EARTH'S CORE

EARTH'S CORE
Geophysics of a Planet's Deepest Interior

VERNON F. CORMIER
Professor of Physics and Geophysics, University of Connecticut, Storrs, CT, United States

MICHAEL I. BERGMAN
Professor of Physics, Bard College at Simon's Rock, Great Barrington, MA, United States

PETER L. OLSON
Adjunct Professor, University of New Mexico, Albuquerque, NM, United States

Elsevier
Radarweg 29, PO Box 211, 1000 AE Amsterdam, Netherlands
The Boulevard, Langford Lane, Kidlington, Oxford OX5 1GB, United Kingdom
50 Hampshire Street, 5th Floor, Cambridge, MA 02139, United States

Copyright © 2022 Elsevier Inc. All rights reserved.

No part of this publication may be reproduced or transmitted in any form or by any means, electronic or mechanical, including photocopying, recording, or any information storage and retrieval system, without permission in writing from the publisher. Details on how to seek permission, further information about the Publisher's permissions policies and our arrangements with organizations such as the Copyright Clearance Center and the Copyright Licensing Agency, can be found at our website: www.elsevier.com/permissions.

This book and the individual contributions contained in it are protected under copyright by the Publisher (other than as may be noted herein).

Notices

Knowledge and best practice in this field are constantly changing. As new research and experience broaden our understanding, changes in research methods, professional practices, or medical treatment may become necessary.

Practitioners and researchers must always rely on their own experience and knowledge in evaluating and using any information, methods, compounds, or experiments described herein. In using such information or methods they should be mindful of their own safety and the safety of others, including parties for whom they have a professional responsibility.

To the fullest extent of the law, neither the Publisher nor the authors, contributors, or editors, assume any liability for any injury and/or damage to persons or property as a matter of products liability, negligence or otherwise, or from any use or operation of any methods, products, instructions, or ideas contained in the material herein.

Library of Congress Cataloging-in-Publication Data
A catalog record for this book is available from the Library of Congress

British Library Cataloguing-in-Publication Data
A catalogue record for this book is available from the British Library

ISBN: 978-0-12-811400-1

For information on all Elsevier publications
visit our website at https://www.elsevier.com/books-and-journals

Publisher: Candice Janco
Acquisitions Editor: Amy Shapiro
Editorial Project Manager: Lena Sparks
Production Project Manager: Bharatwaj Varatharajan
Cover Designer: Mark Rogers

Typeset by STRAIVE, India

Contents

About the authors   vii
Preface   ix

1. Radial structure of Earth's core

1.1 Geophysical evidence   1
1.2 Reference models   3
1.3 Analysis of the seismic wavefield   4
1.4 Viscoelastic attenuation   11
1.5 Scattering   12
1.6 Anisotropy   14
1.7 Viscosity   15
1.8 Summary   17
Appendix 1.1 Moment of inertia   18
Appendix 1.2 Elastic equation of motion   19
Appendix 1.3 Seismic nomenclature   21
Appendix 1.4 Adams-Williamson equation and the Bullen parameter   23
Appendix 1.5 Viscoeleastic attenuation parameterization   24
Appendix 1.6 Birch's law and seismic velocity/density relations   26
Appendix 1.7 Composite elastic moduli and density   27
Appendix 1.8 Elastic anisotropy   28
Appendix 1.9 Equations of state   30
References   31

2. Chemical and physical state of the core

2.1 Composition of the core   33
2.2 Temperature in the core   38
2.3 Transport properties of the core   40
2.4 Thermodynamics of the core   46
2.5 Inner core mineralogy   57
2.6 Summary   64
Appendix 2.1 Construction of phase diagrams and the partition coefficient   64
Appendix 2.2 Thermodynamic relations and the Gruneisen parameter   67
Appendix 2.3 The free electron Fermi gas, phonons, and the Debye model   68
Appendix 2.4 Miller indices and pole figures   69
References   71
Further reading   73

3. Geodynamo and geomagnetic basics

3.1 Preliminaries   75
3.2 The geomagnetic field in the core   78
3.3 The geodynamo process   91
3.4 Geomagnetic images of the core flow   104
3.5 Summary   109
Appendix   109
References   111
Further readings and resources   112

4. Outer core dynamics

4.1 The outer core environment   115
4.2 Dimensionless parameters   118
4.3 Thermochemical transport and buoyancy   121
4.4 Steady laminar flows   124
4.5 Waves in the outer core   135
4.6 Outer core convection   143
4.7 Numerical dynamos   151
4.8 Summary   168
Appendix   169
References   175
Further readings and resources   176

5. Boundary regions

5.1 D″ (lowermost mantle region)   179
5.2 CMB topography   186
5.3 E′ region (uppermost outer core)   187
5.4 F region   189
5.5 ICB topography   190
5.6 Summary   191
References   192

6. Inner core explored with seismology

6.1 Elastic anisotropy   195
6.2 Attenuation and scattering   197
6.3 Hemispherical differences   201
6.4 Differential rotation   204
6.5 Shear modulus, density, and viscosity   207
6.6 Summary   212
References   213

7. Inner core dynamics

7.1 Solidification of the inner core   215
7.2 Deformation in the inner core   223
7.3 Annealing: Recovery, recrystallization, grain growth, and coarsening   234
7.4 Grain size and the deformation mechanism map of the inner core   239
7.5 Inner core viscosity   240
7.6 Inner core elastic anisotropy, attenuation, and isotropic heterogeneity   242
7.7 Summary   244
References   245
Further reading   246

8. Formation and evolution of the core

8.1 Formation of the core	247
8.2 Core evolution	259
8.3 The geodynamo in the deep past	271
8.4 Seeding the early geodynamo	273
8.5 Summary	276
Appendix	276
References	278

9. Future research goals

9.1 Introduction	281
9.2 Seismology	281
9.3 Mineral physics	283
9.4 Core dynamics	286
References	289

Notation tables	**291**
Core properties and parameters	**297**
Glossary	**301**
Index	**309**

About the authors

Vernon F. Cormier is a professor of physics and geophysics at the University of Connecticut. He received a BS from the California Institute of Technology and an MPhil and a PhD from Columbia University. His research has specialized in seismic wave propagation and deep Earth structure, with applications that have included earthquake ground motion, monitoring of underground nuclear tests, and the structure and dynamics of the crust, mantle, and core. He has served as an editor for the *Geophysical Journal International* and *Physics of the Earth's Interior*. Among services to professional societies, he has served the Seismological Society of America as Vice President and the American Geophysical Union as Secretary of the Seismology Section and Chair of the Study of the Earth's Deep Interior (SEDI) interest group.

Michael I. Bergman is the Emily H. Fisher Professor of Physics at Bard College at Simon's Rock, where he teaches courses ranging from introduction to geology to solid-state physics to a field course in volcanology on the island of Montserrat. He earned a BA from Columbia University and a PhD from MIT. His research has focused on the solidification and deformation of metals, alloys, and aqueous solutions, with applications to Earth's core and sea ice. He is the recipient of the Doornbos Memorial Prize for research on the deep Earth by a beginning investigator, has been the long-standing secretary of SEDI, and has served as a guest editor for several journals.

Peter L. Olson is an adjunct professor at the University of New Mexico. He earned a BA from the University of Colorado Boulder; an MA and a PhD from the University of California, Berkeley; and taught earth and planetary science for 37 years at Johns Hopkins University. His research focuses on the dynamics of planetary interiors, including Earth's mantle and core. His service activities include Editor for numerous journals, past President of the Tectonophysics Section of the American Geophysical Union, and a member of the U.S. National Academy of Sciences.

Preface

Earth's core is the most remote part of our planet, recognized as a compositionally distinct region little more than a century ago. It remains the final frontier of our knowledge about Earth. Advances in seismology, computational and experimental mineral physics, geodynamics, geomagnetism, paleomagnetism, and geochemistry have made core research into a truly multidisciplinary effort. Exploration of the solar system has revealed iron-rich cores in all the terrestrial planets and several planetary satellites. Exoplanet exploration will doubtless reveal many more. Most observationally accessible of all cores, Earth's offers great challenges for understanding its composition, structure, and dynamics. Molten iron at over 4000 K churns in the outer core to produce Earth's magnetic field, and is compressed to a solid in the inner core to pressures that exceed three million times the atmospheric pressure. Research of Earth's core requires applying known chemistry and physics to an extreme environment, while being open to new theories to explain surprising phenomena.

The aim of this book is twofold: to give the foundation required to investigate Earth's core and to present the outstanding research questions. Our intended audience includes those at the beginning of their graduate studies. Few of the audience will have studied deep Earth geophysics in detail, having entered the field with disparate backgrounds in physics, geology, chemistry, mathematics, or engineering. While we assume our readers have a background in at least one of these disciplines, we endeavor to broaden their comprehension of other disciplines that bear on the core. This will, for example, enable an established researcher to appreciate more clearly the issues discussed at multidisciplinary meetings such as the Study of the Earth's Deep Interior.

In Chapters 1–3, we present (1) the basic radial structure of the core; (2) its composition, temperature, and material properties, core thermodynamics, and inner core mineralogy; and (3) geomagnetism and elementary dynamo theory. These chapters review the seismology, mineral physics, and geodynamics required to begin a study of the core. Much in these chapters is widely accepted by the geophysical community, except the precise composition of the core, and some of the material properties of iron and its alloys under core conditions, which continue to be active areas of research. Chapters 4–7, while still presenting foundational material, emphasize current research problems. These include (4) details of outer core dynamics and the dynamo process; (5) boundary layers on both sides of the core–mantle boundary and above the inner core boundary; (6) inner core seismology, including elastic anisotropy and attenuation; and (7) inner core solidification and deformation. Some overlap of topics is inevitable because different disciplines often ask the same questions. Chapter 8 delves into the truly speculative—the formation and early evolution of the Earth and its core. We conclude with Chapter 9: a brief chapter on future directions.

There are many excellent monographs that cover the particular disciplines relevant to Earth's deep interior. Except for those on geomagnetism and dynamo theory, few of these specialize in applications to the core and are often more detailed than useful for introducing topics of current research. Stacey's *Physics of the Earth* and Anderson's *New Theory of the Earth* come closest to the spirit of this book, although they do not focus on the core. The choice of topics here reflects those that are of the most pressing interest, tempered by our own expertise. It is for this reason that the book is admittedly short on the contributions of geochemistry. In a book of this broad scope, it is not possible to cover details of experimental techniques, ab initio calculations, or computational methods for seismology or dynamo studies. Many excellent detailed reviews on the research topics introduced in this book can act as supplements and extensions to the book, including Elsevier's *Treatise on Geophysics*.

The question of how to deal with uncertainty is central to the study of Earth's core. Readers will immediately encounter instances of uncertainty, starting in Chapter 1 and increasing as the book progresses. For example, the uncertainty in seismic wave speeds in the core is generally less than 1%, but for density a few percentage. The temperature of the core is uncertain at about 10% level, and it is difficult to put uncertainties on quantities such as alloying elements in the core. Many thermodynamic and geomagnetic field properties inside the core have uncertainties as large as a factor of 2. Certain dynamical properties, such as the outer core convective velocities, are uncertain by almost an order of magnitude, and the outer and inner core viscosities even more so. In many cases, the uncertainty itself is uncertain.

In light of these uncertainties, the challenge becomes how to apply quantitative tools to infer the operation of a system whose properties are known, in many cases, only within broad limits. We respect uncertainties by considering a range of values for each such property. We then do not restrict ourselves to one set of data, one interpretation, or one model. We must be cognizant of the assumptions that go into any interpretation or model, and be open to the possibility that new observations and new ideas may well replace long-accepted ones. It is this combination of imagination guided by rigorous science and sound judgment that gives the exploration of the core its special character.

A number of problems, additional resources, and links to engage the student are available at the following website: https://www.elsevier.com/books-and-journals/book-companion/9780128114001. The instructor site will be available by February 2022. The problems concern fundamental seismology, materials science, and fluid mechanics, and others pertain to the core in particular. They are ranked in order of difficulty by asterisks from one to three, with the most difficult being suitable for discussions leading to research investigations.

This book has been enriched by generous contributions of figures as well as by conversations and debates with research collaborators, students, and colleagues from seismology, mineral physics, and geodynamics. Vernon especially thanks Paul Richards for early collaborations on seismic wave propagation and core structure, and Donald Helmberger for demonstrating the equal importance of source and structure to the interpretation of seismograms. Michael thanks Thierry Alboussiere, Jeremy Bloxham, Dave Cole, Renaud Deguen, David Fearn, Ludovic Huguet, Eric Kramer, Dan Lewis, Ted Madden, Sebastian Merkel, Peggy Shannon, Quentin Williams, and James Yu for their advice, assistance, and encouragement with the book and over the years. Vivian Shi provided much-needed assistance with line figures. Peter acknowledges the contributions to the book by the many graduate students and postdoctoral researchers who worked with him on the core and the deep Earth.

We each thank our families for the thoughtful support that made our book project possible. Vernon thanks his wife Leslie for the early idea of writing this book and his son Kiat for shared sports breaks. Michael thanks Sarah and Ben for forgoing more than a few excursions to kayak through swamps, and his wider family for their general support. Peter is and will forever be grateful for the support of his family, Claudia, David, and Nicholas.

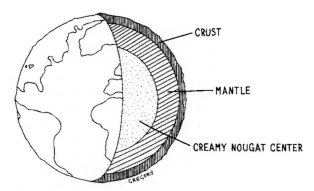

Confectioner's view of Earth (originally published in The New Yorker, September 26, 1999).

CHAPTER 1

Radial structure of Earth's core

1.1 Geophysical evidence

Earth's core, as we know it today, extends to about half of Earth's radius. It consists of a liquid outer core of iron alloyed with nickel and other unknown light elements and a predominantly Fe-Ni solid inner core with lesser amounts of lighter elements. In most parts of this book, we examine Earth's core as a snapshot in time, when it has been possible for humans to measure or estimate its internal properties from its magnetic and gravitational fields, its elastic structure from seismic wavefields, its variations in rotation, and its chemistry from rocks brought to its surface by its internal dynamics or deposited as meteorites. Its current chemistry, state, and dynamics have evolved over a much deeper period of time, starting from the origin of the solar system. We save a glimpse of that deeper time period for Chapter 8.

In this chapter, we briefly review gravity, magnetic, and seismic evidence of gross core structure. We emphasize seismic observations because they are capable of imaging its deep interior in three dimensions at spatial scales approaching 10 km or less. These small scales are those of elastic velocity variations. Elastic velocities, which are combined functions of elastic moduli and density, are only weakly sensitive to *viscosity* and can have complex and often unknown correlations with chemistry and other physical properties important to the geodynamo such as electrical and thermal conductivity. Nonetheless, many useful relations exist for predicting how seismic velocities change due to variations in rheology, composition, temperature, and pressure (see Appendices 1.4–1.9).

1.1.1 Moment of inertia and gravity

Even chemically homogeneous planets can be expected to exhibit some density stratification, with density increasing with depth due to volumetric compression from the effects of increasing pressure toward their centers. The uncompressed bulk density of Earth (4050 kg/m^3), however, is much higher than that predicted for the uncompressed density of mantle silicates (3000 kg/m^3). This inescapably leads to the conclusion that the existence of the density stratification must be also accompanied by a compositional stratification.

A first-order representation of a planet's stratification in density is given by its measured moments of inertia. Two moments of inertia of Earth can be measured from the spatial variation of its gravitational potential, determined from observations of the path of Earth-orbiting satellites, and from the measured precessional period of its axis of rotation about a normal to its orbital plane about the sun (Appendix 1.1). These two measurements are substituted into two equations and for two unknowns, the moment of inertia about Earth's axis of rotation and that for an axis lying in its equatorial plane. The equatorial moment of inertia is slightly larger than the polar, representing the effects of flattening from Earth's rotation. Within two significant digits both moments of inertia are $0.32MR^2$, where M and R are the mass and average radius of the Earth, respectively. Since the moment of inertia of a homogeneous sphere is $0.4MR^2$, it is clear that denser material must exist nearer its center. Since the compression of silicate minerals with increasing pressure at depth is not sufficient to explain this low moment of inertia, this provides evidence for iron concentrated in a core beginning at a depth approximately halfway to its center.

1.1.2 Magnetic field: Spatial spectrum and time variation

The spatial and temporal behavior of Earth's magnetic field provides additional, indirect evidence, for the location, state, and physical properties of Earth's outer core. Downward extrapolation (Chapter 3) of the spatial spectrum of

Earth's magnetic field predicts a whitened spatial spectrum of low-order spatial harmonics about halfway to Earth's center. This agrees with the behavior expected for a dynamo consisting of an electrically conducting fluid in convective motion, driven by vigorous flows over a wide range of scales. From the time variations of Earth's magnetic field, the magnitudes of the convective velocities of this fluid are inferred to be on the order of 10–20 km/year, more than 5 orders of magnitude larger than the convective velocities of Earth's mantle and observed plate motions at its surface (cm/year). From the observations of Earth's gravity, rotational motion, and magnetic field alone, we can thus conclude that it has a dense, electrically conducting, low viscosity fluid, outer core, with a radius about half of Earth's. That it is enriched in iron can be inferred from the known chemistry and density of surface igneous rocks and iron meteorites, in combination with measured solar abundances, stripped of hydrogen and helium.

1.1.3 Seismology: Body waves and normal modes

Solutions to the seismic equation of motion (Appendix 1.2) can establish strong constraints on composition from the elastic moduli and densities of Earth's core from the measurement of elastic vibrations observed as body waves above 0.01 Hz in frequency and normal modes in the millihertz band. Solutions in the high-frequency band propagate as body waves (compressional P and shear S) outward from an earthquake, explosion, or impact in quasi-spherically shaped wavefronts, sampling the elastic wave speed of the Earth along ray paths normal to the wavefront. Solutions in the low-frequency band are described by the elastic normal modes of Earth. Mode spectra are identified and measured in the frequency domain, whereas body wave velocities are inferred from the travel times of waves observed in seismograms recorded in the time domain.

Emil Wiechert in 1896 was the first physicist and seismologist to quantify the existence of a metallic core. Timing of seismic body waves beginning in the early 20th century with the work of Oldham and Lehmann provided estimates of its internal elastic properties and state with depth, adding constraints to its density structure, and more precisely locating the radii of the boundaries of the liquid outer and solid inner cores (Fig. 1.1). By the 1930s, global seismometer coverage was sufficient to reveal the existence of the inner core. With current coverage, the signatures of the sharp discontinuities of the outer and inner cores can be easily observed (Fig. 1.2). Evidence of the outer core came from the observation of a shadow zone, in which no direct P and S waves are observed beginning around a great circle range of 95°. This is consistent with a strong sharp decrease in P wave velocity and the absence of direct S waves beneath a radius of 3480 km, due to the fluidity of the outer core. Evidence for an inner core originated from the identification of a P wave multipath, associated with a *travel time triplication* of P waves (different core waves arrive at the same distance from different angles as explained in Section 1.3.1) arriving in the great circle range of 120–155°. This is predicted by a sharp increase in P wave velocity at radius 1215–1220 km.

FIG. 1.1 The pioneering seismologists who determined the basic structure of the core. Left to right, with date of seminal publications on Earth's core: Emil Wiechert (1896), Richard Oldham (1906), and Inge Lehmann (1936).

Knowledge of the state (liquid or solid) of the core has been obtained from seismology in both the body wave and free-oscillation frequency bands. The lack of S waves propagating through the outer core is inferred from the travel time and polarization of SKS waves, which propagate as S waves in the mantle and compressional waves in the liquid outer core (Fig. 1.3). Direct observation of shear waves propagating in the solid inner core (PKJKP) has been difficult because the amplitude of this wave is quite weak. Weak PKJKP amplitude is due to the small coefficient of conversion of P to S waves at the inner core boundary for possible angles of incidence on the inner core boundary. Further weakening of PKJKP is also possible due to the possibility of shear wave splitting from the *elastic anisotropy* of the inner core, which will widen a single pulse into two interfering PKJKP pulses on both vertical and radial components of motion. Free oscillation *eigenfrequencies* are consistent with a liquid outer core and a solid inner core with a bulk averaged shear velocity of 3.5 km/s. Using correlation wavefields, Tkalčić and Pham (2018) reported a value of 3.42 km/s at the *ICB* rising to 3.58 km/s at Earth's center. These values have been problematic to reconcile with estimates from likely inner core compositions, which are predicted to approach 5 km/s for nearly pure iron at the temperatures and pressures of the inner core (see Chapter 6).

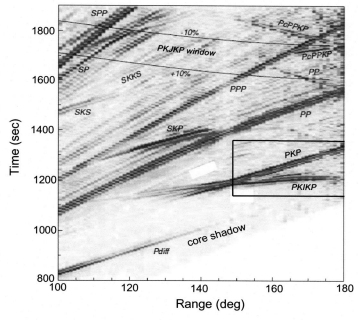

FIG. 1.2 Body wave travel time curves created by stacking 90,673 seismograms from 3648 shallow earthquakes (<50 km depth) from Shearer et al. (2011). (Nomenclature of labeled waves is associated with their ray paths in Earth and is explained in Appendix 1.3.) A signature of the outer core is the P wave shadow, where the direct P wave fades into a wave diffracted (Pdiff) around the core-mantle boundary. A signature of the inner core is the bifurcated PKP curve evident between 150° and 180°, consisting of a faster PKIKP wave, having a near horizontal slope, transmitted through the inner core and a slower PKP wave, having a steeper slope, transmitted through only the outer core. The shear wave traveling through the solid inner core (PKJKP) cannot be easily detected.

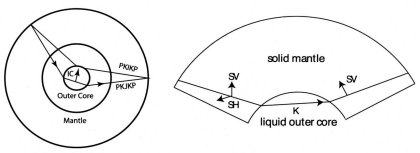

FIG. 1.3 P and S wave polarization sensitivities to the solid/liquid states of the inner/outer cores. Left: the shear wave transmitted through the inner core wave is predicted to exist as only an *SV polarized* S wave within a solid, isotropic inner core. Right: the liquid state of the outer core can be confirmed by the polarization of the SKS wave, in which only the SV polarized component of the S wave incident on the core-mantle boundary can be converted to a P polarized wave in the liquid outer core (K) and then converted to an SV polarized wave on exit.

1.2 Reference models

When determining the detailed structure of Earth's interior, it is useful to seek deviations from reference Earth models, which assume an average, radially symmetric, elastic, and density structure. Two such models are PREM (Dziewonski and Anderson, 1981) and AK135-F (Kennett et al., 1995) as shown in Fig. 1.4. These and other reference models are available as comma-separated (CSV) spreadsheet files from web pages of the Data Management Center of the Incorporated Research Institutions for Seismology (IRIS DMC, 2011). The core model of PREM assumes a chemically homogeneous and neutrally buoyant outer and inner core. PREM is designed to fit a suite of elastic-free oscillations and a catalog of travel times of body waves. Since body waves interacting with the boundaries of Earth's inner and outer core have a relatively complex structure (Figs. 1.5 and 1.6), PREM ignores fitting cataloged travel times in distance ranges where the travel time curves of body waves exhibit this complexity. Hence, body wave studies frequently choose AK135-F rather than PREM as the best reference model for core structure. AK135-F adds details in the form of changes in gradients to elastic velocity and density above the core-mantle boundary and on either side of the inner core boundary. In later chapters, we will see that changes in velocity and density gradient have important consequences for chemistry and stable stratification of the outer core. It is thus not surprising that their existence and lateral variation is still one of the more active areas of core research.

Readers of seismological literature will encounter references to a historical scheme of lettered regions of the Earth A–F developed by K. Bullen to classify depth regions differing in elastic velocity and/or its gradient in depth. Extensions of Bullen's classification scheme have led to identifying a region D″ (the region 200–300 km above the core-mantle boundary) and regions E′ and F of similar thicknesses in the liquid outer core just below the core-mantle boundary and just above the inner core boundary, respectively. Modern research concentrates on interpreting the seismic observations that best constrain smaller scale structures of Bullen's D″, E′, and F regions. Changes in composition and the velocity gradients near the boundaries of the outer and inner cores (CMB and ICB), too small to be represented in Fig. 1.4, and the existence of scatterers in D″ and the uppermost inner core are important for understanding the dynamics and evolution of the inner and outer cores. Evidence for and models of small perturbations near the CMB and ICB are treated in Chapter 5.

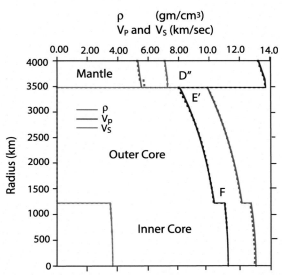

FIG. 1.4 Plotted starting 120 km above the core-mantle boundary (CMB) at radius 3480 km are reference models PREM (solid) and AK135-F (dashed) for P wave velocity (blue, black in print version), density (red, dark gray in print version), and S wave velocity (green, light gray in print version) from the lowermost mantle to the center of Earth's inner core. The inner core boundary (ICB) at radius 1220 km separates the liquid outer core from the solid inner core. The location of Bullen's (Jeffreys and Bullen, 1940) depth regions D″ (lowermost mantle), E′ (uppermost outer core), and F (lowermost outer core) are marked.

1.3 Analysis of the seismic wavefield

Observable elastic vibrations of Earth sensitive to core structure occur over a broad spectrum of frequencies, from less than 1 mHz to greater than 10 Hz. The solution of the equation of motion of these vibrations and their observations are commonly divided into either a representation as body waves or normal modes. Body waves propagate outward from an earthquake, explosion, or impact source as quasi-spherical wavefronts whose normals are called rays. For a normal mode of oscillation, the whole Earth vibrates at a discrete frequency, with an amplitude that exponentially decays with time from its initial excitation. The vibrations form a quilted pattern of a small number of nodes of zero amplitude at its surface, similar to the spatial pattern of the modes of a drum head in 2D or a stringed instrument in 1D.

The representation either as traveling body waves or normal modes of vibration begins with the treatment of the elastic equation of motion (Appendix 1.2). A body wave representation is accurate when gravity and the pseudo-forces of rotation can be neglected in the equation of motion at sufficiently high frequency. At lower frequencies, a normal mode representation is required because the magnitude of the forces of gravity and rotational pseudo-forces (*centrifugal* and *Coriolis*) can be close to the magnitude of the elastic contact forces.

1.3.1 Body waves

Body wave constraints on densities are relatively weak, being only sensitive to the sharp changes in the product of velocity and density or seismic impedance at steep angles of incidence to material discontinuities. The *bulk modulus* or incompressibility obtained from joint P and S wave velocities constrains the density gradient through the Adams-Williamson equation (Appendix 1.4). In the free-oscillation band, however, gravity becomes a restoring force in the equation of motion, and mode eigenfrequencies can establish additional constraints on core density. Density profiles of the core have been determined from combined inversions of body wave velocities, mode eigenfrequencies, and mass and moment of inertia of Earth, together with the assumption of neutrally buoyant stratification.

The body waves commonly used to study core structure are subject to two discontinuous changes in elastic wave velocities and densities, one at the core-mantle boundary and one at the inner core boundary. Any rapid or discontinuous decrease in elastic velocity produces a *caustic* and two paths in a lit zone, each having opposite sign of curvature in their associated travel time curves. A caustic is a surface, line, or point in space where body waves are strongly focused and frequency-independent *ray theory* breaks down. Any rapid or discontinuous velocity increase produces a triplication of the travel time curve. A triplication is a region of distances in which three body waves having different ray paths (multipaths) can arrive at the same seismic station, for example, the travel time curves for P waves interacting with the inner core boundary at distances greater than 145° in Fig. 1.5. The points where the curvature of travel time curves changes sign are the distances at which the caustics intersect the Earth's surface.

The interaction with the core boundaries of P waves, whose polarization direction is coincident with the direction of their rays, provide an example of the waveform complexity induced by a discontinuous velocity decrease at the core-mantle boundary, followed by a discontinuous velocity increase at the inner core-mantle boundary. It is the existence of this complexity in travel time curves that enables an estimate of the depths of both Earth's outer and inner cores to within 10 km or less. A shadow zone and caustic are induced in P waves by a discontinuous P velocity decrease at the core-mantle boundary, and a triplication is induced by a discontinuous P velocity increase at the inner core boundary.

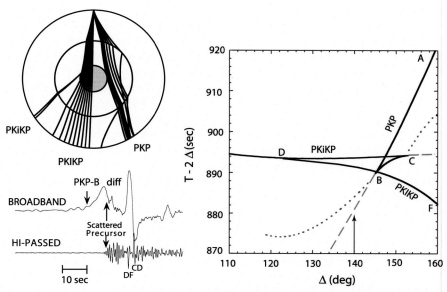

FIG. 1.5 P wave seismograms (left) observed from a deep focus earthquake at 140° emphasize several complex interactions with both the inner core (PKIKP + PKiKP) and the outer core and the core-mantle boundary (PKP, diffraction from the B caustic and C cusp, and CMB scattering). A high-pass filter emphasizes a precursor to PKIKP + PKiKP scattered from the CMB region, while in the broadband recording the low-frequency diffraction from the B caustic becomes visible. The travel time curve (right) shows ray-theoretical waves as black lines, low-frequency diffracted waves as dashed red lines, and high-frequency scattered waves from the CMB as dotted blue lines.

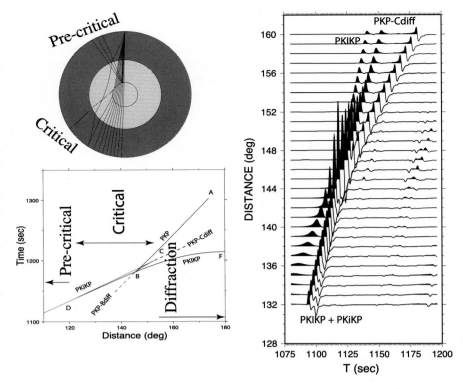

FIG. 1.6 Left: ray paths and travel time curves of core P waves, identifying key distance ranges sensitive to inner core structure: a precritical reflection, where PKiKP amplitudes are small, and a postcritical reflection PKiKP, where P waves are totally reflected. Right: synthetic record section of P wave displacement from an explosive source illustrating how the complexity of the travel time curve is expressed by complexity in the P body waves. Two frequency-dependent waves, not predicted by ray theory, are visible in the wavefield: a diffraction from the inner core boundary (PKP-Cdiff) and the diffraction from the outer core caustic (PKP-Bdiff). The change in shape of PKP-AB wave visible as the latest arriving pulse after 152° is due to a $\pi/2$, phase shift relative to the PKIKP wave.

The triplication is denoted by lines connecting points A, B, and C in the travel time curves shown in Figs. 1.5 and 1.6. The P waves transmitted through the outer core are sometimes denoted as either PKP-AB or PKP-BC according to the branch of their travel time curve.

The velocity decrease at the core-mantle boundary generates a reversal of the travel time-distance curve plotted for a series of increasing vertical takeoff angles. A strong focusing of PKP waves occurs at the caustic distance B. In standard reference Earth models, the core shadow zone starts at 95° and the caustic distance at point B is close to 145°. The discontinuous increase in velocity at the outer-inner core boundary generates the triplication C-D-F. Frequency-dependent *diffraction* occurs along the extension of BC to longer distances. A lower amplitude partial reflection along the dashed segment extends from D to shorter distances. In addition to the effects induced by radially symmetric structure, lateral heterogeneity near the core-mantle boundary can scatter higher frequency body waves in all directions, detectable by a high-frequency (>1 Hz) coda that arrives before PKIKP at distances that precede the B caustic, riding on top of a low-frequency signal diffracted from the caustic. Scattering at the core-mantle boundary can also induce a high-frequency arrival, less-frequently observed, between PKIKP (PKP-DF) and PKP (PKP-AB) at ranges beyond cusp C. The curved dashed line extended to shorter distances from point B in Fig. 1.5 represents the minimum arrival time of these high-frequency PKIKP precursors scattered from either heterogeneity near or topography on the core-mantle boundary. The combined effects of the P wave scattered by heterogeneities near the CMB and the B caustic diffraction are best seen in broadband seismograms spanning a distance range of P wave core interaction between 130° and 145° (Figs. 1.5 and 1.6).

The amplitude of the PKiKP wave varies from small values at short distances and near vertical incidence to a total reflection as it approaches grazing incidence on the inner core. PKiKP corresponding to the range of total (critical) reflection is the travel time branch labeled PKP-CD in Figs. 1.5 and 1.6. The transition in amplitude of PKiKP from *precritical* to *postcritical reflection* is a gradual, frequency-dependent transition surrounding the distance of point D.

Unlike the motion of P waves polarized parallel to the direction of their rays, the motion of S waves is polarized perpendicular to the direction of their rays, similar to the polarization of electromagnetic waves. Assuming elastic isotropy, the orientation of the transverse polarization of an S wave is fixed by the orientation of the slip vectors along an earthquake fault plane. This polarization is commonly decomposed into an SV component in the plane containing the source, receiver, and center of Earth, and an SH component perpendicular to that plane. The SH component of S polarization cannot be converted to a compressional wave in Earth's liquid outer core because SH particle motion has no component in the propagation direction of the converted P wave. Hence, its interaction with the outer core produces only a pure *SH polarized* ScS and a diffracted S. SV polarized waves can convert to P waves and vice versa at the core-mantle boundary. Since compressional waves in the core are faster than the S waves on the mantle side of

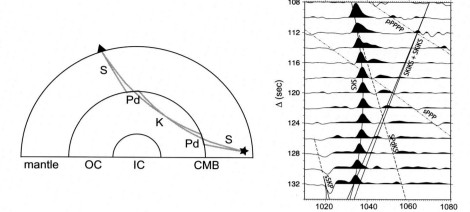

FIG. 1.7 SV interactions with the core-mantle boundary region in which an SPdiffKS wave is excited (Rondenay et al., 2010).

the core-mantle boundary, the travel time curves of SV polarized S, ScS, and SKS waves are examples of a triplication equivalent to a discontinuous velocity increase. The SKS wave should be a purely SV polarized wave (Fig. 1.3). This is indeed the case, except for cases in which significant anisotropy occurs in the mantle along the path exiting the outer core to the receiver. Fig. 1.7 shows an example observation of SV polarized S waves interacting with both the outer and inner core boundaries. It illustrates another important S interaction with the CMB in which at some angles the SV polarized SKS wave can excite a P wave diffracted along the CMB before it converts to a compressional K wave at the outer core. This can happen at either the receiver end, source end, or both ends of a path, and has been useful in research on the structure and lateral variation of structure in the D″ region at the base of the mantle.

Similar to the sequence of underside reflections that are excited by the velocity increase at the inner core boundary (PKIIKP, PKIIIKP+..) shown in Fig. 1.7 travel time curves, there is a sequence of underside reflections of the outer core boundary (SKKS+SKKKS+...) excited by the equivalent increase of S velocity on the mantle side of the CMB to the higher compressional (K) velocity on the outer core side of the CMB. These collections of underside multiples are examples of interference head waves or whispering gallery waves. The amplitudes and travel times of the SmKS multiples are especially sensitive to the compressional wave velocity and its gradient in Bullen's E′ region of the uppermost inner core, and have been used to determine if there exists a stably stratified layer at the top of the outer core that does not participate in the primary convection driving the geodynamo (Chapter 5).

For a given frequency content the sensitivity of body wave travel times and amplitudes to elastic velocity structure can be represented by a banana-shaped volume surrounding their ray paths, the lower the frequency, the wider the waist of the banana (Fig. 1.8). Midway along the ray connecting a source and receiver, the diameter of the zone of

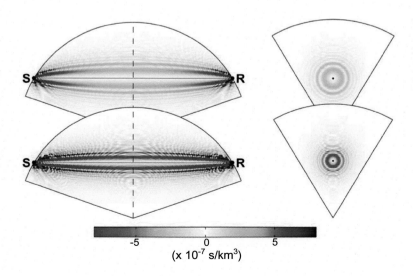

FIG. 1.8 3D sensitivity kernels for an SH wave in a homogeneous sphere at an epicentral distance of 120° calculated in narrow frequency bands centered at 25 mHz (top) and 50 mHz (bottom). In these examples, the sensitivity kernels are calculated using an approximate ray-theoretical approach described in Zhao et al. (2000).

highest sensitivity is roughly the size of the square root of the length of the ray path times wavelength (Nolet and Dahlen, 2000). This zone of sensitivity can more accurately be estimated from the volume of the first Fresnel zone surrounding a ray path. The first Fresnel zone contains all Snell rays connecting a source location, a point scatterer location, and a receiver location, such that the difference between that path and the least-time Snell ray path is less than a half period. More precise measures of a zone of sensitivity can be determined from a superposition of the normal mode eigenfunctions that correspond to a representation of specific body waves. Calculated either from rays or normal modes, the sensitivity kernel K is a function of both frequency and space. It is defined by integrating over volume a 3-D perturbation $\left(\frac{\delta V_{P,S}}{V_{P,S}}\right)$ to seismic velocity that is needed to model the observed difference in travel time $\delta t_{P,S}$ between that predicted from a reference and true velocity structure:

$$\delta t_{P,S} = \iiint K_{P,S}(\omega, \vec{x}) \left(\frac{\delta V_{P,S}}{V_{P,S}}\right) d^3x \tag{1.1}$$

Body waves, whose ray paths turn from downward to upward propagation, are especially sensitive to structure near their turning points. Areas of this enhanced sensitivity are of special interest for structures near the core-mantle boundary and inner core boundary, where phase and chemistry change either discontinuously or over a depth transition whose width is much less than the dominant wavelength of the body wave. Examples include core grazing P and S waves observed before the shadow of the core-mantle boundary and beyond the shadow as frequency-dependent diffractions, Pdiff and Sdiff. Another important core-grazing wave is PKP-AB observed at distances far from the PKP-B caustic.

SKS and SKnKS are strongly sensitive to structure on both sides of the core-mantle boundary. The collection of SKS, SKKS, S3KS, ... at distances greater than 110° has been used as a test for the existence of a chemically distinct region of stable stratification at the top of the outer core. Inner core boundary grazing PKiKP at postcritical incidence at ranges between D and C, together with PKIKP in the same distance range, has been used to infer structure in the uppermost inner core. The distance of the PKP-C cusp, together with the amplitude, travel time, and pulse dispersion of PKP-Cdiff diffracted around the inner core has been important in constraining structure at the bottom of the outer core (Bullen's F region) as well as determining bounds on inner core topography.

Differential travel times measured by cross-correlating the waveforms of two body waves that have nearly identical ray paths in shallower regions have proved to be a powerful approach in removing the effects of shallower heterogeneous structure where the sensitivity kernels of the two waves overlap. Fig. 1.9 shows the effective sensitivity of

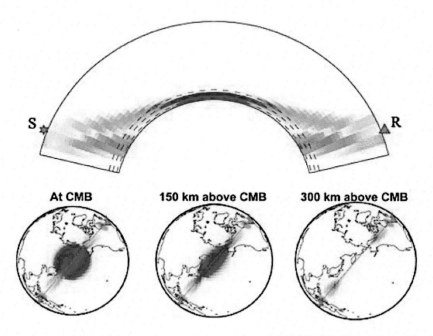

FIG. 1.9 Sensitivity kernel for PKP-DF—Pdiff differential times at low frequency. S and R denote source and receiver locations, respectively. *From Karason, H., van der Hilst, R., 2001. Tomographic imaging of the lowermost mantle with differential travel times of refracted and diffracted core phases. J. Geophys. Res. 106, 6559–6588.*

the difference between the travel time of a P wave transmitted through the inner core and the travel time of a P wave diffracted around the outer core. The effect of the near identical ray paths through the upper mantle is minimized, and the resultant sensitivity is focused on D″ structure near the core-mantle boundary. A similar approach can be used to image structure of the E′ region of the uppermost outer core using the difference between the travel time SKKS and SKS.

With increased density of digitally recorded and telemetered seismic networks, it has become possible to extract weaker body waves sensitive to core structure using techniques of stacking and cross-correlation. Among these are PcP, PKiKP, and PKIIKP at near vertical angles of incidence on the core-mantle and inner core boundaries. At near vertical angles, these P waves are sensitive to only to the P velocity and density changes at discontinuities. Frequency dependence and complexity of their waveforms may reveal the nature of any transition zone structures in the boundary region. The most elusive of these weak waves is the PKJKP wave, which is a compressional wave that converts to an S wave in Earth's solid inner core, and converts back to a compressional wave in the outer core. The conversion coefficient of the compressional K wave in the outer core into the SV polarized wave in the solid inner core is very small for all possible angles of incidence. Making the observation of PKJKP even more difficult is that for some incidence angles and types of elastic anisotropy (Appendix 1.8) the energy of the inner core S wave may be split into two interfering S waves, differing in polarization from either SH or SV, and arriving at slightly different times.

P waves can be observed from reflections by the outer core boundary (PcP) or inner core boundary from either above (PKiKP) or below (PKIIKP). At more grazing angles of incidence (precritical PKiKP) and PKIIKP at ranges less than 180°, their amplitudes are affected by both P and S velocity discontinuities and density discontinuities. For P waves at near vertical angles incidence, the amplitudes of reflected and transmitted P waves depend only on the P velocity and density discontinuities and are nearly independent of the shear velocity jumps at the outer and inner core boundaries. At near vertical incidence, all of the CMB and ICB reflected waves are at a precritical angle of incidence and are partially reflected and hence are weak and relatively difficult to detect in the presence of ambient noise and other scattered waves. When these partial reflections are observable, their amplitudes in combination with P velocity estimates from waveform and travel time modeling can potentially provide better estimates of the density jump at the inner core boundary. This density jump is important to driving a core dynamo powered by compositional convection.

1.3.2 Free oscillations

An earthquake, explosion, or an impact source can excite the free oscillations (*normal modes*) of Earth. They are best observed in the frequency domain, Fourier transforming a seismogram recorded in the time domain. Structure near the boundaries of the outer and inner cores, where chemical heterogeneity and phase changes are most likely, is an important goal of current research. Structure near these boundaries can ideally be inverted from Stoneley type modes, whose energy is concentrated near either the core-mantle boundary or inner the core boundary. Unlike body waves, which provide relatively weak constraints on density gradients and density jumps near discontinuities, normal mode oscillations are measurably affected by a gravitational restoring force. Thus, they can provide constraints on the average density throughout a depth region.

The eigenfrequencies of free oscillations are classified as either *spheroidal* ($_nS_l^m$) or *toroidal* ($_nT_l^m$) modes. The motions of spheroidal modes are perpendicular to Earth's surface; those of torroidal modes are tangent to Earth's surface. Superpositions of spheroidal modes correspond to P-SV body waves and Rayleigh surface waves; those of toroidal modes correspond to SH body waves and Love waves. The subscript l in $_nS_l$ and $_nT_l$ is termed as the angular order number and its value is related to the wavelength λ between peaks in displacement of the mode at the Earth's surface at radius R by the relation $l = \frac{2\pi R}{\lambda}$. The subscript n is termed as the radial order number and also the overtone number. The radial order number n corresponds to the number of nodes with depth in the Earth where the displacement of the mode is zero and undergoes a reversal in the sign of its motion. Each angular order number l can have $2l+1$ possible azimuthal order number m of modes having slightly different eigenfrequencies clustered about a center frequency. Modes having low radial order number n and small angular order number l are most sensitive to upper mantle structure. These modes largely comprise the elastic energy making up Love and Rayleigh surface waves. Modes having large radial order number n and small angular order number l are most sensitive to deep mantle and core structure, comprising energy that makes up deeply penetrating body waves. Mode nomenclature, the senses of motion, and subscripts n and l and superscript m are described in detail in Appendix 1.3.

The $2l+1$ eigenfrequencies possible for different azimuthal order numbers m would all be equal (degenerate) in a spherically symmetric, nonrotating, elastic, isotropic Earth. Degeneracy is removed, making each m have slightly

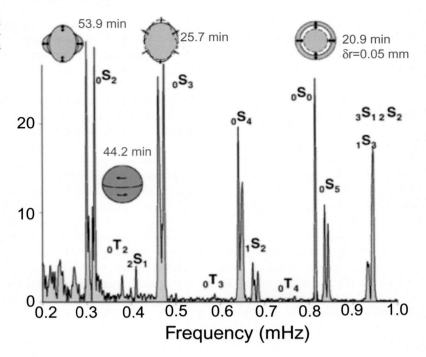

FIG. 1.10 Example spectrum of Earth's normal modes, with illustrations showing senses of displacement for selected spheroidal (yellow) and toroidal modes (blue) (Park et al., 2005).

different frequency, due to the effects of Earth's rotation, ellipticity, lateral heterogeneity, intrinsic attenuation, and anisotropy. The existence of these different eigenfrequencies or singlets is termed as mode splitting and the effect of the superposition of these split modes is to broaden the spectrum of the mode peaks seen in Fig. 1.10 that would otherwise be narrower because each $2l+1$ eigenfrequency would be identical. The broadening of a spectral peak associated with the constructive interference of m's having slightly different frequency for a specific n and l is termed self-coupling. Some mode identifications can be made difficult because of cross-coupling. Strongly cross-coupled modes are those whose n's and/or l's differ but whose eigenfrequencies are sufficiently close together that their observation must be interpreted and modeled jointly.

The structural sensitivity of normal modes can be displayed in various ways. The simplest approach is to display a component of displacement associated with a mode, showing node lines or directions of displacement varying with depth or latitude and longitude at the surface, e.g., Appendix 1.3 and insets in Fig. 1.10.

To provide more insight into which regions of Earth significantly contribute to the observed splitting of a mode one can calculate a structure function. Structure functions can be measured from synthetic seismograms for a reference Earth model and exploited in an iterative, linearized, inversion of observed seismograms to determine perturbations to the structure of the reference model (Woodhouse, 1980). Usually no more than two iterations are performed, starting from a reference Earth model that includes the splitting effects of ellipticity and rotation. The procedure computes a structure coefficient c_{st} and a *splitting function* f_E, where

$$c_{st} = \int_0^{r_e} M_s(r)\delta m_{st}(r) r^2 dr \tag{1.2}$$

$M_s(r)$ is a kernel function depending on the components of motion of the mode eigenfunction and $\delta m_{st}(r)$ is a 3-D structural perturbation with respect to a reference model expanded in *spherical harmonics* $Y_s^t(\theta,\varphi)$.

$$f_E(\theta,\phi) = \sum_{s=2}^{2l}\sum_{t=-s}^{s} c_{st} Y_s^t(\theta,\varphi) \tag{1.3}$$

In Eq. (1.3) the structure coefficients are summed over spherical harmonics to define a splitting function that can be projected on to the surface of the Earth to obtain an image of lateral structural sensitivity of a specific mode. To represent perturbations to velocity discontinuities, an additional term can be added to the definition of the structure

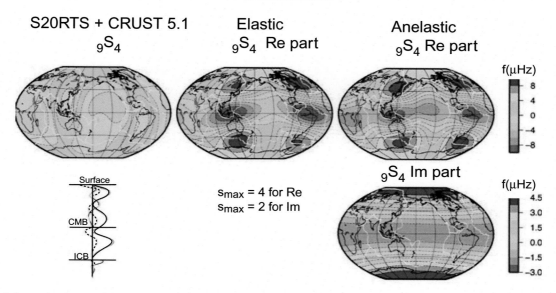

FIG. 1.11 Splitting functions plotted for the spheroidal mode $_9S_4$. Left column: predicted splitting function from a model that includes only the effects of 3D velocity and density variations in the crust and mantle estimated from a tomographic model. Lower left: vertical displacement sensitivity of $_9S_4$ to P velocity (solid black line), S velocity (dashed red line), and density (dotted black line). Middle and right columns: splitting functions obtained from inverting mode data for a model assuming either pure elastic or anelastic structural perturbations in Earth's inner core (Mäkinen and Deuss, 2013).

coefficient c_{st} in Eq. (1.2). An additional function $f_A(\theta,\varphi)$ can also be defined to represent the splitting effects of anelasticity (viscoelastic attenuation), resulting in a complex splitting function,

$$f(\theta,\varphi) = f_E(\theta,\varphi) + if_A(\theta,\varphi) \tag{1.4}$$

A relatively complete sense of both the radial sensitivity of a mode and the lateral perturbations in structure required to match its spectral shape can be obtained by plotting its radial eigenfunction versus depth and its splitting function with latitude and longitude. Fig. 1.11 compares the sensitivity of a mode singlet to that predicted from a 3D model of elastic velocities and density in the crust and mantle with the predictions for the splitting functions obtained from inverting mode data. Anelastic anisotropy in the inner core is required to achieve the best match to observed modal spectra. The anelastic anisotropy has a strong zonal pattern, with fast polar regions correlating with high attenuation compared with slower velocities correlating with lower attenuation in the equatorial region. This sense of correlation is opposite to what is commonly observed in Earth's mantle, which is discussed in more detail in Chapter 6.

1.4 Viscoelastic attenuation

The observed attenuation of the seismic wavefield with time and space and the complexity of body wave pulses assist in understanding the thermal state, microstructure, and larger scale texture of Earth's cores. Elastic waves suffer little or no attenuation in the liquid outer core, similar to the high efficiency of acoustic waves propagating in low-viscosity fluids. In Earth's solid inner core, P waves are observed to suffer much higher attenuation. This attenuation is likely a combination of viscoelastic dissipation due to the internal friction associated with the motions of crystal dislocations or included partial melt and/or scattering attenuation due to the effects of 0.1–50 km scale heterogeneities that redistribute elastic energy into later time windows and locations from observed PKIKP waveforms.

Seismic viscoelastic, or intrinsic attenuation, is the energy lost to heat and internal friction during the passage of an elastic wave. The microscopic mechanisms of intrinsic attenuation have been described in several different ways, including the resistive and viscous properties of oscillator models of the atoms in crystalline lattices, the movement of interstitial fluids between grain boundaries and cracks, and the frictional sliding of cracks.

For elastic waves the existence of viscoelastic attenuation requires that phase and group of elastic body waves are not exactly equal and must be weakly dispersive. This dispersion is difficult to detect in typically narrow band seismograms (Chapter 6) but is important to include in comparing the elastic moduli or velocities determined in the body wave band around 1 Hz with those determined in the mHz band of free oscillations. The reference Earth model

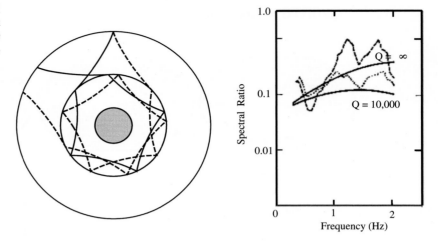

FIG. 1.12 Left: ray paths of P4KP (solid) and P7KP sampling the outer core. Right: estimated bulk Q factors for attenuation in the outer core from the spectral ratio P7KP/P4KP (Cormier and Richards, 1976).

TABLE 1.1 Seismic viscoelastic attenuation estimates reported from Q parameters of reference models at 1 Hz for Earth's lowermost mantle (D″), outer core, and inner core.

Region	AK135				PREM			
	Q_K	Q_μ	Q_P	Q_S	Q_K	Q_μ	Q_P	Q_S
D″	725	273	362	273	57,823	372	840	372
OC	57,822	0	57,822	0	57,823	0	57,823	0
IC	626	85	338	85	1327	85	448	85

Subscripts K and μ refer to bulk and shear attenuation, respectively; P and S to P and S wave attenuation, respectively.

velocities plotted in Fig. 1.4 are shown for 1 Hz elastic waves. In the mHz frequency band, appropriate for free-oscillation observations, the phase velocities and elastic moduli are slightly smaller. A model of viscoelastic attenuation parameters including their frequency dependence, termed as the *relaxation spectrum* (Appendix 1.5), must be known and applied to properly compare the elastic velocities and moduli estimated in the body wave band to those estimated in the free oscillation band.

In estimates of global averages, reference models show increased attenuation in the lowermost mantle D″ region, associated with increased temperature in the thermal boundary layer close to the core-mantle boundary and/or the presence of partial melt. Reference models incorporate evidence for low or near-zero attenuation in the liquid outer core, consistent with the behavior of most low-viscosity fluids. In order to explain the observations of high-frequency PnKP waves (Fig. 1.12) that travel long distances in the outer core, multiply reflecting along the underside of the core-mantle boundary, attenuation in the outer core in both bulk and shear must be very nearly zero (Q's near infinity).

Attenuation increases again in the inner core, consistent with a solid iron close to its melting temperature. Midway into the inner core some models show a sharp decrease in attenuation. Reference models generally agree on the values of P wave attenuation in D″, the inner core, and on the low bulk attenuation of the liquid outer core. The detailed behavior of inner core attenuation, however, is complex, exhibiting anisotropy, depth dependence, and frequency dependence. Because of this, reference models for inner core attenuation have been difficult to construct, illustrate, or be of much use except over broad regions of depth. Reference attenuation models in Table 1.1 are given for Q values in D″, the outer core, and inner core. We will return to a discussion of inner core seismic attenuation in Chapter 6.

1.5 Scattering

The fabric or texture of Earth's solid crust, mantle, and inner cores consists of assemblages of polycrystalline minerals organized into compositionally different patches of organized crystals and crystalline phases, sometimes including lenses of partial melt. If the velocity and/or densities of these patches sufficiently differ across their boundaries, they can scatter elastic waves whose wavelength is on the order of or less than the size of the patch. Seismic observations have found that this is especially true in Earth's lowermost mantle near the CMB and in its inner core, e.g., Shearer (2015). Scattering redistributes energy into different distances and time windows, removing it from the first arriving pulse of a body wave. The shape and frequency content of the scattered coda that arrives following the first break of a body wave pulse can be modeled to determine a statistical description of the heterogeneity scale lengths and percent velocity and density perturbations responsible for the scattered coda.

Scattering effects of body wave coda are quantified by domains defined by the product ka of wavenumber k times scale length a of the scatterer. Realistic fabrics of many heterogeneous materials are well characterized by a power spectrum of velocity and/or density fluctuation ε about a background average value plotted as a function of ka. The scalings between ε fluctuations ($\Delta V_P/V_P$, $\Delta V_S/V_S$, and $\Delta \rho/\rho$) are often assumed from Birch's (1960) empirical observations for the behavior of P velocity and density as a function of mean molecular weight (Appendix 1.6). A commonly assumed shape for the spectrum of heterogeneity is the von Kármán spectrum, which is flat up to a corner of $ka=1$ and then decays at a negative power of ka, termed a Hurst number (Fig. 1.13). The waveform effects of scattering can be divided into different behaviors depending on the magnitude of ka (Sato et al., 2012).

Any sharp (relative to wavelength) contrast in P velocity, S velocity, and/or density and any topography on the inner and outer core boundaries can contribute to scattered coda of body waves. The D″ region core near the mantle boundary is a well-documented region of increased scattering from observations of a high-frequency precursory coda to PKIKP in the 110–140° range (broadband seismogram in Fig. 1.5), confirmed from seismograms synthesized assuming statistical models of topography on and heterogeneity above the CMB. The upper mantle, mid-mantle, lower mantle D″ region, and the inner core can all contribute to the scattered coda observed in high-frequency seismograms of P waves interacting with Earth's outer and inner cores. After correcting for scattering in the mantle, the *heterogeneity spectrum* of the inner core and topography on its boundary can be estimated from modeling the coda of PKiKP, a P wave reflected by the ICB (Fig. 1.14).

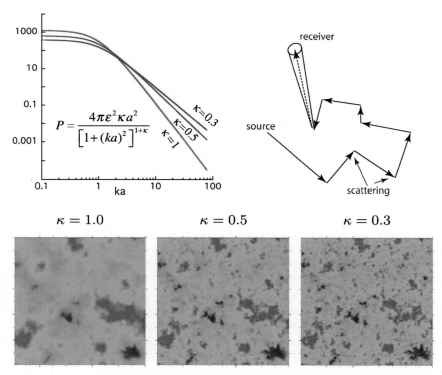

FIG. 1.13 Random walk in which deterministic ray paths conveying packets of elastic energy are interrupted by scattering events in a heterogeneous medium, contributing to the coda (e.g., Fig. 1.14) following the direct arrival predicted from a long-wavelength, equivalent medium. Top left: 2D von Kármán heterogeneity spectrum as a function of correlation wavenumber k; spectral power is flat up to $k_{corner} \sim 1/a$, after which smaller spatial scales decay with falloff rate controlled by Hurst parameter κ. Bottom: example spatial realizations of heterogeneity spectra for various κ at a magnification level of ten scale lengths a. From Sanborn, C.J., Cormier, V.F., Fitzpatrick, M., 2017. *Combined effects of deterministic and statistical structure on high-frequency regional seismograms. Geophys. J. Int.* 210(2), 1143–1159, https://doi.org/10.1093/gji/ggx219.

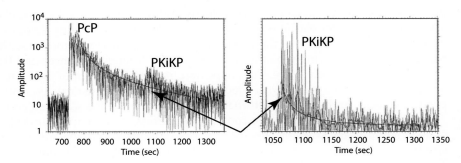

FIG. 1.14 Left: coda envelopes of PcP+PKiKP from an earthquake at 82° (Leyton and Koper, 2007). Right: the smoothed envelope of PcP shown at left is subtracted from the envelope of PKiKP to emphasize scattering from the inner core.

Improved observations and interpretations of the heterogeneity spectrum of the lowermost mantle may eventually aid in discriminating among various processes that have been proposed for increased scattering in D″, including remnant material from subducted slabs, partial melt, a postperovskite phase change, and products of chemical reactions between the silicate mantle and liquid iron and light elements in the outer core. Modeling the heterogeneity spectrum in the inner core may assist in determining whether it is due to lenses of partial melt or boundaries between organized patches of intrinsically anisotropic iron crystals.

1.6 Anisotropy

Only two elastic constants are required to represent an isotropic elastic region in which the stress-strain relation does not depend on the choice of coordinate axes. For comparison with laboratory measurements and the spatially simplest forms of stress, the best choices for reporting these two constants are the bulk modulus K and shear modulus μ. Two types of body waves exist, P and S, with different velocities and polarizations, with S transverse to the ray direction and decomposed into SH and SV components related to the source-receiver geometry. The assumption of an isotropic medium is appropriate for Earth's liquid outer core. Minerals characterized by an arrangement of atoms in a unit cell repeated over a volume in a regular lattice structure, however, are nearly always both elastically as well as electromagnetically anisotropic, requiring up to 21 different elastic moduli to describe the stress-strain relation. In the solid mantle and inner core, the assumption of elastic isotropy may still be appropriate for a polycrystal aggregate when individual silicate or Fe crystals are sufficiently disorganized in orientation or shape. The validity of this assumption weakens in cases where processes align or texture crystallographic axes or where there is a shape anisotropy. Several theories exist for estimating the elastic moduli for composite mixtures of intrinsically anisotropic minerals having a statistically variable orientation. The most advanced of these (Appendix 1.7) incorporate information on the statistics of the orientation of intrinsically anisotropic crystals. Fortunately no more than five separate elastic moduli that can be confidently measured in laboratory crystals and no more than three separate waves with different elastic velocities can theoretically exist in an anisotropic elastic solid (Appendix 1.8). These three waves are a P wave and two quasi-S waves. The quasi-S waves are polarized transverse to their ray but are mutually perpendicular to each other and travel at different velocities. Their polarization is set by the medium elastic constants rather than decomposed into SV and SH directions that are governed by source and receiver positions. With these different speeds and polarizations, the S wave is termed as split.

Weak anisotropy is treated by perturbation theories in the analysis of P waves in the high-frequency band and free oscillations in the low-frequency band. Observation of anisotropic S wave splitting manifests itself as an interference of the two quas-S waves on both SH and SV resolved components of motion. In the case of SKS waves, general and weak anisotropy manifests itself by making observations of the SKS waveform on the transverse (SH) component of motion visible and approximately equal to the time derivative of the SKS waveform observed on the radial (SV) component of motion. Azimuthal variation in P wave velocity due to anisotropy can trade off with lateral variation in isotropy but observation of shear wave splitting is prima facie evidence of anisotropy.

Evidence exists for anisotropy in the region of the lowermost mantle (D″) and in the inner core. In D″ anisotropy may be associated with lenses of anisotropic silicates, orientations of subducted slabs, and convective flow either parallel to the core-mantle boundary or in upward-projecting zones above the CMB from upwelling plumes. Given all these scenarios, it is not surprising that investigations of D″ anisotropy are characterized by strong lateral variations and by difficulties in assessing their resolution.

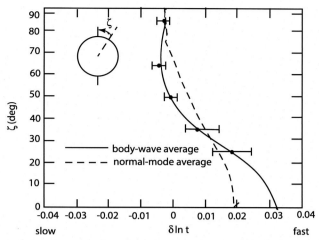

FIG. 1.15 Fractional travel time anomaly as a function of ray angle with respect to Earth's rotation axis, showing fits to an inner core-sensitive travel time anomaly assuming transverse isotropy with an axis of symmetry coincident with the rotation axis. The fits are to differential travel times of body waves PKP-DF-PKP-AB and the predictions to those observations from separate fits to normal mode eigenfrequencies. *Adapted from Song, X., Jordan, T.H., 2017. Stochastic representations of seismic anisotropy: transversely isotropic effective media models. Geophys. J. Int. 209(3), 1831–1850, https://doi.org/10.1093/gji/ggx112.*

Evidence of anisotropy in the inner core has been found both in body wave travel times and free oscillations, consistent with a fast P wave velocity in polar directions compared to equatorial directions (Fig. 1.15). The mechanism for explaining inner core anisotropy is intimately connected to the processes of solidification and deformation of the inner core and the lattice structure of iron. Because of the importance for inner core solidification to a dynamo driven by compositional convection, we will return to this topic in Chapter 6 on observational seismology applied to the inner core and Chapter 7 on inner core dynamics.

1.7 Viscosity

One of the most important physical properties for the operation of a planetary dynamo is the viscosity of the electrically conducting fluid, whose convective motion sustains the dynamo. The form of viscosity that appears in a viscous *rheology* (the equation that defines the relation between stress and strain) is termed the dynamic viscosity. Viscosities are said to be Newtonian if the rheology is such that stress is simply linearly proportional to strain rate, with viscosity η defined as the coefficient of proportionality, i.e.,

$$\sigma = \eta \frac{\partial \varepsilon}{\partial t} \tag{1.5}$$

Note that both the stress and strain are normally assumed to be in shear, and a tensor relation is not needed. The SI units of dynamic viscosity are Pascal-*sec* (Pa s). Kinematic viscosity, which is the form usually shown in the equations for the motion of fluid cores in the descriptions of planetary dynamos, is just the dynamic viscosity divided by density. The MKS units of kinematic viscosity are m^2 s.

The viscosity of the outer core has been estimated both theoretically from ab initio calculations and observationally from the attenuation of body waves, free oscillations, and length of day variations sensitive to the rheology of the uppermost outer core. Ab initio calculations consider the Coulomb force interactions between electrons and atom nuclei to estimate physical properties. Dynamic viscosity η is obtained from the estimate of the self-diffusion coefficient D determined from the theory of Brownian motion for the diffusion of spherical particles through a liquid with low Reynolds number (laminar flow). The relation is known as the Stokes-Einstein equation, with D defined from

$$D = \frac{k_B T}{6 \pi \eta r}, \tag{1.6}$$

where k_B is Boltzmann's constant, T temperature, and r the radius of the spherical particle. Most ab initio calculations have assumed a pure iron liquid outer core and a volume containing an order of about 100 atoms. At the temperature and pressure conditions of Earth's outer core, the ab initio estimates of Alfè and Gillan (1998) predict an outer core

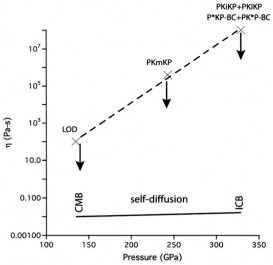

FIG. 1.16 Viscosity estimates for the liquid outer core. Absolute upper bounds are determined from observed length of day (LOD) variations, observed 2 Hz PnKP waves having m ray legs within the outer core, and observed 6 Hz P waves scattered at the core-mantle boundary whose rays turn near the inner core boundary (P*KP-BC + PK*P-BC, where the symbol * indicates scattering at the CMB of the wave preceding the symbol.)

shear viscosity of 10^{-2} Pa s. At first glance estimates of viscosity from geophysical observations present a confusion of results scattered across 11 orders of magnitude (Fig. 1.16). When identified with their depth region of sensitivity and considered as upper bounds, they are consistent with a viscosity that may exponentially increase with increasing pressure or depth in the inner core (Smylie et al., 2009).

At the CMB, Mound and Buffett (2007) estimate the upper bound of 10^2 Pa s from the coupling between the fluid outer core and mantle, which affects Earth's rotation observed in variations of the length of day. The four orders of magnitude difference between Mound and Buffett's upper bound and that obtained from estimates of self-diffusion coefficients in iron melts may arise from the effects of nonlaminar flow in a turbulent outer core. Other upper bounds can be determined from measurements of the attenuation of PnKP body waves sampling the mid-outer core and direct and scattered P waves sampling the lower outer core. These upper bounds can be determined from the theory for bulk sound attenuation Q_K^{-1} in fluids (Lautrup, 2011). These assume that there exists an effective bulk viscosity ς roughly equal to the shear viscosity η. The bulk attenuation is given by.

$$Q_K^{-1} = \frac{\omega}{\omega_o}, \text{where } \omega_o \cong \frac{K}{2\eta} \tag{1.7}$$

Calculating the bulk modulus K from the P velocity and density at mid-outer core depth and substituting 8π for the radian frequency ω for the 2 Hz maximum frequency of observed PnKP waves (Cormier and Richards, 1976) results in an upper bound estimate of η equal to 4×10^6 Pa s. Similarly an upper bound of 10^9 Pa s can best estimated from the attenuation of inner core grazing P waves (Cormier, 2009). Considered as simply upper bounds, however, the not-uncommon observation of even higher frequency (up to 6 Hz) seismic waves transmitted or scattered into the outer core are entirely consistent with an outer core having a near-infinite Q value and a near-zero viscosity from the CMB to the ICB, agreeing with the low estimates from LOD observations and self-diffusion.

In Chapter 4 on core dynamics and the dynamo, we will see that viscosity controls the Ekman number, which is the ratio of viscous to Coriolis forces important for predicting the character of the outer core flow that drives the dynamo. The low ab initio values estimated for outer core viscosity predict Ekman numbers that are consistent with highly turbulent, complex flow. The attenuation of high-frequency P waves in the outer core are also easily consistent with either a viscosity near zero or the higher viscosities within the upper bounds shown in Fig. 1.16. If the upper bounds of viscosity are accepted, however, the Ekman numbers in a large volume of the outer core will be in the range of those commonly accessible by numerical simulations of the dynamo. Since these upper bound viscosities are much higher than ab initio estimates, their values are sometimes termed as hyperviscosities. An unanswered question requiring experimental, observational, and computational advances is whether the assumption of hyperviscosities in the outer core are realistic in their prediction of fluid motions responsible for the magnetic field and its detailed secular variation.

1.8 Summary

The Preliminary Reference Earth Model (PREM) was designed to fit a large catalog of normal modes and free oscillations. In the three broad regions of the lower mantle, outer core, and inner core, it assumes chemical homogeneity and no phase changes within each region. Elastic moduli and density in each region are assumed to obey the predictions of finite strain theory. Density gradients are constrained by the gradient of the bulk modulus. Each region is assumed to be approximately buoyantly neutral, with temperature gradients nearly equal to the predicted adiabatic gradient in each region. Evidence for departure from these assumptions is strongest near the boundaries of each region, originating primarily from the observations of seismic body waves or normal modes that strongly sample the boundary regions. The consequences of these departures from homogeneity of phase and chemistry and neutral buoyancy, however, are critical for understanding the evolution of the inner and outer cores. They are also important for understanding both the slow convective motions of the mantle (0.1–10 cm per year) and the vigorous convective motions of the outer core (0.1–10 km per year) that sustain the Earth's magnetic field.

Models suggesting both simple as well as complex departures in the velocity and density gradients at the top and bottom of a presumed chemically homogeneous liquid outer core require changes in chemistry and mechanisms that sustain these structures in a turbulently convecting liquid outer core. Among these structures is the velocity gradient in the lowermost F region of the outer core in model AK135. Another is a proposed steepening of the compressional velocity gradient at the top of the outer core needed to match the relative travel times and amplitudes of SmKS waves multiply reflected from the underside of the outer core. Such a structure in the outer core E' region also requires a chemistry and mechanism that allow this feature to be sustained with time in an otherwise turbulent, low viscosity, outer core beneath the feature. Numerical dynamos, discussed in Chapter 3, have been found to be sensitive to the thickness and properties of both the E' and F regions. Thicker E' regions decrease field intensity and laterally varying E' regions affect spatial and secular variations of the magnetic field. Some numerical simulations predict strong lateral heterogeneity originating in the F region near the solidifying boundary of the inner core. An increase in the density of seismic observations and new inversion and modeling techniques may yet arrive at consensus theories for the existence for the E' and F regions. These theories are reviewed in greater detail in Chapter 5.

An alternative view of radial anomalies in core structure that can be helpful to consider is that perhaps some or all of the documented structural variations in the E' and F regions may be due to artifacts in the spatial sampling of data, noise contamination, or unconsidered effects of wavefield scattering. With this view, we can instead first seek the simplest models consistent with lower frequency data, such as free oscillations, which are less sensitive to smaller scale heterogeneity, and provide more reliable and stable averages over large volumes. A starting point for determining such a simple model is to assume an *equation of state*. An equation of state is simply an equation that predicts how the density of a material changes with pressure and temperature. Appendix 1.9 briefly reviews the parameters of two equations of state often assumed for planetary interiors. An example is the EPOC outer core model in a study by Irving et al. (2018). From free oscillations and an equation of state for a chemically homogeneous outer core, the authors determine the depth variation of bulk modulus and density in the outer core. With EPOC they obtain better fit to multiple SmKS travel times than models that have assumed a chemically different layer in the outermost core.

In this chapter, we have reviewed current knowledge of the radial structure of Earth's core, primarily the physical properties of elastic structure and density. The things we know well are the boundaries of the inner core and outer core within ±10 km. P and S elastic velocities with depth are known over average depth ranges of 100 km to within several per cent or less. Density models are derived from constraints from moment of inertia, mass of the Earth, elastic velocity models, and free oscillations. Less well known, but vitally important to the function of the geodynamo are the shear velocity structure of the inner core and smaller scale structures close to the inner core and outer core boundaries, the D″, E′, and F regions. These include regions of modified velocity and density gradient, anisotropy in D″ and the inner core, smaller scale structures that scatter seismic waves, and CMB and ICB topography having heights less than 10 km and lateral scales less than 1000 km. These are discussed in greater detail in Chapters 5 and 6.

Although we have reviewed here the constraints on viscosity important for fluid motion of the outer core and geodynamo, other physical properties, including thermal conductivity, electrical conductivity and permittivity, and magnetic permeability are also important to the operation of the geodynamo. Constraints and assumptions for these parameters are discussed in Chapters 2, 3, 4, and 7.

Appendix 1.1 Moment of inertia

The effects of moment of inertia on the torques needed to change the rotation of a body are similar to the effects of mass on the force needed to change linear motion. The moment of inertia for a body rotating about any axis through its center of mass is the essentially the analog of mass for a body in linear motion. The linear momentum of a mass m moving with constant velocity v is the product mv. The angular momentum of a body rotating at constant angular velocity is the product $I\omega$, where I is the moment of inertia about an axis of rotation about its center of mass and ω is the angular velocity of rotation. Expressed similar to Newton's 2nd law of linear motion, where the time derivative of momentum is equal to the sum of forces acting on a mass, the time derivative of angular momentum is equal to the sum of torques acting on the rotating body. The moment of inertia is always defined with respect to a specific axis of rotation. Expressed as either a discrete sum or integral over a continuous distribution of mass, moment of inertia is given by

$$I_{axis} = \sum_i \Delta m_i r_i^2. \tag{1.8}$$

where r_i is the perpendicular distance to the rotational axis measured from the center of the discrete mass m_i.

For a generally shaped body, any rotation can be expressed as a linear combination about three orthogonal principal axes of rotation. These three principal moments of inertia are equal for a spherical body. Planetary bodies of size scale greater than several 100 km tend to assume a spherical shape to minimize their gravitational potential energy. The processes of gravitational differentiation combined with material compressibility will tend to make a compositionally heterogeneous planet form into a series of spherically symmetric layers, with successively higher density layers closer to the center of the planet. Thus in a perfectly spherically symmetric planet, density ρ is only a function of radius r and the moment of inertia about any rotational axis through its center of mass can be replaced by the calculation

$$I = \iiint \rho(r) r^4 \sin(\theta) \, dr \, d\theta \, d\phi \tag{1.9}$$

which is independent of the choice of axis as long as it goes through the center of mass.

A convenient reference moment of inertia is that for a planet having constant density with depth. In this case the expression above evaluates to $I = 0.4 MR^2$, where M is the mass of the sphere and R is its radius.

Even homogeneous materials have some measureable compression as internal pressure increases with depth. Hence all spherically shaped planets will always have denser compressed material closer to their center, nearer to their axis of rotation, and thus always have $I < 0.4 MR^2$. This inequality suggests that measurement of moments of inertia will be an important constraint on both the density distribution and the compressibility of the materials that compose a planet.

Measurement of moment of inertia:

Since planets also tend to rotate about a specific axis after formation, they tend to be ellipsoids of revolution rather than perfect spheres because the total force acting on any volume of mass at depth point in the body will be a combination of the effect of the gravity from the distribution of mass below that point and the effect of the centrifugal pseudo-force of rotation. One of the principal moments of inertia of an ellipsoid of revolution is defined by the rotational axis. For an ellipsoid of revolution the remaining two of the principal axes of moment of inertia are degenerate (equal) and can be defined from any two perpendicular axes in the equatorial plane.

Ellipsoids of revolution are an excellent starting assumption to constrain the density distribution of a rotating planet. Thus most estimates of planetary moments of inertia often attempt to measure the moment of inertia A for an axis through the center of mass in the equatorial plane and an another axis equal to the rotational axis C. A common approach is to solve for the two unknowns C and A using two equations obtained from two types of observations. The first of these is obtained from the observation of the spatial behavior of the gravity or gravitational potential of the planet. The second of these is the observation of the 25,800 year precession of Earth's rotational axis excited by torques from the gravitational forces of the sun and moon acting on the equatorial bulges of Earth's ellipsoidal shape. The Earth's rotational axis is tilted by 23.5° with respect to the normal to the plane of Earth's orbit about the sun and the moon's orbit about Earth. Hence, torques exerted by gravitational force of the sun and moon can act on the bulges. The two equations for C and A are given as follows, the first for gravitational potential ψ and the second for the precessional rate Ω of the planet.

$$\psi(r, \theta) = \frac{-GM}{r^2} + \frac{G}{r^3}(C - A)\left(\frac{3}{2}\cos^2\theta - \frac{1}{2}\right) - \frac{1}{2} r^2 \omega_{spin}^2 \sin^2\theta \tag{1.10}$$

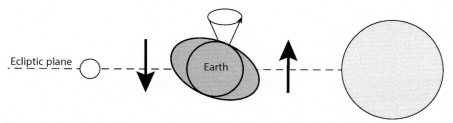

FIG. 1.17 The equatorial bulge of the ellipsoidal shaped Earth is tilted by 23.5° with respect to the plane of its orbit about the sun (ecliptic plane). The gravitational pull of the sun and moon in the ecliptic plane exert torques on this bulge, giving Earth a 25,800-year-period precessional motion about an axis perpendicular to the plane of its orbit.

$$\Omega = \frac{3\omega_{orbit}^2}{2\omega_{spin}} \frac{C-A}{C} \cos\varepsilon \quad (1.11)$$

where θ is the co-latitude, ε the tilt of Earth's rotational axis with respect to a normal to its orbital plane, ω_{spin} the angular velocity of Earth's rotation, and ω_{orbit} the angular velocity of its orbit about the sun (Fig. 1.17).

To within two significant digits C and A are both $0.33MR^2$. Including more significant digits, the dynamic flattening ratio, $(C-A)/C$, can be computed to be $1/306$. This flattening is consistent with the ellipticity due to the changes in the effective acceleration from a faster rotation of Earth closer to the time of its formation. Since C and A are less than 0.4 MR^2 they confirm the existence of a denser mass concentration in its center. With reasonable assumptions for the compositional building blocks of the inner terrestrial planets (silicates and iron nickel alloys), the moments of inertia agree with a Earth model having an iron-enriched core about half of Earth's radius surrounded by a silicate mantle.

For more reading:
Bills and Rubincam (1995), Stacey and Davis (2008).

Appendix 1.2 Elastic equation of motion

Earth responds elastically to infinitesimally small strains excited by earthquakes, explosions, impacts, or tidal forcing. The stresses (forces/area) generated by these sources are assumed to be proportional to the strains, and the rheology (the relation between stress and strain) is said to be linear, often referred to as Hooke's Law. In this regime, the equation of motion for the small motions recorded by seismographs can be written by a form of Newton's 2nd law, in which the time rate of momentum is equal to a sum of forces. When the motion of specific point in a continuum is considered, this equation sets the time derivative of a momentum density equal to the sum of force densities. Omitting the force density of the source, assuming a linear rheology, and a homogeneous Earth of density ρ, this equation can be written as

$$\rho \frac{\partial^2 \vec{u}}{\partial t^2} = \vec{\nabla} \cdot \sigma - 2\rho\, \vec{\Omega} \times \frac{\partial \vec{u}}{\partial t} - \rho \vec{g} \quad (1.12)$$

Note that every term in this equation has dimensions of force/volume. The three components of Earth displacement are contained in the vector $\vec{u}$ and σ is the stress tensor. The remaining two terms on the right-hand side represent the effects of the Coriolis force density due to Earth's rotation and the force density of Earth's self-gravitation. (The effect of the centrifugal force density of Earth's rotation can be absorbed into $\vec{g}$.) $\vec{\Omega}$ is the angular velocity vector of Earth's rotation (2π radians/24h directed along Earth's rotational axis). Knopoff (1955) demonstrated that Lorentz (electromagnetic) forces due to seismically excited motions in the electrically conducting, fluid, outer core are vanishingly small compared to other force densities in Eq. (1.12).

The stress tensor and elastic constants:

With indicial notation the components of the stress tensor σ_{ij} are defined by the linear elastic rheology through a series of elastic moduli c_{ijkl}, expressed as a fourth order tensor, through the equation

$$\sigma_{ij} = c_{ijkl}\varepsilon_{kl} \quad (1.13)$$

where ε_{kl} are strains defined by partial derivatives of displacement with respect to spatial coordinates such that

$$\varepsilon_{ij} = \frac{1}{2}\left(\frac{\partial u_i}{\partial x_j} + \frac{\partial u_j}{\partial x_i}\right) \quad (1.14)$$

Symmetries exist ($\sigma_{ij}=\sigma_{ji}$, $\varepsilon_{kl}=\varepsilon_{lk}$, and $c_{ijkl}=c_{klij}$) so that no more than 21 independent elastic moduli are needed to express the stress-strain relation. With the assumption of isotropy (stress/strain relation independent of the choice of axes), the relation between stress and strain requires no more than two independent elastic moduli. Given in component form, the stress tensor becomes

$$\sigma_{ij} = \lambda \delta_{ij} \frac{\partial u_k}{\partial x_k} + 2\mu \varepsilon_{ij}, \tag{1.15}$$

where λ and μ are elastic constants (moduli) named the Lame constants, $\frac{\partial u_k}{\partial x_k}$ is summed over $k=1$ to 3, and δ_{ij} is the Kroneker delta. Eq. (1.15) is often termed as the elastic constitutive equation. The assumption of a linear elastic rheology by either Eq. (1.13) or Eq. (1.15) is an excellent assumption when the strains (nondimensional quantities $\frac{\partial u_k}{\partial x_k}$) can be considered to be infinitesimal (much less than 1). The smallest detectable nonlinear effects generally vanish for strains less than 10^{-6}. The Lame parameter μ is the same as the shear modulus, where 2μ is the coefficient between shear stress $\sigma_{ij}(i \neq j)$ and shear strain ε_{ij} $(i \neq j)$. The symbol G is also often used for μ.

An elastic modulus K termed as the incompressibility (bulk modulus) is defined as the coefficient between volumetric strain ($\Delta V/V = \varepsilon_{11} + \varepsilon_{22} + \varepsilon_{33}$) and a change in pressure ($\Delta P = \frac{\sigma_{11}+\sigma_{22}+\sigma_{33}}{3}$). The elasticity of an isotropic material is fully defined by either λ and μ or λ and K, but the elastic constitutive relation is simpler to write in terms of the Lame parameters.

Other commonly reported elastic moduli and properties are Young's modulus E and Poisson's ratio ν. E is the coefficient of proportionality between uniaxial stress and uniaxial strain. ν is the negative ratio between lateral and longitudinal strains. (The negative sign in its definition is consistent with the fact that the compression of a material by uniaxial stress applied in a longitudinal direction usually results in extension of the material in the orthogonal, lateral direction.) Poisson's ratio ranges between 0.5 in liquids such as the outer core, for which $\mu=0$, and 0.25 in the solid mantle, for which $\lambda = \mu$ is a good approximation. Setting $\nu = 0.5$ for a liquid is equivalent to assuming conservation of volume or that the liquid is incompressible ($K \to \infty$). Some relations between E, ν, λ, μ, and K are

$$E = \frac{\mu(3\lambda + 3\mu)}{\lambda + \mu} = 2\mu(1+\nu) = 3K(1-2\nu) \tag{1.16}$$

$$\nu = \frac{3K - 2\mu}{6K + 2\mu} = \frac{\lambda}{2(\lambda+\mu)} = \frac{1}{2} - \frac{E}{6K} \tag{1.17}$$

Body wave velocities:

Substituting this rheology into the equation of motion and assuming for now that the effects of rotation and gravity are small relative to the elastic restoring force, the equation of motion becomes

$$\rho \frac{\partial^2 \vec{u}}{\partial t^2} = (\lambda + 2\mu) \vec{\nabla} \vec{\nabla} \cdot \vec{u} - \mu \vec{\nabla} \times \vec{\nabla} \times \vec{u} \tag{1.18}$$

Either the divergence or curl of this equation will give a wave equation. The divergence is equal to the volumetric strain and its wave equation is that for P waves having the velocity $V_P = \sqrt{\frac{\lambda + 2\mu}{\rho}}$. The curl is equal to a rotational strain that propagates with the shear wave velocity $V_S = \sqrt{\frac{\mu}{\rho}}$.

In Earth's mantle, in which the Lame parameters λ and μ are approximately equal, V_P is approximately $\sqrt{3} V_S$, corresponding to Poisson's modulus ν equal to 0.25.

By substituting a plane wave representation of displacement of the type

$$\vec{u} = \vec{A_0} \exp\left[i\left(\vec{k} \cdot \vec{x} - \omega t\right)\right] \tag{1.19}$$

into the equation of motion above and keeping terms highest in order of frequency, it is possible to show that the direction of the vector describing the polarization of displacement is such that the faster P wave is polarized along the direction of propagation k and that the slower S wave is polarized perpendicular to k. At radial discontinuities in Earth, it is helpful to decompose the transverse polarization of S waves into a component SV polarized in the plane containing the source-receiver and center of Earth and a component SH polarized perpendicular to that plane. This decomposition is convenient because it is possible for an SV polarized S wave to generate reflected and transmitted P waves upon incidence on a radial discontinuity. The SH component of an S wave incident a radial solid boundary generates only reflected and transmitted S waves having SH polarization.

To aid in identifying constraints on the compressibility of materials (how their volume changes with increasing pressure), it is often helpful to express the P velocity in terms of the elastic constant of incompressibility K as $V_P = \sqrt{\frac{K + 4/3\mu}{\rho}}$. K can be written in terms of a partial derivative with respect to density ρ or volume V as

$$K = -V \frac{\partial P}{\partial V} = \rho \frac{\partial P}{\partial \rho}. \tag{1.20}$$

Note the dimension of K and all other elastic moduli are pressure. The sign difference in the definition of K can be remembered by recalling that the volume of a material sample decreases and density increases as pressure increases.

For computational feasibility, strategies for solving the elastic equation of motion have naturally evolved into two overlapping but separate domains in frequency band when applied to global seismology, a low-frequency band between 0.001 and 0.1 Hz and a high-frequency band between 0.01 and 10 Hz. A feature that drives this natural separation is the relative importance of the terms on the right-hand side of Eq. (1.12). Keeping all the terms in the original equation of motion and substituting the plane wave solution in Eq. (1.19), it is easy to demonstrate that the Coriolis force and gravitational restoring force are lower order in frequency compared to both the left-hand side and the elastic restoring force and can be neglected in the higher frequency band. Another convenience to this separation into a low- and high-frequency band is that the spectrum of seismic "noise" peaks around 0.1 Hz. Noise around this frequency is primarily excited by ocean waves striking continental margins, which then propagates into the interior of continents as surface waves.

For more reading:
Aki and Richards (2002).

Appendix 1.3 Seismic nomenclature

Body waves:

At the high frequencies of body waves, a ray theoretical approach to the solution of the elastic equation of motion is often a very accurate approximation, in which rays (normals to wavefronts) are bent by changes in elastic velocities according to Snell's law, and amplitudes are predicted by frequency-independent estimates of geometric spreading. Important exceptions to the accuracy of ray theory occur when ray trajectories graze velocity discontinuities or when the formula for geometric spreading becomes singular near surfaces of focusing (caustics). When ray theory is a good approximation, Snell's law predicts that with velocities increasing with depth, ray curvature will be concave upward. Upon incidence on a discontinuity in velocity, rays of transmitted waves will be bent more toward the radial direction for a velocity decrease and more away from the radial for a velocity increase. These effects are seen in the ray paths shown in Fig. 1.18, illustrating the effects of a P velocity decrease from the mantle to the outer core, followed by a P velocity increase from the outer core to the inner core. The nomenclature for specific body waves uses uppercase letters to denote the type of wave along segments in the mantle (P or S), compressional wave segments in the outer core (K) and inner cores (I), and shear wave segments in the inner core (J). Lowercase p and s are used to denote surface reflections from the upgoing rays of P and S waves, close to deep focus earthquakes. Lowercase i and c are used to denote reflections from the outer core and inner core, respectively.

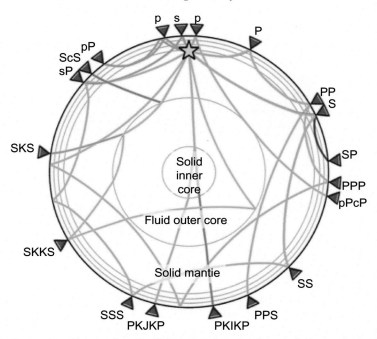

FIG. 1.18 Example body waves from a deep focus earthquake: their nomenclature and ray paths interacting with Earth's surface, outer core, and inner core.

Normal modes:

Subscripts n and l and superscripts m in normal mode nomenclature denote the position of nodes of zero motion in radius and longitude, respectively. Motion reverses sign on either side of the node. The subscript n denotes the radial order or overtone number, with displacement reversing sign n times as a function of radius (depth) in Earth. The angular order number l denotes the number of nodal planes on the surface, and the value of azimuthal order number m affects the ordering of the surface nodal planes as illustrated in Fig. 1.19. Two modes that cannot exist are $_0S_1$, which would require the displacement of the center of gravity, and $_0T_1$, which would require the entire spherical Earth to twist back and forth, violating the conservation of angular momentum (Fig. 1.20).

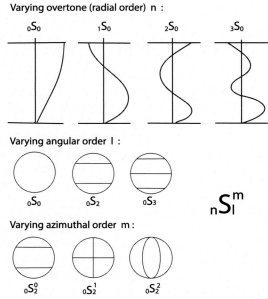

FIG. 1.19 Example spheroidal free oscillations showing effects of nodal patterns in depth and on the surface as n, l, and m are varied. Note: $_1S_0$ can exist but not $_0S_1$.

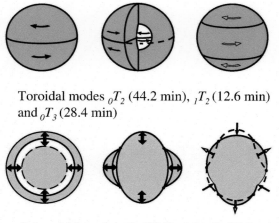

Toroidal modes $_0T_2$ (44.2 min), $_1T_2$ (12.6 min) and $_0T_3$ (28.4 min)

Spheroidal modes $_0S_0$ (20.5 min), $_0S_2$ (53.9 min) and $_0S_3$ (25.7 min)

FIG. 1.20 Sense of displacements and nodes of example toroidal and spheroidal free oscillations. *From Stein, S., Wyssession, M., 2009. An Introduction to Seismology, Earthquakes, and Earth Structure, Wiley-Blackwell Publishing, 512 pp.*

Appendix 1.4 Adams-Williamson equation and the Bullen parameter

Density models of Earth are built from the constraints provided by its mass and moment of inertia. Seismic velocities and normal mode eigenfrequencies provide an additional constraint from their estimate of K/ρ (incompressibility/density).

Hydrostatic equilibrium and the incompressibility of materials can be exploited to predict the density gradient within chemically homogeneous regions. The combined effects enable us to write the density gradient with depth by a product of two partial derivatives:

The derivative $\frac{\partial \rho}{\partial P}$ measures the effects of compressibility, or how density changes (generally increases) with changes in pressure. Using the definition of the bulk modulus (incompressibility) $K = -V\frac{\partial P}{\partial V}$ or $\rho\frac{\partial P}{\partial \rho}$, the derivative $\frac{\partial \rho}{\partial P}$ can be written as

$$\frac{\partial \rho}{\partial P} = \frac{\rho}{K} \tag{1.21}$$

For a point of material in hydrostatic equilibrium in a spherically symmetric planet, the rate of change of pressure with respect to radius is given by

$$\frac{\partial P}{\partial r} = -\rho g \tag{1.22}$$

The density gradient can then be expanded by the chain rule using these two partial derivatives so that

$$\frac{d\rho}{dr} = \frac{\partial \rho}{\partial P}\frac{\partial P}{\partial r} = -\frac{\rho^2 g}{K} \tag{1.23}$$

The ratio $\frac{\rho}{K}$ can be written in terms of the seismic velocities V_P and V_S as

$$\frac{\rho}{K} = \frac{1}{V_P^2 - 4/3 V_S^2} \equiv \frac{1}{\phi_S} \tag{1.24}$$

where ϕ_S is defined as the seismic parameter. Substituting this expression into Eq. (1.23) gives an expression for the density gradient in terms seismic velocities and the gravity acceleration at any radius r by

$$\frac{d\rho}{dr} = -\frac{\rho g}{\phi_S}. \tag{1.25}$$

Eq. (1.25), known as the Adams-Williamson equation, assumes adiabaticity in a homogeneous self-compressed material, which is equivalent to the assumption of neutral buoyancy at any point at which the density gradient satisfies the equation.

In the case of a nonadiabatic temperature gradient, a correction term is added to the Adams-Williamson equation that includes the coefficient of thermal expansion, $\alpha = -\frac{1}{\rho}\left(\frac{\partial \rho}{\partial T}\right)_P$, and the difference between the local temperature gradient and the adiabatic temperature gradient, $\tau = \left(\frac{\partial T}{\partial r}\right) - \left(\frac{\partial T}{\partial r}\right)_S$. With this correction, the density gradient becomes

$$\frac{d\rho}{dr} = -\frac{\rho g}{\phi_S} + \alpha \tau \rho \tag{1.26}$$

Although both the mantle and liquid outer core are convecting, the correction term in the Adams-Williamson equation is estimated to be small enough that it can be usually ignored in constructing Earth density profiles (Stacey and Davis, 2008).

In the most vigorously convecting regions such as Earth's outer core, τ is vanishingly small.

The correction to the Adams-Williamson equation can be better understood by defining another parameter that equivalently measures departures from adiabaticity and homogeneity. This is the Bullen parameter. Expanding $\frac{d\rho}{dr}$ by the chain rule to $\frac{\partial \rho}{\partial P}\frac{\partial P}{\partial r}$ and substituting $-\rho g$ for $\frac{\partial P}{\partial r}$ leads to an equivalent form of the Adams-Williamson equation, which defines a parameter η_B:

$$\eta_B = \phi_S \frac{\partial \rho}{\partial P} = -\frac{\phi_S}{\rho g}\frac{\partial \rho}{\partial r} \tag{1.27}$$

The Bullen parameter η_B essentially measures the ratio of the predicted density gradient from the Adams-Williamson equation to the observed density gradient. $\eta_B = 1$ in regions that are homogeneous and adiabatic; $\eta_B < 1$ in thermal boundary regions that separate regions that do not convectively mix; and $\eta_B > 1$ in regions having either strong chemical heterogeneity or phase changes. Matyska and Yuen (2001) provide an expanded definition of the Bullen parameter that can be calculated in three-dimensionally varying Earth models.

For more reading:
Anderson (2007), Matyska and Yuen (2001).

Appendix 1.5 Viscoeleastic attenuation parameterization

In viscoelastic attenuation, the strain ε induced by elastic displacement lags the stress σ that excited the displacement, resulting in a stress-strain hysteresis curve excited by a cycle of stress. The behavior of stress and strain is analogous to the phase lag in the response to forces in mechanical systems composed of springs and dashpots and the phase lag in current from voltage applied to circuits composed of resistors and capacitors. The theory of this behavior is also equivalent to the theory of the index of refraction for electromagnetic waves in materials, which results in a complex, frequency-dependent index of refraction.

Mathematical representations:

Complete versions of the theory provide a mathematical description that includes the effects of recoverable short-term strain due to an oscillatory application of stress, as well as nonrecoverable strain due to a long-term application of a constant stress. An example of one of the simplest spring/dashpot models of such a rheology is the Burgers model, in which the time-dependent rheology is described by the equation:

$$\varepsilon(t) = \sigma_0 \{ J_U + \delta J \left[1 - \exp(-t/\tau) \right] + t/\eta \} \tag{1.28}$$

From left to right, the three terms on the right-hand side of the equation represent the elastic response of the material with an elastic compliance J_U, the anelastic response of the material due to recoverable strain, and the nonrecoverable strain due to fluid-like deformation of the material having a Newtonian viscosity η. $J(t)$, defined by the terms in the braces {}, is termed the creep compliance function. For short-term applications of stress (small t), the term containing viscosity, representing the nonrecoverable strain due to fluid-like flow, can be dropped. As t approaches 0 the 2nd anleastic term in the braces tends to 0 and material becomes fully elastic with $\varepsilon = J_U \sigma$. The value of time constant τ is a relaxation time that controls whether strain is in phase with the applied stress. For time long compared to the relaxation time, the elastic compliance becomes $J_U + \delta J = J_R$, where J_R is defined as the relaxed compliance. Fourier transforming the equation above defines a complex, frequency-dependent compliance,

$$\widehat{J}(\omega) = J_U + \delta J/(1 + i\omega\tau) - 1/\eta\omega \tag{1.29}$$

where $\widehat{\varepsilon}(\omega) = \widehat{J}(\omega)\,\widehat{\sigma}(\omega)$.

In the frequency domain, the complex elastic modulus is defined by $\widehat{G}(\omega) = 1/\widehat{J}(\omega)$. Phase velocities $\widehat{v}(\omega)$ of body waves become complex and frequency-dependent (dispersive) through the relation $\widehat{v}(\omega) = \sqrt{\frac{G(\omega)}{\rho}}$, where ρ is density. The imaginary part of $\widehat{v}(\omega)$ will give rise to an attenuation of waves as they propagate. Since both real and imaginary parts of velocity depend on frequency; both real velocity and attenuation will depend on frequency. We can see that as frequency approaches zero, the term containing viscosity will dominate and the material behaves like a fluid having a Newtonian viscosity. In the seismic frequency band, the elastic and anelastic terms will dominate. Within the seismic band, and depending on the magnitude of the product $\omega\tau$, the real part of the elastic compliance, elastic modulus, and phase velocity will correspond to either the relaxed state for $\omega\tau < 1$ or the unrelaxed state for $\omega\tau > 1$.

The Q parameter and relaxation spectrum:

A parameter that measures the magnitude of attenuation due to anelasticity is the seismic Q parameter. A viscoelastic Q can be defined by the average energy W per cycle divided by the energy lost or work done per cycle, ΔW:

$$Q = \frac{W}{\Delta W} \tag{1.30}$$

Measured as a depth-dependent property of the Earth, Q is defined from the average energy density and loss per cycle of a complex plane wave. From this definition, it can be shown that $Q = \dfrac{\text{Re}(\widehat{G})}{\text{Im}(\widehat{G})}$, where $\widehat{G}$ is the complex elastic modulus. It is often less confusing to report the reciprocal parameter Q^{-1}, termed as either the attenuation or attenuation parameter, which represents the usually small perturbations to perfect elasticity.

Given a viscoelastic rheology, the attenuation due to anelasticity is always frequency-dependent and dispersive, controlled by the value of a relaxation time. The example discussed thus far has considered only a single relaxation time. Most materials, however, exhibit a spectrum of relaxation times governed by the different anelastic behavior of bonds between different types of atoms and the distribution and types of lattice defects. Hence, a resonant peak of absorption centered on a single frequency, sometimes termed as Debye peak, will not dominate attenuation. Instead, the dependence of attenuation on frequency will be characterized distribution of Debye peaks, each with its own relaxation time. This distribution is termed as the relaxation spectrum. An important restriction on the relaxation spectrum is that the attenuation parameter Q^{-1} cannot increase or decrease faster than a ± 1 power of frequency. Fig. 1.21 is an example of a type of relaxation spectrum commonly assumed in seismic observations, in which the attenuation is nearly constant or slowly varying in a frequency band between two limiting corner frequencies. Maximum velocity dispersion occurs between these two corners. A common assumption is that the width of the zone of near constant attenuation spans at least five decades of frequency.

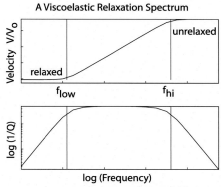

FIG. 1.21 An example viscoelastic relaxation spectrum commonly assumed in seismic investigations.

Practical constraints and applications from Q measurements:

For comparison with laboratory scale measurements or for inclusion in applications of thermodynamic models of Earth mineralogy, the relaxed modulus ($\omega \to 0$) and the unrelaxed modulus ($\omega \to \infty$) may be equated with the isothermal and adiabatic elastic moduli, respectively. The difference between the unrelaxed and relaxed moduli (velocities) is termed the modulus (velocity) defect. Modulus defects in Earth materials are rarely higher than 25% (12.5% in velocity), which places a limit on the product of the width of the relaxation spectrum and the peak attenuation. For a relaxation spectrum five decades wide in frequency, over which attenuation is assumed to be nearly constant or slowly varying, the maximum attenuation conforming to a modulus defect can be estimated from

$$\frac{v_U - v_R}{v_U} = \frac{Q_{\max}^{-1} \ln(\tau_1/\tau_2)}{\pi} \tag{1.31}$$

where v_U and v_R are the unrelaxed and relaxed velocities and τ_1 and τ_2 are the relaxation times corresponding to the low- and high-frequency corners of a relaxation spectrum having near constant attenuation between the corner frequencies. For a 12.5% velocity defect, occurring over five decades of frequency, $Q_{\max}^{-1} = 0.034$ and $Q_{\min} = 30$.

The assumption of a viscoelastic rheology can provide some weak constraints on the viscosity of a region of Earth from estimates of the shear wave Q parameter and the unrelaxed shear modulus G_U. For example, for a frequency band in which the rheology is well approximated by a Burgers solid model having a single anelastic relaxation time, the viscosity is related to the unrelaxed modulus G_U, Q, and frequency by

$$\eta = \frac{G_U Q}{2\pi f} \tag{1.32}$$

where frequency f is assumed to be in the domain $\omega\tau > 1$. For this approximation to be more than minimally useful requires an estimate of the high-frequency corner of the relaxation spectrum, since Q^{-1} will decay as ω^{-1} (Q increasing as ω) when $\omega\tau > 1$.

Relative attenuation of P and S waves:

The attenuation for traveling elastic waves can be decomposed into losses due to *viscoelasticity* in shear or bulk moduli. Since the velocity of a P wave depends on both the bulk and shear moduli, the attenuation Q_P^{-1} of a P wave can be written as a linear combination of the attenuations Q_K^{-1} and Q_S^{-1} defined from complex shear and bulk moduli:

$$Q_P^{-1} = L Q_S^{-1} + (1-L) Q_K^{-1} \tag{1.33}$$

where $L = (4/3)(V_S/V_P)^2$ and V_P and V_S are the compressional and shear velocities, respectively (Anderson, 2007). Although plausible mechanisms for defects in bulk moduli have been found in both laboratory measurements and analytic models, measurements on real data find that bulk dissipation in the deep Earth is small and, in most cases, can be neglected. One exception may occur when the pressure and temperature in a region of the Earth are close to a phase transition, either solid-liquid or solid-solid, possibly in Earth's liquid outer core. Except for these regions, intrinsic attenuation occurs almost entirely in shear, associated with lateral movement of lattice defects, grain boundaries and/or fluids rather than with changes in material volume. Assuming $Q_K^{-1} = 0$ and a Poisson solid, for which $V_P = \sqrt{3} V_S$,

$$Q_P^{-1} = \frac{4}{9} Q_S^{-1} \tag{1.34}$$

Morozov (2015), however, has demonstrated that assuming zero bulk attenuation leads to an auxetic medium, in which Poisson's ratio becomes negative. A number of studies in the frequency band of free oscillations have cited the need for some bulk attenuation in the solid regions of Earth on the order of $Q_K^{-1} = 0.002$ to agree with the observed attenuation of spheroidal modes. Without knowledge or good constraints on the frequency dependence or shape of the bulk relaxation spectrum, it has proved difficult to partition the required bulk attenuation with depth. Plausible arguments can be made for bulk attenuation due to the effects of phase transitions in the upper mantle, outer core, inner core, or all three.

For more reading:
Anderson (2007), Cooper (2002), Jackson (2007), Cormier (2020).

Appendix 1.6 Birch's law and seismic velocity/density relations

Rooted in experimental work leading to the Birch-Murnaghan equation of state (Appendix 1.9), Francis Birch (1960) verified a simple linear relationship between seismic P velocity and density as a function of mean molecular weight at pressures up to 10 kbar, which can be exploited as a constraint on chemistry in at least Earth's upper mantle. In that pressure range, Birch found that P velocity is well approximated by a linear function of density of the form

$$V_P = a + b\rho \tag{1.35}$$

where a is a function of the mean molecular weight of the mineral. For the oxides and silicates composing the mantle for which the mean molecular weight is close to 20, Birch found a good fit to observed P velocities with $a = -1.87$ and $b = 3.05$. By applying Birch's empirical formula in the ranges of pressure and mineralogy where it has been experimentally verified, chemical variations may be detected in Earth models constructed from seismic observations of free oscillations and normal modes.

From Birch's Law, proportionalities between fractional ratios of fluctuations of P velocity, S velocity, and density $\left(\frac{\Delta V_P}{V_P}, \frac{\Delta V_S}{V_S}, \frac{\Delta \rho}{\rho}\right)$ can be estimated and applied to predict the effects of scattering by small-scale chemical heterogeneity. These ratios can also be expressed as derivatives of the natural logarithm of velocities or density, $\left(\frac{d \ln V_P}{dV_P}, \frac{d \ln V_S}{dV_S}, \frac{d \ln \rho}{d\rho}\right)$. Assuming the Birch's values for parameters a and b for upper mantle oxides and silicates and the measured ranges for P and S velocity corresponding to pressures in the upper 400 km of the mantle gives

$$\frac{\Delta \rho}{\rho} \cong 0.8 \frac{\Delta V_P}{V_P}. \tag{1.36}$$

Coupled with separate observations of $\frac{d \ln V_s}{d \ln V_P} = 1.6$ in the same depth region, results in a prediction that

$$\frac{d \ln \rho}{d \rho} = 0.5 \frac{d \ln V_S}{d V_S}. \tag{1.37}$$

Large departures in the assumption of $\frac{d \ln V_s}{d \ln V_P} = 1.6$ have been identified from tomographic images of the mantle below 1000 km and in experiments on mantle minerals at temperatures and pressures at and below the upper mantle phase transitions starting at 400 km depth. These find that $\frac{d \ln V_s}{d \ln V_P} \geq 2$, with some estimates exceeding 3 or 4 near the core-mantle boundary.

Although a linear form of Birch's law performs surprisingly well in the mantle, errors in its linearization have been identified in laboratory and ab initio estimates, especially at pressures approaching those in the inner core. Advances in these experiments provide the best reference constraints on Earth chemistry from seismic velocities. Modifications of Birch's law have been sought for the case of hcp iron, which is a favored candidate for the composition and lattice state of the inner core. At inner core temperatures and pressures, Sakamaki et al. (2016) fit a modified form of Birch's law to incorporate the effects of both temperature and pressure, and Roy and Sarkar (2017) achieve a better fit to the P velocity and density behavior of hcp iron by assuming a linear fit of the product $V_P \sqrt[3]{\rho}$ plotted against ρ.

For more reading:
Birch (1960), Roy and Sarkar (2017), Sakamaki et al. (2016).

Appendix 1.7 Composite elastic moduli and density

Composites of isotropic crystals:

Several schemes are commonly used for estimating the elastic moduli and density of polycrystalline solids. The simplest ones assume that each crystal in the composite is elastically isotropic and occurs in a known volume fraction. The Voight average assumes that the strain is constant throughout the composite. The Voight average estimate for the composite bulk modulus is given by

$$K_V = \sum_i V_i K_i \tag{1.38}$$

where V_i is the fraction of the total volume of material i having bulk modulus K_i.

The Reuss average assumes that stress is constant throughout the composite. The Reuss average estimate for the bulk modulus is given by.

$$K_R = \sum_i \left(\frac{V_i}{K_i}\right)^{-1} \tag{1.39}$$

The Voight-Reuss-Hill estimate is the average between the Voight and Reuss estimates:

$$K_{VRH} = \frac{(K_V + K_R)}{2} \tag{1.40}$$

The shear modulus of the composite is calculated by the same averaging schemes.

In the computation of the elastic velocities estimates of both the composite elastic moduli and composite density must be known. The density of composite is simply given by

$$\rho = \frac{\sum_i \rho_i V_i}{\sum_i V_i} \tag{1.41}$$

The BURNMAN computer code (Heister et al., 2016) has options for computing the Voight, Reuss, and Voight-Reuss-Hill estimates of the composite moduli, their Hashin-Strikhman bounds, and the composite density.

Composites of randomly oriented anisotropic crystals:

In the case of a single compound consisting of intrinsically anisotropic, randomly oriented crystals, the Voight and Reuss averages are defined from the six unique elastic constants of the compound's unit cell. These are given by elements of the fourth order elastic constant tensor c_{ijkl}, but allow only physically realizable media, such that only six unique elastic moduli are required. These are termed the Kelvin stiffness parameters arranged for calculations with the elastic constitutive relation in the form of a 6×6 matrix such that pairs of indices ij or kl in c_{ijkl} map into a single index. The mappings are $11 \to 1$, $22 \to 2$, $33 \to 3$, $23, 32 \to 4$, $13, 31 \to 5$, and $12, 21 \to 6$. For example, $c_{1111} = C_{11}$, $c_{3313} = C_{35}$, $c_{2323} = c_{3223} = C_{44}$. With these definitions, the formulas for the Voight and Reuss averages for bulk modulus are given by

$$K_V = \frac{2C_{11} + 2C_{12} + C_{33} + 4C_{13}}{9}$$
$$K_R = \frac{1}{2C_{11}^{-1} + 2C_{12}^{-1} + C_{33}^{-1} + 4C_{13}^{-1}}$$

(1.42)

The formulas for the Voight and Reuss averages for shear modulus are given by.

$$\mu_V = \frac{2C_{11} - 2C_{12} + C_{33} - 2C_{13} + 6C_{44} + 3C_{66}}{15}$$
$$\mu_R = \frac{15}{2C_{11}^{-1} - 2C_{12}^{-1} + C_{33}^{-1} - 2C_{13}^{-1} + 6C_{44}^{-1} + 3C_{66}^{-1}}$$

(1.43)

Hashin-Strikhman bounds:

The Hashin-Strikhman (HS) bounds provide narrower error bounds than that given by the difference between the Voight and Reuss averages. For a composite composed of two isotropic materials, the Hashin-Strikhman bounds for the bulk and shear moduli are

$$K_{HS} = K_1 + \frac{V_2}{(K_2 - K_1)^{-1} + V_1(K_1 + 4/3\mu_1)^{-1}}$$
$$\mu_{HS} = \mu_1 + \frac{V_2}{(\mu_2 - \mu_1)^{-1} + \frac{2V_1(K_1 + 2\mu_1)}{5\mu_1(K_1 + 4/3\mu_1)}}$$

(1.44)

The HS upper bounds are obtained when the medium 1 has the larger moduli; lower bounds when medium 2 has the larger moduli. Schemes for computing the VHR estimates and Hashin-Strikman bounds on composites of n intrinsically anisotropic crystals having differing types of symmetry and orientations continue to be developed, e.g., Man and Huang (2011).

For more reading:
Heister et al. (2016), Man and Huang (2011), Hashin and Strikhman (1963).

Appendix 1.8 Elastic anisotropy

At the level of a mineral having a uniform arrangement of atoms, termed its lattice, its elastic rheology is naturally anisotropic because the elastic properties of bonds between atoms are different in different directions. In most depth ranges, the effect of the polycrystalline nature of Earth tends to volume-average these differences to zero and elastic isotropy is an excellent approximation. Exceptions are regions where mechanisms exist to preferentially align lattice unit cells along directions related to flow, deposition, or crystallization. In the case of core structure, evidence of such exceptions occur in the mantle D″ region above the outer core, due to either concentrations of intrinsically anisotropic minerals preferentially oriented by mantle convective flow near the core-mantle boundary, and in the inner core, due to either preferential directions of crystallization or slow convective deformation that orients crystals.

To include the effects of elastic anisotropy require a higher number of elastic moduli c_{ijkl} (at least 5). These effects include three possible body waves with different, directionally dependent, phase and group velocities and mutually orthogonal polarizations. The fastest of these waves is still called P waves, but the two slower waves are called quasi S waves. A frequent starting point to demonstrate these properties is to substitute a plane wave representation of the type in Eq. (1.45) for the displacement into the elastic equation of motion (Eq. 1.12 of Appendix 1.2, ignoring Coriolis and gravitational force densities), and assuming the most general form of the elastic constitutive relation for the stress tensor σ_{ij} (Eq. 1.13 of Appendix 1.2):

$$\vec{u} = \vec{g}^m \exp\left[i\left(\vec{k}^m \cdot \vec{x} - \omega t\right)\right] \quad (1.45)$$

These substitutions will define an eigenvector/eigenvalue problem incorporating the tensor of elastic moduli c_{ijkl}, which can solved to determine the three possible eigenvectors of body wave polarization $\vec{g}^m$ ($m = 1, 2, 3$) associated with three eigenvalues v_m^2 equal to the square of the phase velocity. In component form, this eigenvector/eigenvalue problem is expressed as

$$\left(c_{ijkl} n_i n_l - v_m^2 \delta_{jk}\right) g_j^m = 0 \quad (1.46)$$

(Zhou and Greenhalgh, 2008), where the vector $\vec{n}^m = \frac{\vec{k}^m}{|\vec{k}^m|}$ is the normal to the wavefront associated with eigenvector m. Once the eigenvectors $\vec{g}^m$ are determined from the equation above, one can multiply the expression by eigenvector components g_k^m and use the properties of the Kroneker delta δ_{jk} to determine the phase velocities as

$$v_m = \sqrt{c_{ijkl} n_i n_l g_j^m g_k^m} \quad (1.47)$$

For general anisotropy, the group velocity $\vec{U}^m$ (the velocity of energy propagation) is no longer in the same direction as phase velocity $\vec{v}_m$. In component form, a formula for the group velocity is given by

$$U_i^m = \frac{c_{ijkl}}{v_m} n_l g_j^m g_k^m \quad (1.48)$$

For body waveforms, sampling the anisotropic regions of the deep Earth, the differences between phase and group velocity are small enough that they can usually be ignored in interpretations of waveforms.

It is important to note that the polarization of the two quasi S waves depends on material properties (specific elastic moduli) and not on the reference SH (transverse) and SV (radial) polarizations used to resolve the components of the polarization of an S wave in an isotropic material. In general, the polarization of the two quasi S waves are neither SH nor SV polarized and will appear as two separate pulses on both radial and transverse components of motion (Fig. 1.22).

Although shear wave splitting is an unambiguous evidence of elastic anisotropy, most of the evidence for anisotropy in Earth's inner core has come from the directional dependence of P velocities, interpreted from applying a perturbation theory to the solution of the equation of motion. The type of anisotropy analyzed is called transverse isotropy. It assumes five elastic constants. The directional dependence of P and S velocities is specified by three parameters that may be estimated from data. The directional dependence of velocities is given through an angle a ray (normal to a wavefront) makes, an angle θ with respect to an axis of symmetry of the medium. The propagation of body waves

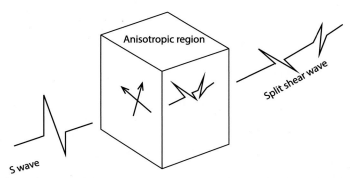

FIG. 1.22 Upon entering a region having elastic anisotropy a shear wave's energy is split into two quasi-shear wave pulses, orthogonally polarized in directions shown by the arrows on the face of the plane of incidence shown above. The orientations of the polarizations depend on the elastic constants of the region. These polarizations are retained after the waves exit the anisotropic region, accumulating a travel time difference that depends on the length of their paths in the anisotropic region.

in materials consisting of planar layers thinner than the wavelength having differing composition and isotropic velocities is well represented by transverse isotropy with an axis of symmetry perpendicular to the planar layers. The directional dependence of velocities in a transversely isotropic material with a vertical axis of symmetry is well approximated by the equations:

$$V_P(\theta) = V_{P0}(1 + \delta \sin^2\theta \cos^2\theta + \varepsilon \sin^4\theta)$$
$$V_{SV}(\theta) = V_{S0}\left[1 + \left(\frac{V_{P0}}{V_{s0}}\right)^2 (\varepsilon - \delta) \sin^2\theta \cos^2\theta \right] \quad (1.49)$$
$$V_{SH}(\theta) = V_{S0}(1 + \gamma \sin^2\theta)$$

(Thomsen, 1986). The velocities V_{P0}, V_{S0} are the P and S velocities in the unperturbed isotropic medium and the parameters ε, δ, γ are nondimensional functions of the five elastic constants, each of which must be $\ll 1$ to maintain the validity of the perturbation approach. The axis of symmetry need not be vertical, in which case the S velocities V_{SV}, V_{SH} are the two quasi S velocities that will arrive and interfere on both the transverse (SH) and radial (SV) components of motion. In application to the measurement of inner core anisotropy, the angle θ is taken to be the angle that a P wave ray makes along its path with respect to an axis of symmetry taken to be a line passing the center of the Earth, often assumed to be the rotational axis.

For more reading:
Thomsen (1986).

Appendix 1.9 Equations of state

An equation of state is any equation that predicts how the volume or density of a material changes as a function of temperature and pressure. In the inner core, equations of state are helpful in predicting and constraining the lattice structure of iron. Important physical properties in equations of state are partial derivatives of volume with respect to temperature and pressure, which are commonly defined by the bulk modulus or incompressibility and the coefficient of thermal expansion. An additional helpful physical property is the Gruneisen parameter because it changes very little over large ranges of pressure and temperature.

Two possible equations of state are the Birch-Murnaghan and the Vinet equations. These are defined using the bulk modulus and its pressure derivative at zero pressure and K_o, K_o', and the volume at V_o at zero pressure. Since the ratio of zero pressure volume to compressed volume, $\overline{V} = V_0/V$, can be significantly larger than 1 as pressure increases, these equations of state are often termed finite strain theory to contrast their behavior from the predictions of elastic strain theory, in which the magnitude of volumetric strains $\left|\frac{V-V_0}{V_0}\right|$, are assumed to be infinitesimal, i.e., $\ll 1$. In the following, the Birch-Murnaghan equation of state is given by Eq. (1.50a) and the Vinet by Eq. (1.50b).

$$P(V) = \frac{3K_0}{2}\left[\overline{V}^{\frac{7}{3}} - \overline{V}^{\frac{5}{3}}\right]\left\{1 + \frac{3}{4}(K_0' - 4)\left[\overline{V}^{\frac{2}{3}} - 1\right]\right\} \quad (1.50a)$$

$$P(V) = 3K_{0S}\overline{V}^{-\frac{2}{3}}\left(1 - \overline{V}^{\frac{1}{3}}\right) \exp\left[\eta\left(1 - \overline{V}^{\frac{1}{3}}\right)\right]$$
$$\text{where } \eta = \frac{3}{2}(K_0' - 1) \quad (1.50b)$$

In Earth's liquid outer core, equations of state together with seismic P velocity can be used to either constrain its chemistry or at least assist in identifying chemical anomalies near its upper and lower boundaries. The Vinet equation of state has been used by Irving et al. (2018) to test the assumption of chemical homogeneity in the outer core when the seismically determined P velocity or bulk sound speed is used to assign the bulk modulus K, while fixing the pressure at the top and bottom of the outer core. That study was able to achieve a good fit to the predicted travel times of SmKS waves reflecting along the underside of the outer core boundary by assuming a chemically homogeneous outer core without the need for a chemically stratified layer in the upper most outer core.

For more reading:
Stacey et al. (1981).

References

Aki, K., Richards, P.G., 2002. Quantitative Seismology. University Science Books.
Alfè, D., Gillan, M.J., 1998. First-principles calculation of transport coefficients. Phys. Rev. Lett. 81, 5161.
Anderson, D.L., 2007. New Theory of the Earth, second ed. Cambridge University Press.
Bills, B.G., Rubincam, D.P., 1995. Constraints on density models from radial moments: application to the Earth, Moon and Mars. J. Geophys. Res. 100, 26305–26315.
Birch, F., 1960. The velocity of compressional waves in rocks to 10 kilobars, Part 1. J. Geophys. Res. 65 (4), 1083–1102. https://doi.org/10.1029/JZ065i004p01083.
Cooper, R.F., 2002. Seismic wave attenuation: energy dissipation in viscoelastic crystalline solids. In: Karato, S.-I., Wenk, H.R. (Eds.), Reviews in Mineralogy and Geochemistry. vol. 51(1), pp. 253–290, https://doi.org/10.2138/gsrmg.51.1.253.
Cormier, V.F., 2009. A glassy lowermost outer core. Geophys. J. Int. 179 (1), 374–380. https://doi.org/10.1111/j.1365-246x.2009.04283.x.
Cormier, V.F., 2020. Seismic viscoelastic attenuation. In: Encyclopedia of Solid Earth Geophysics. Encyclopedia of Earth Sciences Series, Springer, Cham, https://doi.org/10.1007/978-3-030-10475-5_55-1.
Cormier, V.F., Richards, P.G., 1976. Comments on "The Damping of Core Waves" by Anthony Qamar and Alfredo Eisenberg. J. Geophys. Res. 81, 3066–3068.
Dziewonski, A.M., Anderson, D.L., 1981. Preliminary reference earth model. Phys. Earth Planet. In. 25 (4), 297–356.
Hashin, Z., Strikhman, S., 1963. A variational approach to the elastic behavior of multiphase materials. J. Mech. Phys. Solids 11 (2), 127–140. https://doi.org/10.1016/0022-5096(63)90060-7.
Heister, T., Unterborn, C., Rose, I., Cottaar, S., 2016. Burn Man v0.9 [software]. In: Computational Infrastructure for Geodynamics., https://doi.org/10.5281/zenodo.546210. https://zenodo.org/record/546210.
IRIS DMC, 2011. Data Services Products: EMC, A Repository of Earth Models. https://doi.org/10.17611/DP/EMC.1.
Irving, J.C.E., Cottar, S., Lekić, V., 2018. Seismically determined elastic parameters for Earth's outer core. Sci. Adv. 27 (4). https://doi.org/10.1126/sciadv.aar2538.
Jackson, I., 2007. Properties of rocks and minerals—Physical origin of anelasticity and attenuation in rocks. In: Schubert, G. (Ed.), Treatise on Geophysics. vol. 2. Elsevier, Amsterdam, pp. 493–525.
Jeffreys, H., Bullen, K.E., 1940. Seismological tables. British Association Seismological Committee, London.
Kennett, B.L.N., Engdahl, E.R., Buland, R., 1995. Constraints on seismic velocities in the earth from travel times. Geophys. J. Int. 122, 108–124.
Knopoff, L., 1955. Interaction between elastic wave motions and a magnetic field in electrical conductors. J. Geophys. Res. 60 (4), 441–456.
Lautrup, B., 2011. Chapter 15: viscosity. In: Physics of Continuous Matter. CRC Press, Taylor & Francis Grouop, Boca-Raton, pp. 257–259.
Lehmann, I., 1936. Publications du Bureau Central Séismologique International. vol. A14(3), pp. 87–115.
Leyton, F., Koper, K.D., 2007. Using PKiKP coda to determine inner core structure: 2. Determination of Q_C. J. Geophys. Res. 112. https://doi.org/10.1029/2006JB004370.
Mäkinen, A.M., Deuss, A., 2013. Normal mode splitting function measurements of anelasticity and attenuation in Earth's inner core. Geophys. J. Int. 194 (1), 401–416. https://doi.org/10.1093/gji/ggt092.
Man, C.-S., Huang, M., 2011. A simple explicit formula for the Voight-Hill-Reuss average of elastic polycrystals with arbitrary crystal and texture symmetries. J. Elast. 105, 29–45.
Matyska, C., Yuen, D.A., 2001. Bullen's parameter: a link between seismology and geodynamical modeling. Earth Planet. Sci. Lett. 198, 471–483.
Morozov, I., 2015. On the relation between bulk and shear seismic dissipation. Bull. Seismol. Soc. Am. 105, 3180–3188.
Mound, J., Buffett, B., 2007. Viscosity of the Earth's fluid core and torsional oscillations. J. Geophys. Res. 112 (B5). https://doi.org/10.1029/2006JB004426.
Nolet, G., Dahlen, F.A., 2000. Wave front healing and the evolution of seismic delay times. J. Geophys. Res. 105 (B8), 19043–19054. https://doi.org/10.1029/2000JB900161.
Oldham, R.D., 1906. The constitution of the interior of the earth as revealed by earthquakes. Q. J. Geol. Soc. Lond. 62, 459–486.
Park, J., Song, T.-R., Tromp, J., Okale, E., Stein, S., Roult, G., Clevede, E., Laski, G., Kanamori, H., Dvis, P., Berger, J., Braitenberg, C., Van Camp, M., Lei, X., Sun, H., Xu, H., Roasat, S., 2005. Earth's free oscillations excited by the 26 December 2004 Sumatra-Andaman Earthquake. Science 308, 1139–1144. https://doi.org/10.1126/sicence.1112305.
Rondenay, S., Cormier, V.F., Van Ark, E., 2010. SKS and SPdKS sensitivity to two-dimensional ultra-low velocity zones. J. Geophys. Res. 115. https://doi.org/10.1029/2009JB006733, B04311.
Roy, U.C., Sarkar, S.K., 2017. Large dataset test of Birch's law for sound propagation at high pressure. J. Appl. Phys. 121, 225901.
Sakamaki, T., Ohtani, E., Fukui, H., Kamada, S., Takhasi, S., Takahta, A., Sakai, T., Tsutsui, S., Ishikawa, D., Shiraishi, R., Seto, Y., Tsuchiya, T., Baron, A.Q.R., 2016. Constraints on Earth's inner core composition inferred from measurements of the sound velocity of hcp-iron in extreme conditions. Sci. Adv. 2 (2). https://doi.org/10.1126/sciadv.1500802.
Sato, H., Fehler, M.C., Maeda, T., 2012. Seismic Wave Propagation and Scattering in the Heterogeneous Earth, second ed. Springer-Verlag.
Shearer, P.M., 2015. Deep earth structure: seismic scattering in the deep earth. In: Schubert, G. (Ed.), Treatise on Geophysics, second ed. In: Dziewonski, A., Romanowicz, B. (Eds.), Deep Earth Seismology, vol. 1. Elsevier Ltd, Oxford, pp. 759–781.
Shearer, P.M., Rychert, C.A., Liu, Q., 2011. On the visibility of the inner-core shear phase PKJKP at long periods. Geophys. J. Int. 185 (3), 1379–1383.
Smylie, D.E., Brazhkin, V.V., Palmer, A., 2009. Direct observations of the viscosity of Earth's outer core and extrapolation of measurements of the viscosity lf liquid iron. Physics-Uspekhi 52 (1), 79–92.
Stacey, F.D., Davis, P.M., 2008. Physics of the Earth, fourth ed. Cambridge University Press, New York.
Stacey, F.D., Brennan, B.J., Irvine, R.D., 1981. Finite strain theories and comparisons with seismological data. Surv. Geophys. 4 (4), 189–232.
Thomsen, L., 1986. Weak elastic anisotropy. Geophysics 51 (10), 1942–2156.
Tkalčić, H., Pham, T.-S., 2018. Shear properties of Earth's inner core constrained by a detection of J waves in global correlation wavefield. Science 362, 329–332. https://doi.org/10.1126/science.aau7649.
Wiechert, E., 1896. Über die Beschaffenheit des Erdinnern. Sitz.-Ber. Physik.-Ökonom. Ges. Königsberg 37 (4), 4–5.

Woodhouse, J.H., 1980. The coupling and attenuation of nearly-resonant multiplets in the earth's free oscillation spectrum. Geophys. J. Roy. Astron. Soc. 61, 261–283.

Zhao, L., Jordan, T.H., Chapman, C.H., 2000. Three-dimensional Frechet differential kernels for seismic delay times. Geophys. J. Int. 141 (3), 558–576.

Zhou, B., Greenhalgh, S., 2008. Velocity sensitivity of seismic body waves to the anisotropic parameters of a TTI-medium. J. Geophys. Eng. 5 (3), 245–255. https://doi.org/10.1088/174202132/5/3/001.

CHAPTER

2

Chemical and physical state of the core

We first review the composition of the core: primarily Fe, Ni, and the case for light elements. With the assumption that the entire outer core is convecting so that the geotherm lies along the adiabat, and with the inner core boundary temperature fixed at the Fe alloy melting temperature, we derive the core geotherm in terms of reasonably constrained geophysical parameters. We then examine the physics of the important, but not well-constrained, core transport properties. Thermodynamics allows us to estimate the age of the inner core, and the power available for the geodynamo, as well as to frame a discussion of stably stratified layers in the core. We conclude the chapter with a discussion of inner core mineralogy: its stable phase, and elastic and anelastic properties.

2.1 Composition of the core

2.1.1 Fe-Ni

The case for Fe is based in part on the cosmochemical abundance of the elements. Because Fe-56 is the most stable nuclide (it has the lowest mass per nucleon of all the elements), it is produced in great abundance by thermonuclear reactions in main sequence stars; in particular, via fusion of 14 α-particles to form Ni-56, followed by two β$^+$ (positron emission) decays. Because of this abundance of Fe, because the seismically inferred density of the core matches crudely with that of Fe under high pressure, and because of the existence of Fe meteorites that are thought to be the remnants of planetesimal cores, it is widely accepted that Fe is the primary component of the terrestrial planetary cores.

The case for Ni is interesting because it seems to have little effect on core properties, though that gets re-examined periodically. Its density at core pressures is seismically indistinguishable from that of Fe. The cosmic abundance of Ni is 4–5 wt-pct, and Fe meteorites typically contain 5–25 wt-pct Ni. There are two common crystalline *phases* in Fe-Ni meteorites (Fig. 2.1), kamacite with a body-centered cubic (bcc) structure, which is typically 5–10 wt-pct Ni, and taenite, with a face-centered cubic (fcc) structure, which is typically 20–65 wt-pct Ni. As a meteor cools kamacite precipitates from taenite, and the crystallographic relationship between the two leads to the characteristic Widmanstätten pattern of Fe meteorites (Fig. 2.2). The pattern can also be present in steel and other industrial alloys, but on a much smaller lengthscale due to the more rapid cooling rate, and where it can lead to brittleness. This does lead one to wonder whether the presence of Ni might affect the deformation or viscosity of the inner core. Based on its cosmic abundance, its content in Fe meteorites, and its moderately *siderophile* (i.e., Fe-loving) nature, Ni likely comprises at least 6 wt-pct of the core, but it could perhaps be as high as 10 wt-pct. Chapter 8 discusses in more detail the cosmic abundances of the core-forming elements.

2.1.2 Light elements

In 1952 Francis Birch wrote that, beyond Fe-Ni, the core is *"an uncertain mixture of all the elements"* (Birch, 1952). The state of knowledge today is perhaps somewhat better. A comparison of the density of pure Fe (or Fe-Ni) under outer core conditions determined from both high-pressure experiments and *ab initio* (i.e., quantum mechanical; sometimes also referred to as first principles) calculations with the outer core density inferred seismically implies that the outer core is 6%–10% less dense than pure Fe, which we label the outer core density deficit $\Delta\rho_{OC}$ (Fig. 2.3). We also know from seismology that there is a density increase across the boundary between the liquid outer core and the solid inner core. According to the Preliminary Reference Earth Model (PREM), the density jump is $\Delta\rho_{ICB} = 590\,\text{kg/m}^3$, though

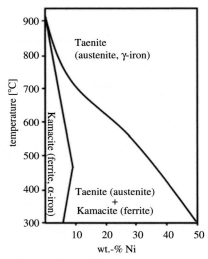

FIG. 2.1 Fe-Ni phase diagram at atmospheric pressure, showing the existence of two stable phases, Ni-poor kamacite, and taenite. See Sections 2.1.3 and 2.5.1 for details on phase diagrams and crystal structures.

FIG. 2.2 Iron meteorite from the Northwest Africa 4710 fall, showing Widmanstätten pattern. *Photo courtesy of Arizona Skies Meteorites.*

FIG. 2.3 Density as a function of radius, for pure Fe (dashed line), experimentally and theoretically determined, and for the core, seismically determined (solid line). $\Delta\rho_{OC}$ is the outer core density deficit from that of pure Fe and $\Delta\rho_{IC}$ the inner core density deficit. The difference equals the seismically determined density difference at the ICB, $\Delta\rho_{ICB}$, minus the density difference due to the phase change, $\Delta\rho_{phase}$.

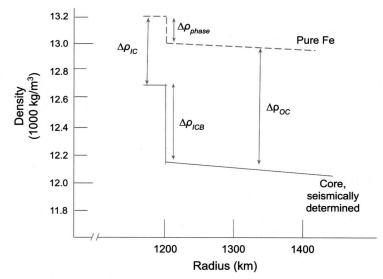

normal mode studies find $\Delta\rho_{ICB} = 820 \pm 180 \, \text{kg/m}^3$, and ICB reflection studies find $\Delta\rho_{ICB} = 850 \, \text{kg/m}^3$. Some studies also find values smaller than that of PREM, $\Delta\rho_{ICB} = 520 \pm 240 \, \text{kg/m}^3$ (for further details on the density discontinuity at the ICB, see Chapter 6). In any case, even the smallest $\Delta\rho_{ICB}$ values exceed both experimental and ab initio estimates of the density jump due solely to the phase change from liquid to solid Fe, $\Delta\rho_{phase}$, which ranges from $200 \, \text{kg/m}^3$ to $240 \, \text{kg/m}^3$. It is thus widely accepted that the outer core contains a higher fraction of less dense alloying components (typically called "light elements") than the inner core, a result of partitioning upon solidification.

However, the inner core is also likely not pure Fe-Ni, with the inner core density deficit $\Delta\rho_{IC} = 2\%$–3%. Assuming the ICB is at the *liquidus* (see the next section) for the composition of the bulk outer core,

$$\Delta\rho_{ICB} - \Delta\rho_{phase} = \Delta\rho_{OC} - \Delta\rho_{IC}, \tag{2.1}$$

but this could be complicated by the possible existence of a stable, dense (i.e., depleted in light elements) layer at the base of the outer core that is not well-mixed with the rest of the outer core.

Aside from their effect on the densities of the inner and outer cores, the presence of less dense alloying components in the inner core may have implications for the stable *crystal structure* of Fe under inner core conditions, which may in turn have implications for the transport and mechanical properties of the inner core. Light elements in the core may likely play a role in the energetics of the core, and also give us clues about core formation and evolution. The light elements most often suggested to explain the density deficit include cosmically abundant Si, S, O, C, and H (see Fig. 8.5). Mg has also been suggested for the core, not primarily to explain the density deficit, but based on an energetics argument, discussed in Section 2.4.4.

Light elements in the core must explain the inner and outer core density deficits, $\Delta\rho_{IC}$ and $\Delta\rho_{OC}$, the density jump at the ICB that exceeds that due to the phase change, $\Delta\rho_{ICB} - \Delta\rho_{phase}$, compressibility consistent with seismic velocity (see Chapter 1), and perhaps the anomalously low inner core *shear modulus* (Section 2.5.4). Arguments for the different light elements in the core also depend on their volatility, how siderophile they are under core conditions, their depletion in the mantle relative to their *chondritic* (expected terrestrial planetary) abundance, and core formation. All of these are tricky issues.

For instance, the Mg/Si ratio in the upper mantle, approximately 1.3, is higher than that of chondritic meteorites, about 1.0, leading to debate about whether the "missing" Si is in the lower mantle or the core. Si isotope differences between terrestrial samples and meteorites suggest that Si went into the core at the time of core formation. However, ab initio calculations (Alfe et al., 2002a) show that the chemical potentials of S in the liquid and solid phases (in an assumed hexagonal close-packed) in Fe under core conditions are equal, so that S partitions equally into the inner and outer cores. The same is likely true of Si because Si and S have atomic sizes comparable to that of Fe. Neither are therefore able to explain $\Delta\rho_{OC} - \Delta\rho_{IC}$. On the other hand, O atoms differ appreciably in size from those of Fe, so O does not fit easily into a solid Fe-rich *lattice*, and likely does not partition into the inner core. Hence, Si or S may be required to explain $\Delta\rho_{IC}$, while O is required to explain $\Delta\rho_{OC} - \Delta\rho_{IC}$ (Fig. 2.4). Given the likelihood that Si condensed as SiO_2, one would expect the dissolution of Si into the core to also result in the presence of large amounts of O. Although at low pressure there is low simultaneous solubility of Si and O in liquid Fe, there is increasing evidence that the solubilities of both increase with pressure. Experiments support that sufficient amounts of Si and O to explain the 10% outer core density deficit can dissolve into an Fe-Ni melt in a deep *magma ocean* (Hirose et al., 2013). It is also possible that the lower mantle and outer core are not in equilibrium, and that Si and O are dissolving into (or out of) the outer core at the core-mantle boundary.

While S is abundant and siderophile, it is also somewhat volatile, and C are H are even more so. The argument for C in the core is that Fe carbides and the graphite phase of C are sometimes found in Fe meteorites, and for H that it may have been incorporated into the core when the Earth's magma ocean may have been surrounded by an H-rich primordial atmosphere.

It seems increasingly clear that the core is likely not a simple binary system and, indeed, there is no reason to think it is. As an example of a possible core composition, Hirose et al. (2013) suggested a model of some 6 wt-pct Si, making the outer core about 5% less dense than pure Fe-Ni, 1–2 wt-pct S, yielding another 1% density decrease from that of Fe-Ni, and 4 wt-pct O, which contributes another 4% to the outer core density deficit (for a total of 10%), the latter also providing the density jump at the ICB beyond that due to the phase change. C and H continue to have their proponents however, and the composition of the light elements in the core continues to be an uncertain mixture of all the elements (Fig. 2.5).

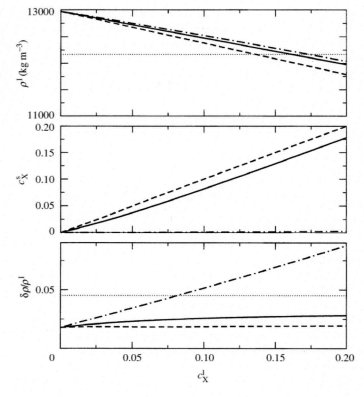

FIG. 2.4 Effect of light element X, S (solid line), Si (dashed line), and O (dot-dashed line), on the liquid density ρ^L at the ICB (*top*), the solid mole fraction of light element C_X^S that results from equal chemical potentials in the solid and the liquid (*middle*), and relative density jump at the ICB, $(\Delta\rho_{ICB} - \Delta\rho_{phase})/\rho^L$ (*bottom*), as a function of liquid mole fraction of light element C_X^L. Dotted line in the top panel is the value from PREM. Dotted line in bottom panel is from normal mode data, $\Delta\rho_{ICB} = 820\,\text{kg/m}^3$. *From Alfe, D., Gillan, M.J., Vocadlo, L., Brodholt, J., Price, G.D., 2002a. The ab initio simulation of the Earth's core. Philos. Trans. R. Soc. Lond. A 360, 1227–1244.*

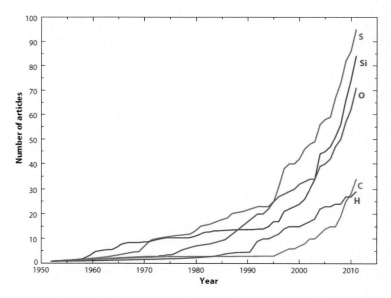

FIG. 2.5 Cumulative number of papers published that study one of the five most commonly proposed light elements in the core, a few years outdated. *From Hirose, K., Labrosse, S., Hernlund, J., 2013. Composition and state of the core. Annu. Rev. Earth Planet. Sci. 41, 657–691.*

2.1.3 Solid-liquid phase diagrams

Because the core is an alloy, its melting temperature varies with composition, thereby requiring construction of a phase diagram. Given the large pressure variation across the radial expanse of the entire core, the phase diagram of the core is three-dimensional, a function of temperature, composition, and pressure. Moreover, as we have seen the core is likely a mixture of at least several elements. Nevertheless, a binary system as a function of temperature and composition captures many of the essential features of solid-liquid phase diagrams, so it is a sensible starting point. We review briefly the thermodynamics that results in solid-liquid phase diagrams, how to interpret such diagrams, and their application to Earth's core, including melting point suppression due to alloying. Understanding the basics of phase diagrams is also essential to understanding core solidification, discussed in Chapter 7. Appendix 2.1 gives a few details on constructing and reading phase diagrams that are not discussed in the main text.

Although the possibility of immiscibility in the liquid core has been raised in regard to a proposed stable layer at the top of the outer core, it seems unlikely enough that we will consider components A and B, say Fe and FeX, where X is your favorite light element, to be miscible in the liquid phase. To simplify further we will assume that each pure solid exists only in a single solid phase (crystal structure), although there have been suggestions of an innermost inner core with differing seismic properties, which is perhaps suggestive of a solid phase transition to a different crystal structure, but this remains highly speculative.

Under these assumptions three types of phase diagrams are possible, the one that occurs is the one that minimizes the Gibbs free energy,

$$G = H - TS, \qquad (2.2)$$

where H is the enthalpy, T is the temperature, and S is the entropy. If the crystal structures of A and B are the same, and their atomic sizes are similar, then there may be miscibility in the solid state as well as in the liquid state ($\Delta H = 0$ upon mixing). Liquid and solid are then both ideal solutions, and only two phases, liquid and solid, are possible in the temperature-composition field (Fig. 2.6). Cu-Sn (bronze) and Cu-Zn (brass) are examples of ideal *solid solutions*, though both exhibit several stable solid phases depending on the concentration, and hence more complex-looking phase diagrams than Fig. 2.6.

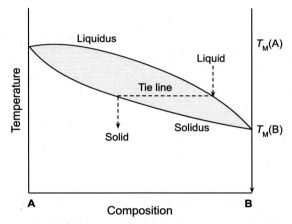

FIG. 2.6 Phase diagram for a binary system (A,B) exhibiting complete solubility in both the liquid and solid phases. $T_M(A)$ is the melting temperature of pure A, and similarly for B. For a given composition, the liquidus is the temperature at which liquid begins to freeze. The composition that freezes out at that temperature is given by drawing a horizontal *tie line* to the *solidus*. The shaded region is thermodynamically unstable.

On the other hand, if A and B have different crystal structures and/or atomic sizes, then $\Delta H > 0$, and the solid is said to be a real solution. In this case, Eq. (2.2) yields a lower Gibbs free energy G if the mixture remains a liquid to a temperature below that of the pure solid melting temperatures, i.e., melting point suppression. Moreover, at sufficiently low temperatures the decrease in the entropic term TS associated with two separate solid phases, an A-rich α-phase and a B-rich β-phase (rather than randomly mixed atoms in a single phase), is compensated by the smaller enthalpy H that results from A and B existing primarily in coexisting separate phases. The miscibility gap at low temperatures results in the two solid phases coexisting. If ΔH is sufficiently positive, the miscibility gap can exist all the way up to the minimum melting temperature, so that liquid and solid phases α and β can coexist.

This results in a *eutectic* phase diagram (Fig. 2.7), with a eutectic composition that has a minimum, or eutectic, melting temperature. Due to the entropic contribution, B can have some nonzero solubility in the A-rich α-phase, and vice versa. Examples of simple eutectic systems include the very familiar NaCl-H₂0, as well as NH₄Cl-H₂0, Pb-Sn, and Zn-Sn. As for ideal solid solutions, eutectic systems may have several stable solid phases depending on the concentration.

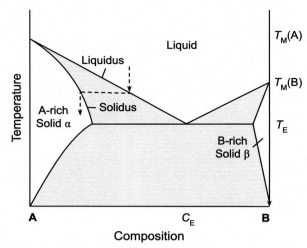

FIG. 2.7 Eutectic phase diagram for a binary system, in which there is complete solubility in the liquid phase, but where $\Delta H > 0$ in the solids, so that the solids are real solutions. Because $\Delta H > 0$ in the solids, the system remains liquid to lower temperatures than $T_M(A)$ and $T_M(B)$. The eutectic temperature T_E is the lowest temperature at which liquid can be in equilibrium with the solid phases A-rich α and B-rich β. At T_E, the composition of the liquid is C_E. The shaded regions are thermodynamically unstable. A horizontal tie line (dashed, not labeled) between the liquidus and the solidus again relates the composition of the liquid to that of the solid that freezes out, in equilibrium.

Last, if $\Delta H < 0$, then it is energetically favorable for a solid phase to form at temperatures above the line connecting the melting temperatures of pure A and B. Such a solid is known as an ordered alloy because the atoms arrange themselves in a particular structure known as a superlattice. Ordered alloys often occur only within a certain compositional range. There is no experimental evidence that any of the possible core alloy components form an ordered alloy with Fe under inner core conditions.

Figs. 2.6 and 2.7 are two-dimensional diagrams; in the Earth's core, where the pressure varies from 136 GPa at the CMB to 364 GPa at the Earth's center, there is a third important axis, pressure. This of course is what allows the inner core to freeze at a higher temperature than the outer core, but it doesn't cause any fundamental change in the type of solidification that can occur.

As an example, we have seen that FeS likely partitions equally between the solid inner and liquid outer cores, which could explain the presence of some light elements in the inner core. This implies that the *phase loop* for the binary Fe-FeS alloy is very narrow. It also suggests that Fe-FeS forms an ideal solid solution (unless a significant fraction of the inner core has not yet cooled to the eutectic temperature). On the other hand, FeO may likely not partition to any appreciable extent into the inner core, so that Fe-FeO might exhibit eutectic behavior with a nearly vertical solidus. It seems probable that the core is at least a ternary system, if not more. Since the composition of the core remains uncertain, the nature of the core solid-liquid phase diagram is also uncertain, but in spite of this a binary phase diagram captures at least many of the salient features required to understand core solidification.

2.2 Temperature in the core

2.2.1 Thermoelastic properties and the adiabatic gradient in the outer core

Because of the existence of the geodynamo, it is generally assumed that the outer core is vigorously convecting, aside from thin conductive *boundary layers* above the ICB and below the CMB (and perhaps thicker *stably stratified layers* above the ICB and below the CMB, whose existence remains uncertain; see Chapter 5). Fluid dynamics predicts that these conductive boundary layers in the outer core are on the order of a few meters thick. Outside of possible stably stratified layers, the outer core temperature gradient $(dT/dr)_{OC}$ is therefore close to the *adiabatic gradient*,

$$(dT/dr)_{ad} = (\partial T/\partial P)_S (dP/dr), \tag{2.3}$$

where P is the pressure and S is the entropy (recall that for an adiabatic process there is no heat flow and the entropy remains constant). Unlike in an incompressible fluid where the adiabatic gradient is zero because $(\partial T/\partial P)_S$ is zero, in a compressible one such as the atmosphere or outer core it is not. If a neutrally buoyant fluid parcel moves radially along the core adiabat it remains neutrally buoyant, and it is only the superadiabatic part of the temperature gradient that can lead to convection. The deviations of T_{OC} from the adiabat, i.e., lateral temperature variations in the outer core, are tiny, on the order of 10^{-6} K (see Chapter 4).

Although the *Gruneisen parameter* was originally defined for solids in terms of lattice vibrations, we often use a thermodynamic Gruneisen parameter (Poirier, 1991; Stacey, 1992; Appendix 2.2).

$$\gamma = \alpha_T V K_T / c_V = \alpha_T V K_S / c_P, \tag{2.4}$$

where α_T is the coefficient of thermal expansivity, V is the volume, $K_T = -V(\partial P/\partial V)_T$ and $K_S = -V(\partial P/\partial V)_S$ are the isothermal and adiabatic bulk moduli (incompressibilities), and c_V and c_P are the specific heats at constant volume and constant pressure. The second equality in Eq. (2.4) uses the thermodynamic relation $K_S/K_T = c_P/c_V$, from Appendix 2.2. Using another thermodynamic relation from Appendix 2.2, $(\partial T/\partial P)_S = \alpha_T VT/c_P$, $(dP/dr) = -\rho g$ (from elementary physics, with g of course gravity), and $\gamma = \alpha_T V K_S / c_P$ from Eq. (2.4), Eq. (2.3) becomes

$$(dT/dr)_{ad} = -\gamma \rho g T_{ad}/K_S. \tag{2.5}$$

Because one can determine the density ρ (or volume), elastic moduli, specific heats, and thermal expansivity seismologically, experimentally, and/or quantum mechanically, one can determine γ. Hence, if one can fix the temperature in the outer core at one point, one can determine the temperature along the outer core adiabat, $(dT/dr)_{ad}$ ($= (dT/dr)_{OC}$), using Eq. (2.5). A constant value of $\gamma = 1.5$ for the outer core seems uncontroversial (Gubbins et al., 2003; Hirose et al., 2013), although γ proportional to $1/\rho$ may be a better approximation (ρ varies by about 20% across the outer core). Because γ varies little across large variations in pressure and temperature across the outer core, it is a particularly useful thermodynamic parameter.

The one point that we have independent information about is the ICB, where the temperature T_{ICB} is the melting temperature of the core Fe alloy at 330 GPa. From Appendix 2.2, $(\partial T/\partial V)_S = -\gamma T/V$, or $(\partial T/\partial \rho)_S = \gamma T/\rho$, so for constant γ the temperature along the adiabat is given by

$$T_{ad} = T_{ICB} (\rho/\rho_{ICB})^\gamma, \tag{2.6}$$

where ρ_{ICB} is outer core density at the ICB. Using $\gamma = 1.5$ and ρ from PREM, $T_{CMB} = 0.74\, T_{ICB}$. This result could, however, be complicated by the possible existence of stably stratified layers at the bottom and/or top of the outer core (not to be confused with much thinner thermal conduction boundary layers), where the temperature need not lie along the adiabat.

2.2.2 T_{ICB} and T_{CMB}

Because the core is an alloy, its melting temperature T_M depends on its composition as well as pressure. The melting temperature of pure Fe at core pressures has been estimated in three ways, using *shock-wave* experiments, static diamond anvil cell experiments, and ab initio calculations. Shock-wave experiments can reach the ICB pressure of 330 GPa, but the experimental results typically have a lot of scatter and T_M can vary by as much as 2000 K, in part because they involve measurements over very small time intervals (Nguyen and Holmes, 2004). The scatter is less for static experiments, but sufficient pressures have not yet been obtained, so some extrapolation is required. It is also difficult to obtain uniform temperatures on the tiny samples required by diamond anvil cells. Ab initio calculations can achieve any pressure, but different methods and approximations have yielded somewhat different results. In essence, the quantum mechanical methods look for the temperature when the solid and liquid phases have the same Gibbs free energy, at 330 GPa. Values for T_M for pure Fe range from about 4900 K to 6370 K, with most results nearer to the high value (Laio et al., 2000; Alfe, 2009). Experiments examining the stable phase of pure Fe under inner core conditions also find solid above 5700 K, lending further credence to a higher value (Tateno et al., 2010).

Whether Fe forms a solid solution or eutectic with the less dense components in the core, T_M will be depressed from that of pure Fe (unless it forms a solid solution with a component that has a higher melting temperature than that of pure Fe, for which there is no evidence). The amount of melting point depression depends, of course, on the light elements present in the core, which we have seen is uncertain. Ab initio calculations incorporating both O and either S or Si into the core show a melting point suppression of 700 K (Alfe et al., 2002b), which some experiments also support. This leads to T_{ICB} in the range 4200 K to 5600 K, with values toward the upper end much more likely. Using this range of values for T_{ICB}, and Eq. (2.6), one finds T_{CMB} ranges between 3150 K and 4200 K. The discovery of a post-perovskite phase in the lowermost mantle requires T_{CMB} to exceed about 3700 K (Hernlund et al., 2005), which again points toward a higher value of T_{ICB}. *If you can't stand the heat, get out of the core.*

It remains unclear if the inner core is convecting, and hence whether its temperature T_{IC} lies along the adiabat as well. We take up this question in Chapter 7. Fig. 2.8 sketches the temperature profile in the core. The increase in the magnitude of the adiabatic gradient with radius in Fig. 2.8 is due primarily to the increase in g across the core, in spite of the decrease in T, Eq. (2.5). This has consequences on where possible stably stratified layers in the outer core are most likely to form.

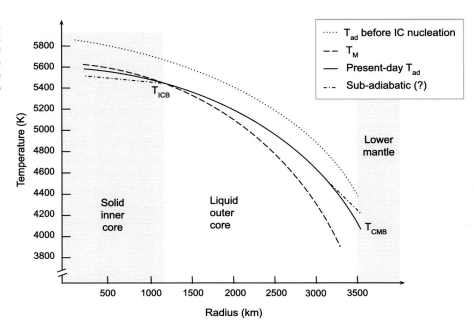

FIG. 2.8 Estimated temperature profiles in the core, showing the adiabat (assumed equal to the geotherm) in the present and before inner core nucleation, the melting temperature of Fe alloys under core pressures, and possible subadiabatic (thermally stratified) layers in the inner core and top of the outer core.

2.3 Transport properties of the core

Transport properties are those that are relevant for the diffusion of some property, be it temperature (thermal conductivity k), magnetic field strength (electrical conductivity σ), atomic or vacancy concentration (known either as self or mass diffusivity D), impurity or solute concentration (compositional or solute diffusivity κ_C, which depends on the species), or momentum (kinematic viscosity ν). Self-diffusivity D or compositional diffusivity κ_C can be studied experimentally or using computational molecular dynamics, and are nearly always small compared with thermal diffusivity $\kappa_T = k/\rho c_P$. This means that heat diffuses much more readily than does solute, a relevant factor when studying convection. In the case of solids, D depends on whether the diffusion is occurring through the lattice or along grain boundaries, and κ_C depends on whether the impurity is substitutional or interstitial. Values for κ_C in liquids are typically of the order 10^{-9} m²/s, values for D in solids near their melting temperature T_M are given in Fig. 2.9. Some experiments have found that $D/(T/T_M)$ is relatively independent of pressure (Porter and Easterling, 1992), but a recent ab initio calculation found that D for hexagonal close-packed (hcp) Fe under inner core conditions is less than 10^{-15} m²/s (Ritterbex and Tsuchiya, 2020), much less than the 10^{-12} m²/s value for hcp metals in Fig. 2.9. This is an area of active research, with self-diffusivity D playing an important role in the *plastic deformation* of solids (Chapter 7). Compositional diffusivity κ_C is particularly important for the style of thermochemical convection in the outer core (Chapter 4). Notice that all diffusivities have the SI units of m²/s.

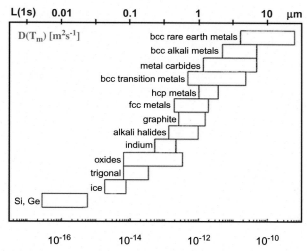

FIG. 2.9 Melting point self-diffusivities $D(T_M)$ in solids (bottom axis). Top axis is the 1-s diffusion lengthscale. *Data from Brown, A.M., Ashby, M.F., 1980. Correlations for diffusion constants. Acta Metall. 28, 1085–1101.*

Unlike the kinematic viscosity in the liquid outer core, solid-state viscosity is not just a material property, but also depends on the way in which the solid deforms, which in turn depends on the dynamics, e.g., the stress level and grain size. We therefore defer until Chapters 6 and 7 discussion of inner core viscosity. Incidentally, in an apparent effort to increase the number of symbols, it is traditional to use the dynamic viscosity $\eta = \nu\rho$ for solid-state viscosity, and also sometimes for liquid viscosity.

2.3.1 Outer core viscosity

The kinematic viscosity ν of the outer core has been inferred from experiments, theoretical considerations, geodynamics, geomagnetism, and seismology, and was reviewed in Chapter 1. Values span at least 11 orders of magnitude. Despite this, there is remarkably little controversy and considerable consensus. This is in part because the larger values may not be measuring the true molecular viscosity, or because they represent an upper bound. Experiments and theory, and some results from geodynamics and geomagnetism, support a small kinematic viscosity ν, on the order of 10^{-6} m^2/s, which is comparable to that of water and liquid metals at atmospheric pressure. Fig. 2.10 gives values for the dynamic viscosity η predicted by different methods.

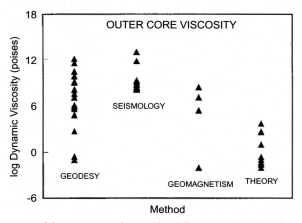

FIG. 2.10 Values for the dynamic viscosity η of the outer core obtained by different methods. To convert to kinematic viscosity ν, divide by the core density ρ, approximately 10^4 kg/m^3. *From Secco, R.A., 1995. Viscosity of the outer core. In Mineral Physics and Crystallography, A Handbook of Physical Constants, AGU.*

Because self-diffusivity and fluid viscosity both involve movement of atoms under the driving force of chemical potential (for diffusion) or shearing force (for viscosity), they are related, by the Stokes-Einstein relation $D = Mk_BT$, where D is the self-diffusivity, M is the atomic mobility (the ratio of atomic drift velocity to shearing force), k_B is Boltzmann's constant, and T is the temperature in Kelvin. The relation has been experimentally verified for liquid metals and also has a theoretical footing. Essentially, both D and ν are due to thermally activated jumps of molecules into "vacancies" in the liquid, and hence follow an Arrhenius law of the form e^{-Q/R_gT}, where Q is activation energy and R_g is the universal gas constant. Interestingly, there are empirical correlations between the melting temperature for liquid metals, T_M, and the activation energy for viscosity, $Q_\nu = 2.6 R_g T_M$, and the activation energy for diffusion, $Q_D = 3.2 R_g T_M$. (Fig. 2.11). Although these correlations are empirical, their nearness to $Q = 3 R_g T_M$ is thought to be due to the Dulong-Petit law for the average molar energy in a solid. These results predict that ν and D stay constant along T_M, independent of the pressure, as seen in Fig. 1.16, with ν on the order of 10^{-6} m^2/s.

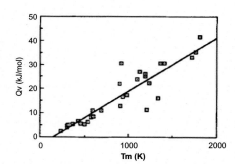 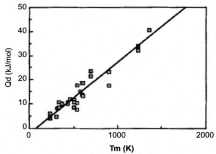

FIG. 2.11 Experimentally determined activation energy for viscosity, Q_ν (left), and for diffusion, Q_D (right), versus the melting temperature of metals. *From Poirier, J.-P., 1988. Transport properties of liquid metals and viscosity of the Earth's core. Geophys. J. 92, 99–105.*

A low viscosity is consistent with vigorous outer core convection and a rotationally dominated geodynamo. Even a viscosity an order or more larger likely makes little difference for the dynamics. Perhaps this latitude is the origin for the lack of controversy concerning ν. A low viscosity also results in the likelihood of turbulent flow in the outer core. Because of the difficulty of numerically simulating the small lengthscales associated with turbulence on a global lengthscale the size of the outer core, many researchers have adopted turbulent (eddy) viscosity values that are many orders of magnitude larger than the true molecular value. Chapter 4 explores the consequences of the small value of ν for outer core dynamics.

Last, some explanations for the seismically inferred F-layer above the ICB (see Chapter 5) involve laterally varying properties that require a viscosity much larger than that of the bulk outer core. Two causes for a high viscosity layer have been suggested. One involves a mixed solid/liquid region, known as a *slurry*, in which solidification is taking place (Gubbins et al., 2008). The effective viscosity of such a region depends on the lengthscales and connectivity of the liquid, but can be larger than the liquid viscosity. However, the metallurgical evidence for a slurry is doubtful (see Chapter 7). The other is that the F-layer is in a glassy state, in which the atoms exhibit frozen-in disorder. This is based on one quantum mechanical study (Xian et al., 2019), but this too is in contradiction with what is usually observed, in that a glassy state, especially in metals, results from rapid solidification, the opposite of what one would expect in the core!

2.3.2 Electrical and thermal conductivities

Because of their importance to the energetics and evolution of the core, and the geodynamo, the thermal and electrical conductivities of Fe under core conditions have been the subject of much research. Both have been computed using ab initio calculations, but whether all of the relevant physics has been included has been controversial. Experimentally, the electrical conductivity has been more reliable to obtain than the thermal conductivity, due to the need to model heat transfer in the latter experiments. However, the electrical and thermal conductivities of metals are generally related due to the role played by free conduction electrons, so if one can obtain the electrical conductivity one can infer the thermal conductivity. The relation, the *Wiedemann-Franz law*, is not entirely straightforward, however. Recently there have been direct measurements of the thermal conductivity of Fe under core conditions.

One of the first great successes of quantum theory was the band theory of solids, which explained the origin of electrical insulators, semiconductors, and conductors. Each band has energy levels that are closely spaced, and the bands are separated by energy gaps. In an insulator, the valence band's energy levels are filled, the conduction band's energy levels are empty, and the gap is large, so there is no mobility of electrons. In a semiconductor, the situation is the same, except that the gap is small, so that electrons can be easily promoted from the valence to the conduction band, and the conductivity increases with temperature. In a metal, the conduction band is partially filled (up to the *Fermi energy level* ε_F), allowing the electrons great mobility between levels. Fig. 2.12 sketches elementary band theory.

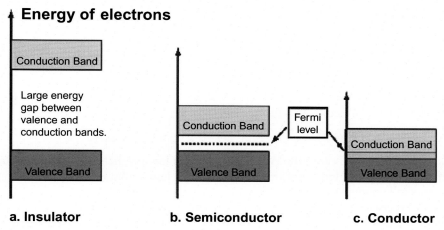

FIG. 2.12 The band theory of solids.

For the most basic understanding of conductivity in metals, to a first approximation, we can ignore the effects of the periodic potential of the crystal that is responsible for the existence of bands, as well as the effects of the Pauli exclusion principle. This is the classical (Drude) model of electrical conductivity. See Appendix 2.3 for a very brief review of the semiclassical (Sommerfeld) free electron *Fermi gas* model that takes into account the exclusion principle. In the classical model, the electric current density $J = -nev_d$, where n is the electron density, e is the electron charge, and v_d is the electron drift velocity due to the electric field E. Even under the influence of E, v_d reaches a steady-state value because of collisions, i.e., scattering events. If τ is the average time between collisions, then $v_d = -eE\tau/m$, where m is the electron mass, so that $J = ne^2E\tau/m$. Ohm's law, $J = \sigma E$, where σ is the electrical conductivity, thus yields $\sigma = ne^2\tau/m$. With $\tau = \Lambda/v$, where Λ is the electron mean free path and v is the total rms velocity (drift plus thermal),

$$\sigma = ne^2\Lambda/mv, \tag{2.7}$$

where v_F is the Fermi velocity, with $\varepsilon_F = \frac{1}{2}mv_F^2$.

The electrical resistivity

$$\rho = 1/\sigma = \rho_{impurities} + \rho_{phonons} + \rho_{electrons} \tag{2.8}$$

is due to the additive effects of electron scattering by impurities $\rho_{impurities}$, lattice *phonons* $\rho_{phonons}$, and other electrons $\rho_{electrons}$ (*Matthiessen's rule*, a similar "rule" holds for the electron contribution to thermal resistivity). At temperatures well above the *Debye temperature* Θ_D (the temperature associated with a crystal's highest normal mode of vibration used in the *Debye model*, discussed in Appendix 2.3), such as is the case in the Earth's core, the temperature dependence of ρ has generally been thought to be primarily from the contribution due to phonons. The scattering probability of electrons by phonons is proportional to the mean square of the displacement of the vibrating atoms $<x^2>$. From the equation of harmonic motion for a vibrating atom, $M d^2x/dt^2 = -bx$, where M is the atomic mass and b is the bond spring constant, so that $b = \omega^2 M$, where ω is the angular frequency of a phonon. As an approximation one can set $\omega = \omega_D = k_B\Theta_D/\hbar$ (Eq. 2.70), where $\hbar$ is the reduced Planck constant and ω_D is the *Debye frequency*, the cutoff frequency of vibrations in the Debye model, corresponding to Θ_D. With the harmonic energy $\frac{1}{2} b <x^2> = \frac{1}{2} k_BT$ (the 1D thermal energy),

$$\rho_{phonons} \propto <x^2> = k_BT/b = (\hbar)^2 T/(k_B \Theta_D^2 M). \tag{2.9}$$

The significance of Eq. (2.9) is that at core temperatures the part of the electrical resistivity that is due to scattering by phonons is proportional to T, and hence σ decreases with temperature. On the other hand, because $<x^2>$ decreases somewhat with pressure, σ increases with pressure.

Although thermal energy can be transported by both phonons and electrons, as opposed to charge, which can be transported only by electrons, the phonon contribution to heat transfer in metals is small compared with that made by electrons, which makes possible the Wiedemann-Franz law. In 1D the electronic contribution to the heat flux $Q_x = -nc\Lambda(dT/dx)v_x$, where n is again the electron density (or linear density, in 1D), c is the specific heat per electron, so that $c\Lambda(dT/dx)$ is the amount of heat transported by an electron over its mean free path Λ, and v_x is the average speed in the x-direction. One can write this as $Q_x = -1/3 \, nc\Lambda(dT/dx) \, v$, where v is the rms velocity. We thus have the thermal conductivity

$$k = 1/3\, ncv\Lambda. \tag{2.10}$$

Dividing Eq. (2.10) by Eq. (2.7), we obtain

$$k/\sigma = 1/3\, mv^2 c/e^2. \tag{2.11}$$

For a monatomic ideal gas of electrons $c = 3/2\, k_B$ and $1/2\, mv^2 = 3/2\, k_B T$, so $k/(\sigma T) = 3/2\, (k_B/e)^2$. A more proper semiclassical treatment yields the Wiedemann-Franz law,

$$k/(\sigma T) = \pi^2/3\, (k_B/e)^2 = L_0 = 2.45 \times 10^{-8}\, W\Omega/K^2. \tag{2.12}$$

L_0 is known as the *Lorenz number*. Although the Wiedemann-Franz law predicts correctly the relation between k and σ for a wide range of metals at different temperatures, it is not a strict law in that many assumptions are built in that may not always hold, as we shall soon see.

Extrapolating measurements of σ in Fe-Si to core pressures and temperatures (e.g., Eq. 2.9), and using the Wiedemann-Franz law, values for k in the range 28–46 W/m/K were typically used for thermal models of core evolution (Stacey, 1992). More recently, however, static measurements of σ for pure Fe using a diamond anvil cell and use of the Wiedemann-Franz law have found values for k of about 100 W/m/K, and even as high as 220 W/m/K (Ohta et al., 2016). For Fe-Si this method has yielded slightly lower values of k, but still two to three times the older values. Ab initio calculations of σ and k have also found larger values for k for both pure Fe and Fe-Si, between 100 and 150 W/m/K (de Koker et al., 2012; Pozzo et al., 2012). On the other hand, direct static measurements of k for pure Fe find a value more in line with the older values (Konopkova et al., 2016; Fig. 2.13). Why the discrepancies?

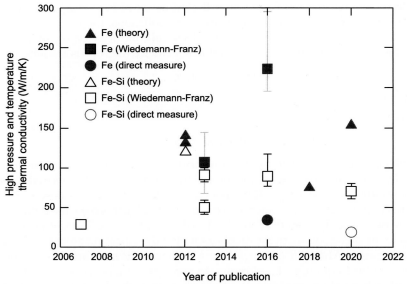

FIG. 2.13 Values for CMB outer core thermal conductivity for pure Fe and Fe-Si alloy, obtained by theory, by experimental measurement of the electrical conductivity and use of the Wiedemann-Franz law, and by direct experimental measurement. *Data from Stacey, F.D., Loper, D.E., 2007. A revised estimate of the conductivity of iron alloy at high pressure and implications for the core energy balance. Phys. Earth Planet. Int. 161, 13–18; De Koker, N., Steinle-Neumann, G., Vlcek, V., 2012. Electrical resistivity and thermal conductivity of liquid Fe alloys at high P and T, and heat flux in Earth's core. Proc. Natl. Acad. Sci. U. S. A. 109, 4070–4073; Pozzo, M., Davies, C., Gubbins, D., Alfe, D., 2012. Thermal and electrical conductivity of iron at Earth's core conditions. Nature 485, 355–358; Seagle, C.T., et al., 2013. Electrical and thermal transport properties of iron and iron-silicon alloy at high pressure. Geophys. Res. Lett. 40, 5377–5381; Gomi, H., et al., 2013. The high conductivity of iron and thermal evolution of the Earth's core. Phys. Earth Planet. Int. 224, 88–103; Konopkova, Z., McWilliams, R.S., Gomez-Perez, N., Goncharov, A.F., 2016. Direct measurement of thermal conductivity in solid iron at planetary core conditions. Nature 534, 99–101; Ohta, K., Kuwayama, Y., Hirose, K., Shimizu, K., Ohishi, Y., 2016. Experimental determination of the electrical resistivity of iron at Earth's core conditions. Nature 534, 95–98; Xu, J., Zhang, P., Haule, K., Minar, J., Wimmer, S., Ebert, H., Cohen, R.E., 2018. Thermal conductivity and electrical resistivity of solid iron at Earth's core conditions from first principles. Cond. Mat. Mtrl. Sci. arXiv:1710.03564v6; Pourovskii, L.V., Mravlje, J., Pozzo, M., Alfe, D., 2020. Electronic correlations and transport in iron at Earth's core conditions. Nat. Commun. doi:10.1038/s41467-020-18003-9; Hsieh, W.-P., Goncharov, A.F., Labrosse, S., Holtgrewe, N., Lobanov, S.S., Chuvashova, I., Deschamps, F., Lin, J.-F., 2020. Low thermal conductivity of iron-silicon alloys at Earth's core conditions with implications for the geodynamo. Nat. Commun. doi:10.1038/s41467-020-17106-7; Zhang, Y., et136al., 2020. Reconciliation of experiments and theory on transport properties of iron and the geodynamo. Phys. Rev. Lett. 125, 078501-1–078501-7.*

It has been suggested that the older, lower temperature measurements of σ, extrapolated to higher temperatures, did not account for what is known as the resistivity saturation effect. As seen in Eq. (2.9), $\rho_{phonons}$ increases with temperature because of the increase in atomic vibration displacement, and hence smaller mean free path before there is an electron-phonon scattering event. However, this effect reaches an upper limit when the mean free path due to phonon scattering becomes as small as the atomic spacing. Thus, $\rho_{phonons}$ and hence ρ cease to increase with temperature, so that predicted σ and k in the core that include resistivity saturation are larger than had previously been predicted.

On the other hand, both indirect measurements (i.e., finding σ and then using the Wiedemann-Franz law to find k) and theory may not have properly accounted for electron-electron scattering $\rho_{electrons}$ in Eq. (2.8). Because charge must be conserved in a scattering event, but thermal energy need not be, if the scattering is inelastic then use of the Wiedemann-Franz law could overestimate k from a measurement of σ. This effect is known to decrease L_0 in Eq. (2.12) to as small a value as 1.59×10^{-8} WΩ/K^2. Some theoretical calculations support that including electron-electron scattering may lower predictions of k, to 77 W/m/K for liquid pure Fe (Xu et al., 2018). Others find only a 20% reduction of k in bcc pure Fe and perhaps a 25% reduction in hcp Fe due to electron-electron scattering, thus still yielding higher values (Pourovskii et al., 2020). Direct measurements of k obviously do not need to correct for electron-electron scattering, though the required thermal modeling of a diamond anvil cell is by no means straightforward.

The derivations of σ and k under the free electron model do not require a crystalline lattice with a periodic potential, but the possibility of electron scattering by phonons implicitly assumes the metal is a solid. The free electron model works surprisingly well for liquid metals, however, likely because of the experimentally confirmed existence of phonons in liquids at frequencies in which the Debye model is a good approximation. Although more work on the connection between solid and liquid conductivities is needed, current estimates are that the electrical conductivity of the solid inner core is likely 10%–20% larger than that of the liquid outer core, perhaps closer to 10%.

The effect of impurities, $\rho_{impurities}$ in Eq. (2.8), also decreases the thermal conductivity of Fe alloys from that of pure Fe, but by how much depends on the impurity and its quantity, and on whether solid or liquid (Fig. 2.14). A very recent (not included in Fig. 2.14) direct measurement of k finds a very low value, about 20 W/m/K, for solid Fe-8 wt-pct Si at 144 GPa and 3300 K (Hsieh et al., 2020). *What goes up, must come down, spinnin' wheel, got to go 'round.* It should be clear that there is not yet a consensus on the value(s) of k in the inner and outer cores, but given its importance for the evolution of the core, it seems likely that work on the thermal conductivity of Fe and its alloys under core conditions will continue to be intense, and values of k will converge.

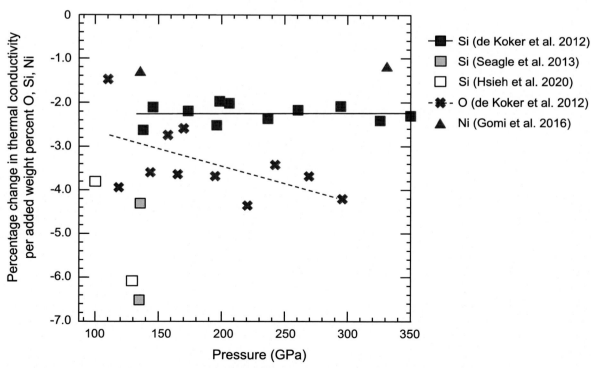

FIG. 2.14 The effect of light elements on the thermal conductivity of Fe under core conditions.

Lower temperature measurements of the electrical resistivity of pure Fe, extrapolated to outer core conditions, yield $\rho = 1.1\,\mu\Omega\,m$, or $\sigma = 9.1 \times 10^5\,S/m$. However, higher temperature measurements and ab initio calculations that include resistivity saturation suggest σ of pure Fe could be two to three times higher, though some have suggested the saturation effect may have been overestimated in calculations, and there could be some error in the temperature measurements. The effects of impurities and the liquid state are almost certain to lower σ, which therefore likely lies in the range between $4 \times 10^5\,S/m$ and $20 \times 10^5\,S/m$ in the outer core where the geodynamo is operating. The value for σ has consequences for the geodynamo and geomagnetism, as we shall see in the following chapters.

2.4 Thermodynamics of the core

Paleomagnetic evidence shows that the Earth has had an internally generated, predominantly dipolar, reversing magnetic field for at least 3.4 billion years (see Chapters 3 and 8). Although reliable intensity measurements become increasingly scarce as the record gets older, there are no obvious changes in its statistical properties over time. However, even the present-day *toroidal* magnetic field is hidden from us at the surface, as are its smaller wavelength components, and it is in these in which there is more Ohmic dissipation. Thus, it is primarily through numerical models and scaling analyses that we have estimates for the power needs of the geodynamo but unfortunately, these have very large error bars. Because the magnetic field is generated through the action of rotating magnetoconvection in the outer core, understanding the energetics of the core is of concern for those who study the geodynamo.

Thermodynamically, the core is a heat engine, with heat removed from the high-temperature reservoir ICB and throughout the outer core, and added to the low-temperature reservoir mantle at the CMB, with work being done in the form of convection, which appears as Ohmic and viscous dissipation in the outer core (Fig. 2.15). There is also another energy term, gravitational potential energy associated with the partition of light elements during solidification at the ICB. This is commonly known as compositional energy, and it drives compositional convection. Compositional energy associated with the partition of light elements is essentially 100% efficient because it is not directly added to the mantle as heat, and because compositional conduction is tiny because of the very small compositional diffusivity κ_C. Thus, all of the compositional energy removed from the core results in useful work, i.e., it is available to drive convection and power the geodynamo. This is unlike the situation for heat, much of which can be conducted up the adiabat. The more heat that is conducted up the adiabat, the less that is available to do work.

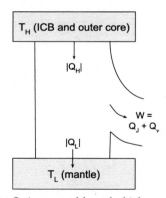

FIG. 2.15 Schematic of the core as a heat engine. Heat Q_H is removed from the high-temperature reservoir, T_H, which is the ICB and nonuniform temperature outer core, and heat Q_L added to the low-temperature reservoir, T_L, which is the mantle. (Although the outer core is the thermodynamic system, it also serves as a nonuniform high-temperature reservoir.) Useful power W is obtained in the process, which appears as Ohmic and viscous dissipation Q_J and Q_ν. In the core Q_J and Q_ν are ultimately added to the mantle, but in order to extract W there must be a temperature difference between T_H and T_L.

The smaller the temperature difference between the high- and low-temperature reservoirs, T_H and T_L, the lower the Carnot efficiency $e = 1 - T_L/T_H$ (Fig. 2.15). For this reason, not all heat removed from the core is equally efficient in terms of its ability to do work. For example, latent heat of solidification, which serves as a Q_H that is removed at the high-temperature ICB, is more efficient than heat removed due to cooling (termed *secular cooling*, meaning temporal, as though lacking a spiritual or permanent basis) of the core, which is removed across the whole volume of the core, and hence at temperatures on average lower than that at the ICB.

Under the approximations that the mantle is electrically insulating relative to the core, and that the CMB is a stress-free boundary condition for outer core flow, Ohmic and viscous dissipation occur completely within the core. Hence the heat produced by them is neither directly removed from the core nor directly added to the mantle and does not enter into the energy balance (conservation of energy) equation. Rather, dissipation redistributes energy in the temperature-varying core, and so appears in the entropy balance equation. Similarly, heat conducted up the adiabat is a net zero energy contribution, and so too does not appear in the energy balance equation. Rather, it also redistributes energy and appears in the entropy balance equation. For this reason, we need both the energy and entropy balance equations to understand the maintenance of the present-day geodynamo, as well as the thermal evolution of the core.

Conservation of energy requires that the total energy removed from the core, whether as heat or compositional (gravitational potential) energy, and whether removed at the ICB or throughout the core, must equal that added to the mantle, i.e., the CMB heat flux Q_{CMB}. This assumes that the compositional flux across the CMB is zero, necessitating that any compositional energy removed from the core is ultimately converted to heat within the core. Likewise, the difference between the entropy removed from the core and that added to the mantle must equal the entropy gain associated with that of conducted heat (and solute, which again, is essentially nil), and Ohmic and viscous dissipation.

Aside from possible stably stratified layers at the bottom and top of the outer core (see Chapter 5), seismic data suggests the core *geotherm* is close to that of the adiabat, and we will assume an isentropic and compositionally well-mixed outer core (the convective state of the inner core is uncertain). We emphasize that the equilibrium thermodynamic model that follows is strictly valid only in the absence of stably stratified layers. The equilibrium model also assumes the ICB is an equilibrium phase boundary, so that $T_{ICB} = T_M$, that the outer core light element concentration is uniform except for small perturbations related to convection, and that the timescale of outer core dynamics is short compared with that of changes to the core thermodynamic structure.

Interestingly, even if Q_{CMB} is less than the heat that is conducted up the adiabat (i.e., Q_{CMB} is *subadiabatic*), it is possible for the outer core to maintain an adiabatic profile, and sustain a dynamo, through compositionally driven convection. Such might be the case, for instance, if the thermal conductivity k is large. In this case compositional convection does work to bring heat downward, from a cooler region to a warmer one, i.e., it acts as a *refrigerator*, as first pointed out by Loper (1978). Of course, there is then less power available for the geodynamo. Fig. 2.16A diagrams this possibility. Incidentally, we use the phrase "heat flow up the adiabat" because although heat of course flows spontaneously from high to low temperature along the adiabat, in the core this means from smaller to larger radius.

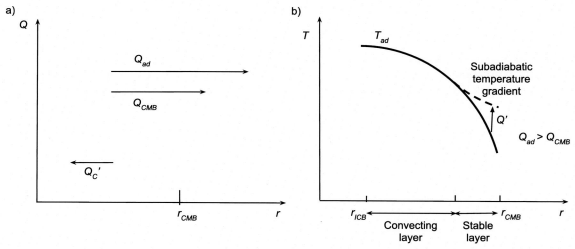

FIG. 2.16 (A) Even if the core-mantle heat flux Q_{CMB} is less than that conducted up the adiabat Q_{ad} throughout the outer core, convection and magnetic field generation can still occur through the action of a fraction of the compositional energy $Q_C' = Q_{ad} - Q_{CMB}$ bringing heat back down into the core to maintain the adiabatic temperature gradient. This is sometimes known as the refrigerator effect. (B) Because T_{ad} is steepest at the top of the outer core, it is possible that the heat conducted up the adiabat Q_{ad} exceeds Q_{CMB} only above some radius, so that a thermally stably stratified layer (i.e., one with a subadiabatic temperature gradient) grows above that radius, as the excess heat $Q' = Q_{ad} - Q_{CMB}$ is conducted into this layer.

In addition to deciphering available power sources for the geodynamo, thermodynamics can also provide constraints on the age of the inner core, which is a key to understanding inner core dynamics. Thermodynamics can also provide insight about the existence of possible stably stratified layers in the outer core, particularly at the top, which has profound consequences for dynamo theory and geomagnetism. Although the detailed thermodynamic analysis that we present is only valid if the entire outer core is well-mixed, we conclude this section with a brief discussion of stably stratified layers.

2.4.1 The energy and entropy balance equations

In this section we develop the thermodynamic framework and expressions for the important contributions Q_i to the energy balance equation and S_i to the entropy balance equation. Although we have simplified the development from more general treatments in an effort to make it more digestible, it might at first still look unappetizing, but the effort to slog through is worth it, as it will give you a nutritious understanding of many issues concerning the core. If you are still hungry for more see detailed treatments (Buffett et al., 1996; Gubbins et al., 2003, 2004; Labrosse, 2003; Nimmo, 2015).

Because many of the geophysical parameter inputs are still under debate, the thermodynamic framework we present is helpful as it is transparent and easy to model the effects of changing the inputs. Although we won't fully explore parameter space, in the next section we will look at geophysical estimates of the various energy and entropy terms, in an effort to understand their relative importance for sustaining the present and past geodynamo. Q_i is a unit of power (rate of energy production), in the core measured in TW; S_i is a unit of rate of entropy production, measured in MW/K.

2.4.1.1 Secular cooling

Secular cooling Q_{SC} is the heat removed from the core as it cools at constant pressure. It is easy to write down,

$$Q_{SC} = -\int \rho c_P \dot{T}\, dV. \tag{2.13}$$

The integral is taken over the entire volume of the core. There is another term that is sometimes denoted as pressure heating, Q_P, which is due to an isothermal increase in pressure during contraction, but it is small and we will not consider it further.

We can recast Eq. (2.13) in a way that is more illuminating, using a simplified core model. This simplified core model is spherically symmetric (thermodynamically a good approximation, even though convection and magnetic field generation are inherently asymmetric), has $T = T_{ad}$, and assumes that some core properties, including incompressibilities K_T and K_S, specific heats c_V and c_P, thermal expansivity α_T, and Gruneisen parameter γ, are constant as a function of radius and time. With $\gamma \rho g / K_S$ approximately constant as a function of time, each side of Eq. (2.5), $dT_{ad}/T_{ad} = -(\gamma \rho g/K_S)\, dr$, is independent of time. With $T = T_{ad}$ this yields

$$\dot{T}/T = \dot{T}_{ICB}/T_{ICB}, \tag{2.14}$$

as we are assuming the ICB lies along the adiabat. Using Eq. (2.14) we can thus write Eq. (2.13) as

$$Q_{SC} = -(\dot{T}_{ICB}/T_{ICB})\left(c_P \int \rho T dV\right), \tag{2.15}$$

where the integral can be evaluated from seismic data and estimates of the geotherm.

Further, we can express the term $\dot{T}_{ICB}$ in terms of the growth rate of the inner core, $\dot{r}_{ICB}$, which will be useful. Fig. 2.17 diagrams how a cooling ICB results in an increasing inner core radius, defined as the intersection of T_{ad} with the melting temperature T_M, by an amount that depends on the relative slopes dT_{ad}/dr and dT_M/dr. Specifically,

$$\begin{aligned}dT_{ICB} &= (dT_M/dr - dT_{ad}/dr)_{ICB} dr_{ICB} \\ &= (dT_M/dP - dT_{ad}/dP)_{ICB}(dP/dr)_{ICB} dr_{ICB} = (dT_M/dP - dT_{ad}/dP)_{ICB}(-\rho g)_{ICB} dr_{ICB},\end{aligned} \tag{2.16}$$

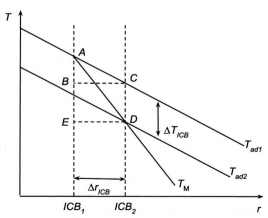

FIG. 2.17 As the core cools from T_{ad1} to T_{ad2} ($\Delta T_{ICB} = CD$), the inner core advances from ICB_1 to ICB_2 (Δr_{ICB}), because the inner core radius is defined by the crossing of T_{ad} with T_M. From the diagram $AE = (dT_M/dr)\Delta r_{ICB}$ and $AB = (dT_{ad}/dr)\Delta r_{ICB}$, so $\Delta T_{ICB} = CD = BE = AE - AB = (dT_M/dr - dT_{ad}/dr)\Delta r_{ICB}$. This derivation assumes the adiabat near the ICB remains parallel for small cooling. *Adapted from Gubbins, D., Alfe, D., Masters, G., Price, G.D., Gillan, M.J., 2003. Can the Earth's dynamo run on heat alone? Geophys. J. Int. 155, 609–622.*

so that

$$\dot{T}_{ICB} = (dT_M/dP - dT_{ad}/dP)_{ICB}(-\rho g)_{ICB}\dot{r}_{ICB}, \quad (2.17)$$

and

$$Q_{SC} = (1/T_{ICB})\left(c_P\int \rho T dV\right)(dT_M/dP - dT_{ad}/dP)_{ICB}(\rho g)_{ICB}\dot{r}_{ICB}. \quad (2.18)$$

Notice that $(dT_M/dP - dT_{ad}/dP)_{ICB} > 0$, so $Q_{SC} > 0$.

Recall that entropy $dS = dQ/T$. In the case of secular cooling, T varies across the core for the heat that is lost from the core as it cools, but $T = T_{CMB}$ for the heat that is gained by the mantle at the CMB. We thus have that the rate of entropy that is gained as the core cools is

$$S_{SC} = Q_{SC}/T_{CMB} + \int (\rho c_P \dot{T}/T) dV, \quad (2.19)$$

where Q_{SC} is from Eq. (2.18). Using Eqs. (2.14) and (2.17),

$$S_{SC} = (1/T_{ICB})c_P(dT_M/dP - dT_{ad}/dP)_{ICB}(\rho g)_{ICB}\dot{r}_{ICB}\left\{(1/T_{CMB})\int \rho T dV - \int (\rho dV)\right\}. \quad (2.20)$$

Because the temperature in the first integral in curly brackets in Eq. (2.20) is always greater than T_{CMB}, $S_{SC} > 0$: entropy increases as heat from the hot core flows across the CMB into the cooler mantle. The second integral, $\int(\rho dV) = M_{core}$, and using Eq. (2.6) one can evaluate the first integral with $T = T_{ad}$. In Chapter 8 we give an example of an analytic model for evaluating this integral.

2.4.1.2 Latent heat

Latent heat Q_{LH} is the heat that must be removed from the core as the liquid core solidifies at the r_{ICB}. It too is easy to write down,

$$Q_{LH} = 4\pi r_{ICB}^2 L\rho_{ICB}\dot{r}_{ICB}, \quad (2.21)$$

where L is the latent heat of solidification of the Fe alloy. Of course, it too is proportional to $\dot{r}_{ICB}$.

Since all latent heat is extracted from the core at the ICB and all added to the mantle at the CMB, the total rate of entropy gain due to latent heat is

$$S_{LH} = 4\pi r_{ICB}^2 L\rho_{ICB}\dot{r}_{ICB}\{1/T_{CMB} - 1/T_{ICB}\}, \quad (2.22)$$

and it is relatively efficient.

2.4.1.3 Radioactive heat

Heat produced by radiogenic elements in the core, Q_R, is one term that does not depend on cooling in the core. It is given by

$$Q_R = \int \rho h\, dV = hM_{core}, \quad (2.23)$$

where h is the rate of radioactive heat produced per mass in the core, assumed constant in a well-mixed core.

By the same line of reasoning as for secular cooling, the rate of entropy gained due to radioactive heat production that then flows across the CMB is

$$S_R = h\left\{M_{core}/T_{CMB} - \int (\rho/T)\, dV\right\}. \quad (2.24)$$

2.4.1.4 Compositional energy

As we have seen, the outer core is enriched in light elements as compared with the inner core. As the inner core solidifies it must therefore release light elements into the base of the outer core, which then tend to rise up, much as does oil beneath vinegar in a tasty salad dressing. This gravitational potential energy, Q_C, is also known as compositional energy. It takes the form

$$Q_C = \int \rho v \cdot g\, dV = -\int \rho v \cdot \nabla\psi\, dV = \int \psi\, \nabla\cdot(\rho v)\, dV = -\int \psi(\partial\rho/\partial t)_{P,T}\, dV$$
$$= -\int \psi(\partial\rho/\partial C)_{P,T}(dC/dt)\, dV, \quad (2.25)$$

where gravity $g = -\nabla\psi$ is the gradient of the gravitational potential ψ, C is the light element mass fraction in the outer core, the third equality uses integration by parts and the boundary condition that the velocity v is zero at the core boundaries, and the fourth equality uses the continuity equation, $\partial\rho/\partial t = -\nabla\cdot(\rho v)$. The compositional expansivity α_C is defined as

$$\alpha_C = -(1/\rho)\,(\partial\rho/\partial C)_{P,T} \approx \left(\Delta\rho_{ICB} - \Delta\rho_{phase}\right)/(\rho\,\Delta C), \qquad (2.26)$$

where recall that $\Delta\rho_{ICB} - \Delta\rho_{phase}$ is the density deficit difference between the outer and inner cores that is due to compositional differences, and ΔC is the compositional difference between the outer and inner cores, each assumed of uniform composition. Using Eq. (2.26) we can write Eq. (2.25) as

$$Q_C = \int \psi\,\rho\,\alpha_C\,(dC/dt)\,dV \approx \int \psi\left(\Delta\rho_{ICB} - \Delta\rho_{phase}\right)(1/\Delta C)(dC/dt)\,dV. \qquad (2.27)$$

Finally, we can relate dC/dt to the inner core growth rate, $\dot{r}_{ICB}$, as for secular cooling and latent heat. The rate of light element mass fraction added to the outer core, dC/dt, equals the rate of solid mass fraction added at the ICB, $4\pi r_{ICB}^2 \rho_{ICB}\,\dot{r}_{ICB}/M_{OC}$, where M_{OC} is the outer core mass, multiplied by the compositional difference between the outer and inner cores. Thus,

$$dC/dt = 4\pi r_{ICB}^2 \rho_{ICB}\,\dot{r}_{ICB}\,(\Delta C/M_{OC}), \qquad (2.28)$$

so that Eq. (2.27) becomes

$$\begin{aligned}Q_C &= \int \psi\left(\Delta\rho_{ICB} - \Delta\rho_{phase}\right) 4\pi r_{ICB}^2 \rho_{ICB}\dot{r}_{ICB}(1/M_{OC})\,dV \\ &= 4\pi r_{ICB}^2 \rho_{ICB}\dot{r}_{ICB}\left(\Delta\rho_{ICB} - \Delta\rho_{phase}\right)(1/M_{OC})\int \psi\,dV,\end{aligned} \qquad (2.29)$$

where the integral can be computed from seismic data.

Because compositional energy is not extracted from the core as heat, but it is added as heat to the mantle across the CMB, the core's rate of entropy gained as compositional energy is released is

$$S_C = Q_C/T_{CMB}. \qquad (2.30)$$

The lack of a term proportional to $-1/T$ or $-1/T_{ICB}$ is what makes compositional energy so efficient from a Carnot standpoint.

2.4.1.5 Heat of chemical reactions

Because the core is an alloy there is the possibility that energy is released or absorbed during chemical reactions between Fe and the light elements. This energy is known as the heat of chemical reaction. Because species must be conserved, whatever energy is released (absorbed) at the ICB, it is absorbed (released) throughout the volume of the outer core, so it does not enter into the energy balance equation, assuming the heat of reaction, H_R, measured in J/kg, is constant. However, because the energy is released (absorbed) at the ICB, and absorbed (released) throughout the volume of the outer core, there is an efficiency factor, leading to a nonzero rate of entropy production associated with the heat of chemical reactions, S_{CR}. The expression for S_{CR} is

$$S_{CR} = \int \rho H_R\,(dC/dt)\,(1/T)\,dV. \qquad (2.31)$$

Using Eq. (2.28), and that the energy is released (absorbed) at the ICB (not at the CMB) and vice versa throughout the outer core, this becomes

$$S_{CR} = 4\pi r_{ICB}^2 \rho_{ICB}\,(H_R \Delta C/M_{OC})\left\{\int (\rho/T)\,dV - M_{core}/T_{ICB}\right\}\dot{r}_{ICB}, \qquad (2.32)$$

which is of course also proportional to the inner core growth rate. If the reaction is exothermic ($H_R > 0$), then $S_{CR} > 0$, because through chemical reactions heat must be removed from the core at the high-temperature ICB and added to it at lower temperatures throughout the outer core.

2.4.2 Geophysical estimates for the energy and entropy balance equations

Eqs. (2.18), (2.21), (2.23), and (2.29) give expressions for the power produced by secular cooling, Q_{SC}, latent heat, Q_{LH}, radiogenic elements, Q_R, and compositional energy, Q_C. By most estimates these are the primary contributors to the energy budget in the core. Others that we therefore won't discuss further include pressure heating and pressure effects on the latent heat. *Elliptical instabilities* driven by tidal forces can feed energy into the core, and result in fascinating fluid dynamics, but are likely a small contributor to the dynamics of the present day core, even as they may have played a role in the early core (Chapter 8).

Conservation of energy thus tells us that

$$Q_{SC} + Q_{LH} + Q_C + Q_R = Q_{CMB}, \tag{2.33}$$

or

$$\left\{ (1/T_{ICB}) \left(c_P \int \rho T dV \right) (dT_M/dP - dT_{ad}/dP)_{ICB} (\rho g)_{ICB} + 4\pi r_{ICB}^2 L \rho_{ICB} \right. \\ \left. + 4\pi r_{ICB}^2 \rho_{ICB} \left(\Delta \rho_{ICB} - \Delta \rho_{phase} \right) (1/M_{OC}) \int \psi \, dV \right\} \dot{r}_{ICB} + h M_{core} = Q_{CMB}. \tag{2.34}$$

Because neither the heat conducted up the adiabat nor the Ohmic and viscous dissipation appear in Eq. (2.34) it tells us nothing about sustaining the geodynamo. This should come as no surprise: the first law of thermodynamics does not preclude heat from spontaneously flowing from the CMB to the ICB, nor molten Fe in the outer core from taking (random) thermal energy and turning it into mechanical or magnetic energy. Hence we need the second law.

In Eq. (2.34) all terms on the left-hand side are proportional to the inner core growth rate $\dot{r}_{ICB}$, except for the radioactive heat $Q_R = hM_{core}$. Thus, for a given Q_{CMB}, the larger the contribution from Q_R, the smaller $\dot{r}_{ICB}$, and the older the inner core. Although Q_R is relatively inefficient because it is removed from the entirety of the core rather than just at the high-temperature ICB (Eq. 2.24), the presence of radiogenic elements in the core could in theory have helped to sustain the geodynamo by increasing the time over which latent heat and compositional energy have been contributing. Before inner core nucleation (ICN), Q_{LH} and Q_C were obviously zero, leaving just Q_{SC} and Q_R. It turns out however, as we will see, that Q_R is likely relatively small, at least at present. While secular cooling has of course been occurring since the core formed, one cannot use Eq. (2.18) before ICN, because one can then not relate the cooling rate at the CMB to the inner core growth rate.

The entropy balance equation is

$$S_{SC} + S_{LH} + S_C + S_{CR} + S_R - S_{ad} = S_J + S_\nu, \tag{2.35}$$

where S_{SC}, S_{LH}, S_C, S_{CR}, and S_R are given by Eqs. (2.20), (2.22), (2.30), (2.32), and (2.24), and represent the rate at which entropy increases due to energy that leaves the high-temperature core and flows into the low-temperature mantle. By subtracting the rate at which entropy is lost due to heat conduction up the adiabat, S_{ad}, and which is therefore not available to do work, we are left with the rate of entropy that can drive convection and power the geodynamo, which appears as Ohmic and viscous dissipation, $S_J + S_\nu$.

The expression for S_{ad} is

$$S_{ad} = \int 4\pi r^2 k \left\{ (dT_{ad}/dr)/T_{ad} \right\}^2 dr. \tag{2.36}$$

A large thermal conductivity k results in a large S_{ad}, and less entropy available for Ohmic dissipation. If S_{ad} exceeds the entropy terms associated with heat removed from the core, $S_{SC} + S_{LH} + S_{CR} + S_R$, convection and magnetic field generation is only possible if S_C is sufficiently large to make up for the requirements of S_J. Unfortunately, S_J cannot be measured or expressed in terms of observables. Rather, it must be determined by models and scaling of magnetic field generation. The parameters k, dT_M/dP, $\Delta\rho_{ICB} - \Delta\rho_{phase}$, h (the radioactive power per mass), and $\dot{r}_{ICB}$ have all been topics of considerable debate as well, making model integration of Eqs. (2.34) and (2.35) uncertain. We also note the uncertainty concerning the latent heat L, though this has perhaps received less attention. Moreover, we do not directly know Q_{CMB}, now or in the past.

2.4.2.1 The CMB heat flux, and the heat and entropy conducted up the adiabat

Ultimately it is the mantle's ability to remove heat, primarily through solid-state convection, that governs the rate at which the core cools. Q_{CMB} is thus a critical parameter. The integrated surface heat flux, Q_{EAR}, is due to mantle and crustal cooling, plus the contributions due to mantle and crustal radioactive heat production, plus Q_{CMB}. Measurements of conducted heat flow taken from different types and ages of crust, and then integrated over the Earth's surface, yield about 35 TW. When added to the estimated 11 TW due to hydrothermal circulation near mid-ocean ridges, Q_{EAR} is about 46 TW. But what fraction of that is due to Q_{CMB}?

Estimates of the present-day core heat flow are based on a variety of indirect methods, including interpretations of lowermost mantle seismic structure, calculations based on *mantle plume* fluxes, and predictions from models of the global circulation of the mantle. In recent years there has been an upward trend in these estimates. Preliminary estimates of core heat flow based on the connection between hotspot activity and mantle plume flux (assuming the heat in mantle plumes comes from the core) yielded low values, in the range of $Q_{CMB} = 2$–5 TW, although later corrections to the method of estimation increased these two-fold. Interpretations of anomalous seismic structure in the D″-region of the lower mantle in terms of a double-crossing of the post-perovskite phase change yield consistently large values for the local (unintegrated) heat flux, in the range $q_{CMB} = 65$–100 mW/m^2 (Wu et al., 2011), equivalent to a total core heat flow of $Q_{CMB} = 10$–16 TW. We note that interpretations of the lateral heterogeneity of the lowermost mantle seismic structure also provide estimates of the lateral heterogeneity in local CMB heat flux in the range of $q_{CMB} = 20$–50 mW/m^2 (Lay et al., 2008). These large lateral variations have important implications for outer, and perhaps inner, core dynamics.

Predictions of the total core heat flow and its time variation can also be obtained from mantle global circulation models (mantle GCMs). The core-to-mantle heat flow in mantle GCMs depends on a number of poorly constrained properties, including the mantle viscosity, thermal conductivity, and temperature profile, the latter depending on the strength of the circulation, compositional heterogeneity, and phase changes in the transition zone and the D″-region, such as that to post-perovskite (Nakagawa and Tackley, 2010). Uncertainties in these mantle properties lead to substantial uncertainties in mantle GCM results. However, mantle GCMs can be tuned to match Earth's present-day surface heat flow and can also be tuned to match the present-day 3D structure of the mantle as revealed by seismology, thereby strengthening their predictive power. For predicting heat flux at the CMB, the structure of dense chemical piles in the lower mantle offers an important geodynamical constraint. It is found that a relatively large total heat flow at the CMB is needed in order to maintain compositionally dense piles the size of the two Large Low Shear Velocity Provinces (LLSVPs) that are seen in the present-day lower mantle (see Chapter 5). In particular, it is found that, to support two dense piles comparable in size and shape to the lower mantle LLSVPs, a total core-mantle boundary heat flow in the range of $Q_{CMB} = 10$–18 TW is needed at the present day (Zhang and Zhong, 2011). In addition, mantle GCMs predict that the time variations in Q_{CMB} over the past several hundred million years are on the order of ± 4 TW or less about the time average value, with no obvious trend. Unfortunately, the uncertainty in Q_{CMB} grows with increasing age, as does the probability of a long-term trend.

The heat conducted up the adiabat just beneath the CMB is given by

$$Q_{ad} = -4\pi r^2 k \, (dT_{ad}/dr), \qquad (2.37)$$

with $r = r_{CMB}$ and dT_{ad}/dr evaluated at the CMB. Using Eqs. (2.5) and (2.6) one finds this to be about 3 TW for a low estimated value of $k = 20$ W/m/K. This is much less than the estimated 10–18 TW for Q_{CMB}. On the other hand, for a higher estimated value of $k = 100$ W/m/K, the heat conducted up the adiabat just beneath the CMB is 15 TW, perilously close to the situation diagrammed in Fig. 2.16A, where Q_{ad} exceeds Q_{CMB} and refrigerator action by compositional convection is required to maintain the adiabat.

Eq. (2.36) for the entropy conducted up the adiabat gives $S_{ad} = 80$ MW/K for $k = 20$ W/m/K and $S_{ad} = 400$ MW/K for $k = 100$ W/m/K.

2.4.2.2 The rate of entropy required by Ohmic and viscous dissipation

Given the very low viscosity of the outer core, $\nu = 10^{-6}$ m^2/s, the viscous dissipation in the outer core is likely to be very small compared with the Ohmic dissipation, even for a turbulent core. Try heating a cup of water by stirring it. We will assume that $S_\nu \ll S_J$.

Turning to the Ohmic dissipation, one can easily enough write down the rate of entropy required by Ohmic dissipation, $S_J = \int (J^2/\sigma)(1/T) \, dV$, where J is the magnitude of the electric current, σ is the electrical conductivity, and T is, as always, the temperature at each point in the core at which the dissipation is occurring. In theory one could scale this as $S_J = Q_J/T^*$, where the Ohmic dissipation $Q_J \approx B^2 V/(\sigma \mu_o^2 l^2)$, B is the strength of the magnetic field, V is the volume of the core, μ_o is the permeability of free space, l is the lengthscale over which B varies, and T^* is a temperature between T_{ICB} and T_{CMB}. However, in the limit that the mantle is electrically insulating the toroidal part of the magnetic field does not leave the core, and in many dynamo models this part can be larger than the poloidal part of the field that we do see. Moreover, because the magnetic field decays with distance as the inverse power of its wavenumber, we can be less certain about the magnitude of the smaller scale components of B where there could be considerable Ohmic dissipation.

Since we do not have direct observations of B and l, we can instead try to estimate Q_J from numerical dynamo scaling laws, as we will explain in Chapter 4. Results vary widely, typically between 0.2 and 4 TW, but tending toward higher values. For a $T^* = 5000$ K, this yields S_J in the range 40–800 MW/K. It is perhaps worth pointing out that numerical dynamos can operate toward the lower end of the Ohmic dissipation range, in case the rate of entropy supplied by the core becomes limited.

2.4.2.3 The rate of energy and entropy change due to radiogenic elements in the core

The primary long-lived radionuclide proposed for the core is K-40, with a half-life of 1.25 Ga. Although U and Th are *lithophile*, they could contribute a small amount of radioactive power. The cosmochemical argument for K in the core is that the crust and mantle are depleted in K relative to its cosmic abundance. On the other hand, it is also somewhat volatile, so the Earth as a whole may not reflect its cosmic abundance. Both theoretical and experimental mineral physics have shown that under high pressure and temperature there may be finite solubility of K in Fe, particularly in the presence of S (Bouhifd et al., 2007). With a plausible core S concentration, the maximum amount of K that is likely present in the core is about 35 ppm (note further that the present-day fraction of K-40/K is about 120 ppm), yielding $Q_R = 0.3$ TW, which is less than 3% of the estimated current Q_{CMB}. Some have argued, however, that larger concentrations of K are possible even without more S, perhaps 100 ppm, yielding $Q_R = 1$ TW. Although radioactive heat likely plays a small role in the current energy and entropy budgets of the core, with its half-life of 1.25 Ga the power production of K-40 would have been 10 times larger 4 Ga ago, and so may have played a more important role before ICN. Chapter 8 examines this possibility in more detail.

Eq. (2.24) yields $S_R = 8$ MW/K for $Q_R = 0.3$ TW. This very small value results from the thermodynamic inefficiency for the already small value of Q_R.

2.4.2.4 The rate of energy and entropy change due to secular cooling, latent heat, heat of chemical reactions, and compositional energy

We now turn to the terms in Eqs. (2.34) and (2.35) that are proportional to the inner core growth rate, $\dot{r}_{ICB}$: secular cooling, latent heat, heat of chemical reactions, and compositional energy. As one way to proceed, using values from the Core Property Tables at the end of the book we will estimate all the parameters for Q_i and S_i except $\dot{r}_{ICB}$, and use Eq. (2.34) with $Q_{CMB} = 10$ TW and 15 TW to bracket the present-day $\dot{r}_{ICB}$. We can then use that range for $\dot{r}_{ICB}$ to estimate the age of the inner core. We can also use it in Eq. (2.35) to examine the entropy budget and the power available for the geodynamo.

The rates of heat removed and entropy gained by secular cooling, Q_{SC} and S_{SC}, are given by Eqs. (2.18) and (2.20). Using $dT_M/dP = 9.4$ K/GPa, so that $(dT_M/dP - dT_{ad}/dP)_{ICB} (\rho g)_{ICB} = (dT_M/dr - dT_{ad}/dr)_{ICB} = 0.17$ K/km, we find $Q_{SC} = (0.17 \times 10^{12}$ TJ/m) $\dot{r}_{ICB}$, and $S_{SC} = (6.2 \times 10^{12}$ MJ/m/K) $\dot{r}_{ICB}$.

The rates of heat removed and entropy gained by latent heat, Q_{LH} and S_{LH}, are given by Eqs. (2.21) and (2.22). With $L = 1.0 \times 10^6$ J/kg, $Q_{LH} = (0.21 \times 10^{12}$ TJ/m) $\dot{r}_{ICB}$, and $S_{LH} = (14.2 \times 10^{12}$ MJ/m/K) $\dot{r}_{ICB}$. Notice the greater efficiency of latent heat than secular cooling.

The net rate of heat produced due to chemical reactions is zero because whatever heat is released in one part of the core is absorbed in another, but S_{CR} is nonzero and is given by Eq. (2.32). There have been limited studies on the heats of reaction, H_R, for different light elements with Fe under core conditions, but if S and Si partition equally into the inner and outer cores, there is obviously no change in their mass fraction. For O, it was computed quantum mechanically that $H_R = -28$ MJ/kg, i.e., the reaction is endothermic. Eq. (2.32) then gives $S_{CR} = -(8.8 \times 10^{12}$ MJ/m/K) $\dot{r}_{ICB}$. It is negative because the reaction absorbs heat at high temperature, the ICB, and removes that heat at lower temperature throughout the core, i.e., it acts as a refrigerator, and is therefore decreasing the power available for the geodynamo. However, more recent work has shown that H_R is likely ten times smaller, so we will use $S_{CR} = -(1 \times 10^{12}$ MJ/m/K) $\dot{r}_{ICB}$, a small negative contribution.

Last, we estimate the rate of compositional energy removed and entropy gained as the inner core solidifies. The density difference between the inner and outer cores that is due to compositional differences, $\Delta\rho_{ICB} - \Delta\rho_{phase}$, is discussed in Section 2.1.2, and further in Chapter 6, but we will adopt here a value of 520 kg/m³, which yields $Q_C = (0.10 \times 10^{12}$ TJ/m) $\dot{r}_{ICB}$ and $S_C = (24.3 \times 10^{12}$ MJ/m/K) $\dot{r}_{ICB}$. The large value of S_C/Q_C as compared with S_i/Q_i for other energy contributions reflects compositional energy's larger thermodynamic efficiency.

2.4.3 Age of the inner core

During a graduate qualifying exam, an examiner asked one of us what one number about the Earth we would most like to know. The answer the examiner was fishing for was the core-mantle heat flux. Although Q_{CMB} is still not known with great precision, the present-day value is likely in the range 10–15 TW, up significantly from estimates in the range of 3 TW at the time of that exam. Q_{CMB} plays a key role in determining the age of the inner core, which in turn plays an important role in inner core dynamics.

Using the values estimated in the previous sections, Eq. (2.34) gives

$$(0.17 \times 10^{12} \text{ TJ/m} + 0.21 \times 10^{12} \text{ TJ/m} + 0.10 \times 10^{12} \text{ TJ/m}) \dot{r}_{ICB} + 0.3 \text{ TW} = Q_{CMB}, \tag{2.38}$$

yielding a present-day inner core growth rate $\dot{r}_{ICB} = 0.63 \times 10^{-3}$ m/yr for $Q_{CMB} = 10$ TW, and $\dot{r}_{ICB} = 0.95 \times 10^{-3}$ m/yr for $Q_{CMB} = 15$ TW. For lack of a better choice, let us assume Q_{CMB} has been constant in time. Consistent with this assumption is that the rate at which mass has been added to the inner core, $\dot{M}_{IC}$, has been constant in time. That rate is given by

$$\dot{M}_{IC} = \left(4\pi r_{ICB}^2\right) \dot{r}_{ICB} \rho_{IC}, \tag{2.39}$$

where values with the subscript $_{ICB}$ refer to the present-day ones, and ρ_{IC} is the average inner core density. For $\dot{r}_{ICB} = 0.63 \times 10^{-3}$ m/yr (corresponding to $Q_{CMB} = 10$ TW), $\dot{M}_{IC} = 1.5 \times 10^{23}$ kg/Ga, and for $\dot{r}_{ICB} = 0.95 \times 10^{-3}$ m/yr (corresponding to $Q_{CMB} = 15$ TW), $\dot{M}_{IC} = 2.2 \times 10^{23}$ kg/Ga. With the age of the inner core, $\tau_{IC} = M_{IC}/\dot{M}_{IC}$, and $M_{IC} = 9.7 \times 10^{22}$ kg, we find $\tau_{IC} = 440$–650 Ma. This range for the age of the inner core is on the young side of some estimates, and even if $Q_R = 1$ TW, ICN is extended only to 690 Ma for $Q_{CMB} = 10$ TW. Nevertheless these results, also shown in Table 2.1, give an idea on how one can use thermodynamics to constrain the age of the inner core. Chapter 8 presents the results of somewhat more detailed calculations.

TABLE 2.1 Estimates for the age of the inner core τ_{IC} for two Q_{CMB}, but assumed constant over time.

Q_{CMB} (TW)	$\dot{r}_{ICB}$ (m/yr)	$\dot{M}_{IC}$ (kg/Ga)	τ_{IC} (Ma)
10	0.63×10^{-3}	1.5×10^{23}	650
15	0.95×10^{-3}	2.2×10^{23}	440

Fig. 2.18A shows an example of a model integration of Eqs. (2.34) and (2.35) that shows the trade-off between k and the maximum age of the inner core, assuming an adiabatic Q_{CMB}. For $k = 100$ W/m/K the maximum inner core age is about 600 million years, but for $k = 20$ W/m/K the inner core could be as old as 3.5 billion years, as old as the oldest record of the paleomagnetic field. A figure such as this is perhaps deceptive, however, because it assumes Q_{CMB} equals the heat conducted up the adiabat, Q_{ad}, and so it appears as though k controls the age of the inner core, i.e., a large k implies a large Q_{CMB} and hence a young inner core. In reality it is the mantle that controls the rate at which the core cools, and hence the age of the inner core. Fig. 2.18A is simply pointing out that there is a maximum inner core age for a given k; the convected heat flux makes up the difference between Q_{CMB} and Q_{ad}.

Fig. 2.18B shows the trade-off between k and the minimum T_{CMB} at the time of ICN, for this model integration. A large k leads to a large initial T_{CMB}, which would likely result in widespread melting of the lower mantle and which may not be in accordance with mantle thermal models.

Ideally, we would have an estimate for ICN that is independent of thermal models. Some geochemical studies had suggested a signal in surface Os isotope ratios that was thought to be due to inner core formation some 3.5 billion years ago, but that cause for the isotope anomaly is no longer widely accepted. Paleomagnetic evidence seems like an obvious place to look for when the inner core commenced solidification, and some studies find evidence for a change in the nature of the geomagnetic field from one that was weaker and more variable to one that was more stable about 560 million years ago, or perhaps 1500 million years ago (Fig. 3.14), but the presence of an obvious signal is perhaps unconvincing. Chapter 8 picks up the search for clues on ICN.

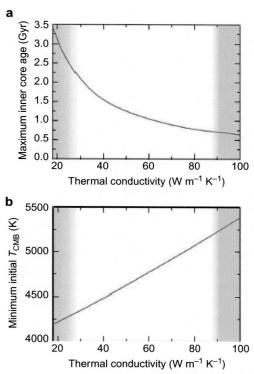

FIG. 2.18 (A) Maximum inner core age and (B) minimum initial T_{CMB} as a function of k, assuming adiabatic CMB heat flow. Blue shaded regions for lowest predicted values of k, pink for moderately high values. From Hsieh, W.-P., Goncharov, A.F., Labrosse, S., Holtgrewe, N., Lobanov, S.S., Chuvashova, I., Deschamps, F., Lin, J.-F., 2020. Low thermal conductivity of iron-silicon alloys at Earth's core conditions with implications for the geodynamo. Nat. Comm. https://doi.org/10.1038/s41467-020-17106-7.

2.4.4 Power for the geodynamo

In Section 2.4.2.1, we found that for k exceeding roughly 100 W/m/K, the heat conducted up the adiabat might exceed even a relatively large $Q_{CMB} = 15$ TW, which is clearly problematic, especially before ICN when compositional energy was not available. However, even a smaller value of k does not guarantee sufficient power for the geodynamo. To examine the power available for the geodynamo we must look at the entropy balance.

Eq. (2.38) yielded a present-day inner core growth rate $\dot{r}_{ICB} = 0.63 \times 10^{-3}$ m/yr for $Q_{CMB} = 10$ TW, and $\dot{r}_{ICB} = 0.95 \times 10^{-3}$ m/yr for $Q_{CMB} = 15$ TW. Using these values for $\dot{r}_{ICB}$ and values for $Q_i/\dot{r}ICB$ and $S_i/\dot{r}_{ICB}$ from Section 2.4.2.4, we can find the contribution for each Q_i and S_i, which are given in Table 2.2.

TABLE 2.2 Present-day contributions to the energy Q_i and entropy S_i balance equations for the core. For all but the smaller value of the entropy gain $S_i = 881$ MW/K (corresponding to $Q_{CMB} = 10$ TW), the larger value of entropy lost to conduction $S_{ad} = 400$ MW/K (corresponding to $k = 100$ W/m/K), and the largest estimated entropy requirement for the geodynamo $S_J = 800$ MW/K, there is currently no power shortage.

	Q_i (TW)	Q_i (TW)	S_i (MW/K) for $Q_{CMB}=10$ TW	S_i (MW/K) for $Q_{CMB}=15$ TW
CMB heat flux	10	15		
Secular cooling (SC)	3.4	5.1	124	187
Latent heat (LH)	4.2	6.3	284	428
Radioactive heat (R)	0.3	0.3	8	8
Compositional energy (C)	2.0	3.0	485	732
Heat of chemical reactions (CR)	0.0	0.0	−20	−30
Sum of entropy gain S_i	−	−	881	1325
Thermal conduction (ad)				
$k = 20$ W/m/K	3	3	80	80
$k = 100$ W/m/K	15	15	400	400
Ohmic heating (J)	0.2–4	0.2–4	40–800	40–800

Eq. (2.35) then gives 881 MW/K $- S_{ad} = S_J$ (neglecting viscous dissipation) for $Q_{CMB} = 10$ TW, and 1325 MW/K $- S_{ad} = S_J$ for $Q_{CMB} = 15$ TW. In Section 2.4.2.1 we found the entropy conducted up the adiabat $S_{ad} = 80$ MW/K for $k = 20$ W/m/K, and $S_{ad} = 400$ MW/K for $k = 100$ W/m/K. With the rate of entropy required by Ohmic dissipation $S_J = 40$–800 MW/K from Section 2.4.2.2, there is clearly ample present-day power for the geodynamo except for the largest values of S_{ad} and S_J, coupled with the smallest values of Q_{CMB}.

If the geodynamo predated the inner core, as seems likely, then one must ask whether secular cooling and radiogenic elements can by themselves provide sufficient entropy, even as the amount of K-40 would have been greater and even if the geodynamo could operate at less than its maximum power requirements.

When estimates of a large k were suggested, a case was made that Mg might help power the geodynamo, especially before ICN (O'Rourke and Stevenson, 2016). The argument is that although Mg is not soluble in liquid Fe at current core temperatures, it was at the higher temperatures during core formation, and that giant impacts may have delivered Mg to the early core. In this scenario, the cooling core is now supersaturated in Mg relative to the mantle, so that Mg is precipitating out at the CMB, a process known as *exsolution*. The resulting liquid beneath the CMB is then enriched in Fe and hence more dense than the bulk outer core, thereby driving compositional convection. Fig. 8.22 shows a cartoon of this scenario, with accompanying further discussion of the geodynamo before ICN. Mg exsolution is akin to the situation at the ICB where solidification results in less dense fluid. Convection driven by Mg exsolution would be very efficient, for the same reasons: it is introduced as gravitational potential energy rather than as heat, so is fully available to do work, and compositional diffusion is very slow. The amount of Mg that has been suggested for the core is about 1–2 wt-pct. Although this scenario is interesting, it remains speculative, because it requires that the solubility of Mg in liquid Fe decreases rapidly with decreasing temperature, in just the right temperature range, and because there is still the issue of getting the Mg into the core in the first place.

2.4.5 Stably stratified layers in the outer core

In Chapter 5 we will see that there is possible seismological evidence for deviations from PREM for the lowermost 150 km of the outer core above the ICB (the so-called "F" layer, or region) and for the uppermost 100–200 km below the CMB (the "E′" layer). The evidence for the former is somewhat less contentious than for the latter. The deviations in both regions have been interpreted in terms of possible stably stratified layers. These layers are much thicker than the thin conductive boundary layers that accompany all convecting systems, but are also thermally conductive and do not follow the adiabat. In Chapter 7 we will look at possible causes for the F-layer, but we will examine briefly here the thermodynamic causes and consequences of the E′ layer. In Chapters 3 and 4 we will see that the existence of a stably stratified layer at the top of the outer core would have profound consequences for the operation of the geodynamo, and the interpretation of geomagnetic signals.

A stably stratified layer at the top of the core could be due to thermal or compositional stratification. If thermal in origin, it would be due to the heat conducted up the adiabat at some radius, Eq. (2.37), exceeding Q_{CMB}. Obviously, this occurs more readily for larger k. Because the magnitude of the adiabatic temperature gradient in the outer core, Eq. (2.5), and plotted in Fig. 2.8, increases with radius, this is most likely beneath the CMB. If Q_{CMB} were subadiabatic everywhere in the outer core, then compositional convection would be required to bring heat back down into the core in order to maintain convection and the geodynamo (Fig. 2.16A) If Q_{CMB} were subadiabatic only at the top of the outer core, however, then in a thermally stably stratified layer could grow there (Fig. 2.16B). Because heat would be transferred across this layer conductively rather than convectively, the temperature gradient in the E′ layer would not follow the adiabat, thereby modifying and complicating the thermodynamic analysis of this section.

If compositional in origin, a stably stratified layer could form by the accumulation over time of light elements partitioned into the outer core during solidification at the ICB. It might also form by dissolution of Si and/or O into the outer core at the CMB, because the present-day Si/O content of the outer core might be less than that needed to be in equilibrium with the base of the mantle (Buffett and Seagle, 2010). The excess Si/O would presumably form a compositionally stratified layer beneath the CMB that grows downward by diffusion. This is in essence the opposite of the Mg exsolution driven geodynamo, although a Si exsolution driven geodynamo has also been suggested! Clearly, we do not yet know the chemical potentials of these species at the CMB. A compositionally stratified layer could also be primordial, a relic of core formation processes.

Unlike a thermally stratified layer that would be both less dense and seismically slow, a compositionally stratified layer would be less dense but, by equations of state such as Birch-Murnaghan or Vinet (see Chapter 1), seismically fast, in contradiction to what is observed. The required changes in the elastic moduli therefore put some constraints on the chemistry of a compositionally stratified layer that is not fully understood.

Because of the large lengthscale and small viscosity of the outer core, horizontal density variations as small as one part in 10^{-7} can drive outer core convection (see Chapter 4). Thus, layers exhibiting as small as 10^{-6} density contrast will tend to be stable against convective disruptions. As should be clear, the existence of these layers, in particular the E′ layer, remains controversial.

2.5 Inner core mineralogy

We now switch gears to examine the mineralogy of the inner core, which is central to understanding inner core dynamics. First, we summarize a few general aspects of mineralogy, in case you missed them along your travels.

A crystalline solid is one whose atoms are arranged in an ordered structure that has a particular, repeating arrangement, known as the crystal structure, or phase. A *grain*, which is a synonym for a single crystal, is a part of that solid whose lattice orientation is the same. Unless it is specially prepared a macroscopic solid is composed of many, many crystals, also known as a *polycrystal*. A *grain boundary* is the surface between two crystals, it is generally a high-energy feature. The orientation of crystals in a macroscopic solid is often random; such solids are untextured. Through processes such as solidification or deformation, however, crystals can tend to align, forming a *textured solid*. A textured solid is said to have a *lattice preferred orientation* (*LPO*). Because individual crystals often exhibit properties that are anisotropic, in particular, elasticity and attenuation, textured solids often exhibit macroscopic anisotropic properties. Elastic and attenuative anisotropies are often observable seismically, and interpreting their patterns in the mantle and, so far, to a lesser extent, in the inner core, have allowed geophysicists to make inferences about Earth's interior's dynamics and evolution.

The *microstructure* of a solid refers to properties of the crystals of which it is composed, aside from the particular crystal structure. These include the grain size and shape (if the grains are elongated they may have a *shape preferred orientation*; if they are not elongated they are *equiaxed*); impurity atom, point defect (vacancy), and line defect (*dislocation*) concentrations; and the presence of such features as *subgrains* that have a small misorientation from the rest of the crystal, or smaller, enclosed volumes of a different phase (perhaps liquid, or of a different crystalline material/phase). Together, the texture and microstructure reflect the processes that formed the solid, and are not simple functions of the state variables temperature, pressure, and composition. In this respect they are like solid-state viscosity. In Chapter 7 we will study these in detail in order to infer clues about the inner core's evolution.

In the following sections, we will examine the likely stable phase of Fe and its alloys, and its elastic and attenuative properties, under inner core conditions. These are intrinsic material properties, like outer core viscosity and thermal conductivity, and hence are more amenable to diamond anvil and ab initio approaches because small lengthscales are sufficient to study them. We first conduct a brief review of crystallography.

2.5.1 Review of crystallography

A lattice is defined by the set of atoms at points

$$r = u_1 a_1 + u_2 a_2 + u_3 a_3, \tag{2.40}$$

where a_1, a_2, and a_3 are the lattice translation vectors, and u_1, u_2, and u_3 are integers. Using group theory one can show that in three dimensions there are 14 lattice types (sometimes known as *Bravais lattices*) distributed amongst 7 *crystal systems* that satisfy symmetry requirements. The crystal systems differ on the magnitudes of and angles between the a_i. The only ones that are thought to be relevant to the inner core are the cubic (a_i equal, a_i perpendicular) and hexagonal ($a_1 = a_2 \neq a_3$, a_1 and a_2 perpendicular to a_3, a_1 forms an angle of 120° with a_2) crystal systems. Incidentally, the macroscopic habit of crystals generally does not reflect the microscopic crystal symmetry, unless crystal growth is unimpeded.

There are three cubic Bravais lattices, simple cubic (sc), body-centered cubic (bcc), and face-centered cubic (fcc). These are shown in Fig. 2.19. For the sc structure, each of 8 corner atoms is shared by 8 *unit conventional cells*, the volume spanned by a_1, a_2, and a_3 (in the case of cubic structures equal to a^3), for a total of one atom per unit conventional cell. For the bcc structure, there is an additional atom at the center, so that there are two atoms per unit conventional cell. For the fcc structure, each face has an additional atom at its center, each shared by another cell, for a total of four atoms per unit conventional cell. The *packing fraction* is the fraction of a unit conventional cell that is occupied by nonoverlapping (hard) spheres representing atoms. The fcc structure has a higher packing fraction than the bcc structure, which has a higher packing fraction than the sc structure.

FIG. 2.19 The lattice point and hard sphere representations for the simple cubic, body-centered, and face-centered lattices.

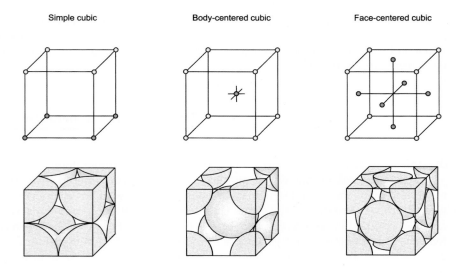

Fig. 2.20 shows the hexagonal close-packed structure. The phase consists of a simple hexagonal lattice, i.e., atoms at the vertices and centers of the bases of a hexagonal prism, each base forming an A layer, plus three atoms in the B layer in between, in the position of "holes" in the over and under lying A layers. The prism bases are known, naturally, as *basal planes*, and the six prism sides as *prismatic planes*. The axis of a prism is called the "*c-axis*," and the line from the center of a hexagon to a vertex the "*a-axis*." For closest packing the c/a ratio is $(8/3)^{1/2} = 1.633$. Even if a particular crystal has a *c/a ratio* that differs somewhat from the "*ideal*," it is still referred to as hcp.

FIG. 2.20 The lattice point and hard sphere representations for the hcp crystal structure. The stacking sequence is ABABAB....

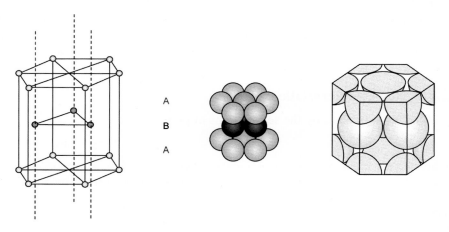

As you can see in Fig. 2.20, in the hcp structure, the spheres in layer B sit in holes in layer A, and in the third layer, the spheres sit in the holes in layer B that are directly above the spheres in layer A, hence the repeating sequence is ABABAB.... In the fcc structure, on the other hand, the spheres in the third layer sit in those holes in layer B that are not directly above the spheres in layer A, hence the repeating sequence is ABCABC.... These two possibilities are sketched together in Fig. 2.21, and Fig. 2.22 shows the ABC *stacking sequence* of planes relative to the fcc unit conventional cell. The hcp and fcc phases have equivalent packing fractions and are *close-packed structures*. Although not an ironclad rule, there is a tendency for high pressure to favor a close-packed phase such as fcc or hcp.

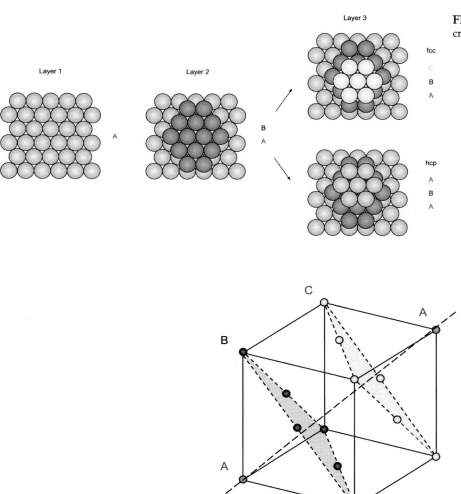

FIG. 2.21 Stacking sequences for fcc and hcp crystals.

FIG. 2.22 Close-packed (111) planes (shaded) relative to the fcc unit conventional cell, showing the ABC stacking sequence. To see the ABC stacking sequence it is easiest to look along a cube diagonal, the dashed line passing between atoms in planes B and C. In a (111) plane, each atom has six nearest neighbors, each a distance $a/2^{1/2}$ away, with the planes spaced a distance $a/3^{1/2}$ apart. In a less close-packed (100) plane (not shaded), each atom is also a distance $a/2^{1/2}$ away from its nearest neighbors, but there are only four of them, so the planes are closer, spaced a distance $a/2$ apart.

We will need to describe crystal planes and directions, which is typically done using *Miller* (or *Miller-Bravais* for hcp lattices) *indices*. Appendix 2.4 gives details on Miller indices, and also on *pole figures*, which are useful for looking at textures.

A *close-packed plane*, as the name suggests, is a plane in which atoms have a high planar density, so that they are relatively tightly bound to each other within the plane. Moreover, because of the high atomic planar density within a close-packed plane, the distance between parallel close-packed planes is correspondingly large. This enables close-packed planes to slide by each other with relative ease. For similar reasons the direction in which those planes tend to slide is in a *close-packed direction*. A *slip system* consists of a family of slip planes {} and slip directions <>. Slip systems are important for the deformation of solids and the development of texture, which we will turn to in Chapter 7.

As an example, the (111) plane is close-packed in fcc crystals, with the directions $[1\bar{1}0]$, $[\bar{1}01]$, and $[01\bar{1}]$ in that plane being close-packed (Figs. 2.22 and 2.23). Because there are three other close-packed planes equivalent to (111), each with three equivalent close-packed directions, there are a total of 12 primary slip systems in fcc crystals. Incidentally, the family of planes {111} that are equivalent, i.e., (111), $(\bar{1}11)$, $(1\bar{1}1)$, and $(11\bar{1})$, are sometimes called octahedral planes. In hcp crystals slip often occurs on the close-packed basal plane (0001), in the $[\bar{1}2\bar{1}0]$ direction (Fig. 2.24). There are two other equivalent close-packed directions, but because of the non-orthogonality of the Miller-Bravais indices, there are only two independent *basal slip* systems. Depending on the c/a ratio, *primary slip* can also occur on a close-packed prism plane, $(10\bar{1}0)$, in the close-packed direction, $[\bar{1}2\bar{1}0]$, but again, there are only two independent *prismatic slip* systems. Although there are other ways for hcp crystals to slip, including *pyramidal slip* and *twinning*, and the criteria for each being active is not always straightforward, in general fcc metals such as Al exhibit greater ductility than do hcp

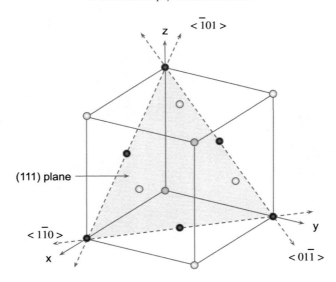

FIG. 2.23 fcc (111) slip systems.

FIG. 2.24 hcp basal $(0001)[\bar{1}2\bar{1}0]$ and prismatic $(10\bar{1}0)[\bar{1}2\bar{1}0]$ slip.

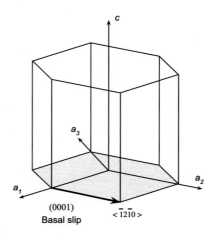
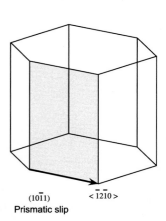

ones such as Zn. Changing the stacking sequence has profound consequences: it changes the symmetry so as to vary the number of close-packed planes and directions, and hence the ability of crystals to deform.

2.5.2 Stable phase of Fe and its alloys under inner core conditions

Fig. 2.7 is a simplified schematic of the solid-liquid part of the Fe alloy phase diagram, showing melting point suppression due to the presence of a light element such as O that exhibits little solid solubility. There should also be a third axis, of course, for pressure. Fig. 2.1 shows the atmospheric pressure phase diagram for Fe-Ni. The left axis, zero wt-pct Ni, i.e., pure Fe, exhibits the well-known transition from the low-temperature bcc α-phase (ferrite) to the higher temperature fcc γ-phase (austenite) at 912°C. Incidentally, Fig. 2.1 also shows that the addition of Ni stabilizes austenite to lower temperatures, which is utilized in making stainless steels. At 1394°C Fe transforms to the high-temperature bcc δ-phase (also called ferrite), before melting at 1538°C.

Our interest is the pressure and temperature dependence of the stable phase of pure Fe and its alloys with Ni and uncertain light elements. It makes sense to begin with Fe and Fe-Ni. Fig. 2.25 is a sketch that compiles all the experimental suggestions for high pressure and temperature phase transitions for pure Fe, as of 2002. At high pressure and up to moderate temperatures the atmospheric pressure phases of Fe transform to an hcp ε-phase. However, at higher temperatures, some diamond anvil cell experiments showed a new phase of Fe, dubbed the β-phase (Saxena et al., 1996). At the SEDI meeting in Tours, France in 1998 one investigator held a birthday party for the β-phase, replete with balloons! The phase was described as having a double hcp (dhcp; stacking order ABACABAC) or orthorhombically distorted structure. Moreover, some shock-wave experiments showed a higher pressure and temperature transformation to a bcc phase.

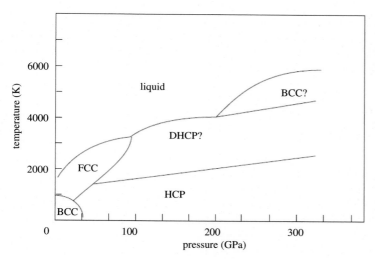

FIG. 2.25 A schematic phase diagram for pure Fe, based on a compilation of experimental results. The high-temperature, atmospheric pressure bcc δ-phase is not shown as it transforms to the fcc γ-phase above about 10 GPa. *From Alfe, D., Gillan, M.J., Vocadlo, L., Brodholt, J., Price, G.D., 2002a. The ab initio simulation of the Earth's core. Philos. Trans. R. Soc. Lond. A 360, 1227–1244.*

However, subsequent experiments have failed to find these phases, and recent experiments by different groups have shown clearly the high pressure and temperature stability of the hcp ε-phase in pure Fe and Fe-10 wt-pct Ni (Fig. 2.26). Ab initio calculations using molecular dynamics to incorporate the effects of high temperature confirm the high pressure and temperature existence of the hcp phase because of mechanical instability of the bcc phase. Although the quantum mechanically calculated Gibbs free energy difference between the phases is small, there is a general consensus that the hcp ε-phase is the stable phase of Fe and Fe-Ni throughout the pressure and temperature range of the inner core, though a few holdouts remain.

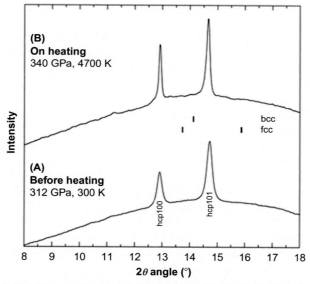

FIG. 2.26 X-ray diffraction patterns for Fe-10 wt-pct Ni, showing the stability of the hcp epsilon-phase at high pressure, for both (A) low and (B) high temperatures. There are no peaks where the bcc or fcc phases would be. *From Tateno, S., Hirose, K., Komabayashi, T., Ozawa, H., Ohishi, Y., 2012. The structure of Fe-Ni alloy in Earth's inner core. Geophys. Res. Lett. 39, L12305.*

The more debated question is whether light elements might stabilize the bcc phase at high temperature because of its greater entropy, thereby lowering its Gibbs free energy, Eq. (2.2). Ab initio calculations have found that 3–6 wt-pct S in Fe stabilizes the bcc phase (Vocadlo et al., 2003), but this has not been studied experimentally. A similar conclusion has been reached using ab initio calculations for Si. Experiments on Si in Fe are somewhat ambiguous, but suggest, on the other hand, stability of the hcp ε-phase (Asanuma et al., 2008). Similarly, small amounts of C in Fe also find stability of the hcp phase. Because O seems not to partition into the inner core, it would seem to be irrelevant for influencing the stable phase.

Clearly more work needs to be done to determine the stable phase of Fe alloys under inner core conditions, though this is made complicated because the inner core's composition remains uncertain. Although hcp is the likely phase of most of the inner core, the apparent small difference in Gibbs free energy between phases suggests the possibility that the stable phase could vary, particularly as a function of depth. This has implications for interpreting inner core seismic elastic anisotropy.

2.5.3 Elastic properties of Fe under inner core conditions

Key questions concerning the elastic properties of Fe under inner core conditions are 1) do the experimentally or quantum mechanically determined isotropic compressional and shear velocities, V_P and V_S, match those found seismically, and 2) what is the single crystal elastic anisotropy? The answer to the latter is central toward understanding the large scale, seismically inferred elastic anisotropy, if it is due to a lattice preferred orientation. The answer to the former lies essentially with finding the combination of geophysically and geochemically plausible light elements present in the inner core to give the seismically inferred values of V_P and V_S, which we discussed earlier in this chapter. However, although the constraints on density can likely be satisfactorily understood, questions remain about the inner core shear modulus μ, which we'll examine briefly in the next section. For further details on elasticity and anisotropy, see Appendices 1.2, 1.7, and 1.8.

For anisotropic materials the general relationship between stress components σ_{ij} and strain components ε_{kl} is

$$\sigma_{ij} = c_{ijkl}\varepsilon_{kl}, \tag{2.41}$$

where c_{ijkl} are the elastic stiffness constants. Because the angular acceleration of an element of solid in equilibrium must vanish, so must its net couple. Hence, $\sigma_{ij} = \sigma_{ji}$, and there are only six independent stress components. Similarly, there are only six independent strain components and thus, 36 elastic stiffness constants. One can show that c_{ijkl} must also be symmetric (Nye, 1985; Kittel, 2005), reducing the number of elastic stiffness constants to 21 for the most general case. By further symmetry conditions, one can show that the number of independent elastic stiffness constants is three in cubic lattices, and five in hcp ones.

By choosing atomic displacement and propagation directions, and using Eq. (2.41) with the appropriate elastic stiffness constants, one can also show that the compressional and shear wave speeds are isotropic in the basal plane of hcp crystals. This cylindrical symmetry about the c-axis is attractive for explaining the large-scale inner core seismic elastic anisotropy of about 3%, which to a first approximation exhibits cylindrical symmetry about the spin axis, as discussed in Chapter 6.

For hcp crystals it is thus sufficient and convenient to plot V_P and the two transverse V_S as a function of angle from the c-axis. At 0 K ab initio calculations show that the c-axis is about 3%–4% faster than the basal plane for compressional waves (Fig. 2.27A). Some ab initio calculations that include the effects of high temperature through particle-in-cell or molecular dynamics methods have found, however, a reversal of the sense of anisotropy, i.e., the basal plane is about 10% faster than the c-axis (Steinle-Neumann et al., 2001). On the other hand, more recent ab initio calculations that include the effects of temperature have not found this reversal, but have found the c-axis to be only about 1%–2% faster than the basal plane, with a 3%–4% V_P minimum at an intermediate direction (Fig. 2.27B). Whether such small single crystal anisotropy can explain the seismic anisotropy seems doubtful. There are suggestions that Si might increase the single crystal elastic anisotropy, but more work needs to be done. Experimental determination of the elastic stiffness constants of Fe at inner core pressures is extremely difficult due to texturing due to nonhydrostatic stresses, and reliable results have so far not been obtained. Experimental studies on hcp metals such as Ti that serve as low-pressure analogs for high-pressure Fe show no reversal of the fast direction, but rather an increase in elastic anisotropy with temperature (Fig. 2.27A).

Because of the possibility that light elements in the inner core might stabilize the bcc phase, mineral physicists have also explored the elastic anisotropy of bcc Fe at inner core conditions. Molecular dynamics show that it could have a single crystal elastic anisotropy as large as 12% (Belonoshko et al., 2008), but the high symmetry of the cubic phases leads to a significant reduction in the anisotropy between the polar axis and its transverse plane, and it is not clear bcc Fe can explain inner core cylindrical anisotropy (Lincot et al., 2015).

Impressively, there have been seismic inferences of inner core shear wave anisotropy. In theory this could be compared with the single crystal anisotropy of V_S as determined by the quantum mechanically calculated elastic stiffness constants, but such work is clearly in its infancy.

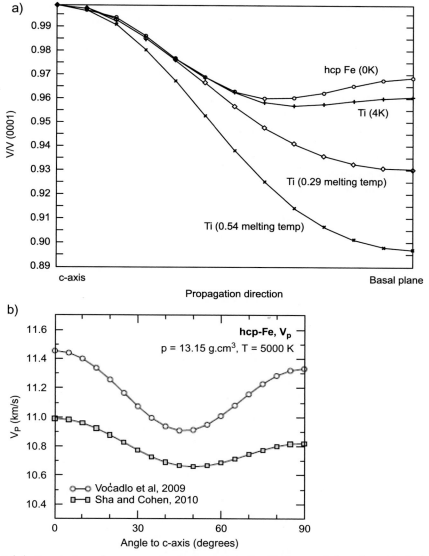

FIG. 2.27 (A) Single crystal elastic anisotropy for compressional waves, normalized to the speed along the *c*-axis, for hcp inner core pressure Fe at 0 K, and for hcp atmospheric pressure Ti for a range of temperatures, experimentally determined. 0 K Fe calculations from Stixrude and Cohen (1995). (B) Ab initio calculations for V_P as a function of propagation direction for hcp Fe under inner core conditions. *Part figure (A) adapted from Bergman, M.I., 1998. Estimates of the Earth's inner core grain size. Geophys. Res. Lett. 25, 1593–1596; (B) From Deguen, R., Cardin, P., Merkel, S., Lebensohn, R.A., 2011. Texturing in Earth's inner core due to preferential growth in its equatorial belt. Phys. Earth Planet. Inter. 188, 173–184.*

2.5.4 Shear modulus of Fe

First, a quick review of relevant *rheologies* (see also Appendix 1.5). For *elastic* behavior, stress σ is proportional to strain ε, with the constant of proportionality an elastic modulus, e.g., $\sigma = \mu \varepsilon$, where μ is the shear modulus. When the stress is removed, the strain returns to zero. There is no loss of elastic energy to heat, and elastic behavior can of course be modeled with a spring. For *anelastic* behavior, stress is proportional to the strain rate $\dot{\varepsilon} = d\varepsilon / dt$, with the constant of proportionality a viscosity η, $\sigma = \eta \dot{\varepsilon}$. The strain is delayed after stress is applied or removed, but the solid or fluid returns to its original state after the stress is removed; there is no permanent deformation. Some non-recoverable elastic energy is lost to heat, equal to the area of the hysteresis loop in a stress-strain diagram. Anelastic behavior can be modeled with a dashpot, the mechanical equivalent of a resistor. For a cyclic load such as a seismic wave, this results in amplitude decay, aside from any geometric spreading. This intrinsic dissipation can be quantified by the inverse *quality factor* Q^{-1}, which represents the fraction of energy lost to heat in each cycle, as discussed in the first chapter.

Anelastic behavior is a special case of *viscous* behavior, where in general the solid or fluid does not return to its original state; there is permanent deformation. Most materials subject to permanent deformation typically exhibit *viscoelastic* behavior, which can be represented by a combination of springs and dashpots in series and/or parallel. There are different models for connecting springs and dashpots to represent different solids' behaviors. Viscoelastic *creep*, also known as just creep or plastic flow, is the response of a solid to a long-term, constant stress, rather than a cyclic one. $\sigma = \eta \dot{\varepsilon}$ for creep, though η need not be constant. We will return to inner core creep in Chapter 7.

As we will see in Chapter 6, the inner core has a low shear modulus μ, between 1.5×10^{11} Pa near the ICB and 1.7×10^{11} Pa at its center. This results in a low shear velocity and high *Poisson's ratio* (see Chapter 1) as compared with values obtained experimentally or computationally, as well as high attenuation Q^{-1} in both the body wave and normal mode bands. Due to the inefficient conversion of compressional to shear wave energy at the ICB, and the high inner core Q^{-1}, even ambiguous observations of inner core shear (J) waves have only been made in the past twenty years. Although the rigidity of the inner core is now well verified by both body waves and normal modes, questions remain about why it appears "soft," even beyond what one might expect for a solid close to its melting temperature.

Several suggestions have been put forth for why the Poisson's ratio might be higher than one would expect for a rigid solid. These include the presence of melt at depth into the inner core, and premelting effects whereby solid Fe begins to show shear softening at temperatures less than T_M (Martorell et al., 2013). However, the smaller than expected value of μ all the way to Earth's center might perhaps suggest these are not the cause, because T/T_M, and the melt fraction, should decrease with depth (though there could be liquid films). The low value of μ has also been ascribed to intrinsic crystalline effects, such as hcp Fe_7C_3 decreasing the shear velocity (but also the density, by too much; Li et al., 2016), or anomalous defects in the bcc phase of Fe (Belonoshko et al., 2007). Recent work has shown that an Fe-C-Si inner core can satisfy both the density and shear modulus constraints, without invoking premelting (Li et al., 2018). On the other hand, others have argued that based on cyclic loading experiments on (cubic) Fe close to its melting temperature, the low seismic value of μ is not anomalously low (Jackson et al., 2000). The inner core shear modulus μ remains somewhat of an enigma.

2.6 Summary

- The core as a whole is Fe-rich, with the outer core 6%–10% less dense than pure Fe under core conditions, and the inner core 2%–3% less dense. There is also some Ni, but that does not explain the density deficit.
- Si, S, O, C, and H have all been proposed to explain the density deficit, but two or more major light elements are likely necessary to explain both the density increase across the ICB beyond that due to the phase change, and the presence of a light component in the inner core.
- Based on the melting temperature of Fe under core conditions, and the effects of light elements, the ICB temperature likely exceeds 5000 K. Under the assumption that the entirety of the outer core is convecting, so that the outer core geotherm lies close to the adiabat, the CMB temperature is then determined to be about 4000 K, which compares reasonably with mantle-side estimates.
- Although some estimates of outer core viscosity have much higher upper bounds, the outer core kinematic viscosity is likely on the order of 10^{-4}–10^{-6} m^2/s. Inner core viscosity is a function of dynamics as well as material properties, so we delay a discussion of inner core viscosity until Chapter 7.
- Experiments, solid-state theory, and first principles calculations on the electrical and thermal conductivities of the core vary widely. Estimates of the thermal conductivity of Fe alloys range from 20 W/m/K to 150 W/m/K, with electrical conductivity ranging between 4×10^5 S/m and 20×10^5 S/m. Thermal conductivity plays a key role in determining whether a thermally stratified layer beneath the CMB can form.
- The CMB heat flux and contributions to the energy budget of the core suggest an age of inner core nucleation between 440 Ma and 650 Ma, but there is considerable uncertainty on input parameters and details of model integrations. Contributions to the entropy budget, including the conducted heat, determine the power available for the geodynamo. There is likely no difficulty powering the geodynamo after inner core nucleation. Chapter 8 examines core energetics of the past.
- Hexagonal close-packed is the likely stable phase of Fe under inner core conditions, although energy differences between it and a cubic phase are small, especially with the presence of light elements. The elastic and anelastic properties of Fe in the inner core remain subjects of research, in an effort to understand seismically inferred elastic anisotropy, and high attenuation.

Appendix 2.1 Construction of phase diagrams and the partition coefficient

Eq. (2.2), $G = H - TS$, expresses the balance between the enthalpic and entropic contributions to the free energy. The phase(s) that exist in stable equilibrium are those that minimize G. At higher temperatures the entropic term dominates, favoring, of course, a well-mixed, high-entropy liquid phase. At lower temperatures the enthalpic term becomes more important, favoring the lower enthalpy of a crystalline solid phase even at the expense of lower entropy. In a two-component system, the components A and B might lower their overall Gibbs free energy by forming two solid phases, A-rich α and B-rich β, depending on the sign of ΔH upon mixing. In particular, if A and B "dislike" each other (ΔH > 0), they can decrease their overall G by maintaining separate phases, thereby decreasing their overall H, even at the expense of also decreasing their overall S.

Fig. 2.28 diagrams the construction of a binary phase diagram for an ideal solid solution (ΔH = 0), Fig. 2.6. Fig. 2.29 diagrams the construction of a binary phase diagram for a eutectic solid system (ΔH > 0), Fig. 2.7. These constructions assume ideal liquids (ΔH = 0 upon mixing in the liquid phase). As T decreases in successive panels, G for the solid phase(s) becomes less than that for the liquid phase, at compositions that depend on the temperature. The chemical potential of component A in phase 1, μ_A^1, is given by

$$\mu_A^1 = \left(\partial G^1 / \partial C_A\right)_{T,P,C_B}, \tag{2.42}$$

and likewise for μ_B^1, and for phase 2. (Truth be told, for the sake of brevity and simplicity we have been a little sloppy in Eq. (2.42) about the distinction between mass, mass fraction, number of moles, and molar fraction for G, C_A, and C_B, but the essential point remains.) The requirement for equilibrium of two phases 1 and 2 (which could be the liquid phase and a solid phase, or two solid phases) is that the chemical potentials of each component A and B in each of the two phases must be equal, i.e., $\mu_A^1 = \mu_A^2$ and $\mu_B^1 = \mu_B^2$. From Eq. (2.42) the slope of the tangents of the free energy curves in Figs. 2.28 or 2.29 must therefore be equal for equilibrium, which determines the equilibrium composition of two phases, for a given temperature.

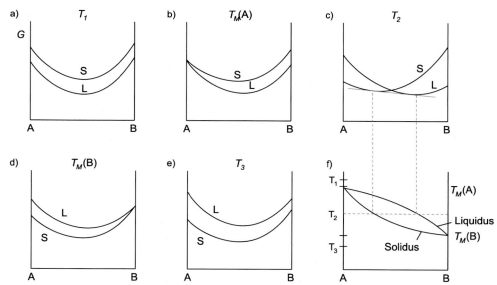

FIG. 2.28 Construction of a solid solution binary phase diagram, panel (F). Panels (A)–(E) show the free energy G for the liquid and solid phases, as the composition varies from A to B, for successively lower temperatures. The stable equilibrium phase is that with the lowest G, but liquid and solid can coexist when the chemical potentials of A and B are equal in each of the two phases, i.e., the slope of the tangents of G_S and G_L are equal, as in (C). The compositions at those two tangent points thus give the composition of the liquidus and the solidus at that temperature (here T_2, panels (C) and (F)), with compositions in between the tangent points thermodynamically unstable. The horizontal line that connects the liquidus and the solidus at a given temperature is the tie line, panel (F) and Fig. 2.6. *Adapted from Porter, D.A., Easterling, K.E., 1992. Phase Transformations in Metals and Alloys, 2nd ed., Chapman and Hall, London.*

A horizontal tie line between the liquidus and solidus (or between any two phase boundaries) therefore gives the compositions of the liquid and solid phases (or between two solid phases) that are in equilibrium. For a eutectic system, Fig. 2.29, either solid phase can be in equilibrium with the liquid phase, depending on the composition, such as at T_2. At T_E, all three phases, solids α, β, and liquid, share the same tangent, hence there is a single equilibrium composition, C_E. Because the equilibrium compositions must lie along a common tangent, other compositions are unstable, for a given T. This is the origin of the shaded, unstable fields in Figs. 2.6 and 2.7. As a system cools, the compositions of the liquid and/or solid(s) therefore move along the liquidus and/or solidus (solidi), in equilibrium.

We delay a discussion of the influence of interfacial energy on equilibrium until we study ICN. However, in anticipation of our discussion on solidification, we briefly review here the *partition coefficient* and the *lever rule*. Fig. 2.30A shows part of a binary phase diagram for a eutectic system, with assumed linear liquidus and solidus. The partition coefficient $k = C_S/C_L$, and C_0 is the initial solute (in the case of the core, the light element) concentration of the liquid. As the system cools, we are interested in the evolution of the composition of the liquid C_L and solid C_S. Two cases are of

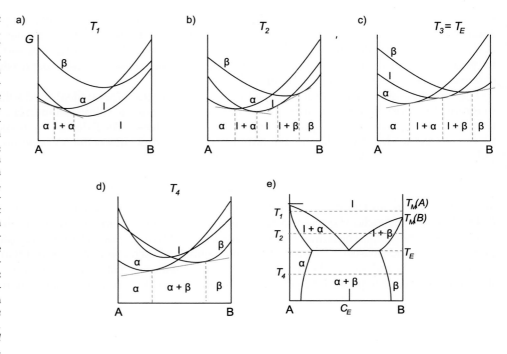

FIG. 2.29 Construction of a eutectic binary phase diagram, panel (E). Panels (A)–(D) show the free energy G for the liquid and solid phases α and β, as the composition varies from A to B, for successively lower temperatures (be careful not to confuse the phase with the composition!). Common tangents again give the equilibrium compositions at a given temperature so, for instance, at T_2 either solid α or β can be in equilibrium with the liquid, with a composition given by its respective solidus, depending on which side of the eutectic composition C_E the liquid is. At $T_3 = T_E$, the eutectic temperature, both solids α and β (with specific compositions) can be in equilibrium with the last drop of liquid (with eutectic composition C_E). Below T_E, at T_4, solids α and β, each with their own composition, can be equilibrium with each other as a solid mixture. *Adapted from Porter, D.A., Easterling, K.E., 1992. Phase Transformations in Metals and Alloys, 2nd ed., Chapman and Hall, London.*

interest: (1) infinitely slow (equilibrium) solidification, and (2) solidification with no diffusion in the solid, but perfect convective mixing in the liquid. As slowly as the core is cooling, with its large size, case (2) is of more interest, but case (1) is useful for comparison.

For case (1), as solidification commences at T_0, a tie line gives $C_S = kC_0$, the initial composition of the solid. Because the system is in equilibrium there is time for C_L and C_S to evolve along the liquidus and solidus during cooling, with $C_S = kC_L$. Fig. 2.30B plots the compositional profile of the liquid and the solid during equilibrium unidirectional solidification of a binary alloy with initial liquid solute concentration C_0. The liquid and solid have uniform but different compositions, but conservation of solute requires that the two shaded areas be equal. The equilibrium lever rule gives the fraction that is solid or liquid. Referring to Fig. 2.30A, at any temperature, say T_1, the liquid fraction is $(C_0 - C_S)/(C_L - C_S)$, and the solid fraction $(C_L - C_0)/(C_L - C_S)$. Solidification is complete at T_2, with the last drop of liquid having a concentration C_0/k, and the solid having a concentration C_0.

For case (2), for which the liquid is convectively well-mixed, but for which compositional diffusion in the solid is too slow to keep it at a uniform solute composition, solidification again commences at T_0 with $C_S = kC_0$. As the system continues to cool, however, the solid that has frozen out retains that solute concentration, and subsequent solid that freezes out has a higher solute concentration, given as always by the tie line between the liquidus and solidus, e.g., at T_1

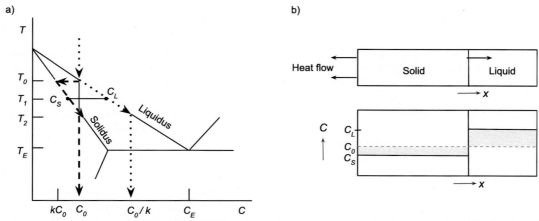

FIG. 2.30 (A) Part of a eutectic phase diagram, showing equilibrium solidification. During cooling the composition of the liquid follows the path of the dotted line, the composition of the solid the dashed line. (B) The composition of the liquid and solid as a function of position x during rightward equilibrium unidirectional solidification.

(Fig. 2.31A). Although the liquid/solid interface is always in local equilibrium, solid away from the interface is not, so that the average solid solute concentration $\overline{C}_S$ is less than the solidus solute concentration C_S at T_1. Fig. 2.31B plots the compositional profile for this case. The solid has nonuniform solute concentration, and conservation of solute, using $\overline{C}_S$ and C_L, results in the liquid becoming more enriched in solute than C_0/k, unlike the case in Fig. 2.30. The liquid can even reach the eutectic concentration C_E.

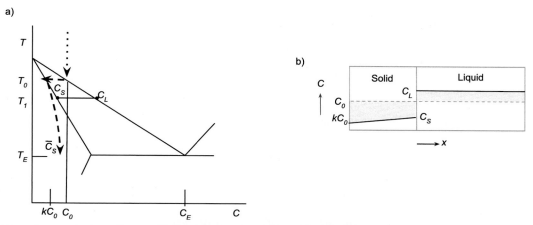

FIG. 2.31 (A) Part of a eutectic phase diagram, for the case where solid-state compositional diffusion is slow. The liquid/solid interface is in local equilibrium, so that C_L follows the liquidus and C_S the solidus, but the average solid solute concentration $\overline{C}_S$ is less than C_S. (B) The composition of the liquid and solid as a function of position x during rightward unidirectional solidification.

For an infinitesimal amount of solute rejected from the solid there is a concomitant increase of solute in the liquid, or

$$(C_L - C_S)\, df_S = (1 - f_S)\, dC_L, \tag{2.43}$$

where f_S is the volume fraction solidified. Using that $C_S = kC_0$ when $f_S = 0$, one can integrate Eq. (2.43) to obtain

$$C_S = kC_0\, (1 - f_S)^{(k-1)}. \tag{2.44}$$

This is known as the nonequilibrium lever (or Scheil) rule, and enables one to calculate the solid solute concentration as a function of fraction solidified.

Appendix 2.2 Thermodynamic relations and the Gruneisen parameter

The Gruneisen parameter γ was originally defined in terms of the change in the vibrational properties of a crystal (i.e., phonons) with a change in the lattice volume. It is thus related to thermodynamic properties, but the exact formulation of a thermodynamic γ can vary depending on the assumptions, though not significantly in practice. The dimensionless Gruneisen parameter for solids plays a somewhat analogous role that c_P/c_V ($\gamma_{ideal\ gas}$) plays for ideal gases. It is particularly useful in solid earth geophysics because it can be related to properties such as the coefficient of thermal expansivity, α_T, the bulk moduli, K_T and K_S, and the specific heats, c_P and c_V, which can be reasonably well determined. It is also relatively constant in the core.

Because at core temperatures the number of phonons at a given frequency is proportional to temperature (this follows from the Planck distribution, Appendix 2.3), a definition for the thermodynamic γ that is consistent with Gruneisen's original definition is

$$\gamma = -(\partial \ln T / \partial \ln V)_S. \tag{2.45}$$

Using the standard Maxwell relations that are derived from the second derivatives of the thermodynamic potentials,

$$(\partial T / \partial V)_S = -(\partial P / \partial S)_V, \tag{2.46}$$

$$(\partial S / \partial V)_T = (\partial P / \partial T)_V, \tag{2.47}$$

$$(\partial T / \partial P)_S = (\partial V / \partial S)_P, \tag{2.48}$$

and

$$(\partial S / \partial P)_T = -(\partial V / \partial T)_P, \tag{2.49}$$

we can relate Eq. (2.45) to material properties, the first equality in Eq. (2.4):

$$\begin{aligned}
\gamma &= -(\partial \ln T/\partial \ln V)_S = -(V/T)\,(\partial T/\partial V)_S = (V/T)\,(\partial P/\partial S)_V = (V/T)\,(\partial P/\partial T)_V(\partial T/\partial S)_V \\
&= (V/c_V)(\partial P/\partial T)_V \quad \text{(using that } c_V = T(\partial S/\partial T)_V) = (V/c_V)\,(\partial S/\partial V)_T \\
&= (V/c_V)(\partial S/\partial P)_T(\partial P/\partial V)_T = -(K_T/c_V)\,(\partial S/\partial P)_T \quad \text{(using that } K_T = -V(\partial P/\partial V)_T) \\
&= (K_T/c_V)(\partial V/\partial T)_P = \alpha_T V K_T/c_V \quad \text{(using that } \alpha_T = (1/V)(\partial V/\partial T)_P).
\end{aligned} \qquad (2.50)$$

Although one often sees Eq. (2.4), $\gamma = \alpha_T V K_T/c_V$, given as the definition of a thermodynamic γ, it ultimately thus comes from a relationship such as Eq. (2.45). Eq. (2.50) also shows $(\partial T/\partial V)_S = -\gamma T/V$, which we used in Section 2.2.1.

The second equality in Eq. (2.4), $\gamma = \alpha_T V K_S/c_P$, derives as follows:

$$dP = (\partial P/\partial S)_T dS + (\partial P/\partial T)_S dT, \qquad (2.51)$$

and for $dP = 0$,

$$(\partial S/\partial T)_P = -(\partial P/\partial T)_S/(\partial P/\partial S)_T. \qquad (2.52)$$

Similarly,

$$dV = (\partial V/\partial S)_T dS + (\partial V/\partial T)_S dT, \qquad (2.53)$$

and for $dV = 0$,

$$(\partial S/\partial T)_V = -(\partial V/\partial T)_S/(\partial V/\partial S)_T. \qquad (2.54)$$

Thus,

$$\begin{aligned}
c_P/c_V &= (\partial S/\partial T)_P/(\partial S/\partial T)_V = \{(\partial P/\partial T)_S(\partial T/\partial V)_S\}/\{(\partial P/\partial S)_T(\partial S/\partial V)_T \\
&= (\partial P/\partial V)_S/(\partial P/\partial V)_T = K_S/K_T.
\end{aligned} \qquad (2.55)$$

Last, in Section 2.2.1 we also used the relation $(\partial T/\partial P)_S = \alpha_T V T/c_P$, obtained as

$$(\partial T/\partial P)_S = (\partial V/\partial S)_P = (\partial V/\partial T)_P\,(\partial T/\partial S)_P = \alpha_T V T/c_P. \qquad (2.56)$$

Appendix 2.3 The free electron Fermi gas, phonons, and the Debye model

In the semiclassical free electron model, the periodic potential of the lattice is ignored, and there are no bands. Valence electrons are thus able to "move about" and participate as conduction electrons, with the infinite square well serving as a suitable potential. They are subject to the Pauli exclusion principle, however, thus behaving as a Fermi gas. In a 3D particle-in-a-box of side L, the energy E_k of an electron orbital with wavenumber k is

$$E_k = \hbar^2 k^2/2m = \hbar^2 \left(k_x^2 + k_y^2 + k_z^2\right)/2m, \qquad (2.57)$$

where m is the electron mass. In the ground state the highest occupied orbital has a wavenumber k_F, with an energy known as the Fermi energy level,

$$\varepsilon_F = \hbar^2 k_F^2/2m. \qquad (2.58)$$

In wavenumber space the filled orbitals form a sphere with volume $4\pi k_F^3/3$. Orbitals with higher energy are unfilled. The spherical boundary between filled and unfilled orbitals is the Fermi surface, at the Fermi energy ε_F. In position space the volume of each orbital is $(2\pi/L)^3$, and with N electrons and 2 electrons per orbital, the total number of filled orbitals is given by

$$\left(4\pi k_F^3/3\right)/(2\pi/L)^3 = N/2, \qquad (2.59)$$

so that

$$k_F = \left(3\pi^2 N/L^3\right)^{1/3} \qquad (2.60)$$

(N/L^3 is thus the electron density), and

$$\varepsilon_F = \hbar^2 \left(3\pi^2 N/L^3\right)^{2/3}/2m. \tag{2.61}$$

From Eq. (2.61), we can solve for N in terms of ε_F,

$$N = L^3/\left(3\pi^2\right)\left(2m\varepsilon_F/\hbar^2\right)^{3/2}, \tag{2.62}$$

so the density of orbitals per energy near the Fermi surface, $D(\varepsilon_F)$, is given by

$$D(\varepsilon_F) = dN/d\varepsilon_F = L^3/\left(2\pi^2\right)\left(2m/\hbar^2\right)^{3/2}\varepsilon_F^{1/2} = 3/2(N/\varepsilon_F), \tag{2.63}$$

which gives the order of magnitude of the density of orbitals at the Fermi energy that participate in electrical or thermal conduction.

Phonons are the quantized elastic vibrations of atoms, and as a means by which energy can be shared, they are thus related to the lattice heat capacity. Like photons, in thermal equilibrium the occupancy n of phonons at frequency ω and temperature T is given by the Planck distribution

$$n = 1/\left[e^{\hbar\omega/(k_B T)} - 1\right], \tag{2.64}$$

with a phonon's energy equal to $\hbar\omega$. Note that for $k_B T \gg \hbar\omega$, $n \approx k_B T/(\hbar\omega)$. The total lattice phonon thermal energy $E_{phonons}$ is thus given by

$$E_{phonons} = \int nD(\omega)(\hbar\omega)d\omega, \tag{2.65}$$

where $D(\omega)$ is the number of phonon modes per frequency, known as the density of states. As for the number of filled electron orbitals in Eq. (2.59) (except without the ½ that is due to 2 electrons per orbital), the number of allowed phonon modes N form a sphere in wavenumber space with radius k:

$$N = \left(4\pi k^3/3\right)/(2\pi/L)^3 = L^3 k^3/\left(6\pi^2\right), \tag{2.66}$$

with N/L^3 the phonon density. The density of states is thus

$$D(\omega) = dN/d\omega = L^3 k^2/\left(2\pi^2\right) dk/d\omega. \tag{2.67}$$

In order to evaluate the integral in Eq. (2.65) to find the lattice phonon thermal energy $E_{phonons}$ and hence the heat capacity $c_V = dE_{phonons}/dT$, one thus needs to know the dispersion relation $dk/d\omega$. A common and useful approximation is the Debye model, in which the sound velocity $V = \omega/k$ is assumed constant for each mode, as is the case for the elastic continuum. With this approximation the density of states becomes

$$D(\omega) = L^3 \omega^2/\left(2\pi^2 V^3\right). \tag{2.68}$$

In the Debye model there thus exists a cutoff frequency, known as the Debye frequency ω_D, given by Eq. (2.66),

$$\omega_D = \left(6\pi^2 V^3 N/L^3\right)^{1/3}. \tag{2.69}$$

In the Debye model only modes with frequencies less than ω_D are assumed to be active, with a maximum wavevector $k_D = \omega_D/V$. The Debye temperature Θ_D is defined as

$$\Theta_D = \hbar\omega_D/k_B. \tag{2.70}$$

$\Theta_D = 470$ K for Fe, so the Earth's core is well above the Debye temperature. One can use Eq. (2.65) and the Debye model to derive the experimentally observed low-temperature limit $c_V \propto T^3$ (as opposed to the classical high-temperature limit in which c_V is temperature-independent), but this is as far as we need to delve into the quantum theory of solids.

Appendix 2.4 Miller indices and pole figures

In crystallography, Miller indices are used to describe a lattice plane (hkl), which intercepts the three points a_1/h, a_2/k, and a_3/l, where a_1, a_2, and a_3 are the lattice translation vectors. The Miller indices are thus proportional to the inverses of the intercepts of the plane. If one of the indices is zero, the plane is parallel to that axis, i.e., the intercept is at infinity. By convention, a negative intercept is written with an overbar, or occasionally with a prime, and the greatest common divisor should be 1. $\{hkl\}$ is the family of planes that are equivalent to (hkl) due to the symmetry of the lattice type. Because for cubic lattices the lattice translation vectors are orthogonal, their directions $[hkl]$ are perpendicular to their planes (hkl).

This is not true for the Miller indices of crystal systems whose lattice translation vectors are not orthogonal, such as hexagonal. ⟨hkl⟩ is the family of directions that are equivalent to [hkl].

FIG. 2.32 The shaded plane has intercepts with the a_1, a_2, and a_3 axes at 3, 3/2, 2, with reciprocals 1/3, 2/3, 1/2, so its Miller indices are (243).

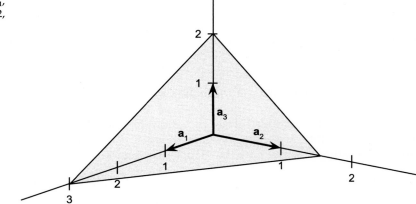

FIG. 2.33 Some low-index planes for cubic lattices.

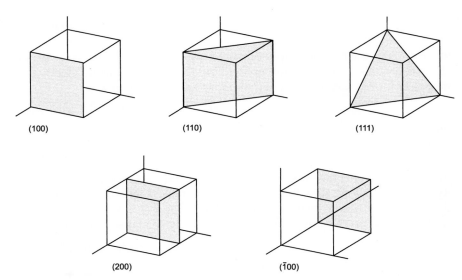

To find the Miller indices (hkl) of a lattice plane, one first finds the intercepts of that plane in terms of the lattice constants a_1, a_2, and a_3, relative to a chosen origin, and then takes their reciprocals. One then reduces these to the three smallest integers in the same ratio. For instance, the shaded plane in Fig. 2.32 has Miller indices (243). Fig. 2.33 shows some important planes in the cubic system. The directions $[hkl]$ are normal to the planes (hkl).

For hexagonal lattices such as hcp, the four index Miller-Bravais nomenclature is useful because planes $(hkil)$ are normal to directions $[hkil]$. Clearly four indices are overdetermined, so the constraint $h+k+i=0$ must be satisfied. Fig. 2.34 gives examples of important planes $(hkil)$ in hexagonal lattices.

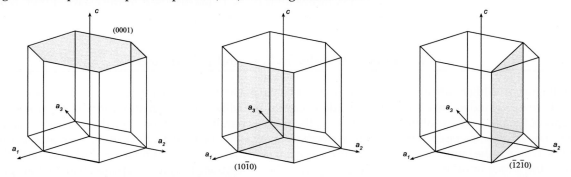

FIG. 2.34 A few low-index planes in the hexagonal system, using the four index Miller-Bravais notation. Notice that the $(10\bar{1}0)$ plane intercepts a_1 at 1 but a_3 at −1, while the $(\bar{1}2\bar{1}0)$ plane intercepts a_2 at ½.

Pole figures are a common and useful way of presenting textural analyses in materials science, although we won't be using them in his book. They can help reveal a lattice preferred orientation such as can occur during solidification or deformation. A pole figure is the stereographic projection of the measured angles (poles) of a chosen crystallographic plane (given by its normal) in a polycrystalline sample, relative to a fixed direction in the sample, which is the center pole. That fixed direction could, for example, be the solidification growth or stress or strain direction. In a spherical sample, such as the Earth, the fixed direction is likely local and the pole figure is most helpful if plotted relative to the local fixed direction. Often one makes pole figures for a few chosen crystallographic planes, and a clumping or pattern to the distribution of poles reveals a lattice preferred orientation. Most crystallographic orientation studies now use electron backscatter diffraction (EBSD), which quickly yields many measurements, so that rather than discrete poles, the pole figures display contours of pole density. Fig. 2.35 gives examples of pole figures. As the name suggests, inverse pole figures are the opposite: measurements of the fixed direction in the sample are plotted as poles relative to a particular crystallographic direction as the center pole. One sometimes displays a slice of the surface of a stereographic projection of an inverse pole figure, which represents a crystallographic plane.

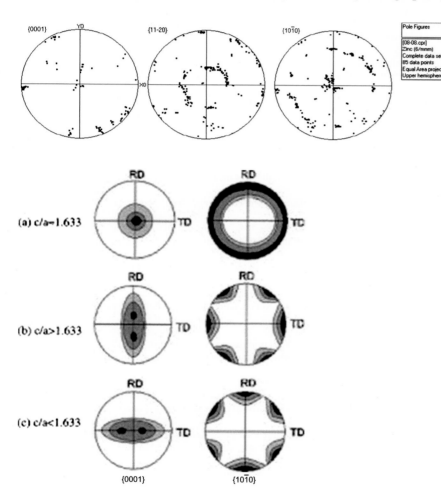

FIG. 2.35 Top: Pole figures for three crystallographic planes, for directionally solidified hcp Zn-3 wt-pct Sn. The center pole has been chosen to be the solidification direction, with the clumping of points at the center of the right panel showing the solidification direction is $[10\bar{1}0]$. Because of the symmetry of the hcp system, $\{10\bar{1}0\}$ also plots 60 degrees away from the center in that pole figure. The circular distribution of poles shows the direction transverse to solidification is randomly oriented. Bottom: Contours of pole density for simulated rolling texture in hcp metals with different c/a ratios (with assumed primary basal $\{0001\}$ $<11\bar{2}0>$ and secondary prismatic $\{10\bar{1}0\}$ $<11\bar{2}0>$ slip for $c/a=1.633$, primary basal $\{0001\}$ $<11\bar{2}0>$ and secondary pyramidal $\{11\bar{2}2\}$ $<11\bar{2}3>$ slip for $c/a>1.633$, and primary prismatic $\{10\bar{1}0\}$ $<11\bar{2}0>$ and secondary basal $\{0001\}$ $<11\bar{2}0>$ slip for $c/a<1.633$; note however that the c/a ratio is not universally recognized as a predictor of the active slip systems; see Section 7.2.2.2). The center pole has been chosen to be normal to the "sample," RD = rolling direction, and TD = transverse direction. Twinning is responsible for the $\{0001\}$ poles deviating slightly from the center. *Top: From Bergman, M.I., Yu, J., Lewis, D.J., Parker, G.K., 2017. Grain boundary sliding in high-temperature deformation of directionally solidified hcp Zn alloys and implications for the deformation mechanism of Earth's inner core. J. Geophys. Res. 123, 189–203. Bottom: From Wang, Y.N., Huang, J.C., 2003. Texture analysis in hexagonal materials. Mater. Chem. Phys. 81, 11–26.*

References

Alfe, D., 2009. Temperature of the inner core boundary of the earth: melting of iron at high pressures from first-principles coexistence simulations. Phys. Rev. B 79, 060101.

Alfe, D., Gillan, M.J., Vocadlo, L., Brodholt, J., Price, G.D., 2002a. The ab initio simulation of the Earth's core. Philos. Trans. R. Soc. Lond. A 360, 1227–1244.

Alfe, D., Gillan, M.J., Price, G.D., 2002b. Composition and temperature of the Earth's core constrained by combining ab initio calculations and seismic data. Earth Planet. Sci. Lett. 195, 91–98.

Asanuma, H., Ohtani, E., Sakai, T., Terasaki, H., Kamada, S., et al., 2008. Phase relations of Fe-Si alloy up to core conditions: implications for the Earth's inner core. Geophys. Res. Lett. 35, L12307.

Belonoshko, A.B., Skorodumova, N.V., Davis, S., Osiptsov, A.N., Rosengren, A., Johans-son, B., 2007. Origin of the low rigidity of the Earth's inner core. Science 316, 1603–1605.

Belonoshko, A.B., Skorodumova, N.V., Rosengren, A., Johansson, B., 2008. Elasticity of Earth's inner core. Science 319, 797–800.
Birch, F., 1952. Elasticity and constitution of the Earth's interior. J. Geophys. Res. 57, 227–286.
Bouhifd, M.A., Gautron, L., Bolfan-Casanova, N., Malavergne, V., Hammouda, T., Andrault, D., Jephcoat, A.P., 2007. Potassium partitioning into molten alloys at high pressure: implications for Earth's core. Phys. Earth Planet. In. 1196 (160), 22–33.
Buffett, B.A., Seagle, C.T., 2010. Stratification of the top of the core due to chemical interactions with the mantle. J. Geophys. Res. 115, B04407.
Buffett, B.A., Huppert, H.E., Lister, J.R., Woods, A.W., 1996. On the thermal evolution of the Earth's core. J. Geophys. Res. 101, 7989–8006.
de Koker, N., Steinle-Neumann, G., Vlcek, V., 2012. Electrical resistivity and thermal conductivity of liquid Fe alloys at high P and T, and heat flux in Earth's core. Proc. Natl. Acad. Sci. U. S. A. 109, 4070–4073.
Gomi, H., et al., 2016. Electrical resistivity of substitutionally disordered hcp Fe-Si and Fe-Ni alloys: chemically-induced resisitivity saturation in the Earth's core. Earth Planet. Sci. Lett. 451, 51–61.
Gubbins, D., Alfe, D., Masters, G., Price, G.D., Gillan, M.J., 2003. Can the Earth's dynamo run on heat alone? Geophys. J. Int. 155, 609–622.
Gubbins, D., Alfe, D., Masters, G., Price, G.D., Gillan, M.J., 2004. Gross thermodynamics of two-component core convection. Geophys. J. Int. 157, 1407–1414.
Gubbins, D., Masters, G., Nimmo, F., 2008. A thermochemical boundary layer at the base of Earth's outer core and independent estimate of core heat flux. Geophys. J. Int. 174, 1007–1018.
Hernlund, J.W., Thomas, C., Tackley, P.J., 2005. Phase boundary double crossing and the structure of Earth's deep mantle. Nature 434, 882–886.
Hirose, K., Labrosse, S., Hernlund, J., 2013. Composition and state of the core. Annu. Rev. Earth Planet. Sci. 41, 657–691.
Hsieh, W.-P., Goncharov, A.F., Labrosse, S., Holtgrewe, N., Lobanov, S.S., Chuvashova, I., Deschamps, F., Lin, J.-F., 2020. Low thermal conductivity of iron-silicon alloys at Earth's core conditions with implications for the geodynamo. Nat. Commun. https://doi.org/10.1038/s41467-020-17106-7.
Jackson, I., Fitzgerald, J.D., Kokkonen, H., 2000. High-temperature viscoelastic relaxation in iron and its implications for the shear modulus and attenuation of the Earth's inner core. J. Geophys. Res. 105 (B10), 23605–23634.
Kittel, C., 2005. Solid State Physics, eighth ed. Wiley.
Konopkova, Z., McWilliams, R.S., Gomez-Perez, N., Goncharov, A.F., 2016. Direct measurement of thermal conductivity in solid iron at planetary core conditions. Nature 534, 99–101.
Labrosse, S., 2003. Thermal and magnetic evolution of the Earth's core. Phys. Earth Planet. In. 140, 127–143.
Laio, A., Bernard, S., Chiarotti, G.L., Scandolo, S., Tosatti, E., 2000. Physics of iron at Earth's core conditions. Science 287, 1027–1030.
Lay, T., Hernlund, J., Buffett, B.A., 2008. Core-mantle boundary heat flow. Nat. Geosci. 1, 25–32.
Li, Y., Vocadlo, L., Brodholt, J., Wood, I.G., 2016. Thermoelasticity of Fe_7C_3 under inner core conditions. J. Geophys. Res. Solid Earth 121, 5828–5837.
Li, Y., Vocadlo, L., Brodholt, J.P., 2018. The elastic properties of hcp-Fe alloys under the conditions of the Earth's inner core. Earth Planet. Sci. Lett. 493, 118–127.
Lincot, A., Merkel, S., Cardin, P., 2015. Is inner core seismic anisotropy a marker of plastic flow of cubic iron? Geophys. Res. Lett. 42 (5). https://doi.org/10.1002/2014GL062862.
Loper, D.E., 1978. The gravitationally powered dynamo. Geophys. J. R. Astron. Soc. 54, 389–404.
Martorell, B., Vocadlo, L., Brodholt, J., Wood, I.G., 2013. Strong pre-melting effect in the elastic properties of hcp-Fe under inner-core conditions. Science 342, 466–468.
Nakagawa, T., Tackley, P.J., 2010. Influence of initial CMB temperature and other parameters on the thermal evolution of Earth's core resulting from thermochemical spherical mantle convection. Geochem. Geophys. Geosyst. 11, Q06001.
Nguyen, J.H., Holmes, N.C., 2004. Melting of iron at the physical conditions of the Earth's core. Nature 427, 339–342.
Nimmo, F., 2015. Energetics of the core. In: Schubert, G. (Ed.), Treatise on Geophysics, second ed. Core Dynamics, vol. 8. Elsevier, Amsterdam, pp. 27–65.
Nye, J.F., 1985. Physical Properties of Crystals. Oxford University Press, Oxford.
O'Rourke, J.G., Stevenson, D.J., 2016. Powering Earth's dynamo with magnesium precipitation from the core. Nature 529, 387–389.
Ohta, K., Kuwayama, Y., Hirose, K., Shimizu, K., Ohishi, Y., 2016. Experimental determination of the electrical resistivity of iron at Earth's core conditions. Nature 534, 95–98.
Poirier, J.-P., 1991. Introduction to the Physics of the Earth's Interior. Cambridge University Press, Cambridge.
Porter, D.A., Easterling, K.E., 1992. Phase Transformations in Metals and Alloys, second ed. Chapman and Hall, London.
Pourovskii, L.V., Mravlje, J., Pozzo, M., Alfe, D., 2020. Electronic correlations and transport in iron at Earth's core conditions. Nat. Commun. https://doi.org/10.1038/s41467-020-18003-9.
Pozzo, M., Davies, C., Gubbins, D., Alfe, D., 2012. Thermal and electrical conductivity of iron at Earth's core conditions. Nature 485, 355–358.
Ritterbex, S., Tsuchiya, T., 2020. Viscosity of hcp iron at Earth's inner core conditions from density functional theory. Sci. Rep. 10, 6311. https://doi.org/10.1038/s41598-020-63166-6.
Saxena, S.K., Dubrovinsky, L.S., Haggkvist, P., 1996. X-ray evidence for the new phase of β-iron at high temperature and high pressure. Geophys. Res. Lett. 23, 2441–2444.
Seagle, C.T., et al., 2013. Electrical and thermal transport properties of iron and iron-silicon alloy at high pressure. Geophys. Res. Lett. 40, 5377–5381.
Sha, X., Cohen, R.E., 2010. Elastic isotropy of ε-Fe under Earth's core conditions. Geophys. Res. Lett. 37, L10302.
Stacey, F.D., 1992. Physics of the Earth, third ed. Brookfield Press, Brisbane.
Steinle-Neumann, G., Stixrude, L., Cohen, R.E., Gulseren, O., 2001. Elasticity of iron at the temperature of the Earth's inner core. Nature 413, 57–60.
Stixrude, L., Cohen, R.E., 1995. High-pressure elasticity of iron and anisotropy of Earth's inner core. Science 267, 1972–1975.
Tateno, S., Hirose, K., Ohishi, Y., Tatsumi, Y., 2010. The structure of iron in Earth's inner core. Science 330, 359–361.
Vocadlo, L., Alfe, D., Gillan, M.J., Wood, I.G., Brodholt, J.P., Price, G.D., 2003. Possible thermal and chemical stabilization of body-centered cubic iron in the Earth's core. Nature 424, 536–539.
Vocadlo, L., Dobson, D.P., Wood, I.G., 2009. Ab initio calculations of the elasticity of hcp-Fe as a function of temperature at inner-core pressure. Earth Planet. Sci. Lett. 288, 534–538.
Wu, B., Driscoll, P., Olson, P., 2011. A statistical boundary layer model for the mantle D″ region. J. Geophys. Res. 116, B12112.

Xian, J.-W., Sun, T., Tscuchiya, T., 2019. Viscoelasticity of liquid iron at conditions of the Earth's outer core. J. Geophys. Res. Solid Earth 124 (14). https://doi.org/10.1029/2019JB017721.

Xu, J., Zhang, P., Haule, K., Minar, J., Wimmer, S., Ebert, H., Cohen, R.E., 2018. Thermal conductivity and electrical resistivity of solid iron at Earth's core conditions from first principles. Cond. Mat. Mtrl. Sci., arXiv:1710.03564v6.

Zhang, N., Zhong, S.J., 2011. Heat fluxes at the Earth's surface and core-mantle boundary since Pangea formation and their implications for the geomagnetic superchrons. Earth Planet. Sci. Lett. 306, 205–216.

Further reading

Composition and temperature of the core:

Alfe, D., Gillan, M.J., Vocadlo, L., Brodholt, J., Price, G.D., 2002. The ab initio simulation of the Earth's core. Philos. Trans. R. Soc. Lond. A 360, 1227–1244 (A review of ab initio results on the temperature and light elements in the core, as well the structure of Fe in the inner core).

Hirose, K., Labrosse, S., Hernlund, J., 2013. Composition and state of the core. Annu. Rev. Earth Planet. Sci. 41, 657–691 (A broad review of the temperature and composition of the core, and the structure of Fe in the inner core).

Poirier, J.-P., 1994. Light elements in the Earth's core: a critical review. Phys. Earth Planet. In. 85, 319–337 (An older but still good read on light elements in the core).

Porter, D.A., Easterling, K.E., 1992. Phase Transformations in Metals and Alloys, second ed. Chapman and Hall, London (An excellent text on phase diagrams, diffusion, and solidification).

Transport properties and thermodynamics:

Gubbins, D., Alfe, D., Masters, G., Price, G.D., Gillan, M.J., 2003. Can the Earth's dynamo run on heat alone? Geophys. J. Int. 155, 609–622 (A very readable treatment of core thermodynamics).

Kittel, C., 2005. Solid State Physics, eighth ed. Wiley (The standard introductory solid-state physics text).

Nimmo, F., 2015. Energetics of the core. In: Schubert, G. (Ed.), Treatise on Geophysics, second ed. Core Dynamics, vol. 8. Elsevier, Amsterdam, pp. 27–65 (A thorough review of core thermodynamics).

Poirier, J.-P., 1991. Introduction to the Physics of the Earth's Interior. Cambridge University Press, Cambridge (Strong focus on the mineral physics that concern transport properties).

Stacey, F.D., 1992. Physics of the Earth, third ed. Brookfield Press, Brisbane (As the title says! Some overlap with this book, but less focus on the core.).

Williams, Q., 2018. The thermal conductivity of Earth's core: a key geophysical parameter's constraints and uncertainties. Annu. Rev. Earth Planet. Sci. 46, 47–66 (A nice overview of the issues concerning thermal conductivity of Fe under core conditions).

Core mineral physics:

Vocadlo, L., 2015. Earth's core: Iron and iron alloys. In: Schubert, G. (Ed.), Treatise on Geophysics, second ed. Mineral Physics, vol. 2. Elsevier, Amsterdam, pp. 117–147 (A review of the mineral physics of the core).

CHAPTER

3

Geodynamo and geomagnetic basics

This chapter summarizes the properties of the geodynamo and the geomagnetic field it produces. It includes discussions of the physical ingredients necessary for planetary dynamos in general, the magnetic induction equation and its application to magnetic field generation by fluid motions in the outer core, the magnetic Reynolds number, the key dimensionless parameter controlling geodynamo behavior, energy pathways in the core that maintain the geodynamo, as well as descriptions of the present-day geomagnetic field and the ancient paleomagnetic field as they pertain to the geodynamo process. We begin with a few basic definitions, some nomenclature, and the rationale for the dynamo mechanism. Next, we examine the geomagnetic field on the core-mantle boundary and its variation over time scales ranging from decades to hundreds of million years. The final parts of this chapter summarize the basics of dynamo theory, including a conceptual model of a fluid dynamo, a simple kinematic dynamo, a laboratory dynamo experiment, estimates of the energy dissipated by the geodynamo, and images of the fluid motions in the outer core derived from measurements of the secular variation of the geomagnetic field.

3.1 Preliminaries

3.1.1 What is a planetary magnetic field?

The following serves as a working definition of a planetary magnetic field. Matter in the neighborhood of a planet experiences a force $\mathbf{F}$ proportional to its electric charge q_e, given by

$$\mathbf{F} = \mathbf{F}_E + \mathbf{F}_M = q_e \mathbf{E} + q_e(\mathbf{v} \times \mathbf{B}). \tag{3.1}$$

This force consists of two parts, an electrostatic part $\mathbf{F}_E$ proportional to the planetary electric field $\mathbf{E}$ and a magnetic part $\mathbf{F}_M$. The magnetic force depends on the product of the electric charge and $\mathbf{v}$, the velocity of the matter relative to the planet. The quantity that transforms this product into the magnetic force is a vector field $\mathbf{B}$, called the *planetary magnetic field*. The magnitude of the vector $\mathbf{B}$, the *magnetic field intensity*, is denoted by B. According to Eq. (3.1), the units of magnetic intensity are

$$[B] = [Fq^{-1}v^{-1}] = \text{N A}^{-1} \text{ m}^{-1}, \tag{3.2}$$

with 1 N A^{-1} m^{-1} = 1 Tesla, or 1 T. Planetary magnetic field intensities are commonly given in terms of milliTesla (1 mT = 10^{-3} T), microTesla (1 μT = 10^{-6} T), or nanoTesla (1 nT = 10^{-9} T). Older references often use gauss, the c.g.s. unit for magnetic field intensity. The conversions are 1 gauss = 10^{-4} T = 0.1 mT = 10^5 nT.

Fig. 3.1 is a map of the geomagnetic field intensity at the Earth's surface at epoch 2014.5, measured by the SWARM satellite constellation. The global root-mean-square (r.m.s.) value of the intensity in Fig. 3.1 is approximately 44,000 nT, but there is a systematic variation with latitude, the subpolar intensities in both northern and southern hemispheres being about twice those near the equator. Although the geomagnetic field is often described as a slightly inclined dipole, substantial deviations from the intensity pattern of an inclined dipole are evident in Fig. 3.1. In particular, the highest magnetic intensities are concentrated in several patches at subpolar latitudes in both the northern and southern hemispheres, and the lowest magnetic intensity is found in a large region in the southern hemisphere, extending across the South Atlantic Ocean and into South America. As shown in Section 3.2, these deviations from simple dipole symmetry become magnified when the surface geomagnetic field is extrapolated down to the core, and they provide important clues to the geodynamo process.

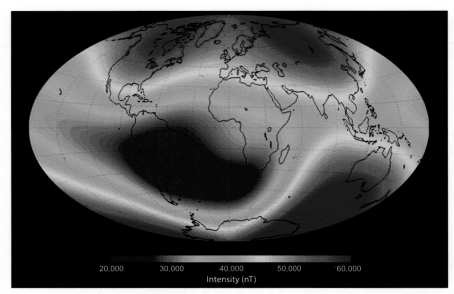

FIG. 3.1 European Space Agency SWARM satellite constellation map of the geomagnetic field intensity at Earth's surface in nanoTesla, nT (10^{-9} Tesla) at epoch 2014.5. *Credit: ESA/DTU Space (Copyrighted).*

3.1.2 What is a planetary dynamo?

In a self-sustaining planetary dynamo, electric currents and magnetic fields are induced by motions of a conducting fluid inside the planet, the conducting fluid in the Earth being the liquid outer core. The term *dynamo* refers to devices or natural systems that convert mechanical energy into electromagnetic energy, inducing electric currents through the relative motion between an electrical conductor and a magnetic field. The term *self-sustaining* implies that the electric currents inside the core and the geomagnetic fields associated with these electric currents persist in the face of Ohmic decay for as long as the fluid motions in the core support them, without the need for external sources of electric current or magnetic field. The term *geodynamo* refers to the set of physical and chemical processes in the Earth that makes this happen.

3.1.3 Rationale for the self-sustaining dynamo mechanism

There are multiple reasons why it is necessary to appeal to a self-sustaining dynamo mechanism in order to explain the origin of the geomagnetic field. First, planetary magnetism is commonplace. Like the Earth, nearly all of the largest objects in the solar system now have, or once had, global-scale magnetic fields of their own. We know that these magnetic fields predominantly have a deep internal origin. Planetary magnetic fields of external origin, such as those produced by the motion of charged particles in the atmosphere or in space, are generally very weak compared to their internal counterparts, and this is true for the Earth. Permanent magnetization occurs on terrestrial planets, but it is generally a weak, near-surface phenomenon because the interior temperatures in planets are usually too high for ferromagnetism to extend to great depths. This is also true for the Earth. And unlike permanent magnetization, the internal geomagnetic field changes rapidly, continuously varying over a wide range of temporal and spatial scales. Furthermore, the internal geomagnetic field has reversed its polarity many hundreds of times in the past, yet excluding the contribution from crustal magnetization, there is no discernible difference between the geomagnetic fields in the two polarities. All of these characteristics point to a dynamo mechanism. The need for a mechanism that is self-sustaining follows from the large discrepancy between the planet age, approximately 4.56 Ga for the Earth, and the time scale for free decay of electric currents in that planet, less than 100 kyr in Earth's interior. Without a regeneration mechanism, any primordial electric currents in Earth's interior and their associated magnetic fields would have long since decayed away.

Exploration of the solar system has shown that planetary magnetic fields differ in many important ways, yet each can be explained as a self-sustaining dynamo in some form. Paradoxically, our sister planet Venus may be the sole exception, showing no evidence of dynamo action at present or in its past. (Pluto may also be in this group, but it is no longer classified as a planet!) Furthermore, dynamo action is not limited to planets. Dynamo action in the

Sun is responsible for the 11-year solar cycle, and there is evidence for dynamo action in the early history of the Moon, in some of the Galilean satellites, and even evidence for dynamo action in meteorite parent bodies. Yet, in spite of this abundance and variety, Earth's core remains the best natural laboratory for deciphering how planetary dynamos work.

3.1.4 Dynamo ingredients

Why are planetary dynamos so common? The reason is that at the most fundamental level, only three ingredients are necessary. The first ingredient is a large volume of electrically conducting fluid. In the Earth, the iron-rich (Fe-rich) liquid outer core is that fluid. Second, there must be an energy source to stir the fluid with enough vigor so that magnetic induction of electric currents balances their resistive decay. The nature of this energy source (or sources) in the Earth's core is a long-standing Geoscience problem, mainly because the geodynamo is a highly dissipative system and requires a supply of power that is both large and long lived.

The first and second dynamo ingredients—the fluidity and conductivity of the outer core and its energetics—are described at length in Chapters 1 and 2. The third ingredient is planetary rotation. Uniform rotation of the Earth does not contribute directly to magnetic induction in the core, and in fact uniform rotation it is not an absolute physical requirement for self-sustaining dynamo action to occur. But rotation is nevertheless a practical necessity, because through its influence on fluid motions it facilitates positive feedbacks between the two otherwise independent magnetic field components—the *poloidal* magnetic field that emerges from core and the *toroidal* magnetic field buried within the core. Furthermore, irregularities in Earth's rotation, such as those produced by tidal forces and by precession, can excite fluid motions in the outer core that contribute directly to the geodynamo.

3.1.5 What we learn from the geodynamo

Global magnetic fields originate deep in planetary interiors, where there is an electrically conducting fluid. This fact already establishes important constraints on the composition, state, and thermal regime of the deep Earth, including the core. The measured structure of the geomagnetic field provides additional information on Earth's interior structure, such as the depth to the dynamo-producing region in the core, its radial extent, the primary energy sources for the geodynamo, and the fluid motions they produce. Crustal magnetization preserves the magnetic history of the core, including the longevity of the geodynamo, its polarity reversal record, and clues to the formation of the inner core and interactions with the mantle over geologic time. And last but not least, Earth's core and its dynamo are unique natural laboratories for studying complex, nonlinear magnetohydrodynamic (MHD), and multiphysics processes, including rotational fluid dynamics, turbulence, convection, solidification, and the interaction of these phenomena with self-generated magnetic fields. Different combinations of these processes are active in planets and stars throughout the cosmos.

The theory of the geodynamo aims to provide answers to questions at two distinct levels. At the first level, we are not so much concerned with the underlying dynamics. Instead, the focus is on imaging the flow in the outer core, in order to determine how these flows produce the observed properties of the core field. In addition, we need to identify the positive feedback mechanisms between poloidal and toroidal magnetic field components that sustain the core field over time. Together these objectives constitute the *kinematic dynamo* problem, and even though it is only part of the picture, the solution of the kinematic problem in the core is not trivial.

For example, consider the difference between an industrial dynamo and the geodynamo. Industrial dynamos are contrived so as to direct electric currents along preferred paths, in order to continuously regenerate a magnetic field. In contrast, the Earth's core is a nearly symmetric spherical conductor, in which no single electric current path is necessarily preferred over others. A system that regenerates magnetic field under these circumstances is called a *homogeneous dynamo*. Obviously, homogeneous dynamos require additional feedback mechanisms beyond the requirements of an industrial dynamo. These mechanisms, along with the techniques used for imaging the flow in the outer core, are described later in this chapter.

Once the kinematic dynamo questions have been answered satisfactorily for the core, the next level involves dynamical questions. How are the electrically conducting fluid motions in the outer core produced? What are the roles of the inner core and mantle? How are the fluid motions in the outer core modified by the presence of the electric currents and magnetic fields they generate? How does the interplay between the available energy sources and the geomagnetic field evolve with time? These are the subjects of *magnetohydrodynamic dynamo theory*, or *MHD dynamo theory* for short, and are treated in Chapter 4.

3.2 The geomagnetic field in the core

3.2.1 The geomagnetic field as a probe of the geodynamo

The geodynamo is a complex three-dimensional (3D) process, directly involving the whole of the core, and indirectly, the mantle. Understanding the geodynamo requires bringing to bear the widest range of geophysical and geochemical tools possible. Unfortunately, some geophysical tools that provide 3D images of crust, mantle, and inner core structure—seismic waves in particular (Chapters 1, 5, and 6) but also gravity and magnetotellurics—are less useful for probing the geodynamo. The limitations of these methods in this context stem from the physical properties of the molten iron compounds of the outer core, particularly their low viscosity. The viscosity of liquid Fe alloys is so low that the outer core fluid cannot maintain appreciable lateral heterogeneity of any property that is closely linked to its density. Because seismic wave speeds have some relationship to density, lateral variations of these properties are very small in the outer core.

Instead, the best probes of the geodynamo now available to us are the structure of the geomagnetic field and its time variations. For example, much information about the geodynamo has been gleaned by interpreting the geomagnetic field and its secular variation in terms of the flow near the top of the outer core. Likewise, short-term geomagnetic variations such as jerks and oscillations offer clues to the mechanics of the geodynamo. Paleomagnetic events such as polarity reversals and excursions provide tests of the stability of the geodynamo, while long-term variations in geomagnetic intensity provide constraints on the evolution of the geodynamo and the core. For these reasons, the geomagnetic field is our clearest window into the geodynamo, and it is worthwhile exploiting this window to its utmost. And in order to do that, we must map the geomagnetic field as close to the dynamo region as possible.

3.2.2 The geomagnetic field in spherical harmonics

The geomagnetic field is continuously measured on Earth's surface at magnetic observatories, above Earth's surface by orbiting satellites as shown in Fig. 3.1, and at lower elevations by aircraft. Here, however, we are primarily interested in the structure of the geomagnetic field in the dynamo region, that is, on the surface of the core, and if possible, within the core itself. This requires a method for projecting near-surface magnetic observations into Earth's deep interior.

The standard method for projecting near-surface magnetic measurements into the interior is to represent the geomagnetic field $\mathbf{B}$ as a potential field satisfying

$$\mathbf{B} = -\nabla \Psi, \tag{3.3}$$

where Ψ is the *geomagnetic potential*. For the part of the surface geomagnetic field originating in Earth's interior, Ψ is written as follows:

$$\Psi = r_{EAR} \sum_{n=1}^{\infty} \sum_{m=0}^{n} \left(\frac{r_{EAR}}{r}\right)^{n+1} P_n^m(\cos\theta)(g_n^m \cos m\phi + h_n^m \sin m\phi), \tag{3.4}$$

where (r, θ, ϕ) are geocentric spherical coordinates (radius, colatitude, and east longitude, respectively); r_{EAR} is the Earth's mean surface radius; P_n^m are the Schmidt quasinormalized associated Legendre polynomials (defined in Appendix 3.1), and g_n^m and h_n^m are the *Gauss coefficients* at spherical harmonic degree n and order m. According to Eq. (3.4), the Gauss coefficients have the same physical units as the geomagnetic field and are usually given in nano-Teslas (nT). The $n=1$ Gauss coefficients represent magnetic dipoles, $n=2$ the quadrupoles, $n=3$ the octupoles, $n=4$ the hexadecapoles, and so on. In theory, the summations for Ψ in Eq. (3.4) extend to infinite n; in practice they are truncated at some maximum value, typically around $n_{\max} \simeq 100$ for satellite data. From Eq. (3.3), the spherical components of $\mathbf{B}$ are

$$(B_r, B_\theta, B_\phi) = -\left(\frac{\partial \Psi}{\partial r}, \frac{1}{r}\frac{\partial \Psi}{\partial \theta}, \frac{1}{r\sin\theta}\frac{\partial \Psi}{\partial \phi}\right) \tag{3.5}$$

and the geomagnetic intensity mapped in Fig. 3.1 is just

$$B = |\mathbf{B}| = (B_r^2 + B_\theta^2 + B_\phi^2)^{1/2} \tag{3.6}$$

evaluated at the Earth's surface.

Two large-scale properties of the geomagnetic field that are useful for characterizing the geodynamo are the geomagnetic dipole moment vector $\mathbf{m}_d$ and the dipole tilt angle θ_d. In physical terms, a dipole field results either from a very small electric current loop or a uniformly magnetized sphere. The geomagnetic dipole moment can be expressed in terms of the Gauss coefficients of degree $n = 1$ as

$$\mathbf{m}_d = (m_x\hat{\mathbf{x}} + m_y\hat{\mathbf{y}} + m_z\hat{\mathbf{z}}) = \frac{4\pi r_{EAR}^3}{\mu_0}(g_1^1\hat{\mathbf{x}} + h_1^1\hat{\mathbf{y}} + g_1^0\hat{\mathbf{z}}) \tag{3.7}$$

where $(\hat{\mathbf{x}}, \hat{\mathbf{y}}, \hat{\mathbf{z}})$ are Cartesian unit vectors with origin at Earth's center (z is the polar axis, aligned with geographic poles, x and y are axes in the equatorial plane, with x through the 0 degree longitude and y through 90 degrees East longitude), (m_x, m_y, m_z) are the dipole moment components along these axes, r_{EAR} again denotes Earth's mean radius, and μ_0 is magnetic permeability, the subscript 0 denoting its free-space value, $4\pi \times 10^{-7}$ N A^{-2}. Defined this way, the dipole moment magnitude is

$$m_d = \frac{4\pi r_{EAR}^3}{\mu_0} g_1 \tag{3.8}$$

where

$$g_1 = [(g_1^1)^2 + (h_1^1)^2 + (g_1^0)^2]^{1/2}, \tag{3.9}$$

and the magnitude of axial dipole moment (ADM) is given by Eq. (3.8) with $|g_1^0|$ replacing g_1. In the SI system, the geomagnetic dipole moment has units of A m^2. At epoch 2020, the dipole Gauss coefficients are $g_1^0 = -29,405$, $g_1^1 = -1451$, and $h_1^1 = 4653$, all in nT, and from Table S1, the geomagnetic dipole moment at epoch 2020 is $m_d = 7.7 \times 10^{22}$ A m^2, or 77 ZA m^2.

The dipole moment is more than the leading order term in a series expansion of the geomagnetic field; it also plays an important role in geodynamo, for several reasons. First, the dipole field dominates the other harmonics in the representation of the geomagnetic field on the core-mantle boundary. In addition, the dipole field represents the largest-scale electric current system within the core. This can be seen by expressing the dipole moment vector in terms of the electric current density $\mathbf{J}$ or the magnetic field $\mathbf{B}$ as

$$\mathbf{m}_d = \frac{1}{2}\int (\mathbf{r} \times \mathbf{J})dV = \frac{3}{2\mu_0}\int \mathbf{B}dV, \tag{3.10}$$

where dV denotes the volume integration over the electrically conducting interior (essentially, the outer and inner core) and $\mathbf{r}$ is the radial position vector. Accordingly, the dipole moment is equivalent to the first moment of the electric current density and it is also a weighted integral of the magnetic field inside the core.

The orientation of the dipole moment vector determines the location of the *geomagnetic pole*. In terms of Gauss coefficients, the colatitude θ_d and east longitude ϕ_d of the geomagnetic pole are given by

$$\theta_d = \cos^{-1}\left(-\frac{g_1^0}{g_1}\right), \quad \phi_d = \tan^{-1}\left(\frac{h_1^1}{g_1^1}\right). \tag{3.11}$$

The present-day (epoch 2020) geographical coordinates of the north geomagnetic pole are 80.6°N and 72.7°E, so the geomagnetic pole colatitude, often referred to as the *dipole tilt*, is now $\theta_d = 9.4$ degrees. However, both θ_d and ϕ_d vary continuously in time, so that the present-day locations of both the north and south geomagnetic poles are just snapshots in time. Indeed, the *best-fitting dipole* is actually eccentric at present, being displaced from Earth's center by several hundred kilometers in the general direction of the Philippine Sea. Nevertheless, paleomagnetic measurements indicate that both the tilt angle and the eccentricity of the geomagnetic dipole tend toward zero when averaged over several thousand years of time. The assumption that the time-averaged geomagnetic dipole is *always* geocentric (i.e., symmetric with respect to the center) and axial (i.e., aligned with Earth's spin axis) is called the *geocentric axial dipole hypothesis*, or GAD hypothesis, and is the basis for most interpretations of the ancient paleomagnetic field. Note that geomagnetic poles are not the same as *magnetic poles*, which are the locations on Earth's surface, where the geomagnetic field vector is oriented vertically. Magnetic poles are superficial effects determined by the local morphology of the surface geomagnetic field in the polar regions, whereas geomagnetic pole locations reflect the orientation of the dipole field produced in the core.

3.2.3 The core field

The internal part of the geomagnetic magnetic field can be evaluated at any depth using Eqs. (3.4), (3.5) along with surface determinations of the Gauss coefficients, provided there are no intervening sources of magnetic field between the surface radius r_{EAR} and the internal radius of interest r. The extent to which this condition is met depends on the internal radius in question and also on the spherical harmonic degree n of the magnetic field. This can be seen by examining the mean-squared magnetic intensity as a function of r and n. In terms of the Gauss coefficients, this is

$$B_{rms}^2(r) = \sum_{n=1}^{n_{max}} \left(\frac{r_{EAR}}{r}\right)^{2n+4} R_n, \qquad (3.12)$$

where

$$R_n = (n+1)\sum_{m=0}^{n}[(g_n^m)^2 + (h_n^m)^2]. \qquad (3.13)$$

Formulas (3.12), (3.13) are called the *Mauersberger-Lowes spectrum*, in honor of its originators P. Mauersberger and F.J. Lowes.

Fig. 3.2 shows the Mauersberger-Lowes spectrum R_n determined from the surface geomagnetic field at epoch 2000 as measured by the Oersted satellite. The dipole term stands above the rest of the spectrum, consistent with the overwhelmingly dipolar content of the surface field, whereas the $n=2$ quadrupole term is somewhat depressed. Beyond the quadrupole, there is more-or-less a uniformly steep decreasing band starting at $n=3$ and extending to approximately $n=14$. This portion of the Mauersberger-Lowes spectrum also exhibits large-amplitude temporal variability on time scales of years to centuries, the *geomagnetic secular variation*.

The portion of the geomagnetic field lying within this band, together with the strong $n=1$ dipole and the relatively weak $n=2$ quadrupole, comprise the *core field*. In contrast, the band consisting of spherical harmonic degrees larger than about $n=16$ is characterized by low magnetic energy, a reverse spectral slope, and relatively low temporal variation. This is the *crustal field* band, and its primary source is the permanent magnetization of crustal rocks.

As noted earlier, Eqs. (3.3)–(3.5) can be used to downward continue the surface geomagnetic field, but only through regions where electric currents and permanent magnetization are negligible. Electric currents are weak enough in the mantle to be considered negligible for this purpose, but this is not the case in the core. Accordingly, the maximum depth for extrapolation of the surface field is the core-mantle boundary. There is also a maximum n-value for downward continuation to the core. Although there is no abrupt separation between the core field and the crustal field in the Mauersberger-Lowes spectrum, there is a transition between the two around $n=14-16$ evident in Fig. 3.2. The location of this transition means that we cannot justifiably downward continue harmonic components of the surface field to the core-mantle boundary beyond $n_{cf} \simeq 14$, approximately. This limitation is called the *crustal filter*. It implies that the core field derived from surface observations is a low-pass filtered image of the magnetic field in the region, where the

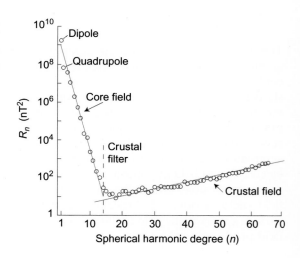

FIG. 3.2 Mauersberger-Lowes geomagnetic spectrum at Earth's surface at epoch 2000 as measured by the Oersted satellite, showing mean-square magnetic intensity R_n in units of nanoTesla squared (nT^2) versus spherical harmonic degree n. The division between the crustal field and the core field, the crustal filter, is indicated by the *dashed vertical line* segment.

geodynamo is active. Because of crustal filtering, the observable part of the core field is effectively truncated at spherical harmonic degree n_{cf} and we only "see" the relatively large-scale magnetic structures produced by the geodynamo process, corresponding to wavelengths of about 30 degrees or more on the core-mantle boundary. The crustal filter renders smaller-scale magnetic structures on the core-mantle boundary invisible to surface magnetic observation, even though these structures might be very important in the operation of the geodynamo.

Fig. 3.3 shows the 150 year time average of the radial component of core field, evaluated on the Earth's surface in panel (A) and on the core-mantle boundary in panel (B). Both maps were constructed using

$$\overline{B}_r = \sum_{n=1}^{14}\sum_{m=0}^{n}(n+1)\left(\frac{r_{EAR}}{r}\right)^{n+2}P_n^m(\cos\theta)(\overline{g}_n^m\cos m\phi + \overline{h}_n^m\sin m\phi), \tag{3.14}$$

in which $\overline{g}_n^m$ and $\overline{h}_n^m$ are the time averages of the Gauss coefficients between epochs 1860 and 2010, $\overline{B}_r$ is the radial magnetic field similarly averaged, and r in Eq. (3.14) is set to the surface radius r_{EAR} in panel (A) and to the core-mantle boundary radius r_{CMB} in panel (B).

Time averaging over multiple decades highlights the more long-lasting geomagnetic structures, although the largest of these structures can also be seen in maps of the core field from individual epochs. The importance of the axial dipole is evidenced by the dominantly negative (radially inward) magnetic field orientation in the northern hemisphere and the dominantly positive (radially outward) magnetic field orientation in the southern hemisphere, most prominently on the surface. Although to a lesser extent, the dipole component also dominates the core field on the core-mantle boundary. For example, the *dipolarity* of the core field, defined as the ratio of the r.m.s. dipole intensity to the total magnetic intensity on the core-mantle boundary, is approximately 0.6 in Fig. 3.3B (Gillet et al., 2013).

Comparison of Fig. 3.3A with Fig. 3.3B reveals how the distance between the core-mantle boundary and Earth's surface modifies the core field. Fig. 3.3B shows that much of the magnetic field that produces the axial dipole is concentrated in four high-intensity, high-latitude flux patches on the core-mantle boundary, two patches in each hemisphere. In the northern hemisphere, for example, these patches lie beneath North America and Northeast Asia. The high-intensity patches in the southern hemisphere are located slightly poleward of their northern hemisphere counterparts, but at longitudes that approximately match those in the north. At the surface, the magnetic field from

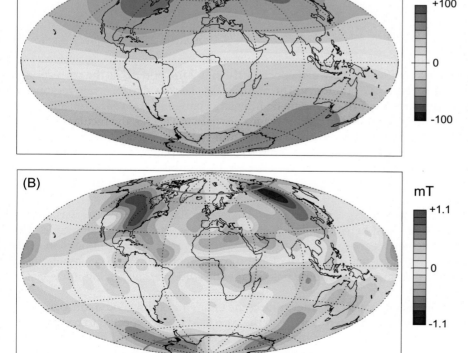

FIG. 3.3 Structure of the core field. Radial component of the geomagnetic field derived from the core, evaluated on the Earth's surface (A) and on the core-mantle boundary (B), both time averaged from 1860 to 2010. Contours of magnetic intensity in microTesla (μT) on the surface in (A) and in milliTesla (mT) on the core-mantle boundary in (B). Continental outlines are shown in both (A) and (B) for reference. Green latitude curves in (B) mark the intersection of the inner core tangent cylinder and the core-mantle boundary. *Data from core field model COV-OBS by Gillet, N., Jault, D., Finlay, C.C., Olsen, N., 2013. Stochastic modeling of the Earth's magnetic field: inversion for covariances over the observatory era. Geochem. Geophys. Geosyst. 14, 766–786. https://doi.org/10.1002/ggge.20041.*

these high-intensity patches merges together, forming lobe-shaped structures, one such structure in each hemisphere, as shown in Fig. 3.3A.

The locations of these high-intensity patches on the core-mantle boundary suggest a possible connection between longitudinally aligned pairs, with magnetic field lines entering the core in the northern patches and emerging from the core at their southern counterparts. Straight-line connectivity between patch pairs across the equator would imply the existence of a columnar structure to the magnetic field structure inside the core. The inference of columnar structures within the core is further supported by the proximity of three of the four high-intensity patches to the cylindrical projection of the solid inner core equator on the core-mantle boundary, as shown in Fig. 3.3B. In Chapter 4, we describe the many effects that this cylindrical projection, called the *inner core tangent cylinder*, has on the dynamics of the outer core.

In addition to the four intense high-latitude magnetic patches, Fig. 3.3B shows that there are a number of lower magnetic intensity regions on the core-mantle boundary that have the opposite polarity for their hemisphere, called *reverse flux patches*. A particularly large concentration of these reverse patches lies beneath the South Atlantic. At and above the Earth's surface, the manifestation of this cluster of reversed-polarity patches is the *South Atlantic Anomaly*, where the intensity of the magnetic field is approximately 30% below average, as seen in the radial component of the surface field in Fig. 3.3A and even more prominently in the surface intensity in Fig. 3.1. There are also smaller reverse flux patches located on the core-mantle boundary very near each pole in Fig. 3.3B, within the inner core tangent cylinder. In spite of their small size, these structures offer important clues to outer core fluid dynamics inside the tangent cylinder, and because of their proximity to the high-intensity patches, clues to the magnetic field structure inside the core in this region.

A third prominent deviation from axial dipole structure in Fig. 3.3B is the swath of high-intensity magnetic field that appears to originate from the high-intensity, high-latitude patch beneath East Antarctica and extends north toward the equator, before turning west beneath the Indian Ocean, south-central Africa, and the equatorial Atlantic. Later in this chapter, we show images of the fluid flow beneath the core-mantle boundary derived from the geomagnetic secular variation that identifies this magnetic structure with a large counter-clockwise circulating gyre in the southern hemisphere of the outer core.

3.2.4 Secular variation of the core field

Fig. 3.4 shows the secular variation of the radial part of the core field on the core-mantle boundary accumulated over the past century. This map is constructed using

$$\Delta B_r = \sum_{n=1}^{14} \sum_{m=0}^{n} (n+1) \left(\frac{r_{EAR}}{r_{CMB}}\right)^{n+2} P_n^m(\cos\theta)(\Delta g_n^m \cos m\phi + \Delta h_n^m \sin m\phi), \quad (3.15)$$

in which Δg_n^m and Δh_n^m are changes in the Gauss coefficients between epochs 1914 and 2014, and ΔB_r is the corresponding change in the radial magnetic field on the core-mantle boundary. The maps in Fig. 3.4 reveal that certain regions on the core-mantle boundary have experienced nearly 1 mT of geomagnetic intensity change in 100 years, reflecting the rapid secular variation of the core field. In comparison with the time average core field in Fig. 3.3, the secular variation is dominated by shorter wavelength structures. Spots of rapid secular variation typically occur in pairs, the spots within a pair having opposite signs. This pairing is usually caused by rapid transport of a patch of high-intensity magnetic flux beneath the core-mantle boundary. As we show later in this chapter, by combining the pattern of secular variation in Fig. 3.4 with the core field structure in Fig. 3.3 we can infer the fluid motions beneath the core-mantle boundary that are responsible for this transport.

Another property of the core field that can be seen in Fig. 3.4B and C is the concentration of the secular variation at low latitudes and mid latitudes in the Atlantic hemisphere, centered more or less beneath 0 degrees longitude. The opposite hemisphere, located beneath the Pacific Ocean, exhibits very little secular variation by comparison. This hemispheric dichotomy in secular variation has persisted ever since global measurements of the geomagnetic field have been available, that is, for 150 years at least, and there is some paleomagnetic evidence suggesting it has lasted far longer.

Establishing the longevity of this dichotomy is important for interpreting the behavior of the geodynamo. According to the classical dynamo theory, azimuthal (east-west) heterogeneity in the core field is not expected to be long lasting, because the strong effects of Earth's rotation tend to homogenize the core field in the east-west direction in time average. However, it is possible that heterogeneities on the boundaries of the outer core—the core-mantle boundary and the inner core boundary—overcome the homogenizing effects of rotation, thereby allowing azimuthal

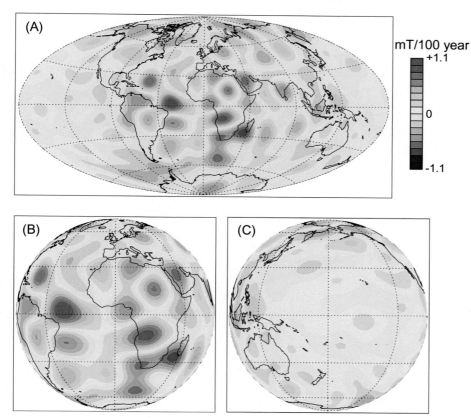

FIG. 3.4 Secular variation of the core field. Changes in the radial component of the geomagnetic field on the core-mantle boundary between 1914 and 2014 contoured in units of milliTesla per century, in global view (A) and in the Atlantic (B) and Pacific (C) hemispheres. Note the difference in the magnitude of secular variation in the Atlantic versus the Pacific hemisphere. Continental outlines are shown for reference. *From geomagnetic field model COV-OBS by Gillet, N., Jault, D., Finlay, C.C., Olsen, N., 2013. Stochastic modeling of the Earth's magnetic field: inversion for covariances over the observatory era. Geochem. Geophys. Geosyst. 14, 766–786. https://doi.org/10.1002/ggge.20041.*

heterogeneity in the core field to persist. The roles that various types of heterogeneity on the core-mantle and inner core boundaries play in structuring the geodynamo are discussed in Chapters 4 and 7.

3.2.5 Historical dipole moment variations

Because magnetic orientation was important for navigation starting in the age of exploration, we can accurately track the motion of the north geomagnetic pole back to about 1600 AD. Fig. 3.5A shows the path of the north geomagnetic pole over 420 years of historic time. Between 1600 and 1840, the dipole tilt angle increased from about 7.5 to 11.4 degrees, while the geomagnetic pole drifted generally westward through about 30 degrees of longitude. Since then, the westward motion of the north geomagnetic pole has continued, while its north-south motion has reversed. From 1840 to 1955, the dipole tilt remained almost constant, but since then it has moved rapidly northward. Over the past few thousand years, the north geomagnetic pole has traced an irregular sequence of prograde and retrograde loops such as shown in Fig. 3.5A, generally within 15 degrees of the north geographic pole. Although the underlying cause of this irregular motion is yet to be firmly established, its kinematic effect is to suppress the equatorial component of the geomagnetic dipole, leaving only the axial part in time average.

Direct measurements of the geomagnetic dipole moment began with the work of C.F. Gauss and date from about 1840. As shown in Fig. 3.5B, the strength of the dipole moment has decreased from about 85 ZA m^2 in 1840 to 77 ZA m^2 in 2020. The average rate of decrease over the intervening 180 years is 1.4×10^{12} A m^2 s^{-1}, an overall rate of nearly 6% per century. The average decrease in the strength of the axial component of the dipole moment—the ADM—over the same 180 years is 1.3×10^{12} A m^2 s^{-1} accounting for more than 90% of the total dipole moment drop during that period. As many authors have noted, if this decrease were to continue forward in time at the same average rate, the geomagnetic dipole moment would be reduced to zero within 2 kyr. Such an extreme outcome is very unlikely because paleomagnetic evidence (see below) indicates that the dipole moment does not completely vanish, even during polarity reversals. However, this may not apply to the equatorial component of the dipole moment. The equatorial moment strength has decreased at an average rate of 5.4×10^{11} A m^2 s^{-1} since 1840, nearly 12% per century. On a percentage

FIG. 3.5 Time variations of the geomagnetic dipole. (A) Motion of the North Geomagnetic Pole since 1600 and (B) dipole moment decrease since 1840 in units of ZA m^2 (10^{21} A m^2). *Data from Jackson, A., Jonkers, A.R.T., Walker, M.R., 2000. Four centuries of geomagnetic secular variation from historical records. Philos. Trans. R. Soc. Lond. A358, 957–990; Olsen, N., Holme, R., Hulot, G., 2000. Oersted initial field model. Geophys. Res. Lett. 27, 3607–3610; and IGRF Main Feld models.*

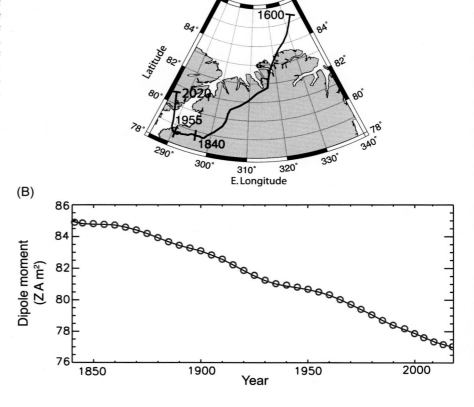

FIG. 3.6 Time variations in δg_0^1, the axial dipole Gauss coefficient with a linear trend removed, in nT (nanoTesla), between 1900 and 2020. *Data from IGRF Main Field models.*

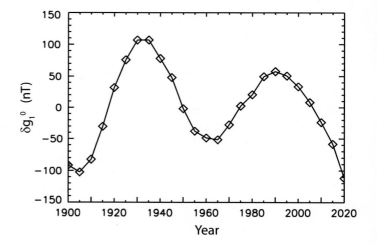

basis, this is twice the rate of the ADM decrease, and it accounts for the recent rapid poleward motion of the geographic pole shown in Fig. 3.5A (Olsen et al., 2000).

Lastly, Fig. 3.5B shows that, although the historical dipole moment decrease is persistent, it is not steady. Instead, there are small-amplitude fluctuations superimposed on a steady decrease. To highlight these fluctuations, Fig. 3.6 shows the time variation of δg_1^0, defined as the axial dipole Gauss coefficient g_1^0 with a linearly decreasing trend removed. The fluctuations in Fig. 3.6, with periods in the range of 60–70 years, have been attributed to wave motion in the E' layer at the top of the outer core. There are also shorter-period fluctuations in the core field with periodicities of 6–8 years that are not evident in Fig. 3.5 or 3.6. These fluctuations correlate with measured changes in the length of day

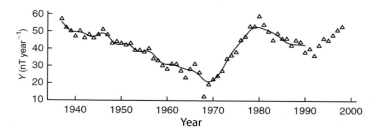

FIG. 3.7 Comparison of the rate of change in the Y (easterly) component of the geomagnetic field from annual means measured at Chambon-la-Foret Observatory in France (*triangles*) with predictions (*solid curve*) from a geomagnetic field model by Jackson et al. (2000). *Reproduced from Fig. 15 in Jackson, A., Finlay, C.C., 2007. Geomagnetic secular variation and its applications to the core. In: Schubert, G. (Ed.), Treatise on Geophysics, vol. 5. Elsevier B.V., pp. 147–193.*

and are attributed to *torsional oscillations* in the interior of the outer core. The various wave motions in the outer core that affect the geomagnetic field are described in Chapter 4.

3.2.6 Geomagnetic jerks

Geomagnetic jerks are defined as discontinuities in the second time derivative of the geomagnetic field (i.e., discontinuities in the geomagnetic acceleration) that produce abrupt changes in geomagnetic secular variation records. Geomagnetic jerks occur irregularly, on average about once per decade. They are seen particularly clearly in the easterly (Y) geomagnetic field component in observatory records from Europe, as shown in Fig. 3.7 (Jackson and Finlay, 2007), although they have been observed at other locations and some have been seen globally. Geomagnetic jerks have been correlated with short time-scale variations in the length of day, suggesting that their underlying cause is linked to irregularities in Earth's rotation.

As yet there is no consensus on the origin of geomagnetic jerks (Pinheiro et al., 2019), although the fact that the geodynamo is capable of abrupt and sometimes global magnetic field changes represents a challenge to both conventional dynamo theory and to our understanding of the dynamics of the outer core. Recent investigations have focused on the possibility that packets of Alfvén waves (propagating magnetic field disturbances illustrated in Fig. 4.15), originating from the deep interior of the core, suddenly perturb the core field and the outer core flow as they approach the core-mantle boundary (Aubert and Finlay, 2019), allowing impulses of angular momentum to be exchanged between the core and mantle.

3.2.7 The secular variation time scale

Just as the magnetic energy in the core field can be expressed in terms of the Lowes-Mauersberger power spectrum R_n, the time variation of magnetic energy in the core field can be expressed as a power spectrum $\dot{R}_n$, with a definition like Eq. (3.13) but involving the time derivatives of the Gauss coefficients, $\dot{g}_n^m$ and $\dot{h}_n^m$. In terms of R_n and $\dot{R}_n$, the correlation time for geomagnetic secular variation at each spherical harmonic degree n is defined as

$$\tau_n = \left(\frac{R_n}{\dot{R}_n}\right)^{1/2}. \tag{3.16}$$

The physical interpretation of Eq. (3.16) is that each τ_n measures the length of time required for the nth spherical harmonic of the core field to either vanish, regenerate, or significantly change its spatial orientation.

Fig. 3.8 shows correlation times τ_n in years from the geomagnetic secular variation at epoch 2005 over the range of the core field, $n = 1-13$. The dipole correlation time $\tau_1 \simeq 1$ kyr corresponds to the wandering motion of the geomagnetic pole described previously in this section. The dipole correlation time lies far above the trend of the rest of the core field in Fig. 3.8 (Lhuillier et al., 2011), primarily because of the very long-term stability of the axial dipole component. In contrast, the quadrupole correlation time τ_2 lies somewhat below the core field trend, consistent with the fact that its axial component has reversed polarity within historical times. The linear fit in Fig. 3.8 defines a geomagnetic secular variation time scale

$$\tau_{SV} = n\tau_n \simeq 415 \text{ years}. \tag{3.17}$$

Eq. (3.17) implies the existence of a characteristic time scale of about four centuries for variability of the geodynamo. What might this characteristic time scale represent? The simplest interpretation is that it reflects geomagnetic

FIG. 3.8 Geomagnetic secular variation correlation times τ_n versus spherical harmonic degree n. Stars: observed correlation times at epoch 2005; solid line: $\tau_{SV} = n\tau_n = 415$ years fit with 90% confidence bars. *Adapted from Fig. 1A in Lhuillier, F., Fournier, A., Hulot, G., Aubert, J., 2011. The geomagnetic secular variation timescale in observations and numerical dynamo models. Geophys. Res. Lett. 38, L09306.*

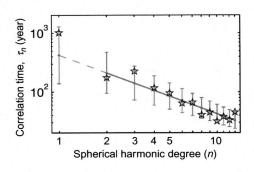

variability produced by uniform transport of magnetic flux beneath the core-mantle boundary. To illustrate this, consider a spatially periodic radial magnetic field at and below the core-mantle boundary of the form

$$B_r = B_0 \cos(n\phi), \tag{3.18}$$

where B_0 is amplitude (magnetic intensity) and ϕ is east longitude. Now suppose this magnetic field is transported beneath the core-mantle boundary in the westward (negative ϕ) direction by a uniform azimuthal flow at a constant (negative) angular speed $\dot{\phi}$. Assuming frozen magnetic flux (see Section 3.4), the resulting geomagnetic secular variation is given by

$$\dot{B}_r = B_0 n \dot{\phi} \sin(n\phi). \tag{3.19}$$

Converting Eqs. (3.18), (3.19) into R_n and $\dot{R}_n$ using Eq. (3.13), substituting the results into Eqs. (3.16), (3.17), and solving for the angular speed of westward transport yields $|\dot{\phi}| = \tau_{SV}^{-1} \simeq 2.4 \times 10^{-3}$ rad year^{-1}, which corresponds to a 2600-year circulation time around the circumference of the outer core.

According to the previous example, the reciprocal of the secular variation time scale, $1/\tau_{SV}$, gives the characteristic angular speed of outer core azimuthal flow. However, a variety of fluid motions in the outer core can produce a constant $n\tau_n$, and hence the same secular variation time scale τ_{SV}. In particular, the combined effects horizontal divergence and convergence associated with large-scale fluid upwellings and downwellings affect $\dot{R}_n$ in much the same way as a uniform horizontal flow, but lead to a different interpretation of the time scale τ_{SV}. For a horizontal convergent/divergent flow, τ_{SV} is proportional to the strength of the upwellings and downwellings. In short, τ_{SV} is a convenient summary parameter for geomagnetic secular variation and it offers an attractive target for numerical models of the geodynamo (see Chapter 4), but it does not provide a unique interpretation in terms of outer core flows.

3.2.8 Dipole variations on millennium times scales

Fig. 3.9 shows the virtual axial dipole moment (VADM) over the past 7.2 kyr determined from the intensity of magnetization of crustal rocks and man-made artifacts. Here, "virtual" refers to the fact that this time series was

FIG. 3.9 Virtual axial dipole moment (VADM) fluctuations during 0–7.2 ka from archeomagnetic field model CALS7K.2. *Dashed line* shows present-day axial dipole moment. *Data from Korte, M., Constable, C.G., 2005. The geomagnetic dipole moment over the last 7000 years—new results from a global model. Earth Planet. Sci. Lett. 236, 348–358*

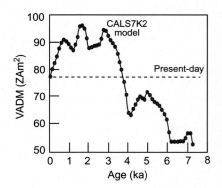

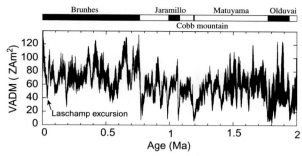

FIG. 3.10 VADM fluctuations over the past 2 Ma recorded in ocean sediments. The geomagnetic polarity chrons (normal polarity Brunhes chron and reverse polarity Matuyama chron) along with several shorter subchrons are shown above the plot. VADM minima mark polarity reversals and excursions. The Laschamp excursion, the most recent such event, is indicated. *Data from the SINT record by Valet, J.P., Meynadier, L., Guyodo, Y., 2005. Geomagnetic dipole strength and reversal rate over the past two million years. Nature 435, 802–805.*

constructed from globally distributed measurements of magnetic intensity, based on the assumption that the geomagnetic field was an axial dipole at all places and times. Magnetic intensity determinations from this time period include data from historical sites plus measurements of rock magnetization at archeological sites as well as recently active volcanic sites. Because they are often based on the magnetization of man-made objects, geomagnetic intensity determinations within this time period are referred to as *archeomagnetic intensity*.

The record of the archeomagnetic VADM extends the historical dipole moment decrease deeper into the past, and provides further confirmation that the geomagnetic dipole moment fluctuates continuously. According to the archeomagnetic field model in Fig. 3.9 (Korte and Constable, 2005), the current episode of dipole moment decrease commenced around 1000 BP and was preceded by an approximately 2.5 kyr interval when the average dipole moment was anomalously high, approximately 90 ZA m^2. Since then, the rate of dipole moment decrease has generally increased with time. Going back further in time, the most recent intensity maximum ended around 3500 BP and was preceded by a longer period of intensity increase that commenced around 6000 BP. This, in turn, was preceded by a shorter period of weak dipole moment extending back in time at least to 7200 BP.

Note that the present-day dipole moment is slightly greater than the 0–7.2 ka average VADM in Fig. 3.9, and it is also greater than the 0–2 Ma average VADM in Fig. 3.10 (Valet et al., 2005). This suggests that the present-day dipole moment is above its long-time-averaged strength, and although it is decreasing rapidly at present, the geomagnetic dipole remains strong relative to the recent geologic past.

3.2.9 Dipole variations on 100 kyr time scales

Longer time-scale fluctuations in the intensity of the core field include transient events in which the intensity of the axial dipole appears to drop to very low values. Fig. 3.10 shows the time variation of the VADM determined from paleomagnetic measurements over the past 2 Ma. The two most recent geomagnetic polarity chrons and several subchrons are shown above the VADM record as black and white bars, indicating the dominant dipole polarity during each chron.

Over the past 2 Ma, the VADM has fluctuated over the range 5–140 ZA m^2. As seen in Fig. 3.10, these fluctuations are broadband, rather than simply periodic, and are suggestive of chaotic behavior of the geodynamo on 100 kyr time scales. Most notably, Fig. 3.10 reveals that the chron and subchron boundaries are characterized by deep minima in the VADM, events that have been interpreted as transient collapses of the ADM followed by ADM recovery. There is some evidence that the dipole moment recovery is, on average, faster than the preceding collapse (Ziegler and Constable, 2011), but both are fast compared to the rate that the dipole moment would change under conditions of free decay (see Section 3.3). In addition, there are multiple VADM minima within polarity the chrons, although these are typically not as deep as those associated with chron boundaries. Nevertheless, many of these additional VADM minima occur at times when the geomagnetic poles moved to low latitudes, in some cases temporarily crossing the equator, only to return to their previous high-latitude positions. These events are called *polarity excursions*.

The paleomagnetic record indicates that the large and rapid dipole moment fluctuations shown in Fig. 3.10 have persisted throughout the history of the geodynamo. We may well ask why they continue to be excited, after billions of years of core evolution. Superficially, these fluctuations imply that core dynamics has two separate effects on the geomagnetic field. During roughly one half of the time, the outer core is in a *dynamo mode*, amplifying the core field (or at least, amplifying its dipolar part). But during the other half of the time, the outer core is in an *antidynamo mode*, actively

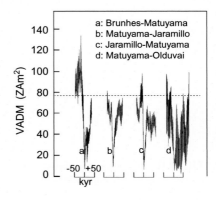

FIG. 3.11 VADM variations during four recent polarity changes, illustrating the dipole collapse and recovery process. Horizontal scale in 50 kyr increments. *Data from the SINT record by Valet, J.P., Meynadier, L., Guyodo, Y., 2005. Geomagnetic dipole strength and reversal rate over the past two million years. Nature 435, 802–805.*

weakening the core field (again, perhaps only its dipolar part). As shown in Chapter 4, it is likely that only very subtle changes in the outer core flow or the configuration of the core field are needed to transition from one state to the other. Regardless, in the effort to understand the geodynamo process, its capacity to rapidly degrade the core field is an often-overlooked property of core dynamics.

The actual intensity of the geomagnetic field during the minima in Fig. 3.10 is not uniquely established because of the assumptions used in constructing VADM from local measurements. However, detailed records of VADM from four recent polarity changes shown in Fig. 3.11 reveal that the VADM drops to as low as 5 ZA m^2 during these events, although the total geomagnetic intensity evidently does not go to zero. In addition, Figs. 3.10 and 3.11 imply no statistical difference in geomagnetic intensity and secular variation between normal and reverse polarity chrons, which is an expected property of a self-sustaining fluid dynamo. Fig. 3.10 also shows that VADM minima occur more frequent than actual polarity reversals, by a factor of 5–10. The fact that they are far more common than reversals suggests that VADM minima with polarity excursions may simply be *failed reversals*, transient perturbations of the geodynamo in which incomplete loss of the axial dipole field inside the core is followed by axial dipole recovery and return to the previous polarity.

3.2.10 The geomagnetic polarity reversal record

Polarity reversals are a signature property of the core field and the geodynamo. For purposes of this discussion, geomagnetic polarity reversals can be defined as transitions between two quasiequilibrium, persistent geomagnetic states that have statistically similar characteristics, apart from the change in sign of the ADM. There have been more than 1000 such polarity reversals during Phanerozoic time (0–540 Ma), and there is abundant evidence for polarity reversals throughout the Precambrian, prior to 540 Ma. Polarity reversals occur at random times for the most part, but these are notable exceptions. Most significantly, the paleomagnetic record includes a number of long polarity *superchrons*, intervals lasting 30–40 Myr in which no (or very few) polarity reversals occurred.

Fig. 3.12 shows the paleomagnetic reversal record from the present day back to about 360 Ma (Ogg, 2012). The bar code at the top is the *geomagnetic polarity time scale*, or GPTS, the yardstick for geologic time based on observed sequences of polarity reversals. The lower part of Fig. 3.12 shows a 5-Myr running average of the reversal rate. Two superchrons are indicated, the *Cretaceous normal superchron*, abbreviated as CNS and the *Kiaman reverse superchron*, the KRS. Here, "Normal" refers to the present-day polarity, and "Reverse" to its opposite. According to Fig. 3.12, the reversal rate varies between essentially zero within a superchron to a maximum of more than 10 reversals per Myr, as exemplified by the so-called *Jurassic hyper-reversals* around 165 Ma. The delineation of the geomagnetic polarity reversal record over Phanerozoic time is among the most important accomplishments in the history of geophysics. The GPTS has not only been instrumental in quantifying plate tectonics by establishing the age distribution of the ocean crust, but also it provides crucial evidence on the history of the geodynamo.

In spite of their abundance in the paleomagnetic record, several important properties of polarity reversals remain poorly understood. One such property is the morphology of the geomagnetic field during polarity change. Two highly idealized conceptual models are often used to interpret paleomagnetic data during reversals. One model consists of a rotation of the geomagnetic dipole axis, starting from a high-latitude position in one hemisphere and ending at a high-latitude position in the opposite hemisphere, this rotation sometimes following a complex path. The other idealized

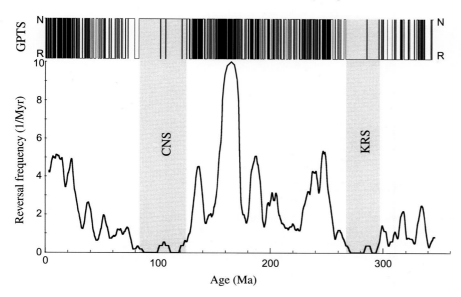

FIG. 3.12 Summary of the paleomagnetic reversal record, 0–340 Ma. Bar code at the top is the geomagnetic polarity time scale (GPTS). *Lower curve* shows a 5-Myr running average of the reversal frequency. Cretaceous normal polarity (CNS) and Kiaman reverse polarity (KRS) superchrons are indicated by shadings. *Data from Ogg, J.G., 2012. The geomagnetic polarity time scale. In: Gradstein, F.M., Ogg, J.G., Schmitz, M., Ogg, G. (Eds.), Geologic Time Scale 2012. Elsevier, pp. 85–114.*

model starts with a fall of the axial dipole intensity to a value below the intensity of the nondipole part of the core field, leading to a transient geomagnetic configuration called the *transition field*. After a few thousand years, the axial dipole reintensifies, but in the opposite polarity. Both of these models are too simple to explain the full range of paleomagnetic observations during polarity transitions, in part because it appears there is a large variation in the behavior of the geomagnetic field from one reversal to another.

Yet despite this variability, there are some characteristics that most reversals share. Typically, polarity reversals are associated with a large drop in the VADM, as seen in Figs. 3.10 and 3.11. This drop generally lasts 10–20 kyr. The directional transition occurs around the mid-point of the VADM minimum, and is usually shorter, typically lasting 2–5 kyr. However, some reversals show more complex behaviors, including intensity and directional changes that precede the actual reversal.

Another unresolved issue is the timing of individual reversals. Unlike the periodic reversals of the solar dynamo, geomagnetic reversals are aperiodic and appear to be largely independent events, but with an average frequency of one in about 240 kyr over the past 5 Ma. The aperiodic character of individual reversals is exemplified by the present-day Brunhes normal polarity chron, which at 775 kyr is more than three times longer than the average chron length over the past 5 Ma.

Still another unresolved issue is the underlying cause of reversals. One thing we do know is that dynamo polarity reversals do not require an external perturbation to occur. As documented in Chapter 4, numerical dynamos commonly exhibit polarity reversals in which the surface field behaves like the paleomagnetic field during reversals. These model reversals are initiated by naturally occurring statistical fluctuations in the velocity and magnetic fields inside the core, and they occur without the help of external perturbations coming from the mantle or other parts of the Earth system. The exact sequence of events in the core that leads to a geomagnetic polarity reversal remains obscure, although reversing numerical dynamos invariably exhibit high levels of time variability, both in the flow and in the magnetic field. In general, polarity reversals in numerical dynamos tend to occur when the dipole moment is weak and the dipole tilt is large and variable, consistent with the known behavior of paleomagnetic reversals. Although individual geomagnetic reversals appear to be largely independent random events, Fig. 3.12 shows the average rate of reversals has changed systematically throughout geologic time, alternating between superchrons and more frequent reversals on a roughly 200 Myr time scale. The geophysical implications of this longer-term variability will be examined in more detail in Chapter 4.

3.2.11 Long-time average core field structure

When interpreting the core field in terms of geodynamo processes, it is important to distinguish shorter transient behaviors such as geomagnetic jerks, excursions, and reversals, from behaviors that are longer lasting. Fig. 3.13 shows a map of the geomagnetic field on the core-mantle boundary derived from paleomagnetic measurements, time

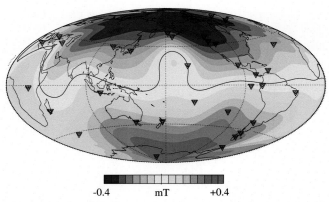

FIG. 3.13 The radial component of the core field on the core-mantle boundary, time averaged 0–5 Ma derived from normal polarity paleomagnetic directions and intensities measured in volcanic rocks worldwide, binned into the regions marked by *triangles*. Continental outlines are shown for reference. *Solid curve* indicates the time-average magnetic equator. *Modified from Fig. 10 model LN3 in Cromwell, G., Johnson, C.L., Tauxe, L., Constable, C.G., Jarboe, N., 2018. PSV10: a global data set for 0–10 Ma time-averaged field and paleosecular variation studies. Geochem. Geophys. Geosyst. 19, 1533–1558.*

averaged over the past 5 million years (0–5 Ma). In this long-time average, the axial dipole field is even more dominant than in the average of the historical field in Fig. 3.3. Nevertheless, significant nonaxial structure persists, especially in the northern hemisphere, where the two high-intensity magnetic flux patches seen in the historical core field are smeared out into high-intensity lobes separated by relatively weak field beneath Africa and the Pacific in the 0–5 Myr average.

The significance of such nonaxial structure for the geodynamo becomes apparent when its 5 Ma persistence is measured against the characteristic time scales of motions in the core versus the mantle, which are centuries in the outer core versus tens to hundreds of million years in the mantle. The persistent deviations from axial symmetry in Fig. 3.13 are therefore far more likely due to the imprint of mantle heterogeneity on outer core dynamics, rather than being intrinsic to the dynamics of the outer core and independent of mantle heterogeneity. If the core-mantle boundary and the inner core boundary were entirely homogeneous, we expect the core field to approach axisymmetry in long-term average. Indeed, paleomagnetic evidence indicates that, to a first approximation, this is what happens. Nevertheless, paleomagnetic evidence does permit small departures from axial symmetry in the time average core field. In Chapter 4, we compare the magnetic field structures produced by numerical dynamos driven by heterogeneous heat flux at the core-mantle boundary with the time average core field shown in Fig. 3.13 (Cromwell et al., 2018).

3.2.12 Geomagnetic intensity in the deep past

Paleomagnetic data from ancient rocks and minerals provide a window on the very long-term history of the geodynamo. Ideally, one would like to have maps of the core field like Fig. 3.13 at many epochs in the deep past, in order to test models of the evolution of the core and for calibration and verification of numerical dynamos. Unfortunately, such reconstructions will require far more paleomagnetic directional data than presently exist. Alternatively, the geomagnetic polarity reversal record could be used for this purpose. Polarity reversals are not limited to Phanerozoic times; there is paleomagnetic evidence for reversals as far back as 2 Ga. However, the geomagnetic reversal record, like the directional data, becomes increasingly fragmented with increasing age.

Because of these limitations, interpretations of the ancient history of the geodynamo mainly rely on determinations of paleomagnetic field intensity. Inferring the paleomagnetic field intensity from the intensity of magnetization in ancient rocks and minerals is an extremely difficult task, so it is no surprise that these measurements provide only a blurry picture of the geodynamo history. That said, two fundamental questions continue to motivate efforts to better resolve the ancient geomagnetic history: first, how old is the inner core? and second, how old is the geodynamo itself?

Fig. 3.14 shows the extent of paleomagnetic field intensity determinations in the deep past (Tauxe and Yamazaki, 2007), expressed as VADM. The quality of each intensity determination is indicated by symbol type. The most obvious property of this data set is its large variance within age bins. Part of this variance stems from the difficulty in converting sample magnetization to magnetizing field intensity, but another other part is due to the inherent time variability of the geodynamo and the core field it produces, as exemplified in Fig. 3.10. Note that the VADM variance is largest where the data coverage is best, in the 0–400 Ma age window. This is a warning: it indicates that the older part of the

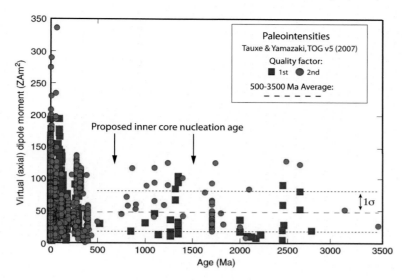

FIG. 3.14 VADM paleomagnetic intensity record from PINT data base, 0–3500 Ma. *Dashed lines* show 500–3500 Ma VADM mean and standard deviation in ZA m^2 (10^{21} A m^2) units. *Arrows* indicate proposed inner core nucleation ages. Quality rankings denoted by *symbols*. Modified from Fig. 18 in Tauxe, L., Yamazaki, T., 2007. Paleointensities. In: Schubert, G. (Ed.), Treatise on Geophysics, vol. 5. Elsevier B.V., pp. 510–557.

paleomagnetic intensity record is grossly undersampled, which means that we cannot confidently answer the questions posed above on the basis of this data alone.

In spite of the limitations in ancient paleomagnetic intensity data, they allow us to draw a few general conclusions about geodynamo history. First, the geodynamo was active at least to 3.45 Ga, and it was quite possibly active before then. Second, there is scant evidence of a lasting trend in paleomagnetic intensity over this time. The 500–3500 Ma time average VADM in Fig. 3.14, 49 ± 30 ZA m^2, is essentially the same as the 46 ± 30 ZA m^2 0–400 Ma time average VADM. Third, in spite of the absence of an obvious trend, there are several time intervals during which the core field may have remained at low intensity for extended periods of time and then recovered. Two such intervals, around 1500 and 700 Ma, are labeled in Fig. 3.14. Both have been proposed as candidates for the time of inner core nucleation. The idea here is that, just when the geodynamo was running low on power and the dipole was weakening, the onset of compositional buoyancy and latent heat release from inner core solidification supplied additional power to the geodynamo, driving a dipole moment increase that has lasted until the present day.

Because the age of inner core nucleation is an important benchmark in the history of the core, there have been numerous attempts to establish its timing, both theoretically on the basis of thermal evolution considerations (Chapter 2) and observationally, using dipole intensity and other paleomagnetic indicators. Not only does the inner core age remains an open question, there are also questions about what inner core nucleation looks like in the paleomagnetic intensity record. These issues are further examined in Chapter 8, using results from thermal evolution models of the core and numerical dynamos that model the formation and growth of the inner core.

3.3 The geodynamo process

3.3.1 How to make a self-sustaining fluid dynamo

The following is a thought experiment on what it takes to make a self-sustaining fluid dynamo. Our purpose here is to define the important variables, physical properties, and necessary conditions for the geodynamo, free of as many complications as possible.

Suppose a cylindrical volume of electrically conducting fluid is placed on a rotating platform, as illustrated in Fig. 3.15. The fluid is initially subjected to a uniform external magnetic field $\mathbf{B}_0$ aligned with the rotation axis (note: if the external magnetic field is uniform, corotation of the platform and magnetic field is not an issue; if it is not uniform, then we suppose that the field corotates with the platform). A propeller is inserted in the fluid, capable of rotating independently of the tank, and a thermometer is inserted to measure temperature. The propeller is the energy (power) source for the dynamo in this experiment.

The relevant physical properties include the platform rotation rate Ω, the fluid depth d, its density ρ, kinematic viscosity ν, electrical conductivity σ, the intensity of the imposed magnetic field $\mathbf{B}_0$, and the propeller angular velocity ω measured relative to the platform rotation rate Ω.

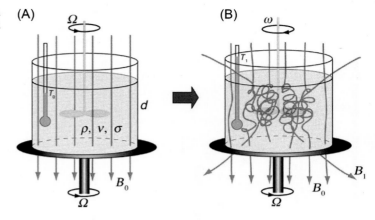

FIG. 3.15 Dynamo Step 1: Apply an external magnetic field to a rotating, conducting fluid (A) and stir to induce electric currents and magnetic fields (B).

The basic *static state*, illustrated in Fig. 3.15A, consists of solid-body rotation produced by rotating both the platform and the propeller at the same rate for a long time, until the fluid is spun up. The thermometer initially reads the lab temperature T_0, the fluid is at rest in the rotating coordinate system of the platform, and the applied magnetic field is unchanged.

The *magnetohydrodynamic state* is illustrated in Fig. 3.15B. This state is produced by rotating the propeller differentially with respect to the platform, thereby inducing motions in the fluid. The fluid motions interact with the applied magnetic field $\mathbf{B}_0$ to generate electric currents in the fluid, which induces additional magnetic fields of their own. These induced magnetic fields are depicted in Fig. 3.15B as tangled field lines within the fluid and as an induced field $\mathbf{B}_1$ outside the fluid. Because of viscous dissipation by the fluid motion and Ohmic dissipation by the electric currents, heat is produced in the fluid and the temperature rises to T_1, as indicted on the thermometer in Fig. 3.15B.

The next step starts from the MHD state, which has been reproduced in Fig. 3.16A. We switch off the externally applied magnetic field $\mathbf{B}_0$ and wait for a while. Free of the imposed magnetic field, the system will evolve toward one of two general states, either *subcritical* (no dynamo) or *supercritical* (dynamo). These two states are illustrated in Fig. 3.16B and C, respectively.

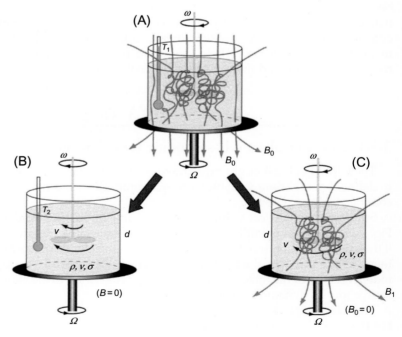

FIG. 3.16 Dynamo Step 2: Remove the external magnetic field in (A) and wait to see if the dynamo is subcritical (B) or supercritical (C).

In the subcritical state, the electric currents and the induced magnetic field $\mathbf{B}_1$ decay away, leaving only the fluid motion driven by the propeller. There is still heating from viscous dissipation, but without electric currents this is less than the heating in the previous state, so the temperature T_2 in Fig. 3.16B is less than T_1 in Fig. 3.16A. In the supercritical state, the electric currents and the induced magnetic field $\mathbf{B}_1$ remain, as illustrated in Fig. 3.16C, although perhaps in different forms than before the imposed magnetic field $\mathbf{B}_0$ was removed.

The threshold condition that separates these two states is called the *critical state*. It depends on the magnitude of the key dimensionless parameter, the magnetic Reynolds number

$$Rm = \mu_0 \sigma v d, \tag{3.20}$$

where again μ_0 is the free-space magnetic permeability, σ is the electrical conductivity of the fluid, v is the r.m.s. velocity of the fluid measured in the rotating system, and d is the conducting fluid depth. The magnetic Reynolds number is the single most important dimensionless parameter for all fluid dynamos, including the geodynamo. The quantity

$$\kappa_B = (\mu_0 \sigma)^{-1} \tag{3.21}$$

has units of length squared over time and is the *magnetic diffusivity*, a fundamental MHD property that governs the behavior of all self-sustaining dynamos, including the geodynamo. In the core, $\sigma \simeq 4\text{--}20 \times 10^5$ S m^{-1}, so κ_B is in the range 0.4–2 m^2 s^{-1}.

In geophysical terms, the magnetic Reynolds number defined by Eq. (3.20) can be thought of as the ratio of two time scales: the free decay time of electrical currents and their associated magnetic fields due to electrical resistance in the core,

$$\tau_{mag} = \frac{d^2}{\kappa_B} = \mu_0 \sigma d^2, \tag{3.22}$$

divided by the characteristic time for circulation of the outer core fluid

$$\tau_{circ} = \frac{d}{v}, \tag{3.23}$$

so that

$$Rm = \tau_{mag}/\tau_{circ} = \frac{vd}{\kappa_B}. \tag{3.24}$$

Self-sustaining dynamo action requires that the magnetic diffusion time scale far exceeds the circulation time scale, so that Rm must be large compared to one. There are multiple ways in which a large magnetic Reynolds number can be achieved. Laboratory fluid dynamos achieve large magnetic Reynolds numbers using good conductors and very high fluid velocities (large v), whereas planetary dynamos reach this regime with modest fluid velocities but very deep conducting fluids (large d). Theory, laboratory experiments, and numerical simulations have convincingly shown that the critical magnetic Reynolds number for dynamo action in Earth's core is of order $Rm_{crit} \simeq 40$, provided the fluid motions in the outer core conform to what is expected for buoyancy-driven convection (see Chapter 4). In other words, if $Rm < 40$ in the outer core, the electric currents and the geomagnetic field in the core decay with time, but if $Rm > 40$ in the outer core, the electric currents and the geomagnetic field in the core are able to amplify with time. This amplification process continues until an equilibrium is reached between magnetic energy added and lost.

3.3.2 The magnetic Reynolds number in the core

The value of the magnetic Reynolds number in the outer core can be estimated from Eq. (3.20) using the electrical conductivity of core alloys at high pressure and temperature (Chapter 2 and Table S2), inverting the geomagnetic secular variation to determine the r.m.s fluid velocity below the core-mantle boundary (Section 3.4 and Table S3), and using the spherical shell thickness of the outer core (Table S1) for d. This approach yields $Rm = 2000 \pm 1000$, the range of values primarily reflecting the uncertainty in core electrical conductivity, and secondarily, ambiguity in estimating the outer core fluid velocity based on the geomagnetic secular variation. But even allowing for these uncertainties, the geophysical implication is abundantly clear: the present-day geodynamo is far beyond critical, with a magnetic Reynolds number in the range $Rm = (25\text{--}75)Rm_{crit}$. The fact that the present-day geodynamo is strongly supercritical has multiple implications for the energetics of the core, both past and present. It also key to understanding the origin of the geomagnetic phenomenon described earlier in this chapter, including the structure of the core field, the geomagnetic secular variation, polarity reversals, and excursions.

3.3.3 Magnetic induction in the core

All of the time variations in the core field described previously in this chapter are governed by the *magnetic induction equation*. The derivation of the magnetic induction equation starting from Maxwell's equations of electromagnetism is given in Appendix 3.2. In its simplest form, assuming uniform electrical conductivity and incompressible (divergence-free) fluid motion, the magnetic induction equation is

$$\frac{\partial \mathbf{B}}{\partial t} + (\mathbf{v} \cdot \nabla)\mathbf{B} = (\mathbf{B} \cdot \nabla)\mathbf{v} + \kappa_B \nabla^2 \mathbf{B} \tag{3.25}$$

where $\mathbf{v}$ is the vector field of the fluid velocity in the outer core and $\kappa_B = 1/\mu_0 \sigma$ is the magnetic diffusivity in the outer core, assumed constant because the electrical conductivity σ is assumed constant.

The physical meanings of the terms in Eq. (3.25) are as follows. The first term on the l.h.s. represents the time rate of change of the magnetic field at a point in the core. The second term on the l.h.s. represents the advection or transport of magnetic field by the outer core fluid velocity. The first term on the r.h.s. represents the stretching of magnetic field lines by gradients in the outer core fluid velocity. Field-line stretching by velocity gradients is the primary MHD mechanism for increasing the density of magnetic energy, and lies at the heart of dynamo action. The second term on the r.h.s. of Eq. (3.25) represents magnetic diffusion. This term tends to dampen the effects of magnetic field-line stretching and reduce the magnetic energy density. Denoting the characteristic fluid velocity by v and the outer core thickness by d, the magnetic Reynolds number Rm defined by Eqs. (3.20), (3.24) is obtained from Eq. (3.25) by taking the ratio of either the advection term or the stretching term to the diffusion term.

The induction equation has important simplifications in the two extreme limits of Rm. For $Rm \ll 1$, Eq. (3.25) reduces to a vector diffusion equation

$$\frac{\partial \mathbf{B}}{\partial t} = \kappa_B \nabla^2 \mathbf{B}. \tag{3.26}$$

With $Rm \ll 1$, according to Eq. (3.26), the geomagnetic field would decay with time at a rate inversely proportional to Eq. (3.22), the magnetic diffusion time in the core. No dynamo action is possible in this limit. However, we have already shown that Rm is of order 1000–3000 in the core, so $Rm \gg 1$ is a more relevant limit. In the extreme case that Rm approaches infinity, Eq. (3.25) reduces to

$$\frac{\partial \mathbf{B}}{\partial t} + (\mathbf{v} \cdot \nabla)\mathbf{B} = (\mathbf{B} \cdot \nabla)\mathbf{v}, \tag{3.27}$$

which is the *perfect conductor* limit of the induction equation.

Although the core is far from being a perfect electrical conductor, the fact that Rm in the core is very large makes Eq. (3.27) useful for interpreting some of the behaviors of the core field, especially its secular variation. In particular, consider a material patch of outer core fluid with surface area A oriented by a normal unit vector $\hat{n}$. Integrating Eq. (3.27) over this patch yields an important condition on the total magnetic flux Φ that lies within that patch:

$$\frac{d\Phi}{dt} = \frac{d}{dt}\int (\mathbf{B} \cdot \hat{n}) dA = 0. \tag{3.28}$$

Eq. (3.28) implies that the total magnetic flux Φ within the fluid patch is conserved in time (i.e., has a zero material time derivative, $d/dt = 0$). This result is called *Alfvén's theorem*: In a perfect conductor, magnetic flux moves as if it were attached to, or frozen into, the conductor. Frozen magnetic flux implies a close connection between the secular variation of the core field and the fluid motions in the outer core. In addition, the concept of *frozen flux*, as it is called, provides a simple geometrical way to visualize the basic process of magnetic field amplification in the core. Fig. 3.17 shows this process in terms of the stretching of a cylindrical volume of core fluid containing magnetic field. Assuming the fluid motion is incompressible, the magnetic field intensity B changes in proportion to changes in the cross-sectional area A of the cylindrical volume and its length L according to Eq. (3.28), so that

$$\frac{B_2}{B_1} = \frac{A_1}{A_2} = \frac{L_2}{L_1}, \tag{3.29}$$

where the subscripts 1 and 2 refer to before and after the stretching, respectively.

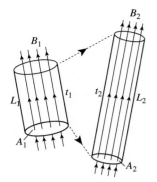

FIG. 3.17 Illustration of magnetic field intensification by fluid motion in the frozen flux (perfect conductor) limit. Stretching of an incompressible perfectly conducting cylinder containing magnetic flux $\Phi = A_1 B_1$ at t_1 and conserves Φ at t_2 but intensifies the magnetic field according to $B_2/B_1 = L_2/L_1 = A_1/A_2$.

3.3.4 Poloidal and toroidal magnetic field components

The next step is to represent the geomagnetic field inside the core as the sum of *toroidal* and *poloidal* parts, defined in terms of scalar potentials $\mathcal{T}$ and $\mathcal{P}$ according to

$$\mathbf{B} = \mathbf{B}_T + \mathbf{B}_P = \nabla \times (\mathbf{r}\mathcal{T}) + \nabla \times \nabla \times (\mathbf{r}\mathcal{P}), \tag{3.30}$$

where $\mathbf{r}$ is the radial vector with origin at the center of the core. According to Eq. (3.30), a toroidal magnetic field $\mathbf{B}_T$ has zero radial component, only horizontal (θ, ϕ) components. Furthermore, this part of the magnetic field is limited to the interiors of electrical conductors. Therefore, the toroidal geomagnetic field lies almost exclusively in the core and is practically zero in the mantle and crust. A poloidal magnetic field $\mathbf{B}_P$, in contrast, has a radial component according to Eq. (3.30), as well as horizontal components. This part of the geomagnetic field comprises the core field on the core-mantle boundary shown in Figs. 3.3 and 3.4, as well as the contribution of the core field to the geomagnetic field measured at Earth's surface.

Substituting Eq. (3.30) into Eq. (3.26), the low magnetic Reynolds number limit of the induction equation breaks into separate diffusion equations for the scalar potentials $\mathcal{T}$ and $\mathcal{P}$:

$$\frac{\partial \mathcal{T}}{\partial t} = \kappa_B \nabla^2 \mathcal{T} \tag{3.31}$$

and

$$\frac{\partial \mathcal{P}}{\partial t} = \kappa_B \nabla^2 \mathcal{P}. \tag{3.32}$$

In this low Rm limit, there is no toroidal-poloidal magnetic field interaction. The toroidal and poloidal magnetic fields diffuse independently, and as mentioned previously, self-sustaining dynamo action does not occur. Nevertheless, it is useful to examine solutions to these equations, in order to determine the structure of freely decaying magnetic fields in the core as well as their rates of decay.

The slowest decay mode is the dipole term in Eq. (3.32), which is also the most important decay mode for the geodynamo. Assuming the mantle has negligible electrical conductivity and the core has uniform electrical conductivity throughout, the solution to Eq. (3.32) for the axial dipole is

$$\mathcal{P} = \mathcal{P}_0 j_1(\pi r/r_{CMB}) \cos\theta (e^{-t/\tau_{dip}}) \tag{3.33}$$

where $\mathcal{P}_0$ is the amplitude of the poloidal magnetic potential at time $t = 0$, j_1 is the spherical Bessel function of order 1, and

$$\tau_{dip} = \frac{r_{CMB}^2}{\pi^2 \kappa_B} \tag{3.34}$$

is the decay time scale of a dipole field in a conducting sphere with radius r_{CMB} and magnetic diffusivity κ_B. Magnetic field lines corresponding to Eq. (3.33) are shown in Fig. 3.18, with the present-day geomagnetic dipole tilt applied, along with contours of the associated electric current density, $\mathbf{J}$.

FIG. 3.18 Magnetic field lines of a freely decaying dipolar field $B_\mathcal{P}$, from Eq. (3.33). Dipolar magnetic field lines oriented as in the present-day (i.e., normal) geomagnetic dipole polarity with the present-day inclination denoted by dip angle θ_d and the location of the North Geomagnetic Pole (NGP). *Filled contours* represent contours of the associated electric current density within the core. Earth's surface, core-mantle boundary, and inner core boundary radii are indicated.

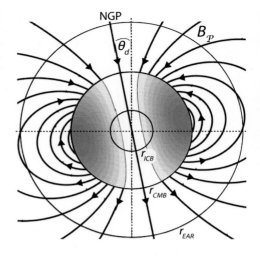

It can be shown that the dipole moment vector in free decay obeys an equation like Eq. (3.33), so that

$$\mathbf{m} = \mathbf{m}_0 e^{-t/\tau_{dip}}. \qquad (3.35)$$

Substituting the outer core radius $r_{CMB} = 3480$ km in Eq. (3.34) and using an electrical conductivity for the core of $\sigma = 4 \times 10^5$ S m^{-1} (corresponding to a magnetic diffusivity of $\kappa_B = 1/\mu_0 \sigma = 2$ m^2 s^{-1}) yields $\tau_{dip} = 19.6$ kyr for the free decay time of the geomagnetic dipole. This is for electrical conductivity at the low end of recent estimates (see Chapter 2 and Table S2). Using a higher electrical conductivity of $\sigma = 10 \times 10^5$ S m^{-1} yields a free decay time of $\tau_{dip} = 49$ kyr for the geomagnetic dipole.

It is instructive to compare the observed rate of dipole moment decrease shown in Fig. 3.5B with these theoretical rates of free decay. Fitting the data in Fig. 3.5B to an expression like Eq. (3.35) yields 1.9 kyr for the exponential time scale of the historical moment decrease. Therefore, the historical dipole moment decrease rate is approximately a factor of 10 faster than theoretically expected for free decay in the core, assuming the lower value of electrical conductivity applies. For the higher electrical conductivity, the historical moment decrease exceeds the free decay rate in the core by more than 25 times! This comparison demonstrates that the measured dipole decrease during historical times cannot be interpreted as the geodynamo simply shutting off. Instead, dynamical processes must be at work in the core, extracting energy from the dipole field and causing it to decrease much faster than it would by free decay alone. In Section 3.4, the flows in the outer core that is responsible for the secular variation of the core field are considered. It turns out that these flows are very energetic, fully capable of rapidly intensifying the dipole field, and likewise, equally capable of rapidly diminishing it.

Returning to the case of finite but large Rm, substituting Eq. (3.30) into Eq. (3.25) yields the following equation for the interaction between the poloidal and toroidal potentials:

$$\frac{\partial}{\partial t} r B_r + (\mathbf{v} \cdot \nabla) r B_r = \kappa_B \nabla^2 r B_r + (\mathbf{B}_\mathcal{T} + \mathbf{B}_\mathcal{P}) \cdot \nabla (r v_r), \qquad (3.36)$$

where

$$r B_r = \mathbf{r} \cdot \mathbf{B} = \mathbf{r} \cdot \nabla \times \nabla \times (\mathbf{r} \mathcal{P}) \qquad (3.37)$$

and $r v_r = \mathbf{r} \cdot \mathbf{v}$, B_r and v_r being the radial components of the magnetic field and the fluid velocity $\mathbf{v}$, respectively.

Eq. (3.36) is mathematically akin to a heat transport equation, with a magnetic field production term, the second on the r.h.s, which involves both poloidal and toroidal magnetic potentials. This equation signifies that poloidal magnetic field is transported and diffused much like temperature, but with a production term that is proportional to both magnetic potentials and also proportional to the fluid radial velocity v_r. The toroidal magnetic field obeys an equation with terms similar to Eq. (3.36).

Two points are worth emphasizing about the magnetic production term in Eq. (3.36). First, it shows that toroidal-poloidal coupling can amplify the poloidal magnetic field through interaction with the fluid motion (the same goes for toroidal magnetic field amplification). Second, because the production term in Eq. (3.36) is proportional to v_r, it vanishes where $v_r = 0$, that is, wherever the radial component of the outer core flow vanishes. This means that radial fluid motions are necessary to generate the poloidal magnetic field in the core. It is not sufficient to drive the geodynamo with horizontal fluid motions, although horizontal motions do affect the structure of the geomagnetic field. Accordingly, questions about the driving force and energy supply for the geodynamo, the subjects of Chapters 2, 4, and 7, naturally focus on processes that are capable of sustaining radial fluid velocities in the outer core. This explains why convective motions in the core, which transport heat and chemistry by radial flow, are considered to be instrumental in maintaining the geodynamo.

3.3.5 Toroidal-poloidal magnetic feedback

Figs. 3.19–3.21 illustrate schematically how frozen magnetic flux concepts can be used to visualize two elementary induction mechanisms that produce positive feedback between toroidal and poloidal magnetic fields, thereby promoting dynamo action. These mechanisms are called the ω-effect and the α-effect, respectively.

The ω-effect transforms poloidal magnetic field into toroidal magnetic field through the shearing action of a large-scale toroidal flow. As illustrated in Fig. 3.19, the toroidal flow is often an azimuthal (i.e., east-west directed) flow, with shear in both the radial and meridional (i.e., north-south) directions. Poloidal magnetic lines of force are dragged from their meridional planes by this flow, generating toroidal magnetic fields oriented in the azimuthal direction. Very intense toroidal magnetic fields can be produced this way. As demonstrated in the example here, in the ω-effect, the ratio of induced toroidal magnetic field to inducing poloidal magnetic field scales with the magnetic Reynolds number of shear flow. For the outer core, this implies that toroidal magnetic fields could be many times more intense than poloidal magnetic fields in places where strong, large-scale shear flows are present.

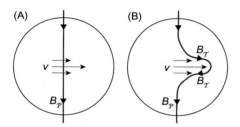

FIG. 3.19 Production of toroidal magnetic field from a poloidal magnetic field by a shear flow. *Horizontal arrows* denote the shear flow with velocity v. *Heavy curves* show the deformation of an initially straight poloidal magnetic field line (labeled B_P in A) by the shear flow, generating two toroidal magnetic field segments with opposite orientations, labeled B_T in B.

3.3.5.1 Toroidal magnetic field generation in a shear flow

To quantify toroidal magnetic field generation by the ω-effect, we consider a classic MHD problem, the generation of a horizontal magnetic field by a linear (uniform shear) Couette flow acting on an imposed uniform vertical magnetic field in a horizontally infinite fluid layer with uniform magnetic diffusivity κ_B. A relevant geometry for the core is shown in Fig. 3.20. In this example, the upper surface of the shear flow (denoted by subscript o, for outer) represents the core-mantle boundary (the very thin viscous boundary layer is ignored here), while the lower surface represents the depth in the outer core where the shear vanishes. In this geometry, B_z represents the poloidal magnetic field in the outer core and B_x represents the toroidal magnetic field.

In terms of the Cartesian coordinates in Fig. 3.20, the z-component of the induction Eq. (3.25) reduces to

$$\left(\frac{\partial}{\partial t} - \kappa_B \frac{\partial^2}{\partial z^2}\right) B_z = 0 \tag{3.38}$$

and the x-component of the same equation reduces to

$$\left(\frac{\partial}{\partial t} - \kappa_B \frac{\partial^2}{\partial z^2}\right) B_x = \frac{\partial}{\partial z}\left(\frac{B_z v_o z}{d}\right). \tag{3.39}$$

FIG. 3.20 Geometry of toroidal magnetic field induction by a linear Couette shear flow.

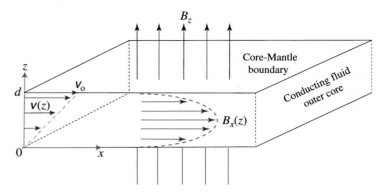

Eqs. (3.38), (3.39) are like heat conduction equations, but only Eq. (3.39) has a source term, meaning that the shear flow affects B_x but not B_z. Accordingly, the steady-state solution to Eq. (3.38) is just the imposed magnetic field, $B_z = B_o$, and in steady-state Eq. (3.39) reduces to

$$\frac{d^2}{dz^2}B_x = -\frac{B_o v_o}{\kappa_B d}. \tag{3.40}$$

Boundary conditions are $B_x = 0$ at $z = d$ and $z = 0$, corresponding to the toroidal magnetic field vanishing at the core-mantle boundary and at the base of the shear flow in the outer core.

The solution to Eq. (3.40) satisfying these boundary conditions is

$$B_x = \frac{B_o v_o d}{8\kappa_B}\left(1 - \frac{(z - d/2)^2}{(d/2)^2}\right). \tag{3.41}$$

The variation of B_x with z is shown in Fig. 3.20. The maximum in the induced toroidal magnetic field occurs at mid-depth in the shear layer, and its intensity relative to the imposed magnetic field is given by

$$\left(\frac{B_x}{B_o}\right)_{max} = \frac{Rm}{8}, \tag{3.42}$$

where $Rm = v_o d/\kappa_B$ is the magnetic Reynolds number of the shear flow. According to Eq. (3.42), the induced toroidal magnetic field intensity is proportional to the product inducing poloidal magnetic field intensity and the magnetic Reynolds number of the shear flow, and the ratio of the toroidal to poloidal magnetic energy is proportional to Rm^2. Because $Rm \gg 1$ in the outer core, very intense toroidal magnetic fields are expected wherever large-scale shear flows are present.

3.3.5.2 Poloidal magnetic field generation by helical flow

Transformation of toroidal magnetic field back into poloidal magnetic field is a bit more complicated to visualize and more difficult to quantify because it involves 3D flows. Fig. 3.21 illustrates the α-effect mechanism, in which a convective eddy interacts with a bundle of toroidal magnetic flux, inducing poloidal magnetic field. The important property of this eddy is its kinematic *helicity*, defined as the correlation between the fluid velocity **v** and its vorticity $\nabla \times \mathbf{v}$:

$$\overline{H} = \overline{\mathbf{v} \cdot (\nabla \times \mathbf{v})}, \tag{3.43}$$

the overbar denoting an average over one eddy or an array of such eddies. There is a related quantity called the *magnetic helicity*, defined as the correlation between the magnetic field and electric current density. In Chapters 3 and 4, we only use Eq. (3.43), referring to it as "helicity."

In many laminar flows, such as the shear flow described earlier, the fluid velocity and vorticity vectors are uncorrelated and the helicity is zero. However, 3D flows affected by Earth's rotation are often strongly helical. Convective eddies influenced by the Coriolis acceleration are particularly relevant examples of this. The flow illustrated in

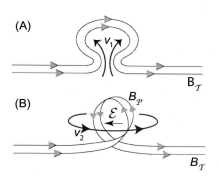

FIG. 3.21 Production of poloidal magnetic field B_P from a toroidal magnetic field B_T by a helical convective eddy. Flow in the eddy consists of two parts, a convective upwelling $\mathbf{v}_1$ in (A) and a quasigeostrophic circulation around the upwelling $\mathbf{v}_2$ in (B) with vorticity in the same direction as the upwelling, generating positive kinematic helicity. Positive helicity induces an e.m.f. $\mathcal{E}$ antiparallel to B_T, and a loop of poloidal magnetic field with the polarity indicated by the *filled arrows*.

Fig. 3.21A and B consists of a localized convective upwelling $\mathbf{v}_1$ plus a circulation around the upwelling $\mathbf{v}_2$ resulting from the Coriolis acceleration due to Earth's rotation. The vorticity $\nabla \times \mathbf{v}_1$ correlates with the fluid velocity $\mathbf{v}_2$ and the vorticity $\nabla \times \mathbf{v}_2$ correlates with the fluid velocity $\mathbf{v}_1$, so that positive helicity is produced in both interactions. In terms of frozen magnetic flux, the upwelling $\mathbf{v}_1$ bends the toroidal magnetic flux bundle into a loop, while the circulation $\mathbf{v}_2$ twists the loop out of its original plane. The result is a poloidal magnetic field loop, ideally at right angles to the toroidal magnetic flux bundle, with a component of the electric current density $\mathbf{J}$ and a component of the induced e.m.f. $\mathcal{E}$ that are aligned with the toroidal magnetic flux bundle. In Fig. 3.21B, the alignment is antiparallel.

The α-effect illustrated here can be parameterized in the form of a linear relationship between the helicity-induced e.m.f. and the toroidal magnetic field:

$$\mathcal{E} = \overline{\mathbf{v} \times \mathbf{B}} = \alpha \mathbf{B}_T, \tag{3.44}$$

with α being the coefficient in this relationship. The flow shown in Fig. 3.21 has positive helicity ($\overline{H} > 0$) and corresponds to $\alpha < 0$ in Eq. (3.44). Without helicity, without the circulation $\mathbf{v}_2$, for example, the electric field induced by the upwelling $\mathbf{v}_1$ would be perpendicular to $\mathbf{B}_T$, and α in Eq. (3.44) would be zero.

Convection in the outer core is strongly influenced by the Coriolis acceleration and, therefore, is expected to be helical and capable of generating an α-effect. Similarly, there is evidence for large-scale shear flows in the outer core in the geomagnetic secular variation (Section 3.4), and these flows are capable of generating ω-effects. Accordingly, the basic mechanisms for producing toroidal-poloidal feedbacks illustrated in Figs. 3.19–3.21 are likely present in the outer core, and we need to ask: what magnetic field configurations do these mechanisms produce?

3.3.6 A simple kinematic dynamo

An illustrative example of kinematic dynamo action consists of equal α-effects acting on both the toroidal and the poloidal magnetic fields in a uniformly conducting sphere. Although highly oversimplified, this example has a number of implications for the core and the geodynamo.

First, we rewrite induction Eq. (3.25) in terms of the induced e.m.f. $\mathcal{E}$:

$$\frac{\partial \mathbf{B}}{\partial t} = \nabla \times \mathcal{E} + \kappa_B \nabla^2 \mathbf{B}, \tag{3.45}$$

in which $\mathbf{B}$ now denotes the large-scale (spatially averaged) magnetic field. Substituting Eq. (3.44) into Eq. (3.45) and representing $\mathbf{B}$ in terms of the magnetic potentials (Eq. 3.30) yields, for constant α

$$\left(\frac{\partial}{\partial t} - \kappa_B \nabla^2\right) \mathcal{P} = \alpha \mathcal{T}, \tag{3.46}$$

and assuming a relationship like Eq. (3.44) but involving $\mathcal{P}$,

$$\left(\frac{\partial}{\partial t} - \kappa_B \nabla^2\right)\mathcal{T} = -\alpha \nabla^2 \mathcal{P}. \tag{3.47}$$

Eqs. (3.46), (3.47) along with their boundary conditions define an eigenvalue problem for the critical α needed to maintain the intensity of the magnetic field in the presence of magnetic diffusion. In this example, the critical value of α determined from the steady-state solution. Setting their time derivatives to zero and eliminating $\mathcal{P}$ yields

$$\nabla^2 \mathcal{T} + \left(\frac{\alpha}{\kappa_B}\right)^2 \mathcal{T} = 0. \tag{3.48}$$

Spherical harmonic solutions to Eq. (3.48) can be written using notation similar to the geomagnetic potential Eq. (3.4), that is,

$$\mathcal{T}_n^m(r,\theta,\phi) = j_n(kr) P_n^m(\cos\theta)(C_n^m \cos m\phi + S_n^m \sin m\phi) \tag{3.49}$$

where $j_n(kr)$ is a spherical Bessel function of order n, P_n^m is a Legendre polynomial (see the Appendix), C_n^m and S_n^m are amplitude coefficients, and $k^2 = \alpha^2/\kappa_B^2$. Applied to the core, the simplest boundary conditions assume the mantle has zero electrical conductivity and the toroidal field vanishes at Earth's center, so that $\mathcal{T}_n^m = 0$ at $r = 0$ and at r_{CMB}, the core-mantle boundary. To satisfy the core-mantle boundary condition, we set kr_{CMB} equal to x_n, the first root of j_n. The critical α is then given by

$$\alpha_{crit}^2 = \left(\frac{x_n \kappa_B}{r_{CMB}}\right)^2. \tag{3.50}$$

The smallest value for x_n is $x_1 = 4.49,\ldots$ for which P_1^m corresponds a dipolar magnetic field and $\mathcal{T}_1^m$ corresponds to a toroidal field consisting of a single axisymmetric magnetic flux bundle symmetric about the dipole equator. For this magnetic field configuration, Eq. (3.50) yields

$$\alpha_{crit} \pm \frac{4.49 \kappa_B}{r_{CMB}}. \tag{3.51}$$

We can go one further step, and express Eq. (3.51) in terms of the equivalent critical magnetic Reynolds number and the characteristic length scale of the convection. Dimensional considerations reveal that $|\alpha| \simeq l^2 \overline{H}/\kappa_B$, where l is the length scale of the helicity-containing flow, the size of the convective eddy in Fig. 3.21. Combining this with Eqs. (3.43), (3.51) implies, for the length scale of the convective eddies $l \simeq r_{CMB}/Rm_{crit}^2$. For $Rm_{crit} \simeq 40$, this implies $l \simeq 2$ km, which is more than two orders of magnitude below the crustal filter.

Several important points can be made here. First, because the smallest eigenvalue corresponds to $n = 1$, dipolar magnetic fields are the easiest to excite in this dynamo. Dipole preference is also a property of the geodynamo. The axial dipole term is by far the largest spherical harmonic in the core field, and in addition, the core field invariably resumes a dipole-dominant configuration following reversals and excursions. Second, back substitution of Eq. (3.49) reveals that the poloidal magnetic field intensity in this dynamo is comparable to the toroidal magnetic field intensity. This is a general property of α^2-dynamos, and it contrasts with dynamos having strong ω-effects, in which the toroidal magnetic field is typically more intense. This begs the following question: How strong is the toroidal magnetic field in the outer core—Is it comparable to the poloidal field intensity or far stronger? An answer to this question places strong constraints on the dynamo mechanism in the outer core. Third, because α in this model is a parameterization of the induction effects of the fluid velocity, its critical value given by Eq. (3.51) is a parameterized version of the critical magnetic Reynolds number for onset of dynamo action. How does this compare with the critical magnetic Reynolds number based on the actual fluid velocity in the outer core? Fourth, this simple α^2-dynamo is isotropic, in that there is no dependence on spherical harmonic order m. For the dipolar solutions, this means that the dipole moment vector has no preference for any particular direction. The core field, in contrast, shows a strong preference for an axial (rotationally aligned) dipole. This begs yet another question: What properties are missing from this kinematic dynamo that give the geodynamo its preference for an axial dipole? And lastly, this kinematic model is valid only at dynamo onset, when the magnetic field is too weak to affect the fluid motion. Accordingly, it does not answer the question of what determines the equilibrium intensity of the core field. In Chapter 4, we examine the dynamics of the outer core, seeking answers to these questions and the others posed elsewhere in this chapter.

3.3.7 The VKS dynamo experiment

The von Karman Sodium (VKS) experiment is perhaps the best-known laboratory fluid experiment to achieve dynamo generation (Berhanu et al., 2010). Strictly speaking, it is not a homogeneous fluid dynamo, but nevertheless

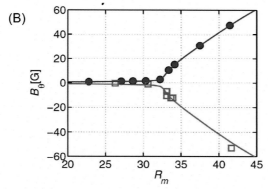

FIG. 3.22 The von Karmen Sodium (VKS) experimental dynamo. (A) Design of the apparatus. A cylinder is filled with liquid sodium at 120°C with circulation driven by two impellers rotating at frequencies f_1 and f_2. The axisymmetric toroidal flow produced by impeller counterrotation is indicated by *tube arrows*; the helical poloidal flow around each impeller is indicated by *ribbon arrows*. (B) Magnetic field intensity in the azimuthal θ-direction measured in gauss versus magnetic Reynolds number. *Squares* and *circles* denote experiments with impeller rotation directions reversed. Bifurcation (dynamo onset) occurs for $Rm_{crit} \simeq 32$. From Figs. 3a and 17b in Monchaux, R., Berhanu, M., Aumaitre, S., 2009. *The von Karman Sodium experiment: turbulent dynamical dynamos*. Phys. Fluids 21, 035108. Credit: VKS team.

the VKS experiment generates a variety of magnetic field configurations, including some with polarity reversals. In addition, it combines the important kinematic properties for facilitating dynamo action described in this chapter.

As illustrated in Fig. 3.22A (Monchaux et al., 2009), the VKS apparatus consists of a cylindrical copper vessel roughly one-half meter in length and diameter, filled with liquid sodium. The flow is driven by twin coaxial impellers mounted at the cylinder ends, each impeller able to rotate independently. When the impellers are counter rotated, the velocity field in the liquid sodium includes helical motion around each impeller, the poloidal flow component, plus a large-scale differential rotation, the toroidal flow component. Both an α-effect and an ω-effect are present in this experiment. The ambient geomagnetic field serves as the seed field.

The major limitation of the VKS experiment is that sustained dynamo action only occurred when one or both sets of impeller blades were made of soft iron with a large magnetic permeability. Impellers made of nonmagnetic materials failed to produce dynamo action. The magnetic Reynolds number in the VKS experiment is defined as

$$Rm = \pi k_e \mu_0 \sigma r_c^2 (f_1 + f_2) \tag{3.52}$$

where r_c is the cylinder radius, $(f_1 + f_2)$ is the sum of the two impeller rotation frequencies, σ is the electrical conductivity of the liquid sodium, μ_0 is again free-space permeability, and $k_e \simeq 0.6$ is an empirical coefficient that measures the efficiency of kinetic energy transfer from the impellers to the fluid.

Fig. 3.22B shows that dynamo onset at a critical value of this magnetic Reynolds number, $Rm_{crit} \simeq 32$. At dynamo onset, the induced magnetic field consists primarily of an axial dipolar field plus an axially symmetric toroidal magnetic field lying within the sodium. Other, more complex magnetic field configurations are found at larger Rm. Polarity reversals in the VKS experiment are observed after slowly decreasing the frequency of one impeller, starting from a stationary dipolar dynamo state. Polarity reversals occupy narrow slices in the space of the experimental control parameters, and it is found that only small perturbations to the experimental control parameters are needed to move from reversing to nonreversing dynamo states and back. The variety of dynamo states observed in the VKS experiment and their sensitivity to the control parameters raises the possibility that the geodynamo might have progressed through a variety of magnetic field states and behaviors in the course of the evolution of the core, as external conditions changed with time.

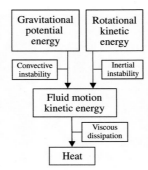

FIG. 3.23 Energy pathways in a hydrodynamic system driven by convective and inertial instabilities.

3.3.8 Dynamo energy pathways

The energy and entropy balances in the core analyzed in Chapter 2 identify the major power sources for convection in the outer core and also provide estimates of how much power is needed to sustain the geodynamo. In essence, energy and entropy balances establish conditions that are *necessary* for the geodynamo, but they do establish the *sufficient* conditions. In this chapter, we are mainly concerned with identifying the latter. So far, we have identified three such conditions: (1) a supercritical magnetic Reynolds number, (2) fluid motion with helicity to produce an α-effect, and (3) fluid motion with shear (an ω-effect). Why are these additional conditions needed? In other words, why are energy and entropy balances not sufficient for predicting dynamo action in the Earth's core, or for that matter, the cores of other terrestrial planets? The answer lies in the fact that self-sustaining fluid dynamos require energy to flow along specific pathways, and these pathways are not defined by global balances of energy or entropy.

To make this point, it is instructive to compare and contrast the energy pathways in a generalized hydrodynamic system with the energy pathways in an MHD dynamo system. Fig. 3.23 depicts the energy pathways in hydrodynamic systems powered by release of gravitational potential energy by convection, or alternatively, conversion of rotational kinetic energy to kinetic energy of fluid motions through an inertial instability. The energy pathway in the convective system converts the gravitational potential energy stored in adverse density gradients into kinetic energy via a convective instability, which produces kinetic energy of fluid motion. The kinetic energy of the fluid motion is ultimately dissipated by viscous heating. The energy pathway in the inertial systems converts some of the kinetic energy in solid-body rotation into kinetic energy of fluid motion through an inertial instability, which is then dissipated by viscous heating.

For comparison, Fig. 3.24 depicts the energy pathways in an MHD dynamo system powered by the same two forcings, convective and inertial. Initially, the pathways in the MHD dynamo system are the same as the hydrodynamic system pathways: convective dynamos convert the gravitational potential energy stored in adverse density gradients into kinetic energy via a convective instability, and inertial systems convert the kinetic energy stored in solid-body rotation into kinetic energy of fluid motions through an inertial instability. But at this point, two new pathways open for MHD dynamos that are not found in hydrodynamic systems. These are first, conversion of kinetic energy into magnetic energy, and second, conversion of magnetic energy back to kinetic energy. Both of these pathways are labeled by horizontal arrows in Fig. 3.24. In addition, MHD dynamo systems acquire an additional mechanism for dissipating energy, Ohmic dissipation. As shown in Chapter 2 and the Appendix, Ohmic dissipation is proportional to the square

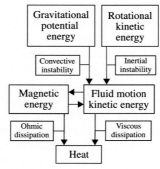

FIG. 3.24 Energy pathways in a self-sustaining dynamo system driven by convective and inertial instabilities.

of the electric current density, and in an MHD dynamo, Ohmic dissipation can be thought of as conversion of magnetic energy into heat, as depicted in Fig. 3.24.

Existence of the two additional energy pathways and the additional dissipation mechanism has first-order implications for the core and the geodynamo. Without these additions, nearly all of the kinetic energy contained in outer core fluid motions would be dissipated by viscous friction. Instead, very little of the kinetic energy in outer core motions is dissipated this way. According to the theory, lab experiments, and geomagnetic observations, most of the kinetic energy contained in outer core motions is lost by Ohmic dissipation. But in order for this to occur, enough kinetic energy in the fluid motion must be converted into electric currents, so that Ohmic dissipation exceeds viscous dissipation. In other words, in the outer core, the pathway between kinetic and magnetic energy illustrated in Fig. 3.24 must be open and well-traveled. Opening such a pathway requires a supercritical magnetic Reynolds number; fluid helicity and fluid shear allow that pathway to be well traveled.

3.3.9 Ohmic dissipation in the core

In principle, it is possible to estimate the Ohmic dissipation in the core by extrapolating the core field part of the Mauersberger-Lowes spectrum to very high spherical harmonic degrees, and converting that extrapolated spectrum into a spectrum for the electric current density inside the core. In practice, however, the accuracy of this approach is severely limited by the crustal filter, along with uncertainties in the structure of the magnetic field inside the core.

To demonstrate this, we downward continue the core field part of the Mauersberger-Lowes spectrum defined by Eq. (3.12) and shown in Fig. 3.2 to the core-mantle boundary and fit the resulting spectrum between spherical harmonic degrees $n = 2$ and 13 to an exponential law, yielding

$$B_{rms}^2(r_{CMB}) = B_0^2 \sum_{n=1}^{13} e^{-\beta n} \tag{3.53}$$

with $B_0^2 \simeq 1.2 \times 10^{10}$ nT^2 and $\beta \simeq 0.1$. Because the electric current density and core field are related by $\mu_0 \mathbf{J} = \nabla \times \mathbf{B}$, the Ohmic dissipation spectrum corresponding to Eq. (3.53) is expected to vary with spherical harmonic degree n like $n^2 \exp^{-\beta n}$, implying convergence at very large n, far beyond the crustal filter $n_{cf} \simeq 14$. And even if this extrapolation could be done accurately, it applies only to the poloidal magnetic field component (i.e., it applies to toroidal electric currents), and does not account for the toroidal magnetic field inside the core (i.e., it ignores poloidal electric currents). Nevertheless, this approach forms the basis of many observational estimates. For example, assuming simultaneous free decay of all the spherical harmonic terms in Eq. (3.53) implies an Ohmic dissipation of $Q_J \simeq 2$ GW (0.002 TW), which is far too small to be reliable or useful. Alternatively, extrapolation of an electric current spectrum of the form $n^2 \exp^{-\beta n}$ to infinite n and assuming isotropic structure for the electric currents inside the core yields a much larger Ohmic dissipation, $Q_J \simeq 100$ GW (0.1 TW). This latter value is more realistic, but is nevertheless at the lower end of the range of Ohmic dissipation implied by the energy and entropy balances in Chapter 2.

The Ohmic dissipation in the core can also be calculated theoretically, using the results of numerical dynamos. These estimates are based on simplified global balances for electromagnetic energy and kinetic energy in the outer core taken from the Appendix and Chapter 4, assuming that the geodynamo is maintained by thermochemical convection in the outer core. Although these simplified energy balances are approximate (they ignore effects of compression, among other things), their analysis is more compact and transparent than the formalism in Chapter 2, which uses more complete but more complex energy and entropy balances. Another advantage of using the simplified energy balances is that they are directly applicable to numerical dynamo results, because they are derived from the same set of governing equations on which most numerical dynamos are based.

We begin with the global kinetic energy balance for the outer core derived in the Appendix of Chapter 4:

$$\frac{\partial}{\partial t} E_{kin} = \mathcal{B} - (Q_\nu + W_L), \tag{3.54}$$

where E_{kin} is the total kinetic energy due to fluid motions, $\mathcal{B}$ is the production of kinetic energy by buoyancy forces, Q_ν is the viscous heating, and W_L is the rate of work done against the Lorentz force. The corresponding global magnetic energy balance derived in the Appendix is

$$\frac{\partial}{\partial t} E_{mag} = W_L - Q_J, \tag{3.55}$$

where E_{mag} is the total magnetic energy and Q_J is Ohmic dissipation. Adding the kinetic and magnetic energy balances (3.54), (3.55) eliminates W_L and yields

$$\frac{\partial}{\partial t}\left(E_{kin} + E_{mag}\right) = \mathcal{B} - \left(Q_\nu + Q_J\right). \tag{3.56}$$

For convection in the outer core, it is usual to assume that Ohmic dissipation greatly exceeds viscous dissipation, so that $Q_J \gg Q_\nu$. In addition, the buoyancy production equals the average thermochemical buoyancy flux F (Section 4.3 of Chapter 4) multiplied by the outer core mass, so that for outer core convection, $\mathcal{B} = M_{OC}F$. Lastly, we assume equilibrium conditions apply to long-time averages of the energy balance, so that $E_{kin} + E_{mag}$ remains roughly constant in such long-time averages. Taken together, these assumptions reduce Eq. (3.56) to the following simple-looking balance:

$$Q_J = M_{OC}F. \tag{3.57}$$

Using $F \simeq 2 \times 10^{-12}$ m^2 s^{-3} from Chapter 4 and Table S3, along with $M_{OC} = 1.84 \times 10^{24}$ kg from Table S1, Eq. (3.57) predicts $Q_J \simeq 3.7$ TW for the total Ohmic dissipation in the outer core. This value is toward the upper end of the range of the estimates obtained in Chapter 2. Tighter constraints on the Ohmic dissipation using this method requires an improved estimate of the outer core buoyancy flux F.

It is possible to estimate the average buoyancy flux F in the outer core by inverting dynamo model scaling laws, but in their current forms these scaling laws are only accurate enough to provide loose constraints on Ohmic dissipation Q_J. For example, inverting scaling laws derived from numerical dynamos (Christensen and Aubert, 2006) for the r.m.s. convective velocity v_{rms} and internal magnetic field intensity B_{rms} for the buoyancy flux F (see Section 4.7 of Chapter 4) yield, respectively,

$$F \simeq \Omega^{1/2} v_{rms}^{5/2} d^{-1/2} \tag{3.58}$$

and

$$F \simeq c_{alfv}^{10/3} \Omega^{-1/3} d^{-4/3}, \tag{3.59}$$

where $c_{alfv} = B_{rms}/(\mu_0 \rho)^{1/2}$ is the characteristic Alfvén wave speed in the outer core (Section 4.5 of Chapter 4). Using $v_{rms} = 2.6 \times 10^{-3}$ m s^{-1} from Chapter 4 for the convective velocity and $c_{alfv} = 4 \times 10^{-2}$ m s^{-1} based on the r.m.s. magnetic intensity, Eqs. (3.58), (3.59) both predict $F \simeq 2 \times 10^{-12}$ m^2 s^{-3}, the same value as above, and according to Eq. (3.57), approximately 3.7 TW of Ohmic dissipation. Note the strong dependence on c_{alfv} and v_{rms} in Eqs. (3.58), (3.59) and the assumption that the coefficients in each relationship are unity; it is the ambiguities in these factors that limit the accuracy of these scaling laws for predicting Ohmic dissipation in the core. Nevertheless, the results here are fundamentally consistent with the conclusions in Chapter 2: geodynamo dissipation is measured in terawatts, and thermochemical convection in the outer core supplies enough power to maintain the present-day geomagnetic field.

Finally, it is important to reemphasize that global energy balances define conditions that are *necessary* for geodynamo maintenance, by thermochemical convection in this case. They do not define *sufficient* conditions, for several reasons. For one, there may be other forcing mechanisms such as rotational instabilities that contribute power toward maintaining the geodynamo. For another, global energy balances alone do not ensure the existence of proper energy pathways. In order to establish sufficient conditions for maintaining the geodynamo, we must delve more deeply into the geodynamo process, including the fluid dynamics in the outer core. These subjects are treated in the following section and in Chapter 4.

3.4 Geomagnetic images of the core flow

The secular variation of the core field, a property of the geomagnetic field that has been observed for centuries, is now understood to be primarily due to advection of the geomagnetic field by outer core fluid motions, an interaction that occurs near the top of the core, within a few hundred kilometers of the core-mantle boundary. On the basis of this interpretation, maps of the core field and its secular variation at different epochs have been used to image the pattern of flow below the core-mantle boundary, thereby providing an observational approach to understand the fluid dynamics of the geodynamo.

3.4.1 Frozen flux tracing: Methods, assumptions, and limitations

Images of the core flow using the geomagnetic secular variation are based on inverting the MHD induction equation, Eq. (3.25) in its perfect conductor or frozen flux limit, that limit given by Eq. (3.27). By ignoring magnetic diffusion, the radial component of Eq. (3.27) just below the core-mantle boundary can be written

$$\frac{\partial B_r}{\partial t} + (\mathbf{v}_H \cdot \nabla) B_r = -B_r (\nabla_H \cdot \mathbf{v}), \qquad (3.60)$$

where B_r is the radial component of the core field, $\mathbf{v}$ is the vector fluid velocity in the outer core below the core-mantle boundary, and the subscript H refers to the horizontal (i.e., tangential) spherical coordinates (θ, ϕ). Here, we have assumed that close to the core-mantle boundary the radial component of $\mathbf{v}$ can be neglected, but not necessarily its radial derivative. In mathematical terms, Eq. (3.60) is a tracer equation, in which B_r is the tracer, $\partial B_r / \partial t$ is its local rate of change, and $\mathbf{v}$ is the velocity field to be imaged. The second term on the l.h.s. of Eq. (3.60) represents advective transport of B_r by the outer core flow, while the term on the r.h.s. represents B_r intensity changes produced by the horizontal divergence or convergence of the outer core flow, as illustrated in Fig. 3.17. Given maps of the radial core field B_r and its secular variation $\partial B_r / \partial t$, it is mathematically possible to invert Eq. (3.60) for the components of $\mathbf{v}$.

Images of the outer core flow derived by frozen flux tracing involve a number of assumptions and limitations, some obvious, others subtle. One set of limitations arises from the absence of magnetic diffusion. Using Eq. (3.60), the effects of magnetic diffusion are either ignored entirely or, in some cases, applied in an ad hoc manner. The justification for entirely neglecting magnetic diffusion in Eq. (3.60) is based on the fact that magnetic Reynolds number Rm is large in the outer core. However, this justification is questionable for the region below the core-mantle boundary, where it is likely that the radial length scale of magnetic field variations may be compressed, creating magnetic boundary layers. Effects of magnetic diffusion will be enhanced within such boundary layers, particularly magnetic diffusion in the radial direction. Unfortunately, the radial dependence of the core field is very poorly constrained by observations, so we are unable to say precisely how important radial diffusion is to the magnetic field structure and the secular variation below the core-mantle boundary. Yet, this is precisely where frozen flux inversions attempt to image the outer core flow.

A second set of limitations arises because frozen flux imaging suffers from problems of nonuniqueness. Even in its most reduced form, strictly tangential flow, $\mathbf{v}_H$ contains two unknowns at each point, v_θ and v_ϕ, related through a single equation, Eq. (3.60). This means that one additional constraint needs to be applied to the flow, in order to reduce the number of unknowns from two to one. Without an a priori knowledge of the outer core flow to serve as a guide, this additional constraint is arbitrary. Third, frozen flux images are necessarily incomplete. Flows that are parallel to the B_r-contours of the core field and are horizontally nondivergent (i.e., satisfy $\nabla_H \cdot \mathbf{v} = 0$) produce zero secular variation in Eq. (3.60) and so cannot be imaged by frozen flux methods. Similarly, flows with length scales much smaller than the crustal filter cannot be imaged this way.

And last but not least, a final ambiguity: precisely at what depth or depths beneath the core-mantle boundary do these methods resolve the flow? This question has been addressed by many studies, and the general conclusion is that they do not apply to the top of the outer core; instead they apply in the so-called *free stream* of the outer core, below the stack of boundary layers that lie underneath the core-mantle boundary. As discussed earlier, the thickest of these boundary layers is due to radial diffusion of the magnetic field. In order of magnitude, the thickness of this boundary layer is given by

$$\delta_B \sim (\kappa_B \tau_{SV})^{1/2}. \qquad (3.61)$$

For magnetic diffusivities $0.4 \leq \kappa_B \leq 1$ m s^{-2} and secular variation time scale $\tau_{SV} = 415$ years, Eq. (3.61) predicts $\delta_B = 70\text{--}115$ km for the magnetic boundary layer thickness beneath the core-mantle boundary. Frozen flux inversions apply to outer core flow at or below this depth range.

In spite of all the limitations and ambiguities just described, frozen flux inversion remains the best tool for the imaging outer core flow. The conventional inversion procedure, originated by P. Roberts and S. Scott (Roberts and Scott, 1965), goes as follows: first, express the fluid velocity $\mathbf{v}$ in terms of toroidal and poloidal potentials and expand each velocity potential in terms of spherical harmonics, just as is done for the magnetic field in Eqs. (3.30), (3.49). These expansions are then substituted into Eq. (3.60), along with the spherical harmonic expansions of B_r and $\dot{B}_r$ derived from core field models at various epochs. The result is a system of linear equations for the spherical harmonic coefficients of the velocity field, which when inverted yields the core flow image. Alternatively, Eq. (3.60) can be solved for the stream function and horizontal divergence of the velocity field on a spherical finite-difference grid.

3.4.2 Velocity constraints

A variety of options for the additional constraint needed for imaging the outer core flow have been investigated, including the assumptions of steady-state flow, purely toroidal flow, tangentially geostrophic or quasigeostrophic flow, columnar flow, and helical flow. The steady-state flow constraint is applied by inverting the secular variation at a number of epochs with the velocity field **v** fixed, or alternatively, by integrating Eq. (3.60) between two epochs using a steady flow. Either approach can account for a large fraction of the secular variation, although steady flows do not explain persistent time variations of the core field, such as the approximately 60–70 years oscillation in the axial dipole.

Toroidal flows are characterized by an absence of radial motion, that is, $v_r = 0$, implying that the horizontal divergence of the velocity $\nabla_H \cdot \mathbf{v}$ is also zero. One motivation for using this assumption is the recent seismic evidence that the E′ layer at the top of the outer core may be stably stratified (see Chapters 1 and 5). In general, toroidal flows fit the geomagnetic secular variation less well than flows having both toroidal and poloidal flow components, although this is expected because the former have fewer degrees of freedom than the latter.

Perhaps the most widely used assumption is *tangential geostrophy*. This assumption applies a geostrophic force balance to the tangential components of the outer core velocity field (see Chapter 4) and leads to the following constraint:

$$\nabla_H \cdot (\mathbf{v}\cos\theta) = 0. \qquad (3.62)$$

A closely related constraint termed *columnar flow* constrains the tangential divergence of the velocity to obey

$$\nabla_H \cdot \mathbf{v} = \frac{2\tan\theta}{r_{CMB}} v_\theta. \qquad (3.63)$$

Lastly, *helical flow* assumes that the tangential divergence of the flow $\nabla_H \cdot \mathbf{v}$ is spatially correlated with the radial component of vorticity $(\nabla \times \mathbf{v})_r$. This assumption is motivated by the properties of the α^2-dynamo (Section 3.3) as well as numerical and laboratory models of convection in the core, in which the flow is strongly helical due to the influence of the Coriolis acceleration.

3.4.3 Core flow images

Fig. 3.25 shows the streamlines of the toroidal part of the flow beneath the core-mantle boundary from an inversion of the core field and its secular variation at epoch 2005 that uses the columnar flow assumption. Arrows indicate the flow directions, and the spacing between streamlines is inversely proportional to the magnitude of the fluid velocity there. The r.m.s. fluid velocity in this inversion is approximately 15 km year^{-1}, although in a few places the fluid velocities are three times larger than this.

The most robust feature in Fig. 3.25 is the large counterclockwise gyre in the southern hemisphere, with its center of circulation located beneath the Southern Ocean, south of Africa. The strongest limb of this gyre is a belt of westward azimuthal flow in the southern hemisphere extending westward from beneath the Indian Ocean across the South Atlantic to beneath South America. This flow structure is common to virtually all frozen flux inversions and it appears to persist through the entire historical record of the geomagnetic secular variation. Other prominent flow structures in Fig. 3.25 are the two large, high-latitude vorticies in the northern hemisphere. These two vortices correlate with the high-latitude,

FIG. 3.25 A frozen flux image of flow near the top of the outer core at epoch 2005 based on the columnar flow constraint. Streamlines of the horizontal flow with *directional arrows* are shown overlaying contours of the radial magnetic field on the core-mantle boundary at the same epoch. Continental outlines are shown for reference. *Black cross* and *circle* denote the location of the North and South Geomagnetic Poles, respectively. *Credit: H. Amit.*

high-intensity magnetic flux patches in the northern hemisphere of the core field, seen in this figure and in Fig. 3.3. The locations of the two high-latitude vortices and their relationship to magnetic intensity centers are suggestive of quasigeostrophic flow in the outer core (described in Chapter 4). There are also several sets of smaller vortices at lower latitudes. However, these smaller vortices are only imaged in some frozen flux inversions, and therefore are not as robust as the southern hemisphere gyre or the larger vortices at higher latitudes in the northern hemisphere.

3.4.4 Length-of-day variations

Because they are subject to multiple assumptions and ambiguities, images of the core flow based on frozen flux inversions require independent verification. Fortunately, a few such independent verifications exist. In particular, in the record of Earth's rotation, there are variations amounting to a couple of millisecond fluctuations in the length of day that are not accounted for by near-surface Earth processes. These changes in the length of day (LOD) are shown in Fig. 3.26 and imply that the angular momentum of the crust-mantle system varies with time. Changes in the angular momentum of the core, transmitted to the mantle, are a likely source for these variations. In order to test this mechanism, it is necessary to project frozen flux images of the core flow through the whole core, calculate the total angular momentum content of these projected flows, and determine their time variations. The method of choice for projection is to assume that the axisymmetric part of the outer core azimuthal velocity v_ϕ is purely geostrophic, that is, uniform on cylinders concentric with the Earth's rotation axis. With this assumption, changes in angular momentum of the core involve only the axisymmetric azimuthal outer core flow at spherical harmonic degrees $n = 1$ and 3.

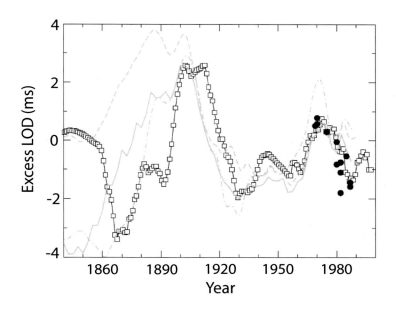

FIG. 3.26 Comparison of the observed excess length-of-day variations (in milliseconds) that is attributed to the deep Earth (*squares and filled circles*), with predictions from various frozen flux images of core flow. *Reproduced with permission from Ponsar, S., Dehant, V., Holme, R., Jault, D., et al., 2003. The core and fluctuations in the Earth's rotation. In: Dehant, V., Karato, S., Zatman, S. (Eds.), Earth's Core: Dynamics, Structure, Rotation. AGU Geodynamics Series v31.*

Fig. 3.26 compares the part of the observed LOD attributed to the deep Earth with LOD predictions from frozen flux core flow models based on various velocity constraints, but all using the same cylindrical projection in their angular momentum calculations (Ponsar et al., 2003). The cross-model agreement with the observations is quite good in Fig. 3.26 back to about 1900, particularly for the models invoking the tangential geostrophy constraint. Prior to 1900, the amplitude and the periodicity of the LOD predicted by the frozen flux models are generally satisfactory, although there is a phase error in most of the model results, compared to the observations. Overall, the correspondence between the core flow-based LOD predictions and the LOD observations provides some confidence that frozen flux imaging does indeed capture the main properties of the core flow at global length scales.

3.4.5 Polar vortices

In addition to imaging core flow on global scales, frozen flux tracing can resolve flow structures in more limited regions, providing constraints on the regional-scale dynamics in the outer core, as well as offering points of comparison with results from numerical dynamos and other predictive tools. An example of regional-scale frozen flux tracing is the

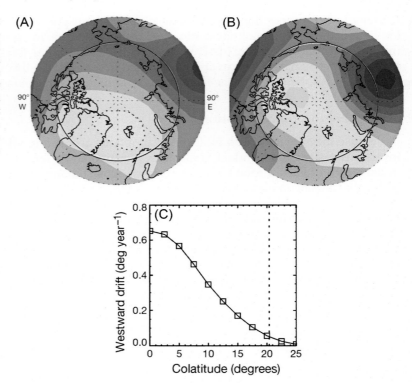

FIG. 3.27 Frozen flux evidence for a north polar vortex in the outer core. Maps of the radial component of the geomagnetic field on the core-mantle boundary at 1870 (A) and 1990 (B) from core field model *ufm* by Bloxham and Jackson (1992). *Dashed contours* mark westward drifting reverse magnetic flux patch. *Solid circle* marks the edge of the inner core tangent cylinder. (C) Angular velocity distribution inferred by frozen flux tracing; *dashed line* marks the tangent cylinder edge. Continental outlines shown for reference. *Modified from Figs. 1a, c and 2a in Olson, P., Aurnou, J., 1999. A polar vortex in the Earth's core. Nature 402, 170–173.*

character of outer core flow within the inner core tangent cylinder. The tangent cylinder region of the outer core is of particular interest because of the possible existence of convective vortices and thermal winds in this region (Chapter 4) that might offer clues about anomalous rotations of the inner core (Chapters 1 and 6).

Fig. 3.27A and B shows the geomagnetic secular variation over 120 years in the northern polar region of the core-mantle boundary using data from a time-dependent core field model by Bloxham and Jackson (1992). The generally westward drift of magnetic structures in this region, particularly evident in the trajectory of the patch of reversed magnetic flux marked by dashed contours, suggests the presence of retrograde (westward) vortex motion. Frozen flux tracing applied to this region, under the restrictive assumption that the flow is axisymmetric, yields a pattern of westward azimuthal flow with the angular velocity distribution shown in Fig. 3.27C that is consistent with the polar vortex interpretation (Olson and Aurnou, 1999). Additional support for polar vortices comes from global frozen flux modeling of more recent satellite-derived magnetic data by Hulot et al. (2002) and Holme and Olsen (2006), and suggests that a similar flow structure may be present in the southern tangent cylinder region of the outer core.

3.4.6 Remarks on frozen flux

Considering all the assumptions that go into frozen flux imaging, it is perhaps surprising (and certainly encouraging!) that most inversions recover the same large-scale structures in the core flow, irrespective of the modeling approaches used. Based on this, several general conclusions can be drawn. First, the fact that frozen flux core flows are broadly consistent with observed length-of-day variations indicates that a substantial portion of the time-varying part of the outer core flow is being imaged correctly. No comparable test is available for the steady component of core flow, but some large-scale flows such as the southern hemisphere gyre are found in most frozen flux global flows, irrespective of modeling assumptions, indicating these structures are also properly imaged. Another confidence building result is that similar r.m.s. velocities of the outer core flow, on the order of 15–20 km year^{-1}, are found across the various imaging methods. In addition, there has also been progress in identifying individual structures in the core flow such as high-latitude and polar vortices, offering the prospect that frozen flux imaging will eventually produce an observationally based interpretation of the entire large-scale flow beneath the core-mantle boundary.

In summary, frozen flux inversions of the core field and its secular variation reveal a pattern of global scale and regional scale recirculating flows below the core-mantle boundary, with speeds exceeding 10^{-3} m s^{-1} in places. Missing from this picture is the nature of smaller-scale flows, which are hidden from view by the crustal filter. Frozen flux also does not

3.5 Summary

- A self-sustaining dynamo in the core is the only explanation we have for the source of the geomagnetic field that is consistent with the structure and composition of the Earth's interior, the physical properties and energy balance of the core, and the behavior of the geomagnetic field.

- Three basic ingredients are needed for a self-sustaining planetary dynamo: an electrically conducting fluid, a source of power for fluid motions, and planetary rotation. Earth's core has all three.

- The magnetic Reynolds number is the most important dimensionless parameter governing the geodynamo. The magnetic Reynolds number in the outer core is 25–75 times greater than the minimum needed for dynamo action.

- The core field, the large-scale part of the surface geomagnetic field projected to the core-mantle boundary, consists of a slightly inclined eccentric dipole plus a weaker nondipole part. Much of the nondipole part has an overall symmetry that is consistent with strong rotational constraints on the geodynamo process.

- The secular variation (i.e., the temporal change) of the core field displays an east-west dichotomy, with higher activity in the Atlantic hemisphere compared to the Pacific hemisphere. The core field also displays abrupt changes called geomagnetic jerks.

- The ongoing decrease in the geomagnetic dipole moment is characteristic of the large amplitude intensity fluctuations that permeate the paleomagnetic record.

- Magnetic polarity reversals are generally random, having occurred about twice per million years on average. Exceptions are magnetic superchrons, intervals with few or no reversals lasting up to 40 million years. Yet, the geomagnetic field is remarkably persistent over geologic time, with little evidence of major long-lasting trends.

- The fundamental interaction that causes dynamo regeneration is positive feedback between the toroidal and poloidal magnetic field components, an interaction facilitated by transverse shear flows and by helical convective flows. Laboratory fluid dynamos operate by generating flows that produce this interaction.

- Because the magnetic Reynolds number is high, magnetic field lines are nearly "frozen" in the outer core fluid. Analyses of the geomagnetic secular variation based on frozen magnetic flux concepts reveal a pattern of recirculating flows in the outer core with characteristic velocities of 0.5–1 mm s^{-1}, including a massive counterclockwise gyre in the southern hemisphere.

Appendix

Appendix 3.1 Schmidt Legendre polynomials for geomagnetism

The Schmidt (quasinormalized) associated Legendre polynomial P_n^m as used in geomagnetism is defined in terms of the standard associated Legendre polynomial $P_{n,m}$ according to Winch et al (2005)

$$P_n^m = \sqrt{(2 - \delta_m^0) \frac{(n-m)!}{(n+m)!}} P_{n,m}, \tag{3.64}$$

where n and m are polynomial degree and order, respectively. $P_{n,m}$ is defined as

$$P_{n,m} = \frac{1}{2^n n!} (1-x^2)^{m/2} \frac{d^{n+m}}{dx^{n+m}} (x^2 - 1)^2 \tag{3.65}$$

with $x = \cos\theta$ for use in spherical coordinate systems. Recurrence relations for generating different polynomials are, for varying n

$$(n-m+1)P_{n+1,m} = (2n+1)xP_{n,m} - (n+m)P_{n-1,m}, \tag{3.66}$$

and for varying m

$$\sqrt{1-x^2}P_{n,m+1} = (n-m)xP_{n,m} - (n+m)P_{n-1,m}. \tag{3.67}$$

The following are the $n = 1$ (dipole), $n = 2$ (quadrupole), and $n = 3$ (octupole) Schmidt associated Legendre polynomials:

$$P_1^0 = \cos\theta \tag{3.68}$$
$$P_1^1 = \sin\theta \tag{3.69}$$
$$P_2^0 = 3/2(\cos^2\theta - 1/3), \tag{3.70}$$
$$P_2^1 = \sqrt{3}(\cos\theta\sin\theta), \tag{3.71}$$
$$P_2^2 = \sqrt{3}/2(\sin^2\theta), \tag{3.72}$$
$$P_3^0 = (5/2)(\cos\theta)(\cos^2\theta - 9/15), \tag{3.73}$$
$$P_3^1 = (5/2)\left(\sqrt{3/2}\right)\sin\theta(\cos^2\theta - 3/15), \tag{3.74}$$
$$P_3^2 = (\sqrt{15}/2)(\cos\theta)\sin^2\theta, \tag{3.75}$$
$$P_3^3 = \left(\sqrt{5/2}/2\right)\sin^3\theta. \tag{3.76}$$

Appendix 3.2 Magnetohydrodynamic induction and magnetic energy equations

To derive the MHD induction equation for the outer core, we start from Maxwell's equations, including Gauss' law

$$\nabla \cdot \mathbf{B} = 0, \tag{3.77}$$

in which B is the magnetic field, Faraday's law

$$\nabla \times \mathbf{E} = -\frac{\partial \mathbf{B}}{\partial t}, \tag{3.78}$$

in which E is the electric field, and Ampere's law

$$\nabla \times \mathbf{B} = \mu_0 \mathbf{J}, \tag{3.79}$$

in which J is the electric current density and μ_0 is the free space permeability. In Eq. (3.79), we have ignored displacement currents, which eliminate electromagnetic radiation; this is a standard low-frequency approximation used in MHD. To this set of equations we add a constitutive relationship, Ohm's law for a moving conductor

$$\mathbf{J} = \sigma(\mathbf{E} + \mathbf{v} \times \mathbf{B}), \tag{3.80}$$

in which σ is electrical conductivity and $\mathbf{v}$ is the velocity of the conductor. The first term on the r.h.s. of Eq. (3.80) represents conduction electric currents, the second induction electric currents.

Elimination of $\mathbf{J}$ and $\mathbf{E}$ from Eqs. (3.77)–(3.80) yields the general form of the induction equation

$$\frac{\partial \mathbf{B}}{\partial t} = \nabla \times (\mathbf{v} \times \mathbf{B}) + \kappa_B \nabla^2 \mathbf{B}, \tag{3.81}$$

where $\kappa_B = 1/\mu_0\sigma$ is the magnetic diffusivity, here assumed constant. For incompressible fluid motion in which $\nabla \cdot \mathbf{v} = 0$, Eq. (3.81) can be written

$$\frac{\partial \mathbf{B}}{\partial t} + (\mathbf{v} \cdot \nabla)\mathbf{B} = (\mathbf{B} \cdot \nabla)\mathbf{v} + \kappa_B \nabla^2 \mathbf{B}, \tag{3.82}$$

or alternatively, using the relationship between partial and full (i.e., material) time derivatives $\partial/\partial t + \mathbf{v} \cdot \nabla \equiv d/dt$,

$$\frac{d\mathbf{B}}{dt} = (\mathbf{B} \cdot \nabla)\mathbf{v} + \kappa_B \nabla^2 \mathbf{B}. \tag{3.83}$$

From left to right, the terms in Eq. (3.83) represent transport, stretching, and diffusion of magnetic field lines, respectively.

To obtain an equation for the magnetic energy, the left side of Ohm's law (3.80) is dotted with **J** and the right side with $\frac{1}{\mu_0}\nabla \times \mathbf{B}$. Using standard vector identities and applying Faraday's law (3.78) to eliminate $\nabla \times \mathbf{E}$ yields

$$\frac{(\mathbf{J}\cdot\mathbf{J})}{\sigma} = -\frac{1}{\mu_0}\nabla\cdot(\mathbf{E}\times\mathbf{B}) - \frac{\partial}{\partial t}\left(\frac{B^2}{2\mu_0}\right) - (\mathbf{v}\cdot\mathbf{F}_L), \tag{3.84}$$

where

$$\mathbf{F}_L = \mathbf{J}\times\mathbf{B} \tag{3.85}$$

is the *Lorentz force* in the conductor, here with units of force per unit volume.

Integration of Eq. (3.84) over the whole of space accounts for the total variation in the magnetic energy. The first term on the r.h.s. of Eq. (3.84) is zero after integration because both **E** and **B** decrease faster than r^2, and thus both vanish at infinity. In addition, both **J** and **v** are assumed to be zero outside of the core. Therefore, the terms that contribution to the change in magnetic energy involve integration over the volume of the core, denoted by subscript *core*:

$$\frac{\partial}{\partial t}\int_\infty \frac{B^2}{2\mu_0}dV = -\int_{core}(\mathbf{v}\cdot\mathbf{F}_L)dV - \int_{core}\frac{(\mathbf{J}\cdot\mathbf{J})}{\sigma}dV. \tag{3.86}$$

The term on the l.h.s of Eq. (3.86) is the time rate of change in E_{mag}, the total magnetic energy. The first term on the r.h.s. is the rate of work done by the fluid against the Lorentz force W_L, which can be positive or negative. The second integral on the right-hand side represents dissipation of magnetic energy due to Joule heating, and is called the *Ohmic dissipation*, Q_J. The integral is always positive and so this term always acts as a magnetic energy sink. In shorthand form, therefore, the magnetic energy balance (3.86) can be written

$$\frac{\partial}{\partial t}E_{mag} = W_L - Q_J. \tag{3.87}$$

References

Aubert, J., Finlay, C.C., 2019. Geomagnetic jerks and rapid hydromagnetic waves focusing at Earth's core surface. Nat. Geosci. 12, 393–398.

Berhanu, M., Verhille, G., Boisson, J., Gallet, B., et al., 2010. Dynamo regimes and transitions in the VKS experiment. Eur. Phys. J. B 77, 459–468.

Bloxham, J., Jackson, A., 1992. Time-dependent mapping of the magnetic field at the core-mantle boundary. J. Geophys. Res. 97, 19537–19563.

Christensen, U.R., Aubert, J., 2006. Scaling properties of convection driven dynamos in rotating spherical shells and application to planetary magnetic fields. Geophys. J. Int. 166, 97–114.

Cromwell, G., Johnson, C.L., Tauxe, L., Constable, C.G., Jarboe, N., 2018. PSV10: a global data set for 0–10 Ma time-averaged field and paleosecular variation studies. Geochem. Geophys. Geosyst. 19, 1533–1558.

Gillet, N., Jault, D., Finlay, C.C., Olsen, N., 2013. Stochastic modeling of the Earth's magnetic field: inversion for covariances over the observatory era. Geochem. Geophys. Geosyst. 14, 766–786. https://doi.org/10.1002/ggge.20041.

Holme, R., Olsen, N., 2006. Core surface flow modeling from high-resolution secular variation. Geophys. J. Int. 166, 518–528.

Hulot, G., Eymin, C., Langlais, B., Mandea, M., Olsen, N., 2002. Small-scale structure of the geodynamo inferred from Oersted and MAGSAT satellite data. Nature 416, 620–623.

Jackson, A., Finlay, C., 2007. Geomagnetic secular variation and its applications to the core. In: Schubert, G. (Ed.), Treatise on Geophysics. vol. 5. Elsevier B.V, pp. 147–193.

Jackson, A., Jonkers, A.R.T., Walker, M.R., 2000. Four centuries of geomagnetic secular variation from historical records. Philos. Trans. R. Soc. Lond. A358, 957–990.

Korte, M., Constable, C.G., 2005. The geomagnetic dipole moment over the last 7000 years—new results from a global model. Earth Planet. Sci. Lett. 236, 348–358.

Lhuillier, F., Fournier, A., Hulot, G., Aubert, J., 2011. The geomagnetic secular variation timescale in observations and numerical dynamo models. Geophys. Res. Lett. 38, L09306.

Monchaux, R., Berhanu, M., Aumaitre, S., 2009. The von Karman Sodium experiment: turbulent dynamical dynamos. Phys. Fluids 21, 035108.

Ogg, J.G., 2012. The geomagnetic polarity time scale. In: Gradstein, F.M., Ogg, J.G., Schmitz, M., Ogg, G. (Eds.), Geologic Time Scale 2012. Elsevier, pp. 85–114.

Olsen, N., Holme, R., Hulot, G., 2000. Oersted initial field model. Geophys. Res. Lett. 27, 3607–3610.

Olson, P., Aurnou, J., 1999. A polar vortex in the Earth's core. Nature 402, 170–173.

Pinheiro, K.J., Amit, H., Terra-Nova, F., 2019. Geomagnetic jerk features produced using synthetic core flow models. Phys. Earth Planet. Inter. 291, 35–53.

Ponsar, S., Dehant, V., Holme, R., Jault, D., et al., 2003. The core and fluctuations in the Earth's rotation. In: Dehant, V., Karato, S., Zatman, S. (Eds.), Earth's Core: Dynamics, Structure, Rotation. AGU Geodynamics Series v31.

Roberts, P.H., Scott, S., 1965. On analysis of the secular variation, 1, A hydromagnetic constraint: theory. J. Geomagn. Geoelectr. 17, 137–151.

Tauxe, L., Yamazaki, T., 2007. Paleointensities. In: Schubert, G. (Ed.), Treatise on Geophysics. vol. 5. Elsevier B.V, pp. 510–557.
Valet, J.P., Meynadier, L., Guyodo, Y., 2005. Geomagnetic dipole strength and reversal rate over the past two million years. Nature 435, 802–805.
Winch, D.E., Ivers, D.J., Turner, J.P.R., Stening, R.J., 2005. Geomagnetism and Schmidt quasi-normalization. Geophys. J. Int. 160, 487–504.
Ziegler, L.B., Constable, C.G., 2011. Asymmetery in growth and decay of the geomagnetic dipole. Earth Planet. Sci. Lett. 321, 300–304.

Further readings and resources

In addition to the references cited in the text of this chapter, the following are some additional readings and online resources, organized by subject matter. These include general texts and monographs, review papers, a selection of older classic studies, links to geomagnetic data sources, plus some recent papers that offer new perspectives on the topics presented in this chapter.

Useful links to geomagnetic and paleomagnetic data

INTERMAGNET for magnetic observatory data: http://www.intermagnet.org.
NOAA Geophysical Data Center for IGRF geomagnetic field models: http://www.ngdc.noaa.gov.
SWARM magnetic satellite home page: https://earth.esa.int/eogateway/missions/swarm.
Paleointensity (PINT) database: http://earth.liv.ac.uk/pint/.
Magnetics Information consortium: https://www2.earthref.org/MagIC.

Geomagnetic field characterization

Backus, G., Parker, R., Constable, C., 1996. Foundations of Geomagnetism. Cambridge University Press, Cambridge.
Merrill, R.T., McElhinny, M.W., McFadden, P.L., 1996. The Magnetic Field of the Earth: Paleomagnetism, the Core and the Deep Mantle. Academic Press, San Diego, CA.
Stacey, F., Davis, P.M., 2008. Physics of the Earth, fourth ed. Cambridge University Press, Cambridge.

Excellent review papers on geomagnetic field structure and measurement

Hulot, G., Finlay, C.C., Constable, C.G., Olsen, N., Mandea, M., 2010. The magnetic field of planet Earth. Space Sci. Rev. 152, 159–222.
Olsen, N., Stolle, C., 2012. Satellite geomagnetism. Annu. Rev. Earth Planet. Sci. 40, 441–465.

Classic papers on geomagnetic field spectra

Langel, R.A., Estes, R.H., 1982. A geomagnetic field spectrum. Geophys. Res. Lett. 9, 250–253.
Lowes, F.J., 1974. Spatial power spectrum of the main geomagnetic field, and extrapolation to the core. Geophys. J. R. Astron. Soc. 36 (3), 717–730.
Winch, D.E., Ivers, D.J., Turner, J.P.R., Stening, R.J., 2005. Geomagnetism and Schmidt quasi-normalization. Geophys. J. Int. 160, 487–504.

Geomagnetic secular variation, forecasting, and jerks

Amit, H., Terra-Nova, F., Lézin, M., et al., 2021. Non-monotonic growth and motion of the South Atlantic Anomaly. Earth Planets Space 73, 38.
Finlay, C.C., Kloss, C., Olsen, N., et al., 2020. The CHAOS-7 geomagnetic field model and observed changes in the South Atlantic Anomaly. Earth Planets Space 72, 156.
Fournier, A., Hulot, G., Jault, D., Kuang, W., et al., 2010. An introduction to data assimilation and predictability in geomagnetism. Space Sci. Rev. 155, 247–291.
Olson, P., Amit, H., 2006. Changes in earth's dipole. Naturwissenschaften 93, 519–542.
Qamili, E., De Santis, A., Isac, A., Mandea, M., Duka, B., Simonyan, A., 2012. Geomagnetic jerks as chaotic fluctuations of the Earth's magnetic field. Geochem. Geophys. Geosyst. 14, 839–850.
Ziegler, L.B., Constable, C.G., Johnson, C.L., Tauxe, L., 2011. PADM2M: a penalized maximum likelihood model of the 0–2 Ma palaeomagnetic axial dipole moment. Geophys. J. Int. 184, 1069–1089.

Time average geomagnetic field: Contrasting perspectives

Carlut, J., Courtillot, V., 1998. How complex is the time-averaged geomagnetic field over the past 5 Myr? Geophys. J. Int. 134, 527–544.
Johnson, C.L., Constable, C.G., 1997. The time-averaged geomagnetic field: global and regional biases for 0–5 Ma. Geophys. J. Int. 131, 643–666.

Geomagnetic polarity reversals: When, how, and so what?

Bogue, S.W., Merrill, R.T., 1992. The character of the field during geomagnetic reversals. Ann. Rev. Earth Planet. Sci. 20, 181–219.
Carbone, V., Alberti, T., Lepreti, F., et al., 2020. A model for the geomagnetic field reversal rate and constraints on the heat flux variations at the core-mantle boundary. Sci. Rep. 10, 13008.
Coe, R.S., Glatzmaier, G.A., 2006. Symmetry and stability of the geomagnetic field. Geophys. Res. Lett. 33, L21311.
Hoffman, K.A., Mochizuki, N., 2012. Evidence of a partitioned dynamo reversal process from paleomagnetic recordings in Tahitian lavas. Geophys. Res. Lett. 39, L06303.

Merrill, R.T., McFadden, P.L., 1999. Geomagnetic polarity transitions. Rev. Geophys. 37, 201–226.
Ogg, J.G., 2012. The geomagnetic polarity time scale. In: Gradstein, F.M., Ogg, J.G., Schmitz, M., Ogg, G. (Eds.), Geologic Time Scale 2012. Elsevier, pp. 85–114.

Dynamo theory and experiments

Moffatt, H.K., 1978. Magnetic Field Generation in Electrically Conducting Fluids. Cambridge University Press, Cambridge.
Rudiger, G., Hollerbach, R., 2004. The Magnetic Universe: Geophysical and Astrophysical Dynamo Theory. Wiley, Weinheim.
Backus, G.E., 1986. Poloidal and toroidal fields in geomagnetic field modeling. Rev. Geophys. 24, 75–109.
Jones, C.A., 2011. Planetary magnetic fields and fluid dynamos. Ann. Rev. Fluid Mech. 43, 583–614.
Roberts, P.H., King, E.M., 2013. On the genesis of the Earth's magnetism. Rep. Progr. Phys. 76, 096801.
Lathrop, D.P., Forest, C.B., 2011. Magnetic dynamos in the lab. Phys. Today 64, 40–45.
Rojas, R.E., Perevalov, A., Zürner, T., Lathrop, P., 2021. Experimental study of rough spherical Couette flows: increasing helicity toward a dynamo state. Phys. Rev. Fluids 6, 033801.

Core Ohmic dissipation

Christensen, U.R., Tilgner, A., 2004. Power requirement of the geodynamo from Ohmic losses in numerical and laboratory dynamos. Nature 429, 169–171.
Roberts, P.H., Jones, C.A., Calderwood, A., 2003. Energy fluxes and Ohmic dissipation. In: Jones, C.A., Soward, A.M., Zhang, K. (Eds.), Earth's Core and Lower Mantle. Taylor & Francis, London, pp. 100–129.
Verhoogen, J., 1979. Energetics of the Earth. National Academy of Science, Washington, DC.

Frozen magnetic flux: Theory and applications

Amit, H., Olson, P., 2006. Time-average and time-dependent parts of core flow. Phys. Earth Planet. Inter. 155, 120–139.
Backus, G.E., 1968. Kinematics of secular variation in a perfectly conducting core. Philos. Trans. R. Soc. Lond. 263, 239–266.
Holme, R., Olsen, N., 2006. Core-surface flow modeling from high resolution secular variation. Geophys. J. Int. 166, 518–528.
Jault, D., Gire, C., Le Mouel, J.L., 1988. Westward drift, core motions and exchanges of angular momentum between core and mantle. Nature 333, 353–356.

CHAPTER

4

Outer core dynamics

We begin this chapter with a description of the physical environment of the outer core, focusing on its most distinctive attributes. We then characterize the outer core environment using dimensionless parameters defined in terms of the physical and chemical properties that govern its dynamics. Special emphasis is given to thermochemical buoyancy, the primary driver for outer core convection. Next, we describe the multiple types of laminar and wave-like flows in the outer core, and what their observations imply about the structure and state of the core. We then apply the results of laboratory and numerical experiments to infer the properties of rotating convection in the outer core. Finally, we show how numerical dynamo models are used to link outer core convection to the state of the core described in Chapter 2, the seismic structure of the core described in Chapters 1, 5, and 6, the dynamics of the inner core described in Chapter 7, and the evolution of the core and the geodynamo described in Chapters 3 and 8.

4.1 The outer core environment

The dynamics of the outer core are analogous to those of a giant heat engine with multiple interacting components. As the Earth cools, heat is extracted from the outer core by the mantle, which creates the conditions for convection and other types of fluid motions to occur in the outer core. As described in Chapter 2, the sources of buoyancy for convection and for wave motion in the outer core include thermal, due primarily to the slow cooling of the core, and compositional, produced by light elements released at the inner core boundary as the inner core progressively solidifies. Additional outer core buoyancy comes from latent heat released during inner core solidification and possibly from chemical exchange with the mantle.

The evidence presented in the previous chapters strongly implicates (but does not absolutely prove) thermochemical convection as the main source of the radial fluid motions that homogenize the outer core and maintain the geodynamo. Other types of fluid dynamics likely contribute to these processes, including a wide variety of possible wave motions, as well as instabilities driven by tidal forces and by precession. But according to our present understanding, thermochemical convection, broadly defined, remains the leading candidate for the most important outer core motions.

A major obstacle we face in deciphering its dynamics is that we have no first-hand experience with an environment quite like the outer core. A deep, rapidly rotating, highly compressed molten iron alloy at temperatures above 4000 K, pressures above 100 GPa, containing a massive electric current system and surrounded by the solid mantle and inner core, both of which exchange energy with the outer core. What characteristics determine the dynamics in this environment? Phrasing the question another way, what attributes of the outer core make its dynamics similar to other parts of the Earth system, and what attributes make it different? Based on the information in the preceding chapters, we suggest the following.

First, the outer core is a very deep fluid, four times greater in depth than the upper mantle, comparable in depth to the lower mantle, and 500 times deeper than the ocean. Yet in spite of its prodigious depth, most of the outer core is well mixed. Consistent with the seismic and geomagnetic evidence, fluid parcels enriched or depleted in heat and light elements continually rise or sink 2000 km or more through the outer core, over time scales of a century or two.

Second, because it is a liquid metal, the outer core fluid has a low viscosity and high electrical and thermal conductivities. Although there is considerable uncertainty in these transport properties because of the high-pressure, high-temperature environment (see Chapter 2), there is general agreement that the outer core viscosity may not be very

FIG. 4.1 The fluidity of molten iron. At Earth's surface pressure, liquid iron and liquid iron alloys flow like water.

different from that of molten iron at low pressure. If this is the case, then outer core fluid is able to flow like water, as depicted in Fig. 4.1. In addition, the thermal conductivity of the outer core is far higher than the mantle, and its electrical conductivity is many times higher than the mantle (and far higher than ocean water, for that matter). This latter property ensures that most of the electric currents generated in the outer core remain there.

Third, the outer core fluid is subject to very strong rotational constraints. In particular, the viscous forces experienced by fluid parcels in the interior of the outer core are orders of magnitude smaller than the effects those same parcels experience from the Coriolis acceleration due to Earth's spin. In addition, the Coriolis acceleration far exceeds the inertial acceleration throughout most of the outer core (mathematical forms of the Coriolis and inertial accelerations are derived in Appendix 4.1). This dominance of the Coriolis acceleration means that all large- and medium-scale motions in the outer core are approximately *geostrophic*. In this regard, we expect that outer core dynamics exhibit some of the properties of large- and medium-scale dynamics seen in the ocean and the atmosphere.

Fourth, the outer core fluid is subject to some very delicate thermochemical and mechanical balances. As an example, we can compare the magnitude of lateral heterogeneity in the outer core to that in the mantle, and the magnitude of the forces derived from the lateral heterogeneity in both regions. The lateral thermal heterogeneity in the mantle is measured in hundreds of degrees Kelvin and mantle buoyancy forces exceed 100 Newtons per cubic meter in places. What about the outer core? Although we are unable to measure in situ temperature heterogeneity and buoyancy in the outer core, we can estimate their order of magnitude as follows. The local convective heat flux in the outer core is given by

$$q_{conv} = \rho_0 c_P \overline{v_r \Delta T}, \tag{4.1}$$

where v_r is the radial fluid velocity, c_P is the specific heat, ρ_0 is a reference density, ΔT is the horizontal variation in outer core temperature we seek, and the overbar denotes a horizontal average that includes several upwellings and downwellings. In general, radial velocity and temperature heterogeneity are highly correlated in thermal convection, so we can approximate the magnitude of the temperature heterogeneity by rearranging Eq. (4.1) to give $\Delta T \simeq q_{conv}/\rho_0 c_P v_r$. A reasonable local convective heat flux in the outer core is $q_{conv} = 0.02$ W/m^2. Using 1.1×10^4 kg/m^3 and 850 J/kg per K for the reference density and specific heat along with $v_r = 5 \times 10^{-4}$ m/s from the geomagnetic secular variation implies $\Delta T \simeq 4$ μK, in other words, lateral temperature heterogeneity on the order of micro-Kelvins, producing thermal buoyancy forces on the order of a few micro-Newtons per cubic meter. A parallel analysis made for lateral heterogeneity due to variations in composition leads to qualitatively similar results, even if we allow for slower radial velocities. The overall implication is clear: in the outer core, lateral density heterogeneity and the buoyancy forces it produces are minuscule.

Taken together, these four attributes imply that the dynamics of the outer core are characterized by a combination of huge force mismatches at some scales, coexisting with exceedingly delicate force balances at other scales. Some of the force mismatches can be quantified. Consider, for example, the relative sizes of the accelerations experienced by outer

TABLE 4.1 Outer core dimensionless parameters.

Parameter	Notation	Definition	Value/range
Thickness scale	$d*$	$(r_{CMB} - r_{ICB})/r_{CMB}$	0.649
Magnetic Reynolds number	Rm	vd/κ_B	1000–3000
Prandtl number	Pr	ν/κ_T	0.1–1
Magnetic Prandtl number	Pm	ν/κ_B	10^{-4}–10^{-6}
Ekman number	Ek	$\nu/\Omega d^2$	1–100×10^{-15}
Rayleigh number, ΔT	Ra_T	$\alpha_T g \Delta T d^3 / \kappa_T \nu$	$\sim 10^{27}$
Rayleigh number, q	Ra_q	$\alpha_T g (q - q_{ad}) d^4 / k \kappa_T \nu$	$\sim 10^{28-29}$
Rayleigh number, F	Ra_F	$F d^4 / \nu^2 \kappa_\chi$	$\sim 10^{29}$
Elsasser number	Λ	$\sigma B_{rms}^2 / \rho \Omega$	10–20
Elsasser number, CMB	Λ_{CMB}	$\sigma B_{CMB}^2 / \rho \Omega$	~ 1
Reynolds number	Re	vd/ν	$\sim 10^7$
Rossby number	Ro	$v/\Omega d$	2–5×10^{-6}
Local Rossby number	Ro_l	$vl/\Omega d^2$	≤ 0.1

core fluid parcels. The dimensionless parameter that measures the relative importance of inertial accelerations to the Coriolis acceleration due to Earth's rotation is the Rossby number Ro. For large-scale motions, in which the applicable length scale is the outer core thickness d,

$$Ro = \frac{v}{\Omega d}, \qquad (4.2)$$

where v is the r.m.s. fluid velocity and Ω is the angular velocity of rotation. Numerical values for these properties from Tables S1 and S3 from Core Properties and Parameters yield $Ro = 2-5 \times 10^{-6}$, as given in Table 4.1. The small size of the Rossby number defined in Eq. (4.2) demonstrates the assertion we made previously, that the effects of Earth's rotation, specifically the Coriolis acceleration, far and away dominate the inertial accelerations experienced by the outer core fluid, at least for large-scale motions.

What about the effects of viscosity versus Earth's rotation in the outer core? Here, the key dimensionless parameter is the Ekman number Ek, the ratio of the viscous force to the Coriolis acceleration. For large-scale motions, the Ekman number is defined as

$$Ek = \frac{\nu}{\Omega d^2}, \qquad (4.3)$$

where ν is the dynamic viscosity of the outer core fluid. Inserting the numerical values from Tables S1 and S2 from Core Properties and Parameters into Eq. (4.3) yields $Ek = 10^{-13}-10^{-15}$—a range of absurdly small numbers!

However, these two numerical examples should not be taken to mean that inertial and viscous effects are dynamically unimportant throughout the outer core, because they presume the flows in question vary on length scales commensurate with the outer core thickness d. As a counter example, the thickness predicted for Ekman boundary layers (Section 4.4) is such that an Ekman number like Eq. (4.3) based on that boundary layer thickness is of order one, implying that viscous and Coriolis effects balance in those layers. Similarly, the cross-sectional dimension of convective eddies in the outer core may be small enough so that a Rossby number based on that length scale is orders of magnitude larger than Eq. (4.2), which would imply that inertial effects are significant in outer core convection at some reduced scale.

In short, in order to thoroughly understand outer core dynamics, we need to determine how the various force mismatches and delicate balances translate into fluid motions, both the motions we can observe and those we cannot. We need to know how each type of fluid motion affects the state of the core, its evolution, and the operation of the geodynamo. Given what we already know about its structure and present-day state from Chapters 1 and 2, we anticipate that a great variety of dynamical processes are active in the outer core. Fig. 4.2 summarizes the spatial and temporal relationship of the dynamical processes we focus on in this chapter.

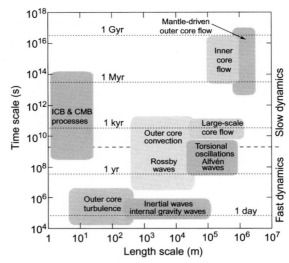

FIG. 4.2 Characteristic time scales and length scales of the outer core dynamical processes described in this chapter.

4.2 Dimensionless parameters

Dynamics in the outer core depend on a very large number of physical and chemical properties. When modeling outer core dynamics, we are forced to choose numerical values for these properties, in spite of the fact that numerical values for many of them are poorly constrained. This means that, in order to make progress, we have to run our numerical models or laboratory experiments many times, varying one or more of the properties while holding the others fixed. To make matters worse, we are generally unable to run models or experiments using realistic values for all of the important physical properties. Finally, there are the issues of validation and comparison: how do researchers, working independently, compare their model or experimental results on the core when so many different property choices are involved?

For all of these reasons, it is imperative to reduce the number of *independent* model parameters to a minimum, and to standardize their definitions. This is best done by working in terms of *dimensionless parameters*, combinations of physical and chemical properties that are without physical dimension. Unfortunately, as we will see, this approach still allows for a multitude of dimensionless parameters, including multiple versions of the same parameter. In this chapter, we focus on a minimal set of dimensionless parameters that are essential for describing outer core dynamics, although we also make use of some dimensionless parameter variants where it is necessary.

The first question to answer is: what constitutes a minimum set of dimensionless parameters for characterizing outer core dynamics, including the geodynamo? Here, we must distinguish between *control parameters* (i.e., system inputs) and *response parameters* (i.e., system outputs). Dimensionless control parameters consist of prescribed properties that define the physical and chemical state of the core. In contrast, dimensionless response parameters include variable properties we wish to determine, such as the fluid velocity and the magnetic field inside the outer core.

For purposes of clarity, we limit the following discussion to convection as the only forcing mechanism. Input properties include structural parameters such as the outer core thickness d, which we adopt as the basic the length scale, the Earth's rotation rate Ω, as well as material properties such as the outer core fluid density ρ, kinematic viscosity ν, electrical conductivity σ, magnetic permeability μ_0, and thermal diffusivity κ_T. The convective forcing is prescribed in terms of the fluid buoyancy g' (reduced gravity), which can be thermal, compositional, or both (i.e., thermochemical). The output properties of greatest interest include the geomagnetic field intensity B in the core and the outer core fluid velocity v, although there are others.

This list amounts to a total of 10 physical properties. As a group, these properties are made up of four fundamental physical units: mass, length, time, and electric current. The *Buckingham Pi theorem* dictates the minimum number of independent dimensionless parameters that we need in order to describe this system is $10-4=6$.

4.2.1 Characteristic time scales

There are formal matrix methods for determining independent sets of dimensionless parameters, but an informal method that also provides useful insight into the dynamics consists of listing the characteristic time scales defined by these 10 physical properties, and then forming dimensionless parameters by taking ratios of these time scales.

For the outer core, the physical properties listed earlier fall into three categories: diffusive, dynamical, and forcing. In the diffusive category, we have three time scales, namely

$$\tau_{visc} = \frac{d^2}{\nu} \quad \tau_{temp} = \frac{d^2}{\kappa_T} \quad \tau_{mag} = \frac{d^2}{\kappa_B} = \mu_0 \sigma d^2, \tag{4.4}$$

where τ_{visc} is the viscous diffusion time scale, τ_{temp} is the thermal diffusion time scale based on thermal diffusivity κ_T, and τ_{mag} is the magnetic diffusion time scale based on magnetic diffusivity $\kappa_B = 1/\mu_0\sigma$. Note that we can also define a time scale based on chemical diffusion, but for now we lump this into the thermal diffusion time scale. In the dynamical category, we also have three time scales. These are

$$\tau_{rot} = \Omega^{-1} \quad \tau_{circ} = \frac{d}{v} \quad \tau_{alfv} = (\rho\mu_0)^{1/2} \frac{d}{B}, \tag{4.5}$$

where τ_{rot} is the time scale for rotation, τ_{circ} is the circulation time, and as shown later in this chapter, τ_{alfv} is the time scale based on the Alfvén wave travel time across the outer core. And finally, we need a time scale for the convective forcing. In order to streamline our analysis, we form the average product of the convective buoyancy $g' = -g\rho'/\rho_0$ and the radial component of velocity v_r, giving the convective buoyancy flux $F = \overline{g'v_r}$ with SI units m^2/s^3, which can be thermal, compositional, or both. An independent time scale involving F is the buoyancy time scale

$$\tau_{buoy} = d^{2/3} F^{-1/3}. \tag{4.6}$$

4.2.2 Control versus response parameters

We can now form six independent dimensionless parameters by taking ratios of the seven characteristic time scales just defined. There are an infinite number of possibilities here, but the following are a standard set that are commonly used for outer core dynamics.

Control parameters defined in terms of diffusive properties include the *Prandtl number*

$$Pr = \frac{\tau_{temp}}{\tau_{visc}} = \frac{\nu}{\kappa_T} \tag{4.7}$$

and the *magnetic Prandtl number*

$$Pm = \frac{\tau_{mag}}{\tau_{visc}} = \frac{\nu}{\kappa_B}. \tag{4.8}$$

A control parameter that ratios diffusive and dynamical time scales is the previously defined *Ekman number*

$$Ek = \frac{\tau_{rot}}{\tau_{visc}} = \frac{\nu}{\Omega d^2}. \tag{4.9}$$

The two Prandtl numbers consist of ratios of the three diffusivities. In the outer core, $Pr \sim 0.1$–1 and $Pm \sim 10^{-5}$–10^{-6} (see Table 4.1). As noted in the previous section, the Ekman number, measuring the relative importance of viscous to Coriolis effects, lies in the range $E \sim 10^{-13}$–10^{-15}, the uncertainty being due to a possible pressure-induced increase in the dynamic viscosity of liquid iron alloys.

The control parameter that characterizes the convective forcing is the *Rayleigh number*, denoted by Ra. There are many definitions of the Rayleigh number now in use for outer core dynamics. Some of these Rayleigh are defined in terms of the specific type of forcing being considered, such as the superadiabatic thermal gradient, the heat flux, the light element flux, or their combination. In addition, there are hybrid Rayleigh numbers, formed by multiplying Ra with one or more of the dimensionless parameters listed earlier. In this chapter, subscripts will be used to identify the various forms of Ra. Their definitions are given in Table 4.2, along with range estimates of their numerical values in the outer core. The Rayleigh number used in this section is based on the buoyancy flux F and is given by

TABLE 4.2 Outer core Rayleigh numbers.

Forcing	Notation	Definition	Range
Temperature difference	Ra_T	$\alpha_T g_{CMB} \Delta\Theta d^3 / \kappa_T \nu$	$\sim 10^{27}$
CMB heat flux	Ra_q	$\alpha_T g_{CMB}(q_{CMB} - q_{ad}) d^4 / k\kappa_T \nu$	$\sim 10^{28-29}$
Codensity flux	Ra_F	$Fd^4/\nu^2 \kappa_\chi$	$\sim 10^{29}$
Codensity change	$Ra_{\dot\chi}$	$g_{CMB} \dot\chi_0 d^5 / \nu^2 \kappa_\chi$	$\sim 10^{29}$
Core cooling	Ra_Θ	$-\alpha_T g_{CMB} \dot\Theta d^5 / \nu^2 \kappa_T$	$\sim 10^{28}$
Radioactive heat	Ra_h	$\alpha_T g_{CMB} h d^5 / c_P \nu^2 \kappa_T$	$< 10^{27}$
Modification	**Notation**	**Definition**	**Range**
$Ek Ra_T / Pr$	Ra_Ω	$\alpha_T g_{CMB} \Delta\Theta d / \Omega\nu$	$\sim 10^{13}$
$Ek^3 Ra_q / Pr^2$	Ra_Q	$\alpha_T g_{CMB}(q_{CMB} - q_{ad}) / \rho_0 c_P \Omega^3 d^2$	$\sim 10^{-13}$
$Ek^3 Ra_\chi / Pr$	$Ra_{\dot m}$	$g_{CMB} F_{\dot m} / 4\pi \rho_0 \Omega^3 d^4$	$\sim 10^{-13}$

Notes: $\Theta = T - T_{ad}$; $F_{\dot m}$ = mass flux.

$$Ra_F = \frac{\tau_{visc}^2 \tau_{temp}}{\tau_{buoy}^3} = \frac{Fd^4}{\nu^2 \kappa_T}. \qquad (4.10)$$

Response parameters include the ratio of the magnetic diffusion time to the circulation time, the *magnetic Reynolds number* from Chapter 3

$$Rm = \frac{\tau_{mag}}{\tau_{circ}} = \frac{vd}{\kappa_B} \qquad (4.11)$$

and the *Elsasser number*

$$\Lambda = \frac{\tau_{mag} \tau_{rot}}{\tau_{alfv}^2} = \frac{\sigma B^2}{\rho \Omega}, \qquad (4.12)$$

which is the ratio of the Lorentz force divided by product of the Coriolis acceleration times *Rm*, and is a widely used dimensionless measure of the magnetic field intensity in the outer core. A final dimensionless parameter we should add to our list is the scaled outer core depth $d^* = d/r_{CMB}$. Models of the outer core rarely have problems with duplicating this parameter, so except for studies of the evolution of the core that includes temporal changes of the outer core thickness d, this is a fixed parameter.

The obvious questions to ask next are: (1) How large are these dimensionless parameters in the outer core? and (2) How closely can we match them using our primary modeling tools—numerical models and laboratory experiments? Figs. 4.3 and 4.4 provide answers to both of these questions. For the control parameters, the answer to question (2) looks very discouraging. According to Fig. 4.3A and B, the ranges that are currently accessible to numerical or laboratory investigation are orders of magnitude away from the outer core for three out of the four control parameters. This would seem to spell doom for any attempt to realistically model outer core dynamics. But the situation is not that bad, and Fig. 4.4 shows why. Both numerical models and lab experiments produce *results* that are in fact *very comparable* to the outer core and to the geodynamo, as measured by the response parameters *Rm* and Λ.

What is going on here? How can model inputs be so wrong, yet those same models give sensible results? One explanation is that the modeling tools we are now using provide good answers but for the wrong dynamical reasons. If true, that will be hard to fix. Alternatively, perhaps there are particular *combinations* of the control parameters that actually govern outer core dynamics, and by simply choosing control parameter values that rank the various forces in their proper order according to their size, our modeling tools yield good results. These are in fact serious issues, and finding which of these alternatives (if either) is true has motivated much of the research effort to better model the dynamics of the outer core and the geodynamo.

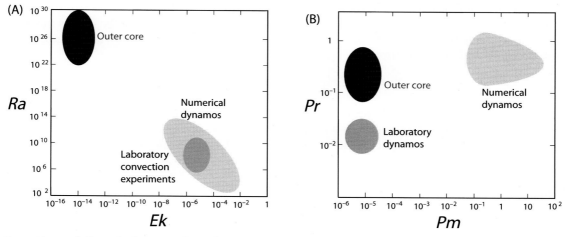

FIG. 4.3 Comparisons of dimensionless control parameters: outer core dynamics versus numerical dynamos and laboratory experiments. (A) Rayleigh-Ekman number regime; (B) Prandtl-magnetic Prandtl number regime. Parameters defined in text.

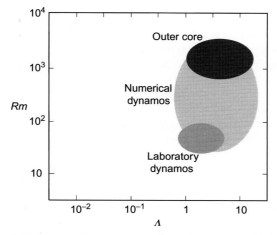

FIG. 4.4 Comparisons of dimensionless response parameters: outer core dynamics versus numerical dynamos and laboratory experiments in magnetic Reynolds-Elsasser number regime. Parameters defined in text.

4.3 Thermochemical transport and buoyancy

The secular cooling of the Earth over geologic time results in a slow reduction in the gravitational potential energy of the core. This potential energy reduction is understood to be the major source of power for outer core convection and for maintaining the present-day geodynamo. In idealized form, the system works as follows. It is usually assumed that prior to the nucleation of the inner core, light elements were mixed more-or-less uniformly through the bulk of the fully liquid core (the term "bulk" signifies that we are ignoring boundary region heterogeneity, such as the proto-E′ layer). As the temperature of the core decreased to the point where inner core solidification began, liquid solutions enriched in light elements were rejected from the inner core, a process that continues to this day (see Chapters 2, 7, and 8). At a common pressure, these enriched solutions are less dense than the outer core bulk fluid, and are, therefore, convectively unstable. This instability is enhanced by thermal buoyancy derived from latent heat release associated with solidification at the inner core boundary and sensible heat loss at the core-mantle boundary. As quantified in Chapter 2, thermochemical convection supplies power to homogenize the outer core and drive the present-day geodynamo in proportion to the rate of inner core growth.

4.3.1 Light element and heat transport in the outer core

In order to model thermochemical convection and the evolution of the core just described, conservation equations for the outer core heat and the light element budgets are needed. We begin by developing a transport-diffusion equation governing the time-dependent concentration of light elements, treating light elements as the solute and liquid metal as the solvent. For simplicity, we further assume the material properties are uniform and the flow is incompressible.

Denoting the light element concentration in the liquid metal by C (with units of kg/kg) and assuming Fick's Law of diffusion, the transport-diffusion equation is

$$\frac{\partial C}{\partial t} + (\mathbf{v} \cdot \nabla)C = \kappa_C \nabla^2 C, \tag{4.13}$$

where κ_C is the diffusivity of the light elements in the liquid metal. We now represent C in the outer core as the sum of a volume average part $C_0(t)$ that increases slowly with time as the inner core grows, plus a convective perturbation C' that varies more rapidly in time and space, that is, on the time and length scales of the dynamics:

$$C = C_0(t) + C' \tag{4.14}$$

with C_0 defined by

$$C_0(t) = V_{OC}^{-1} \int_{V_{OC}} C dV, \tag{4.15}$$

where V_{OC} is the outer core volume. Substituting Eq. (4.14) into Eq. (4.13) yields

$$\frac{\partial C'}{\partial t} + (\mathbf{v} \cdot \nabla)C' = \kappa_C \nabla^2 C' - \dot{C}_0, \tag{4.16}$$

where $\dot{C}_0$ denotes the secular (i.e., slow) time rate of change of C_0.

A parallel derivation for heat conservation in the outer core begins with the heat transport-diffusion equation written in terms of temperature T. Ignoring radioactive heat sources for now, the heat transport-diffusion equation is

$$\frac{\partial T}{\partial t} + (\mathbf{v} \cdot \nabla T) = \kappa_T \nabla^2 T, \tag{4.17}$$

where κ_T is the thermal diffusivity. As we did for light element concentration, we represent temperature as the sum of a slowly decreasing part plus a convective perturbation. For temperature, however, we identify the slowly decreasing part with an adiabat tied to the core-mantle boundary, so that $T = T_{ad}(r, t) + T'$. Substitution into Eq. (4.17) yields

$$\frac{\partial T'}{\partial t} + (\mathbf{v} \cdot \nabla)T' = \kappa_T \nabla^2 T_{ad} + \kappa_T \nabla^2 T' - \dot{T}_{ad}, \tag{4.18}$$

where in this case $\dot{T}_{ad}$ denotes the time rate of change of the outer core temperature along an adiabat.

We now average Eqs. (4.16), (4.18) over spherical surfaces. For light elements, we then multiply each term by a reference density ρ_0 and integrate with respect to radius, yielding

$$\rho_0 \overline{v_r C'} = \frac{r_{CMB}^3 - r^3}{3r^2} \rho_0 \dot{C}_0, \tag{4.19}$$

where we have assumed that the convective light element flux vanishes at r_{CMB}. For temperature, we multiply each spherical average by $\rho_0 c_P$ and perform the same integration, yielding

$$\rho_0 c_P \overline{v_r T'} = -\frac{\rho_0 c_P r \dot{T}_{ad}}{3} - q_{ad}, \tag{4.20}$$

where $q_{ad} = -k(dT_{ad}/dr)$ is the local heat flux conducted along the adiabatic thermal gradient and k is the thermal conductivity, and we have assumed an adiabatic heat flux at the inner core boundary. The quantities in Eq. (4.19), (4.20) are the outer core convective fluxes of light element mass and heat, respectively.

4.3.2 Codensity

To fully describe thermochemical convection in the outer core, we need to separately track the light element concentration C and the temperature T, using the separate transport-diffusion equations derived earlier for each variable. Because two independent diffusivities are involved, this form of thermochemical convection is called *doubly diffusive convection*. Doubly diffusive convection in the outer core is very challenging to simulate because of the large difference in the two diffusivities, thermal versus compositional. An alternative approach is to lump the effects of temperature and light element concentration together by introducing a hybrid variable called the *codensity*. Codensity, denoted here by χ, is usually defined as

$$\chi = \rho_0(\alpha_T T + \alpha_C C), \tag{4.21}$$

where α_T and α_C are the expansion coefficients for temperature and light element concentrations, respectively. With the reference density ρ_0 appearing as a coefficient in Eq. (4.21), the units of codensity are the same as density, kg/m^3. Because both expansion coefficients are positive in the outer core, perturbations in T and C perturb the codensity χ in the same way.

In order to make the codensity variable genuinely useful, we need to assume that the mass and thermal diffusivities are effectively equal in the outer core. As discussed in Chapter 2, this is certainly not true for the molecular value of the light element diffusivity κ_C, which in the outer core is far smaller than the thermal diffusivity κ_T. Here, however, we will make the (very shaky!) assertion that small-scale turbulent motions in the outer core enhance chemical mixing more than thermal mixing, so that the effective light element diffusivity is actually a larger turbulent diffusivity, rather than the smaller molecular diffusivity. We then further assume that the numerical value of this turbulent light element diffusivity roughly equals the molecular thermal diffusivity κ_T.

Provided this logic chain holds, we can combine Eqs. (4.16), (4.18) into a single codensity transport equation with the form

$$\frac{\partial \chi'}{\partial t} + (\mathbf{v} \cdot \nabla)\chi' - \kappa_\chi \nabla^2 \chi' = \epsilon, \tag{4.22}$$

where $\chi' = \rho_0(\alpha_T T' + \alpha_C C')$ is the codensity perturbation, κ_χ is the codensity diffusivity (usually equated to κ_T), and the term on the r.h.s.

$$\epsilon = -\dot{\chi}_0 = -\rho_0(\alpha_T \dot{T}_0 + \alpha_C \dot{C}_0) \tag{4.23}$$

consists of secular variations of the background temperature and light element concentration, in which $\dot{T}_0 = \dot{T}_{ad} - \kappa_T \nabla_r^2 T_{ad}$.

The ϵ-term in Eq. (4.22) functions as a source or a sink of outer core codensity, depending on its sign. Note that cooling reduces temperature in the outer core and tends to make ϵ positive, whereas light element enrichment increases codensity and tends to make ϵ negative. Most models of core energetics and evolution, such as those in Chapters 2 and 8, predict that light element enrichment dominates over cooling in the present-day outer core, so that ϵ is negative at present, on average. Negative ϵ in the outer core is referred to as the *codensity sink*. But in the deep past, prior to inner core nucleation, the sign of ϵ in the outer core was likely different. If secular cooling dominated the buoyancy production, ϵ would have been positive, a codensity source. In addition, if removal of light elements from the outer core dominated the buoyancy production, ϵ would have been positive then as well. This predicted change in the sign of ϵ might have observable expressions in the paleomagnetic field, which would allow us to date inner core nucleation and to identify buoyancy sources when the outer core was entirely liquid. We examine this possibility in Chapter 8 using numerical dynamo models.

4.3.3 Thermochemical buoyancy flux

Thermochemical buoyancy in the outer core, denoted here by g', can be written in terms of codensity and gravity g as

$$g' = -g\rho'/\rho_0 = g\chi'/\rho_0 = g(\alpha_T T' + \alpha_C C'). \tag{4.24}$$

As discussed in Section 4.2, a convenient variable with which to measure the strength of thermochemical convection in the outer core is the *codensity buoyancy flux*. In terms of thermochemical buoyancy (4.24) and radial component of fluid velocity v_r, the codensity buoyancy flux is given by

$$\overline{g'v_r} = g\overline{\chi'v_r}/\rho_0 = g(\alpha_T \overline{v_r T'} + \alpha_C \overline{v_r C'}), \tag{4.25}$$

where the overbars again denote horizontal (or spherical surface) averages.

The unit of codensity buoyancy flux in Eq. (4.25) is m^2/s^3, involving length and time only. Because it is independent of other physical units (in particular, independent of mass, temperature, and concentration) and because it combines both thermal and compositional buoyancy forces, codensity buoyancy flux is a particularly handy variable for characterizing convection in the outer core.

For example, the buoyancy flux Rayleigh number Ra_F introduced earlier in this chapter involves just buoyancy flux and three other properties, all of which are comprised of the same two physical units (length and time). In addition, the codensity buoyancy flux in the outer core can be constrained within roughly an order of magnitude from the heat flux at the core-mantle boundary and the inner core growth rate. Although far from perfect, these constraints on Ra_F are better than for Ra_T, the Rayleigh number based on the superadiabatic temperature gradient, for which even the proper sign in the outer core (positive or negative) remains uncertain! Lastly, Ra_F can be translated into other definitions of the Rayleigh number. Accordingly, we make extensive use of codensity buoyancy flux and Ra_F for scaling numerical simulations and laboratory experiments to outer core convection and the geodynamo.

One difficulty with using Ra_F defined by Eq. (4.25) for this purpose is that both the thermal and the compositional buoyancy fluxes vary with radius in the outer core, whereas they should appear as constant properties when used in a control parameter. The conventional way around this problem is to average the terms in Eq. (4.25) through the outer core. Assuming both gravity g and the local adiabatic heat flux q_{ad} increase linearly with radius, substitution of Eqs. (4.19), (4.20) into Eq. (4.25) and radial averaging yields

$$F = F_T + F_C = \frac{3\alpha_T g_{CMB}}{5\rho_0 c_P}(\bar{q} - q_{ad})_{CMB} + \frac{3\alpha_C g_{CMB} r_{CMB}}{10} \dot{C}_0. \tag{4.26}$$

Here, F is the volume average codensity buoyancy flux, $\bar{q}$ is the spherical average heat flux, the subscripts T and C denote the individual thermal and compositional (light element) buoyancy fluxes, the subscripts CMB and ICB denote the core-mantle and inner core boundaries, and terms of order $(r_{ICB}/r_{CMB})^2$ or smaller have been ignored. In deriving Eq. (4.26), we have also ignored contributions from latent heat release and light element loss at the top of the core.

In order for outer core convection to occur, the average codensity buoyancy flux F needs to be positive. However, as discussed in Chapter 2, it is not necessary that both its thermal and compositional parts be positive everywhere. For example, if the total heat conducted along the outer core adiabat exceeds the total core heat flux, then F_T in Eq. (4.26) is negative but F can still be positive if F_C is sufficiently large and positive. In that case, F_C is the kinetic energy source for the convection and F_T is the kinetic energy sink. In Chapter 2, this arrangement is referred to as the refrigerator effect. If both F_T and F_C are positive, then both are kinetic energy sources for the convection.

We can estimate in order of magnitude the size of F in the outer core, as well as the size of the buoyancy flux-based Rayleigh number Ra_F. For thermal buoyancy, a reasonable upper limit is $\bar{q}_{CMB} - q_{ad} = 40$ mW/m^2 at the core-mantle boundary. Using the property values in Tables S1 and S2 from Core Properties and Parameters, Eq. (4.26) gives $F_T \simeq 5 \times 10^{-13}$ m^2/s^3. Likewise, assuming $\dot{C}_0 = 1.5 \times 10^{-19}$ s^{-1} and using property values from the same tables gives $F_C \simeq 1.5 \times 10^{-12}$ m^2/s^3, three times larger than F_T. By these estimates, present-day convection in the outer core is thermochemical, a mixture of thermal and light element buoyancies, with compositional buoyancy the stronger of the two. Finally, we can use these values of F_T and F_C to estimate F and Ra_F in the outer core. Combining these with values for the other properties from Tables S1 and S2 from Core Properties and Parameters, Eq. (4.10) gives $F \simeq 2 \times 10^{-12}$ m^2/s^3 and $Ra_F \simeq 10^{29}$—another astronomically large number!

4.4 Steady laminar flows

The analysis in Section 4.2 based on the dimensionless parameters Rossby number Ro and Ekman number Ek implicates the Coriolis acceleration as having primary control on outer core dynamics. Thermochemical buoyancy and the Lorentz force from the geomagnetic field must also play important roles, insofar as heat transfer, mass transfer, and magnetic field generation are concerned. In addition, we also need to consider the effects of fluid inertia and the viscous force, these becoming more significant as the length scale of the motion shrinks.

In this section, we examine each of these in turn, focusing on idealized laminar flow examples, starting with flows dominated by the Coriolis acceleration. Table 4.3 summarizes the main properties of these flows. Although they are treated separately here, it is important to keep in mind that in the outer core these flows are apt to be superimposed on one another, so it is likely their dynamics are closely intertwined. General equations of motion appropriate to the rotating coordinate system of the outer core are derived in Appendix 4.2. The laminar flows and waves described in the following two sections are special cases of those more general equations.

4.4.1 Geostrophic flow

The term *geostrophy*, literally "Earth turning," refers to flows that result from a balance between the Coriolis acceleration and the pressure gradient force. Geostrophy, or more precisely, the approach to geostrophy, characterizes large-scale flows in Earth's ocean and in the atmospheres of most planets. In its idealized form, the geostrophic balance for the outer core can be written

TABLE 4.3 Outer core laminar flows.

Flow type	Force balance	Character
Geostrophic	Coriolis vs. fluid pressure	Columnar flow
Magnetostrophic	Coriolis vs. magnetic pressure	Follows B^2-contours
Magnetic wind	Magnetic tension	Columnar shear flow
Thermal wind	Coriolis vs. buoyancy	Azimuthal flow with shear
Reynolds stress	Inertia	From correlated turbulence
Ekman boundary layer	Coriolis vs. viscous	Induces Ekman pumping

$$2\mathbf{\Omega} \times \mathbf{v} = -\frac{1}{\rho_0}\nabla P', \tag{4.27}$$

where $\mathbf{\Omega}$ is Earth's angular velocity of rotation vector, P' is the reduced pressure in the fluid (with a hydrostatic component removed), and the other variables and properties have their previous meanings. The reference density ρ_0 appears in Eq. (4.27) because in its idealized form, geostrophic flow does not involve density variations.

Taking the curl of Eq. (4.27) reveals that geostrophic flow satisfies

$$\frac{d\mathbf{v}}{d\zeta} = 0, \tag{4.28}$$

where ζ is the axial coordinate, parallel to the rotation vector $\mathbf{\Omega}$. Eq. (4.28) is called the *Taylor-Proudman constraint*. It states that geostrophic flow is invariant along coordinates parallel to the Earth's rotation axis. In naturally occurring systems, flows that exactly satisfy Eqs. (4.27), (4.28) are rare, and quite possibly, nonexistent. However, there are many natural flows that approximately conform to Eqs. (4.27), (4.28), making these very useful for the purpose of characterizing the fluid motions in those systems.

As an example, consider geostrophic fluid motions extending through the bulk of the outer core. The Taylor-Proudman constraint (4.28) implies that columns of fluid parallel to ζ (i.e., parallel to the rotation vector $\mathbf{\Omega}$) move uniformly along cylinders of constant height, as illustrated in Fig. 4.5. To the extent that the outer core can be approximated as an oblate spheroidal shell (or even more simply, as a spherical shell), geostrophic cylinders are the infinite set of nested circular cylinders with the Earth's rotation axis as their symmetry axis. According to this definition, the cylinder tangent to the inner core equator is a geostrophic cylinder. The curves defined by the intersection of the geostrophic cylinders and the core-mantle boundary (idealized in Fig. 4.5 as a smooth surface) are called *geostrophic contours*. Assuming hypothetically smooth, axisymmetric core-mantle, and inner core boundaries, latitude circles are geostrophic contours. Geostrophic contours, supposing any exist in the outer core, represent the streamlines of purely geostrophic flow.

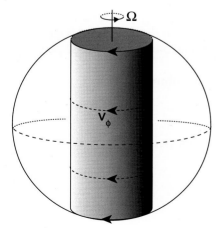

FIG. 4.5 Idealized geostrophic flow along a geostrophic contour with uniform column height in the outer core. v_ϕ is azimuthal velocity, Ω is angular velocity of rotation. *Long-dashed curve* marks the outer core equator.

In contrast to flow along geostrophic contours, core flow along any trajectory that crosses core-mantle boundary topography or is eccentric with respect to Earth's rotation axis implies stretching of the fluid column, as illustrated in Fig. 4.6. Because of column stretching, these flows are called *quasigeostrophic*. This distinction is important because purely geostrophic flow, as we have defined it, has no radial component of velocity. And because radial velocity is necessary for geodynamo maintenance (see Chapter 3), outer core flow cannot be purely geostrophic. This begs the following questions: (1) How does the outer core flow overcome the constraints of geostrophy, and (2) Is outer core flow in fact quasigeostrophic, or does it conform to some other force balance?

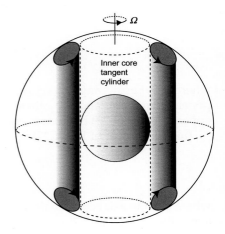

FIG. 4.6 Quasigeostrophic flow in the outer core, illustrating columnar flow with column stretching. Inner core tangent cylinder indicated by *short dashes* and outer core equator by *long dashes*. Ω is angular velocity of rotation.

Providing answers to these questions lies at the heart of outer core dynamics. It turns out there are multiple ways the motions of a rotating fluid can violate the constraints imposed by Eqs. (4.27), (4.28), and it is probable that the actual fluid motions in the outer core are a complex mixture of all of these. In the following sections, we describe some of the possible deviations from pure geostrophy, focusing on steady laminar flows. In Section 4.5, we examine deviations from pure geostrophy in the form of wave motions.

4.4.2 Thin-layer geostrophy

As a first example of relaxing the full impact of the Taylor-Proudman constraint, consider the properties of geostrophic flow confined to a thin layer, such as the stratified E′ layer that has been proposed for the region below the core-mantle boundary, evidence for which is presented in Chapter 5. The combination of its thinness and its stratification implies that radial fluid velocities are much less than horizontal fluid velocities in the layer. In this situation, only the radial r-component of Ω is of dynamical consequence to the flow, the horizontal θ-component of Ω playing a secondary role except in the narrow region very close to the outer core equator.

We can express the horizontal velocity in the thin layer in terms of a toroidal streamfunction ψ

$$\mathbf{v} = \nabla \times (\mathbf{r}\psi). \tag{4.29}$$

Substituting Eq. (4.29) into Eq. (4.27) and solving for the toroidal velocity components yields

$$v_\theta \equiv \frac{1}{\sin\theta}\frac{\partial \psi}{\partial \phi} = \frac{-1}{\rho_0 f \sin\theta\, r\partial\phi}\frac{\partial P'}{} \tag{4.30}$$

and

$$v_\phi \equiv -\frac{\partial \psi}{\partial \theta} = \frac{1}{\rho_0 f r} \frac{\partial P'}{\partial \theta}, \tag{4.31}$$

where $f = 2\Omega \cos\theta$, the *Coriolis parameter*, equals twice the radial component of Ω at colatitude θ. The presence of $\cos\theta$ in the denominators of the pressure force terms makes the velocity components singular at $\theta = \pi/2$, and serves as a reminder that Eqs. (4.30), (4.31) are invalid in the immediate neighborhood of the equator.

A further simplification, appropriate for flow structures in the thin layer that are smaller than global in scale, is to treat f as a constant equal to $f_0 = 2\Omega \cos\theta_0$, where θ_0 is the colatitude of the center of the flow structure in question. This simplification is called the *f-plane* approximation, and is often used in applications of geostrophy to subglobal scale flows the atmosphere and ocean. In the *f-plane* approximation, Eqs. (4.30), (4.31) imply

$$P' = -\rho_0 f_0 r \psi, \tag{4.32}$$

which signifies that isobars and streamlines coincide, and allows the streamfunction of the flow to be directly converted into the reduced pressure variations within the layer. A subtle distinction between thin stratified layer geostrophy and global geostrophy is the internal structure of the flow. The internal flow in the former is invariant along the radial coordinate r, whereas in the latter it is invariant along the axial coordinate ζ. There is little difference between these in high-latitude regions, but the difference becomes more pronounced at lower latitudes.

As an application, consider thin-layer geostrophy as a possible explanation for the large gyre located in the southern hemisphere of the outer core seen in frozen flux images of that derived in Chapter 3 from the geomagnetic secular variation. A plausible geometry is sketched in Fig. 4.7A. The layer is characterized by a uniform density ρ_1 smaller than ρ_0, the density of the underlying fluid. In addition, the layer has thickness h that varies laterally, with the reduced pressure in the layer varying as $P' = (\rho_0 - \rho_1)gh$, where g is gravity at the top of the layer, representing the core-mantle boundary. The velocity vectors in Fig. 4.7A show the thin layer geostrophic response to a positive thickness anomaly of the layer in the southern hemisphere. The implied circulation is counterclockwise (anticyclonic) like the southern hemisphere gyre described in Chapter 3. The streamlines of this flow correspond to the contours of the layer thickness h or alternatively, the contours of the reduced pressure P'.

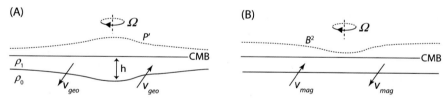

FIG. 4.7 Dynamics in a thin, low-density layer at the top of the outer core. Southern hemisphere geometry. (A) Geostrophic flow produced by lateral variation in layer thickness; *dashed curve* indicates reduced pressure variation in the layer and (B) magnetostrophic flow produced by lateral variation in geomagnetic field intensity; *dashed curve* indicates the magnetic pressure variation. Symbols defined in text.

We can estimate the layer thickness anomaly necessary to support this flow by geostrophy. Again with reference to Fig. 3.25 of Chapter 3, the gyre center is near latitude 60°S, the velocity difference between the gyre margin and its center is of order $\Delta v \simeq 1.5 \times 10^{-3}$ m/s, and the average radius of the gyre is approximately $\Delta\theta \simeq 30$ degrees on the core-mantle boundary. Replacing partial derivatives with differences in Eq. (4.31) and using the outer core radius and the outermost core density from Table S1 from Core Properties and Parameters, we obtain $\Delta P' \simeq \rho_0 f_0 r_{CMB} \Delta\theta \Delta v \simeq 3.4 \times 10^3$ Pa. Assuming $\Delta\rho = \rho_0 - \rho_1 \simeq 50$ kg/m^3 in the E' layer (see Chapter 5) implies a layer thickness variation across the gyre of $\Delta h = \Delta P'/\Delta\rho g \simeq 7$ m. This is a very small lateral heterogeneity—unfortunately too small for direct seismic detection, but certainly plausible in light of the extreme heterogeneity of the core-mantle boundary region.

In short, it is relatively easy to account for the magnitude and geometry of outer core flows inferred from geomagnetic secular variation in terms of thin-layer geostrophy. The main problem is the lack of direct observations of the lateral heterogeneity that is responsible for driving these flows. Because we cannot detect such minute variations in the layer thickness h or the dynamic pressure P', we need to consider indirect methods, including considerations of the thermal interaction between the mantle and the outer core.

4.4.3 Magnetostrophic flow

The term *magnetostrophy* refers to a balance between the Coriolis acceleration and the Maxwell pressure part of the Lorentz force (see Appendix 4.4 for definitions of the Lorentz force and the Maxwell pressure), and *magnetostrophic* is the term used to describe the resulting flows. This terminology is a bit misleading, because magnetostrophic flow is in fact an "Earth turning" flow like geostrophic flow, but with the Maxwell pressure derived from the magnetic field replacing the ordinary fluid dynamic pressure in the driving force.

For large-scale flows in the outer core, a magnetostrophic force balance requires very large geomagnetic field gradients, unrealistically large in most instances. To demonstrate this, we start with an equation expressing the force balance just described:

$$2\mathbf{\Omega} \times \mathbf{v} = -\nabla \left(\frac{\mathbf{B} \cdot \mathbf{B}}{2\rho_0 \mu_0} \right) \tag{4.33}$$

and repeat the steps in Eqs. (4.29)–(4.31), leading to expressions for the magnetostrophic velocity components in a thin layer:

$$v_\theta = \frac{-1}{2\rho_0 \mu_0 f_0 \sin\theta} \frac{\partial B^2}{r \partial \phi} \tag{4.34}$$

and

$$v_\phi = \frac{1}{2\rho_0 \mu_0 f_0} \frac{\partial B^2}{r \partial \theta}, \tag{4.35}$$

where B^2 denotes the square of the magnetic field intensity.

We can repeat the analysis from the preceding section, this time assuming that the prominent anticyclonic gyre in the southern hemisphere of the outer core includes a magnetostrophic component driven by Maxwell pressure gradients, and then ask how large the magnetic field gradient must be in order to make an appreciable contribution to the circulation. The situation is illustrated in Fig. 4.7B. Because the core field is weaker in the gyre center, compared to the field on the gyre rim, the Maxwell pressure force is directed toward the gyre center, and the resulting flow is cyclonic, that is, opposite to Fig. 4.7A and opposite the circulation direction inferred from frozen flux images. In this sense, the magnetostrophic component of the flow acts as a braking mechanism.

Using Eq. (4.35), the relationship between the magnetostrophic velocity and the difference in magnetic field intensity across the gyre is given by $(\Delta B)^2 \simeq 2\rho_0 \mu_0 f_0 r \Delta \theta \Delta v \simeq 9 \times 10^{-3}$ T^2. Using the same gyre properties in this formula as in thin-layer geostrophy example, a significant magnetostrophic braking velocity would, therefore, require an intensity variation of $\Delta B \simeq 100$ mT across the gyre. With reference to the core field map (Fig. 3.3), the difference in the radial geomagnetic field intensity between the center and the edge of the gyre is only about 1 mT, two orders of magnitude less than what is needed for significant braking. Accordingly, we conclude that Maxwell pressure gradients are too weak to appreciably affect the prominent gyre in the southern hemisphere of the outer core, and therefore it is unlikely that this flow includes a major magnetostrophic component.

The previous example should not be understood to mean that magnetically driven flows are entirely absent from the outer core. In the following section, we describe how shearing forces derived from the tangential components of Maxwell stress are involved in a type of flow called magnetic wind. Furthermore, the multiyear length-of-day variations referred to in Chapter 3 appear to be due to torsional oscillations, wave motions in the outer core in which the restoring force is derived from tangential Maxwell stresses (see Section 4.4).

4.4.4 Magnetic wind

In addition to normal forces exerted by the Maxwell pressure, the Maxwell stresses exert tangential forces, capable of producing large-scale shearing motions in the outer core called *magnetic winds*. To illustrate this, consider the azimuthal ϕ-component of the Lorentz force from Appendix 4.4, integrated over the area of the geostrophic cylinder shown in Fig. 4.5. The result is

$$\int_A (\mathbf{J} \times \mathbf{B})|_\phi dA = \frac{1}{\mu_0} \int_A (\nabla \times \mathbf{B}) \times \mathbf{B}|_\phi dA. \tag{4.36}$$

Similar integrations of the Coriolis acceleration and pressure force vanish, and in addition, the buoyancy force lacks a ϕ-component so its integral is zero as well. Accordingly, if the effects of fluid inertia are neglected, then conservation of momentum (given in Appendices 4.2 and 4.3) dictates that Eq. (4.36) should be zero (or nearly zero) on any geostrophic contour. This condition is called the *J.B. Taylor constraint* after plasma physicist J.B. Taylor, not to be confused with G.I. Taylor, coauthor of Eq. (4.28). For several decades following initial publication (in 1963), it was widely thought to offer a recipe for determining the axisymmetric geomagnetic field structure inside the core, and perhaps a shortcut for modeling the geodynamo. Much research has gone into these questions, and it now seems that, in spite of its appealing simplicity, Eq. (4.36) may only provide very loose constraints on the geomagnetic field inside the core.

More specifically, if we multiply the ϕ-component of the Lorentz force by the perpendicular distance from the rotation axis s, integration over the volume of a geostrophic cylinder yields Γ, the electromagnetic torque about the rotation axis exerted on the cylinder. If we then approximate the magnetic field inside the core as axisymmetric, this electromagnetic torque Γ reduces to the following:

$$\Gamma = \frac{2\pi s^2}{\mu_0} \int_{\zeta_1}^{\zeta_2} B_\phi B_s d\zeta, \tag{4.37}$$

where $\zeta_2 - \zeta_1 = H(s)$ is the height of the geostrophic cylinder at cylindrical radius s and the subscripts s and ϕ denote components of the magnetic field intensity normal and tangential to the cylinder, respectively. The condition $\Gamma \neq 0$ implies nonzero electromagnetic torques. If Γ is nonzero and steady in time, the result is steady shearing motions between concentric cylinders, a *magnetic wind*; if Γ is unsteady, the result is a time-dependent shearing motion between nested geostrophic cylinders. The time-dependent forms of this motion, called *torsional oscillations*, are described in Section 4.5.

As an example of a steady magnetic wind, consider the generally westward drift in the outer core imaged in Chapter 3 using frozen flux methods. What are the implications for the geometry of the geomagnetic field inside the outer core if this flow is a magnetic wind driven by electromagnetic torques? The structure of the core field is such that B_s is mostly positive in the southern hemisphere and mostly negative in the northern hemisphere. Therefore, in order that the observed westward drift be a magnetic wind, the azimuthal toroidal magnetic field B_ϕ below the core-mantle boundary would need to be anticorrelated with B_s on average, in order that the electromagnetic torque Γ in Eq. (4.37) be negative. Anticorrelation with the core field B_s implies that on average B_ϕ is positive in the northern hemisphere and negative in the in the southern hemisphere. Establishing the direction of the toroidal field inside the core based on notion that the geomagnetic westward drift is a magnetic wind would be a significant piece of geophysical inference. An important caveat to this interpretation is that there are other ways to generate the westward drift, as shown in the next section. Nevertheless, it is found that B_s and B_ϕ are negatively correlated on average in many numerical dynamos, as exemplified by the numerical dynamos described in Section 4.7. This suggests that magnetic winds may indeed play a role in structuring the geomagnetic field inside the outer core.

Another interesting dynamical effect from magnetic winds that may be important in the outer core is the phenomenon of superrotation. In *magnetic superrotation*, Lorentz forces overcome the constraints of geostrophy and produce local angular velocities in the outer core fluid that is larger than the angular velocities of both the mantle and inner core.

Fig. 4.8 shows the results of a superrotation experiment by Brito et al. (2011), in which a differentially rotating spherical shell of liquid sodium is subjected to a strong axial dipolar magnetic field originating from a magnet located at the center of the sphere. Differential rotation between the inner and outer spherical boundaries generates a spherical Couette flow, which interacts with the dipolar magnetic field to produce a strong toroidal magnetic field in the highly conducting liquid sodium, in the manner described in Chapter 3. The Lorentz forces produced by the interaction of the imposed dipolar and induced toroidal magnetic fields reduce the angular velocity of the fluid in some places, but close to the inner boundary equator, they accelerate the fluid into superrotation, as shown in Fig. 4.8B.

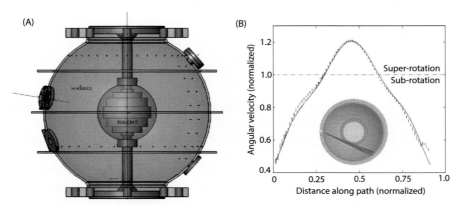

FIG. 4.8 Laboratory experiment on super-rotation driven by magnetic winds. (A) Experimental device consisting of a differentially rotating spherical shell filled with liquid sodium with an applied axial dipolar magnetic field produced by a central magnet. (B) Experimental results. Angular velocity of the fluid is measured along the path marked in the *inset* diagram. Superrotation and subrotation regions (normalized by the fast-rotating inner sphere) are separated by the *dashed line*. Modified from Figs. 3 and 11 of Brito, D., Alboussière, T., Cardin, P., Jault, D., LaRiza, P., Masson, J.-P., Nataf, H.-C., Schmitt, D. 2011. Zonal shear and super-rotation in a magnetized spherical Couette-flow experiment. Phys. Rev. E 83, 066310. Courtesy H.-C. Nataf.

There is as yet no compelling evidence for appreciable superrotation of the fluid in the outer core close to the inner core, as the experiment predicts. However, as shown in Fig. 3.4, there is strong evidence for a marked difference in the geomagnetic secular variation between the Atlantic and Pacific hemispheres, where frozen flux images indicate a large difference in westward flow between these regions. It is possible that in places, subrotation from magnetic winds augments the westward flow due to thermal winds, such that the apparent westward drift is much greater in those places.

4.4.5 Thermal wind

Wherever lateral heterogeneity in density is dynamically significant in the outer core, we need to modify the geostrophic relation by adding a buoyancy force to the r.h.s. of Eq. (4.27). For density heterogeneity produced by temperature, this force (per unit mass) is $-\alpha_T \mathbf{g} T'$, where α_T is the thermal expansion, $\mathbf{g}$ is the gravity vector, and T' is the perturbation temperature. Writing gravity in the outer core as $\mathbf{g} = -g_{CMB}\mathbf{r}/r_{CMB}$ and assuming constant thermal expansion, the curl of Eq. (4.27) with this thermal buoyancy force included becomes

$$\frac{d\mathbf{v}}{d\zeta} = \frac{\alpha_T g_{CMB}}{2\Omega r_{CMB}}(\mathbf{r}\times\nabla)T'. \tag{4.38}$$

This equation is called the *thermal wind* balance, because of its long-standing use in atmosphere dynamics. In the outer core, the $\hat{\phi}$-component is the most important part of Eq. (4.38), as it governs the structure of azimuthal (east-west) flows driven by zonal (north-south) temperature variations:

$$\frac{dv_\phi}{d\zeta} = \frac{\alpha_T g_{CMB}}{2\Omega r_{CMB}}\frac{\partial T'}{\partial \theta}. \tag{4.39}$$

This particular form of the thermal wind balance is applicable to thermal heterogeneity that extends deep into the outer core. The thin layer version of Eq. (4.39), applicable to thermal heterogeneity limited to the region near the top of the outer core, is just

$$\frac{dv_\phi}{dr} = \frac{\alpha_T g_{CMB}}{f_0 r_{CMB}}\frac{\partial T'}{\partial \theta}. \tag{4.40}$$

Note that both of these thermal wind balances can be generalized to include compositional heterogeneity by including the term $\alpha_C \partial C'/\partial \theta$.

Regardless of the source of the density heterogeneity, Eqs. (4.39), (4.40) can be used to estimate the magnitude of the shear in the azimuthal flow in terms of zonal (i.e., north-south) density gradients. Two possible axisymmetric thermal winds in the outer core are illustrated schematically in Fig. 4.9A and B. Fig. 4.9A shows the azimuthal thermal wind resulting from an idealized axisymmetric latitudinal thermal gradient in which the perturbation temperature T' near the equator is above the perturbation temperature at middle latitudes, a possible consequence of heterogeneous inner core growth (see Chapters 6 and 7). In this situation, $\partial T'/\partial \theta$ is positive in the northern hemisphere and negative in the

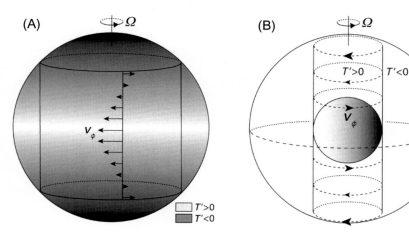

FIG. 4.9 Thermal wind flow outside (A) and inside (B) the inner core tangent cylinder driven in (A) by meridional (north-south) temperature variations T' and in (B) by a temperature contrast $\Delta T'$ across the tangent cylinder boundary. Variations in thermal wind velocity v_ϕ indicated by arrow lengths, sizes, and directions.

southern hemisphere. If v_ϕ is negative (westward-directed) on the equatorial plane, then according to Eq. (4.39), v_ϕ becomes increasingly positive with distance from the equator in both the northern and southern hemispheres. The result is the flow pattern shown in Fig. 4.9A, with a strong retrograde (westward, i.e., negative in ϕ) flow near the equator and weaker flows at higher latitudes, possibly becoming prograde (eastward) far from the equator. Because this flow pattern is aligned along the axial direction ζ, the strongest retrograde motion near the equator lies at depth beneath the core-mantle boundary. This deep, westward-flowing low-latitude thermal wind in the outer core is an example of an *equatorial undercurrent*. Equatorial undercurrents are often found in numerical simulations of outer core convection and dynamo action (see Sections 4.6 and 4.7) and offer yet another explanation for the westward drift observed in the geomagnetic secular variation.

Fig. 4.9B shows another azimuthal thermal wind, this one resulting from outer core temperature differences across the *inner core tangent cylinder*, the geostrophic cylinder circumscribing the inner core equator. The situation depicted in Fig. 4.9B corresponds to higher temperatures inside the tangent cylinder compared to immediately outside. The same dynamical effects would result from a slightly higher concentration of light elements inside the tangent cylinder compared to the surrounding outer core (see Chapter 7). In Fig. 4.9B, $\partial T'/\partial \theta$ is negative in the northern hemisphere and positive in the southern hemisphere, so that according to Eq. (4.39), v_ϕ decreases with distance from the equator in both the northern and southern hemisphere tangent cylinder segments. The result is that v_ϕ becomes increasingly negative with distance from the equator, and the strongest retrograde (westward) flow occurs near the polar caps of each hemisphere.

In Chapter 3, we described evidence from the high-latitude geomagnetic secular variation for a westward circulating polar vortex in the northern hemisphere tangent cylinder region of the outer core. The thermal wind in the northern polar cap in Fig. 4.9B is an example of such a polar vortex, as is the thermal wind in the southern polar cap. The thermal wind relation (4.39) predicts patterns of shear (i.e., $\partial v_\phi / \partial \zeta$) that tends to produce prograde (eastward) motion with increasing depth in both northern and southern tangent cylinder segments, as depicted in Fig. 4.9B. Prograde circulations at the base of each tangent cylinder segment result in prograde tangential forces on the inner core boundary, and would tend to drag the inner core into superrotation. The seismic evidence for a steady-state superrotation of the inner core, once strongly in favor, is now more doubtful (see Chapter 6). Nevertheless, thermal winds inside the tangent cylinder offer a simple mechanism for producing time-variable anomalous rotations of the inner core.

We can estimate the magnitude of the lateral temperature differences needed to support thermal winds in the outer core. Replacing the derivatives in Eq. (4.39) with differences and rearranging yields $\Delta T' \simeq 2\Omega r_{CMB}\Delta\theta \Delta v_\phi / \alpha_T g_{CMB} d$. For the tangent cylinder region $\Delta\theta \simeq 0.4$ radians and $\Delta v_\phi \simeq 10^{-4}$ m/s. Using values for the other parameters from Tables S1 and S2 from Core Properties and Parameters yields $\Delta T' \simeq 10^{-6}$ K—a tiny thermal anomaly! In short, miniscule lateral differences in temperature or light element concentration, if coherent on a sufficiently large scale, can drive vigorous thermal wind flows in the outer core.

4.4.6 Reynolds stress

Another forcing mechanism for large-scale laminar flows in the outer core arises from nonlinear interactions between small-scale motions, particularly small-scale, quasigeostrophic convective motions. The most important

FIG. 4.10 Illustration of Reynolds shear stress generation by an array of small-scale convective vortices. σ' is the Reynolds stress, v_0 is an example of the resulting large-scale flow.

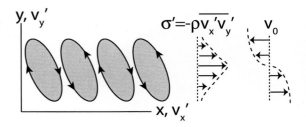

nonlinear interactions of this type involve correlations between the two horizontal components of the small-scale fluid velocity. These correlations generate *Reynolds shear stresses* that can drive large-scale motions.

Consider a horizontal flow in the outer core consisting of motion on two very different length scales, a large-scale velocity field $\mathbf{v}_0$, and superimposed on this a much small-scale velocity field $\mathbf{v}'$, representing quasigeostrophic convection, for example. Averaging these velocities over a length scale intermediate between the length scales of the two flow types implies $\overline{\mathbf{v}}_0 = \mathbf{v}_0$ and $\overline{\mathbf{v}'} = 0$, where the overbar denotes this average. It also implies that products of the large-scale and small-scale velocities average to zero, so that $\overline{\mathbf{v}_0 \cdot \mathbf{v}'} = 0$, for example. The inertial term in the equation of motion (see Appendix 4.1), written in terms of a reference density ρ_0 and these velocities and then averaged this way, yields

$$\overline{\rho_0 \mathbf{v} \cdot \nabla \mathbf{v}} = \rho_0 \mathbf{v}_0 \cdot \nabla \mathbf{v}_0 + \rho_0 \overline{\mathbf{v}' \cdot \nabla \mathbf{v}'}. \tag{4.41}$$

The second term on the r.h.s. of Eq. (4.41) can be written using indicial notation as

$$\rho_0 \overline{v'_j \frac{\partial v'_i}{\partial x_j}} = -\frac{\partial \sigma'_{ij}}{\partial x_j}, \tag{4.42}$$

where σ'_{ij} is the *Reynolds stress tensor*, defined as

$$\sigma'_{ij} = -\rho_0 \overline{v'_i v'_j}. \tag{4.43}$$

Reynolds stresses are present when the components of the small-scale velocity are correlated on average. An example of this correlation is illustrated in Fig. 4.10, which shows a linear array of small-scale vortices, each vortex tilted in the same direction and slightly elongated. The velocity components v'_x and v'_y are inversely correlated, particularly along the line that bisects the vortex array. Because of this inverse correlation, the vortex array generates a positive Reynolds shear stress $\sigma'_{xy} = -\rho_0 \overline{v'_x v'_y} > 0$, as shown in Fig. 4.10. The character of large-scale flows driven by Reynolds stresses depends on the spatial distribution of the Reynolds stress as well as the nature of the resisting force. Fig. 4.10 illustrates a possible large-scale flow v_0 driven by this Reynolds stress.

The atmospheres of the gas-giant planets Jupiter and Saturn have strong azimuthal flow patterns consisting of eastward and westward jets that alternate with latitude. These jets are often attributed to zonal patterns of Reynolds stresses established by small-scale convective vortices. In contrast, flow in the outer core does not show a strong alternating azimuthal jet pattern, except possibly near the inner core tangent cylinder. However, the core flow does include strong azimuthal motions in some regions, such as in parts of the Atlantic hemisphere (see Fig. 3.25 in Chapter 3). Reynolds stresses derived from convective vortices oriented as shown in Fig. 4.10 could provide the force that maintains the preference for westward over eastward azimuthal motion in that region.

4.4.7 Ekman boundary layers

A variety of boundary layers must be present in the outer core, immediately below the core-mantle boundary and above the inner core boundary. Most of these boundary layers are predicted to be very thin, and many are thought to affect outer core flows only locally. Others, however, have global effects on the flow in spite of being very thin. Ekman layers are best example of this latter type. Ekman layers involve a balance between the Coriolis acceleration and viscous shear; they form in order to adjust the velocity and the vorticity of the free stream flow to match the velocity and vorticity of the nearby solid boundary.

Fig. 4.11 shows profiles of the velocity components in an idealized Ekman boundary layer, located between the solid core-mantle boundary and the free stream flow in the northern hemisphere of the outer core. The free stream flow

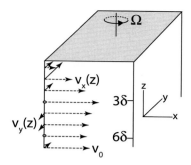

FIG. 4.11 Velocity components versus depth below the core-mantle boundary in a viscous Ekman boundary layer produced by uniform eastward flow v_0 at depth. v_y and v_x are north (y) and east (x) velocity components. Boundary layer thickness scale δ defined in text.

consists of a steady uniform velocity in the x-direction, denoted by v_0 in this example. In the local Cartesian coordinate system shown in Fig. 4.11, the x- and y-components of motion in the Ekman boundary layer are

$$-fv_y = \nu \frac{d^2 v_x}{dz^2} \tag{4.44}$$

and

$$fv_x = \nu \frac{d^2 v_y}{dz^2} + fv_0, \tag{4.45}$$

where ν is kinematic viscosity. Boundary conditions are $v_x = v_0$ and $v_y = 0$ at depth ($-z \to \infty$), plus the no-slip condition $v_x = v_y = 0$ on the solid core-mantle boundary. Solutions to Eqs. (4.44), (4.45) satisfying these conditions are

$$v_x = v_0(1 - \exp(z/\delta)\cos(z/\delta)) \tag{4.46}$$

and

$$v_y = -v_0(\exp(z/\delta)\sin(z/\delta)), \tag{4.47}$$

where $\delta = \sqrt{2\nu/f}$ is the *Ekman boundary layer thickness*. The profiles of v_x and v_y are shown in Fig. 4.11. Because v_x and v_y are out of phase in z, the velocity vector $\mathbf{v} - \mathbf{v}_0$ attenuates and rotates with depth, producing the so-called *Ekman spiral*. In addition, there is a volume transport in the positive y-direction in the Ekman layer, to the left of the primary flow, given by $v_0 \delta/2$.

Larger-scale influences of Ekman boundary layers arise from the effects of *Ekman pumping*. Suppose the magnitude (but not the direction) of the free stream flow v_0 in Fig. 4.11 varies slowly in the y-direction. By "slowly" we mean that $dv_0/dy \ll (dv_x/dz, dv_y/dz)$, so that Eqs. (4.46), (4.47) remain valid locally. Substituting Eqs. (4.46), (4.47) into the continuity equation $\nabla \cdot \mathbf{v} = 0$ (see Appendix 4.2), integrating downward from the core-mantle boundary and using the local f-plane representation yields an expression for the radial velocity induced at depth:

$$v_r = \frac{\delta}{2} \frac{dv_0}{dy}, \tag{4.48}$$

in which the Ekman boundary layer thickness is given by $\delta = \sqrt{2\nu/f_0}$.

Fig. 4.12 shows Ekman pumping below the core-mantle boundary associated with geostrophic flow around circular high- and low-pressure centers in the northern hemisphere of the outer core. The primary geostrophic circulations are depicted with large velocity arrows, the secondary Ekman circulations with small velocity arrows. Around the low-pressure center, for example, the primary flow is a clockwise circulation, while the secondary transport in the Ekman boundary layer is to the left of that motion, that is, inward toward the low-pressure center. The convergence of the Ekman boundary layer flow is balanced by downward motion beneath the low pressure, with radial velocity $v_r < 0$ as shown in Fig. 4.12. The resulting flow structure is called a *cyclone* and its primary geostrophic motion is *cyclonic*. Around the high-pressure center in Fig. 4.12, the flow directions and the pumping actions are reversed, and the radial motion induced at depth is upward, that is, $v_r > 0$. This flow structure is called an *anticyclone* and its primary geostrophic motion is *anticyclonic*. Around high- and low-pressure centers in the southern hemisphere, the directions of the primary circulations in Fig. 4.12 are reversed, but the nomenclature (cyclone around low pressure and anticyclone around high pressure) and the directions of the secondary flows remain the same.

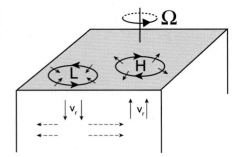

FIG. 4.12 The Ekman pump. Primary geostrophic flows around low-pressure cyclone and high-pressure anticyclone indicated by *large arrows*. Secondary transports due to Ekman boundary layers indicated by *small arrows*. Upwellings and downwellings denoted by v_r and indicated by *vertical arrows*. Deep lateral transport indicated by *horizontal arrows with dashed lines*.

Another consequence of Ekman pumping is horizontal mass transport at depth below cyclones and anticyclones. The direction of this transport, from beneath a cyclone to beneath an adjacent anticyclone, is indicated by the dashed arrows in Fig. 4.12. One effect of this deep transport is to even out irregularities in outer core angular velocity, by transporting high angular velocity fluid from the cyclone to the anticyclone, where the angular velocity of the fluid is lower.

Ekman pumping and deep mass transport also affect the magnetic field structure in the outer core, as sketched in Fig. 4.13. Ekman boundary layer convergence and downwelling tend to concentrate magnetic flux on the core-mantle boundary into cyclones, producing patches of intense magnetic field there. Conversely, Ekman boundary layer divergence and upwelling tend to disperse core-mantle boundary magnetic flux away from anticyclones, producing patches of low-intensity magnetic field at such places. The situation is reversed at depth beneath the core-mantle boundary, where the deep transport disperses magnetic flux from beneath cyclones and concentrates magnetic flux beneath anticyclones.

It is tempting to associate high and low magnetic flux concentrations seen in the core field with the Ekman boundary layer pumping just described. However, other types of core flows can produce similar magnetic field effects. In particular, quasigeostrophic convection involves similar types of motions and produces similar magnetic flux concentrations. In Section 4.6, we show that quasigeostrophic convection induces secondary downwellings and upwellings with transport characteristics that are similar to those of Ekman pumping and are likely to be far stronger in the outer core. Indeed, it remains unclear how significant Ekman transport is in the outer core. Applying the property values in Tables S1 and S2 from Core Properties and Parameters, the outer edge of a laminar Ekman layer measured at a depth of 6δ, is predicted to lie only 1–10 m below the core-mantle boundary. Such a thin boundary layer implies exceedingly weak secondary flows and transports. Possible caveats, which we have not examined here, include the action of turbulence in the Ekman boundary layer or a much larger kinematic viscosity for the outer core fluid (see Chapters 1 and 2), both of which would enhance the strength of Ekman pumping. But regardless of their pumping strength, the kinematic effect of Ekman boundary layers illustrated in Figs. 4.12 and 4.13 serves as a useful guide for visualizing the relationship between a variety of quasigeostrophic core flows and the structure of the geomagnetic field they produce.

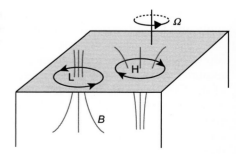

FIG. 4.13 Effects of the Ekman pumping in Fig. 4.12 on magnetic fields. On the core-mantle boundary, magnetic flux (denoted by B) is concentrated in low-pressure cyclones (labeled L) and dispersed from high-pressure anticyclones (labeled H). At depth, the magnetic flux is dispersed below cyclones and concentrated below anticyclones.

4.5 Waves in the outer core

A wide range of unsteady, wave-like motions are likely present in the outer core because the fluid is subject to so many types of restoring forces. In addition to the elastic bulk modulus that supports seismic P-waves (Chapter 1), outer core restoring forces include buoyancy in stably stratified regions, the Lorentz force from the magnetic field, and also the Coriolis acceleration, which can act as a restoring force. Some of the wave types supported by this set of restoring forces have been observed in geomagnetic field behavior, while others have yet to be detected. However, numerical and laboratory simulations of outer core dynamics strongly implicate these wave types. Likewise, their ubiquitous presence in the other geophysical fluid systems—the atmosphere, the ocean, and the magnetosphere—suggests that they are present in the outer core as well. Table 4.4 summarizes the main properties of the wave motions described in this section.

TABLE 4.4 Outer core waves.

Wave type	Restoring force(s)	Period range	Observable
P-wave	Elastic	0.1–20 s	Seismic
Normal modes	Elastic	20–1200 s	Seismic
Alfvén	Magnetic	<10 years	Torsional oscillations
Inertial	Coriolis	>12 h	Tidal amplitudes
Internal gravity	Buoyancy	<24 h	Time-dependent gravity
Rossby	Coriolis, topography	~1 year	Geomagnetic secular variation
MAC	Magnetic, buoyancy, Coriolis	60–70 years	LOD, geomagnetic secular variation

4.5.1 Alfvén waves

As discussed in the previous section in connection with magnetic winds, the geomagnetic field imparts a small rigidity to the outer core fluid due to magnetic field line tension. When distorted by fluid motions, bent magnetic field lines apply a transverse restoring force to the fluid, through the tension part of the Maxwell stress tensor (defined in Appendix 4.4). The action of this restoring force is to straighten the magnetic field lines, in a manner analogous to the action of elastic strings. The inertial response of the fluid to this restoring force consists of both propagating and standing transverse waves. The propagating forms are called *Alfvén waves* in honor of their 1942 discovery by H. Alfvén. Their standing forms in the outer core are referred to as *torsional oscillations*.

Transverse Alfvén waves involve a balance between fluid inertia and the tension part of the Lorentz force, given by

$$\rho_0 \frac{\partial \mathbf{v}}{\partial t} = \frac{1}{\mu_0}(\mathbf{B} \cdot \nabla)\mathbf{B}. \tag{4.49}$$

To describe simple transverse Alfvén waves, we represent the magnetic field as the sum of a uniform background part $\mathbf{B_0}$ plus a wave part $\mathbf{B}'$, while the fluid velocity is represented only with a wave part $\mathbf{v}$. Assuming the wave part of the magnetic field is small in amplitude compared to the background part, Eq. (4.49) reduces to

$$\rho_0 \frac{\partial \mathbf{v}}{\partial t} = \frac{1}{\mu_0}(\mathbf{B_0} \cdot \nabla)\mathbf{B}', \tag{4.50}$$

while in the frozen flux limit the induction equation from Chapter 3 reduces to

$$\frac{\partial \mathbf{B}'}{\partial t} = (\mathbf{B_0} \cdot \nabla)\mathbf{v}. \tag{4.51}$$

Combining Eqs. (4.50), (4.51) to eliminate the fluid velocity yields the Alfvén wave equation

$$\frac{\partial^2 \mathbf{B}'}{\partial t^2} = c_{alfv}^2 \nabla^2 \mathbf{B}', \tag{4.52}$$

where

$$c_{alfv} = B_0/\sqrt{\rho_0 \mu_0} \tag{4.53}$$

is the Alfvén wave propagation speed. In deriving Eq. (4.52), we have assumed that the wave perturbations are perpendicular to $\mathbf{B_0}$, with the fluid velocity and the magnetic field disturbances polarized in the same plane. Fig. 4.14 shows the relationship between the fluid velocity and magnetic field lines in an Alfvén wave packet.

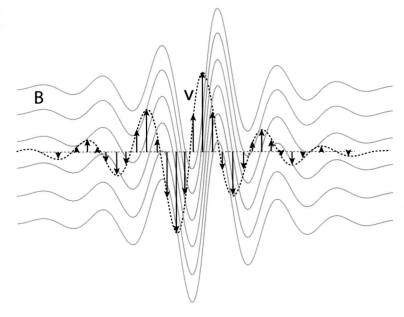

FIG. 4.14 Sketch of an Alfvén wave packet. Magnetic field B-lines in *solid gray curves*, fluid velocity v shown by *arrows* and *dashed curve*.

Alfvén wave propagation speeds in the outer core are rather low. Using $B_0 = 3$ mT as a representative intensity of the radial component of the geomagnetic field in the outer core, Eq. (4.53) yields $c_{alfv} \simeq 2.6 \times 10^{-2}$ m/s. At this speed, it takes an Alfvén wave about two-and-one-half years to propagate from near the inner core boundary to near the core-mantle boundary. Nevertheless, this travel time is short compared to the transport time associated with outer core convection, which is of order 100–200 years along the same path. With outer core travel times on the order of a few years, Alfvén waves are viable candidates for some of the short-term variability seen in the core field, such as magnetic jerks, torsional oscillations, and other sudden changes in geomagnetic intensity or direction.

4.5.2 Torsional oscillations

The slow propagation speed of Alfvén waves in the outer core, just a few centimeters per second, means that the rigidity imparted to the outer core fluid by the geomagnetic field is rather weak. Even so, this magnetic rigidity is strong enough to support global-scale toroidal-type oscillations in the outer core with periodicities shorter than a decade, called *torsional oscillations*.

In the idealized forms shown in Fig. 4.15, torsional oscillations consist of time-variable azimuthal flow, uniform on geostrophic cylinders. The normal modes of torsional oscillations are standing waves between the inner core tangent cylinder and the core-mantle boundary, with a periodicity that depends on the wave number of the standing wave and the Alfvén wave speed. The restoring force for these oscillations comes from the Lorentz force, specifically, the bending of the component of the geomagnetic field $\mathbf{B_0}$ that is normal to the geostrophic cylinders. Therefore, measurements of torsional normal-mode periods offer a means for determining the total geomagnetic flux through each geostrophic cylinder (Wicht and Christensen, 2010).

For example, consider one-dimensional (1D) normal-mode torsional oscillations with variation along the s-direction and subject to a uniform background magnetic field with intensity B_0, also in the s-direction. At $s = 0$, representing the inner core tangent cylinder, we specify that the wave part of the magnetic field satisfies $\partial B'/\partial s = 0$, and at $s = d$, representing the core-mantle boundary, we set $B' = 0$, simulating the vanishing of toroidal magnetic field in the poorly conducting mantle. Solutions to Eq. (4.52) satisfying these boundary conditions have the form

$$B' \propto \cos(k_n s) \sin(\omega_n t) \tag{4.54}$$

with wave numbers given by $k_n = n\pi/2d$ for mode numbers $n = 1, 3, 5, \ldots$ and angular frequency given by $\omega_n = k_n c_{alfv}$. Combining these with Eq. (4.53) and solving for the background magnetic field intensity yields

$$B_0 = \frac{4d\sqrt{\rho_0 \mu_0}}{n t_n}, \qquad (4.55)$$

where $t_n = 2\pi/\omega_n$ denotes the period of the nth normal mode.

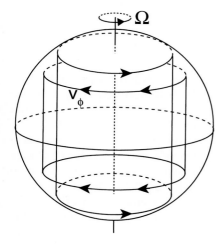

FIG. 4.15 Sketch of torsional oscillations on concentric cylinders in the outer core. v_ϕ is azumuthal fluid velocity indicted by *arrows*; Ω indicates angular velocity of rotation.

Evidence for normal-mode torsional oscillations with periodicity $t_n \simeq 6$ years has been found by Gillet et al. (2010) in both the geomagnetic secular variation and length-of-day records. Using this periodicity in Eq. (4.55) yields $B_0 = 5.3$ mT for the fundamental $n = 1$ mode, and 1.8 mT for the first overtone $n = 3$ mode. More complete modeling by Gillet et al. (2010) that includes spherical geometry of the core plus radial variations of the geomagnetic field in the outer core yields $B_0 \simeq 3$ mT for the r.m.s. intensity of the s-component of the magnetic field. This converts to an r.m.s. magnetic field intensity of about 5 mT in the outer core. We note that this intensity is in basic agreement with an independent estimate by Buffett (2010) based on Ohmic dissipation of the free nutation of the core.

It is worth pointing out that the 60–70-year periodicities seen in the length-of-day record and in the geomagnetic secular variation (Fig. 3.26 of Chapter 3) had previously been interpreted as torsional oscillations. Applied to Eq. (4.55), oscillations in this period range would imply r.m.s. magnetic field intensities in the outer core of only 0.3–0.6 mT, barely different from the r.m.s. intensity of the observable part of the geomagnetic field on the core-mantle boundary. This had long been a problem for dynamo theory, which predicts the r.m.s. magnetic field inside the core interior is a factor of 10 or more stronger than on the core-mantle boundary. Now, it appears that these 60–70-year periodicities are probably not torsional oscillations, but instead may be due to a complex wave type called a MAC wave, propagating in the E' layer beneath the core-mantle boundary. The properties of MAC waves are described later in this section.

4.5.3 Inertial waves

Inertial waves are probably the most common dynamical mechanism by which a rotating flow deviates from the strict constraints of geostrophy. Inertial wave motion often dominates the spectrum of velocity fluctuations in laboratory experiments and in many places in the ocean and atmosphere. Because rotational effects are so strong in the outer core, inertial waves are expected to be present there as well.

Inertial waves use the Coriolis acceleration as a restoring force. To show this, we modify the geostrophic balance from Section 4.4, adding the local fluid acceleration. The resulting equation of motion is

$$\rho_0 \left(\frac{\partial \mathbf{v}}{\partial t} + 2\boldsymbol{\Omega} \times \mathbf{v} \right) = -\nabla P'. \qquad (4.56)$$

Combining the curl and the double curl of Eq. (4.56) yields

$$\frac{\partial^2}{\partial t^2}\nabla^2\mathbf{v} + 4\Omega^2\frac{\partial^2\mathbf{v}}{\partial \zeta^2} = 0, \qquad (4.57)$$

which is known as the *Poincaré equation*, first derived by H. Poincaré in 1910. Propagating plane waves of the form

$$\mathbf{v} = v_0\cos(\mathbf{k}\cdot\mathbf{x} - \omega t), \qquad (4.58)$$

where v_0 is the wave amplitude and $\mathbf{k}$ is the wave vector are solutions to Eq. (4.57) provided the wave frequency satisfies

$$\omega = 2\Omega|\cos\phi|, \qquad (4.59)$$

where ϕ is the angle between the wave vector $\mathbf{k}$ and the rotation vector $\mathbf{\Omega}$.

According to Eq. (4.59), inertial wave frequencies vary within the limits $0 < \omega < 2\Omega$. In addition, they exhibit *anisotropic dispersion*, meaning that the wave speed depends on the propagation direction. In their high-frequency limit, for example, the propagation direction, denoted by the orientation of the wave vector $\mathbf{k}$, is parallel or antiparallel to the rotation vector $\mathbf{\Omega}$. In their low-frequency limit, in contrast, their propagation direction is perpendicular to $\mathbf{\Omega}$.

Inertial waves have several properties that make them dynamically significant in the outer core. One such property involves their energy transmission. The group velocity vector of an inertial wave is perpendicular to its phase velocity vector, that is, perpendicular to $\mathbf{k}$. A low-frequency inertial wave in the outer core propagates subparallel to the equator, while it transmits kinetic energy in the direction of the group velocity, that is, nearly parallel to the rotation axis. This property bears directly on the question of how the rigid constraints of geostrophy can be enforced in a fluid body as large as the outer core. Low-frequency inertial wave transmission provides a mechanism for building geostrophic cylinders and elongated columnar vortices, and it also provides a mechanism whereby small irregularities in uniform geostrophic motion can be radiated away, thereby stabilizing these elongated structures.

A second property of inertial waves that make them significant for magnetic field generation is their helicity. From Eq. (4.58), the inertial wave helicity is given by

$$\overline{H} = \overline{(\nabla\times\mathbf{v})\cdot\mathbf{v}} = -k_\Omega v_0^2, \qquad (4.60)$$

where k_Ω is the component of the wave vector parallel to $\mathbf{\Omega}$. According to Eq. (4.60), inertial waves propagating in a generally northward direction (with positive k_Ω) carry negative helicity, whereas those with a generally southward propagation direction (with negative k_Ω) carry positive helicity.

Suppose there exists, on average, a different propagation direction for inertial waves in the northern and southern hemispheres of the outer core. In that situation, each hemisphere will acquire helicity but with a difference in sign, one positive, the other negative. Helicity that is antisymmetric about an equator offers a simple (perhaps overly simple) explanation for the stability of the axial component of the geomagnetic dipole moment. According to this explanation, the stability of the geomagnetic pole is not so much influenced by the stability of the rotation axis as by the stability of the *helicity equator* in the outer core. We shall revisit this question later in this chapter in the context of the helicity produced by outer core convection.

4.5.4 Internal gravity waves

As discussed in Chapters 1 and 2, the Adams-Williamson equation is a commonly used equation of state for interpreting radial density variations in the outer core in terms of seismic properties. In its simple form, it applies to well-mixed layers with uniform composition and uniform entropy, where the radial density gradient is isentropic (or, adiabatic). In such layers, the adiabatic density gradient can be written as given in Chapters 1 and 2 as

$$\frac{d\rho_{ad}}{dr} = \left(\frac{\partial\rho}{\partial P}\right)_S\frac{dP_0}{dr} = -\frac{\rho g}{\phi_S}, \qquad (4.61)$$

where P_0 is hydrostatic pressure,

$$\phi_S = \frac{K_S}{\rho} = V_P^2 - \frac{4}{3}V_S^2 \qquad (4.62)$$

is the seismic parameter, V_P and V_S are the seismic P-wave and S-wave propagation speeds, K_S is the isentropic bulk modulus, and elsewhere the S subscript denotes constant entropy.

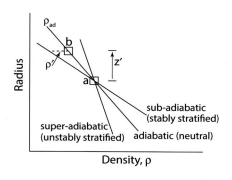

FIG. 4.16 Density perturbation ρ' due to an upward adiabatic displacement z' from radial position a to position b in the presence of a background subadiabatic (stable) radial density profile. ρ_{ad} denotes the adiabatic (neutral) density profile; a superadiabatic (unstable) density profile is shown for comparison.

Even small departures from adiabatic density stratification are dynamically significant in the outer core. For example, consider the buoyancy force exerted on a small fluid parcel displaced vertically from its equilibrium position, as shown in Fig. 4.16. The parcel initially at position a will move along the adiabatic density curve to position b if the displacement is "fast," that is, if the parcel loses no heat and thereby conserves its entropy. Suppose the actual density profile in the outer core is subadiabatic. Then, the displaced parcel will be denser than its surroundings by the amount indicated by ρ' in Fig. 4.16. In the gravity field of the outer core, the buoyancy force on the upward-displaced parcel is downward, that is, stabilizing. Conversely, if the actual density profile in the outer core is superadiabatic, the displaced parcel will be less dense than its surroundings and the buoyancy force on it will be upward, that is, destabilizing.

With reference to Fig. 4.16 and Eqs. (4.61), (4.62), the buoyancy force per unit volume on the parcel in the vertical $\hat{\mathbf{z}}$-direction is given by

$$-\rho' g \hat{\mathbf{z}} = g\left(\frac{d\rho}{dr} - \frac{d\rho_{ad}}{dr}\right) z' \hat{\mathbf{z}} = g\left(\frac{d\rho}{dr} + \frac{\rho g}{\phi_S}\right) z' \hat{\mathbf{z}}, \tag{4.63}$$

where z' is the parcel vertical displacement. If we ignore viscous, rotational, and magnetic effects, the parcel motion is governed by a balance between inertia and this buoyancy force, that is,

$$\frac{d^2 z'}{dt^2} + N^2 z' = 0, \tag{4.64}$$

where

$$N^2 = -\frac{g}{\rho}\left(\frac{d\rho}{dr} + \frac{\rho g}{\phi_S}\right). \tag{4.65}$$

The parameter N has units of 1/s and is called the *buoyancy frequency*. Note that if the density profile is adiabatic, the bracketed term in N^2 consists of the terms in Adams-Williamson equation (4.61) and the buoyancy frequency equals zero.

Plane-wave solutions to Eq. (4.64) are oscillatory for $N^2 > 0$ (stable stratification) and represent horizontally propagating *internal gravity waves* with frequency $\omega = N$ and transverse, vertically polarized particle motion. For an arbitrary propagation direction, generalization of Eq. (4.64) yields $\omega = N \sin\gamma$, where γ is the angle between $\hat{\mathbf{z}}$ (the vertical) and the propagation direction. Therefore, the frequencies of internal gravity waves lie in the range $0 \leq \omega \leq N$. As the wave frequency ω approaches the frequency of rotation Ω, internal gravity waves take on characteristics of inertial waves, and are called *inertial gravity waves*. Like inertial waves, inertial gravity waves possess some helicity, meaning that this type of wave motion can contribute to dynamo action, even in stably stratified portions of the outer core.

An alternative definition of N^2 for thermal stratification involves the difference between the actual temperature gradient in the outer core and the adiabatic temperature gradient, the temperature gradient that corresponds to adiabatic density stratification (see Chapter 1):

$$N^2 = g\alpha_T \left(\frac{dT}{dr} - \frac{dT_{ad}}{dr}\right). \tag{4.66}$$

In the outer core, the bracketed terms in Eqs. (4.65), (4.66) nearly but not exactly cancel, and in those regions $N^2 \neq 0$ over some depth intervals. According to the density and seismic velocity profiles in Fig. 1.4 of Chapter 1, the absolute value of N^2 is generally less than about 10^{-8} rad^2/s^2 through the interior of the outer core. However, according to those profiles, N^2 appears to be slightly negative in some regions. For negative N^2, the solutions to Eq. (4.64) are exponentials in time, implying that this stratification would be dynamically unstable. But because of convective mixing, N^2 cannot be finite and negative over any extended depth range of the outer core. This suggests that the negative N^2 regions in some seismic models of the outer core may not reflect the true structure there. In contrast, the seismic evidence that N^2 is large and positive in the E' layer beneath the core-mantle boundary and in the F layer above the inner core boundary (see Chapter 5) is dynamically consistent with stable stratification in those regions.

4.5.5 Rossby waves

Rossby waves are yet another way in which a rotating fluid overcomes the restrictions of pure geostrophy. They are particularly significant in the outer core because of their role in the onset of quasigeostrophic convection. Here, we examine the simpler case of *barotropic* Rossby waves, as a prelude to the discussion of convective *baroclinic* Rossby waves later in this chapter. For barotropic Rossby wave propagation, density differences are ignored, and the fluid is assumed to be homogeneous.

A convenient starting point is the compound variable *potential vorticity*, commonly used to describe quasigeostrophic flow. For our purposes, we define potential vorticity as

$$Z = \frac{\varpi + 2\Omega}{\rho_0 H}, \tag{4.67}$$

where ϖ is the component of the local fluid vorticity parallel to the rotation vector $\mathbf{\Omega}$ and H is the height of the fluid column with that vorticity. The average core density ρ_0 appears in the denominator of Eq. (4.67) because we are not considering density variations here. The numerator in Eq. (4.67) is the sum of the local fluid vorticity plus the background, or *planetary vorticity* 2Ω. The numerator is, therefore, a measure of the *absolute vorticity* of the fluid, as would be seen from space, for example. Ignoring viscous, buoyancy, and magnetic forces, it can be shown that the potential vorticity is a conserved property of rotating fluid motion, so that

$$\frac{dZ}{dt} = 0 \tag{4.68}$$

for each fluid column.

In Fig. 4.17, two quasigeostrophic columns are shown, one outside the inner core tangent cylinder labeled H_o and located at colatitude θ_o, the other inside the tangent cylinder labeled H_i. Consider how the potential vorticity of each is conserved as these columns move toward or away from the rotation axis. For the column outside the tangent cylinder, northward motion from θ_o increases the column length H, so in order to conserve Z, ϖ must increase, according to Eq. (4.67). Conversely, southward motion of the same column decreases its length, so conservation of Z requires ϖ to decrease in proportion. As the column moves northward, increasingly positive ϖ acts as a restoring mechanism, rotating the column back toward its equilibrium colatitude θ_0. Similarly, as the column moves southward, the increasingly

FIG. 4.17 Geometry for Rossby wave propagation in the parts of the outer core lying outside and inside the inner core tangent cylinder.

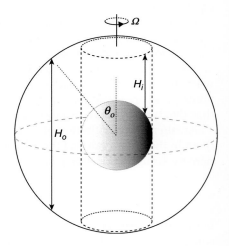

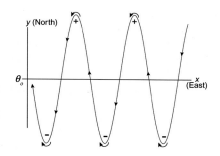

FIG. 4.18 Sketch of a streamline of fluid parcel motion in a Rossby wave superimposed on a steady westward flow. *Plus and minus signs* indicate vorticity perturbations in the wave.

negative ϖ is also restorative, rotating the column back toward its equilibrium colatitude. The resulting motion of the column is, therefore, a periodic oscillation in time and is called a *Rossby wave*.

The kinematics of this action can be visualized by superimposing an azimuthally (east-west) propagating Rossby wave on a steady uniform azimuthal flow, the speed of the steady flow chosen to be equal and opposite to the Rossby wave propagation speed. Under these conditions, the Rossby wave has zero *apparent velocity* and the fluid motion is steady with respect to geographical coordinates and consists of the streamline pattern shown in Fig. 4.18. The restoring vorticity at the Rossby wave crests and troughs is indicated by the plus and minus signs, and the direction of the circulation at those places is indicated by arrows in Fig. 4.18.

To quantify Rossby wave behavior in the outer core, we assume that the local vorticity ϖ is much smaller than the planetary vorticity 2Ω, that changes in column height from wave motion are small compared to the equilibrium height H_0 at colatitude θ_0, and that the north-south velocity v_y is much larger than the east-west velocity v_x in the wave. Under these conditions, Eq. (4.67) reduces to

$$Z \simeq \frac{\varpi + 2\Omega[1 - (y/H_0)dH/dy]}{\rho_0 H_0} \tag{4.69}$$

with $dH/dy = 2\sin\theta_0$ and $\varpi \simeq \partial v_y/\partial x$ outside the tangent cylinder. Substituting Eq. (4.69) into Eq. (4.68) and invoking these simplifications yields

$$\frac{\partial^2 v_y}{\partial x \partial t} = \left(\frac{4\Omega \sin\theta_0}{H_0}\right) v_y. \tag{4.70}$$

Simple plane waves of the form $v_y \propto \sin(kx - \omega t)$ where k is wave number and ω is wave frequency are solutions to Eq. (4.70) provided the phase speed of the Rossby wave satisfies

$$c_{Ro} = \frac{\omega}{k} = \frac{2\Omega \tan\theta_0}{k^2 r_{CMB}}, \tag{4.71}$$

where we have used $H_0 = 2r_{CMB}\cos\theta_0$, the relationship between the height of a geostrophic column outside of the inner core tangent cylinder and the core-mantle boundary radius.

Positive c_{Ro} in Eq. (4.71) signifies that Rossby waves outside the inner core tangent cylinder propagate to the east. According to Eq. (4.71), the propagation speed of a 10^3 km wavelength Rossby wave at 45 degrees colatitude in the outer core is rather fast, $c_{Ro} \simeq 1$ m/s, whereas a 30-km wavelength Rossby wave at the same latitude propagates at about 10^{-3} m/s, quite comparable to core flow speeds inferred from the geomagnetic secular variation (see Chapter 3). Accordingly, we can expect that, when superimposed on generally westward azimuthal core flows, long Rossby waves appear to propagate rapidly to the east, intermediate wavelengths appear to be nearly stationary (i.e., they have either a small or zero apparent velocity) and very short Rossby waves are advected by the flow and have westward apparent velocities.

The above analysis is applicable to Rossby waves outside the inner core tangent cylinder. As shown in Fig. 4.17, inside the tangent cylinder dH/dy is negative because the inner core boundary has greater curvature than the core-mantle boundary. This means that Rossby waves within the inner core tangent cylinder propagate westward, as is the case in Earth's atmosphere. At very high latitudes, dH/dy is numerically small, so that in addition to westward propagation, phase speed of Rossby waves in the polar regions is generally lower than their phase speeds at lower latitudes.

4.5.6 Magnetic Rossby waves

The propagation characteristics of the hydrodynamic Rossby waves described in the previous section change markedly where the Lorentz force is strong. In particular, a strong azimuthal magnetic field may counteract the restoring force from column stretching, altering both the wave frequency and its propagation direction.

In a classic paper, Acheson (1978) derived the properties of slow magnetic Rossby waves, contrasting them with hydrodynamic Rossby waves described above. In a spherical fluid shell containing a uniform azimuthal magnetic field of intensity B_0, the azimuthal propagation speed of a magnetic Rossby wave at colatitude θ_0 is given by

$$c_{mRo} = -\frac{k^2 c_{alfv}^2 r_{CMB}}{2\Omega} \tan\theta_0, \tag{4.72}$$

where c_{alfv} is the Alfvén wave speed given by Eq. (4.53).

Compared to the hydrodynamic Rossby wave propagation speed in Eq. (4.71), the magnetic Rossby wave propagation speed in Eq. (4.72) is far slower, particularly for small wave numbers k, that is, for large wavelengths. In addition, the negative sign in Eq. (4.71) indicates that magnetic Rossby wave propagation is westward in the region outside the tangent cylinder. Taken together, these properties suggest that magnetic Rossby waves may be good candidates for explaining some of the azimuthal propagation behavior observed in the geomagnetic secular variation. Indeed, Finlay and Jackson (2003) have identified similarities between magnetic Rossby wave characteristics and certain features in the secular variation of the core field at low latitudes.

4.5.7 MAC waves

At places in the outer core where the effects of stable stratification, Earth's rotation, and the geomagnetic field act in concert, additional types of wave motions are possible. The best-studied example of these are decadal time-scale *MAC waves*, the abbreviation referring to the action of *M*agnetic, *A*rchimedean (in honor of Greek mathematician Archimedes of Syracuse who formulated the hydrostatic principle, a.k.a., buoyancy), and *C*oriolis effects in these waves.

An idealized model for MAC waves that are broadly applicable to the E' layer of the outer core consists of axisymmetric oscillations in a thin spherical layer with uniformly stable stratification characterized by a constant buoyancy frequency $N > 0$, the layer permeated by a uniform radial magnetic field and rotating at constant angular frequency Ω. S.I. Braginsky, a pioneer in dynamo theory, coined the MAC acronym (Braginsky, 1993). He showed that the frequency of decadal time-scale MAC oscillations in this system is given by

$$\omega_{MAC} = (n(n+1))^{1/2}\left(\frac{c_{alfv} N}{2\Omega r_o}\right), \tag{4.73}$$

where n is the spherical harmonic degree of the oscillation, r_o is the outer radius of the layer, and c_{alfv} is the Alfvén wave propagation speed associated with the uniform radial magnetic field, as given by Eq. (4.53).

Because they involve oscillations of the fluid velocity and axisymmetric part of the geomagnetic field, MAC waves in the E' layer of the outer core are expected to be observable in the geomagnetic secular variation as well as in length-of-day variations and in the outer core flow imaged by frozen flux applied to the geomagnetic secular variation. And because their frequency is proportional to the buoyancy frequency N, they can provide constraints on the stratification in the E' layer.

For example, Buffett (2014) and Buffett et al. (2016) have inferred the strength of the stratification in the E' layer by interpreting the 60–70-year cycles in the excess length-of-day (LOD) and dipole moment fluctuations (Chapter 3) as the results of MAC waves in the outer core. The calculation proceeds by noting that even values of n are capable of producing dipolar and length-of-day variations. Based on the radial part of the present-day core field and Eq. (4.53), the Alfvén wave speed is $c_{alfv} \simeq 4 \times 10^{-3}$ m/s in the E' layer. For $N/2\Omega = 1$, and $n = 2$, Eq. (4.73) predicts $\omega_{MAC} = 2.8 \times 10^{-8}$ rad/s, equivalent to an oscillation period of 71 years. Alternatively, LOD and dipole fluctuation periodicities of 65 years can be matched either by slightly increasing c_{alfv} to 4.4×10^{-3} m/s, or alternatively, by increasing $N/2\Omega$ to 1.1.

In either case, the qualitative implication is the same: assuming the decadal time-scale oscillations are indeed MAC waves, then the E' layer is stably stratified. Quantitatively, however, the inferred stratification is rather weak. Thermal stratification with $N \simeq 2\Omega$ only requires a small deviation from an adiabatic temperature gradient, or alternatively, a very small departure from uniform light element concentration. In the following section, we describe some of the effects this amount of stable stratification is expected to have on outer core convection.

4.6 Outer core convection

Before delving into the subtleties of outer core convection, it is worthwhile to briefly summarize what is known about dynamical conditions in the outer core based on the evidence given in this and other chapters, and to frame the discussion of convection by posing some basic questions. First, we are confident about the following:

- The bulk of the outer core closely adheres to the Adams-Williamson relationship and thus appears to be reasonably well mixed. Boundary regions are exceptions to this rule.
- Pervasive radial fluid velocities are needed for dynamo action. In the outer core, convective motions can provide that radial velocity.
- The geomagnetic secular variation implies fluid velocities of order 10^{-3} m/s in the outer core.
- Heat flux at the core-mantle boundary is probably high, the total heat loss lying in the range 6–18 TW, of the same order as, and possibly larger than, the total heat flow conducted down the outer core adiabat. Core-mantle boundary heat flux is also laterally heterogeneous.
- Cooling of the core produces inner core solidification and light element partitioning into the liquid outer core, adding compositional buoyancy. Overall, this compositional buoyancy is probably stronger than the thermal buoyancy in the outer core.
- Earth's rotation exerts a controlling influence on outer core convection through the Coriolis acceleration. The geomagnetic field also affects convection in the outer core through the Lorentz force, although to a lesser extent.

In light of these, we seek answers to the following questions:

- Are the fluxes of heat and composition large enough to drive convection everywhere in the outer core, overcoming the stabilizing effects of rotation and the geomagnetic field?
- What is the planform of outer core convection? What are its characteristic length and time scales?
- What is the relationship between outer core convection and the core flows imaged by frozen flux applied to the geomagnetic secular variation?
- How does lateral heterogeneity near the core-mantle and inner core boundaries affect outer core convection?
- What properties of convection in the outer core are most important for magnetic field generation?
- What are the connections between outer core convection and geomagnetic polarity reversals?

4.6.1 Onset of outer core convection

In analyzing the onset of convection in the outer core, it is vital to distinguish between the regions inside the inner core tangent cylinder versus the region outside. Inside the tangent cylinder, the gravity vector is nearly aligned with the rotation axis, whereas outside the tangent cylinder the most important component of the gravity vector for dynamics is the component perpendicular to the rotation axis. In addition, the angle between the radial direction **r** (representing the local vertical) and the angular velocity vector Ω is large outside the tangent cylinder, especially at middle and low latitudes. All of these geometrical factors play key roles in the onset of convection in the outer core, because of the strong geostrophic constraints imposed by Earth's rotation.

The term *rotating convection* refers to buoyantly driven convection influenced by solid-body rotation of the fluid, mainly through the Coriolis acceleration. Although it does not include magnetic field effects, rotating convection is nevertheless a meaningful dynamical system for investigating outer core dynamics. Many of the fundamental properties of the flow outside the inner core tangent cylinder can be understood using a simplified model of convection in a rotating sphere that was originated by Busse (1970) and is illustrated in Fig. 4.19.

This model assumes that at the onset of convection, the fluid motion is quasigeostrophic and restricted to a thin, cylindrical annular domain of width d_a, as shown in Fig. 4.19. Within this thin annulus, the spherical core-mantle boundary is modeled in terms of uniformly sloping top and bottom annular caps. The driving force for convection in this model is the component of the buoyancy force perpendicular to the rotation axis, that is, perpendicular to the cylindrical annulus. The combined effects of rotation and the sloping boundary caps of the annulus induce propagating Rossby waves (Section 4.5). According to the theory, convective onset under these conditions consists of columnar baroclinic Rossby waves driven by buoyancy forces directed perpendicular to the rotation axis and resisted by viscous friction between adjacent columns.

The control parameters for thermal convection include the Prandtl number Pr and the Ekman number Ek_a, slightly modified here for the annular geometry in Fig. 4.19:

$$Ek_a = \frac{H\nu}{4\Omega d_a^3 \tan\theta_0}, \tag{4.74}$$

FIG. 4.19 Thin annulus geometry used for onset of convection outside the inner core tangent cylinder. H is the annulus height, d_a is the annulus thickness, and $\mathbf{g}_s$ denotes the component of gravity in the s-direction, perpendicular to the cylindrical annulus.

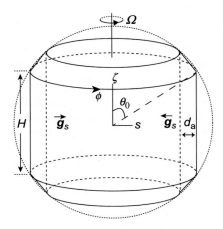

where H is the annulus height, d_a is its width, and $\tan\theta_0$ is the slope of the annular caps. There are two important limiting situations in this geometry. First, if the slopes of the annular caps are dynamically unimportant because the fluid motion does not extend to the core-mantle boundary, then the preferred form of convection consists of two-dimensional (2D) cylindrical columns arrayed around the annulus and aligned with the rotation axis. Because this type of convection is columnar and is restricted to geostrophic contours, the constraints of rotation vanish. In this situation, the critical Rayleigh number for thermal convective onset is the same as for nonrotating (Rayleigh-Bénard) convection. For stress free, isothermal annular boundaries the critical Rayleigh number is given by (Chandrasekhar, 1961)

$$Ra_{crit} = \frac{\alpha_T g_s d_a^3 \Delta T_{crit}}{\kappa_T \nu} = \frac{27\pi^4}{4}, \tag{4.75}$$

where ΔT_{crit} is the critical temperature difference applied across the annulus, g_s is the component of gravity perpendicular to the annulus, and the other symbols have their previous meanings. The wave number of the columns measured in the azimuthal direction (the ϕ-direction in Fig. 4.19) is $k_\phi = \pi/\sqrt{2}d_a$. At convective onset, this flow is steady.

More realistically, if the convection extends to the core-mantle boundary, the slopes of the annular caps in Fig. 4.19 become important. In this situation, the convection is forced to depart from strictly 2D columnar structure. It becomes quasigeostrophic and exhibits Rossby wave propagation. In the regime of rapid rotation ($Ek_a \ll 1$) and for $Pr < 1$, the critical Rayleigh number is approximately

$$Ra_{crit} \simeq \frac{3}{2^{2/3}} \left(\frac{Pr}{Ek_a}\right)^{4/3}, \tag{4.76}$$

while the Rossby wave propagation frequency and azimuthal wave number are approximately

$$\omega \simeq \frac{\nu}{d_a^2} \left(Ek_a^2 Pr\right)^{-1/3}, \quad k_\phi \simeq \frac{1}{d_a}\left(\frac{Pr}{Ek_a}\right)^{1/3}. \tag{4.77}$$

The planform corresponding to Eqs. (4.76), (4.77) is illustrated schematically in Fig. 4.20. These *Busse rolls*, as they are called, propagate in the prograde direction, that is, to the east. Note that the effects of viscosity are felt in the critical Rayleigh number, the propagation frequency (which is low) and in the azimuthal wave number (which is large), even if the Ekman number is very small.

The critical Rayleigh number for Busse rolls (4.76) is enormously increased over columnar Rayleigh-Bénard convection (4.75) because Busse rolls impinge on the spherical core-mantle boundary, and this allows the stabilizing effects of rotation to be felt, via column stretching. According to Eqs. (4.74), (4.76), for an annular gap width of $d_a = 100$ km, a kinematic viscosity of $\nu = 1 \times 10^{-6}$ m/s^2, and a Prandtl number of $Pr = 0.2$, the annular Ekman number for a necklace of Busse rolls arrayed at 30 degrees colatitude is $Ek_a = 3.7 \times 10^{-11}$, and the critical Rayleigh number is $Ra_{crit} \simeq 1.8 \times 10^{13}$, nearly 10 orders of magnitude larger than Eq. (4.75). For the same properties, the frequency and azimuthal wave number of Busse rolls are, from Eq. (4.77), $\omega \simeq 1.5 \times 10^{-9}$ rad/s and $k_\phi \simeq 1.8 \times 10^{-2}$ rad/m, equivalent to a wavelength of about 350 m.

Armed with these results, we are now able to determine the critical superadiabatic temperature gradient needed to initiate Busse roll convection in the outer core. Solving for $\Delta T_{crit}/d_a$ in Eq. (4.76), using the value above for Ra_{crit} along with the thermodynamic and transport properties in Table S2 from Core Properties and Parameters, we find that only

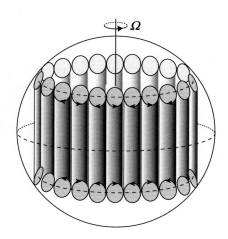

FIG. 4.20 Sketch of Busse rolls, the preferred mode at onset of convection in a rotating cylindrical annulus with sloping end caps. Fit into a spherical shell, Busse rolls are a model for the onset of outer core convection outside of the inner core tangent cylinder.

an imperceptible superadiabatic temperature gradient is needed to initiate thermal convection outside the inner core tangent cylinder, 10^{-10} K/km or less. This is vanishingly small compared to the adiabatic temperature gradients derived in Chapter 2 and given in Table S2 from Core Properties and Parameters, which are in the range of 0.5–1 K/km, and supports our assertion that the temperature gradient in the outer core with convection present is essentially the same as the adiabatic temperature gradient.

All of the previous considerations in this section apply to convection outside the inner core tangent cylinder. Within the inner core tangent cylinder, the gravity vector is nearly aligned with the rotation axis and the effects of boundary slope, while substantial, are less severe than in the region outside. Accordingly, thermal convection in a planar layer with gravity and rotation vectors aligned perpendicular to the layer is a reasonable starting model for convection inside the tangent cylinder.

Rotating convection in a fluid layer heated from below and cooled from above is well studied, experimentally, theoretically, and numerically. The critical Rayleigh number for onset of convection in this geometry always depends on the global Ekman number $Ek = \nu/\Omega d^2$, where d is now the full fluid layer depth. For isothermal stress-free boundaries and stationary onset, the critical Rayleigh number and horizontal wave number in this system are approximately (Chandrasekhar, 1961)

$$Ra_{crit} \simeq 20 Ek^{-4/3}, \quad k \simeq 1.6 d^{-1} Ek^{-1/3}. \tag{4.78}$$

With stationary (rather than wave-like) convective onset, dependence on the Prandtl number does not appear in Eq. (4.78).

Fig. 4.21 shows the path of a fluid parcel as it circulates through a square-shaped *Veronis cell*, the name given to cellular flows in this type of convection, after G. Veronis, who investigated the finite-amplitude properties of these flows. As Fig. 4.21 illustrates, fluid parcels rise and fall under the action of buoyancy forces and simultaneously circulate in a clockwise or counterclockwise sense in response to the Coriolis acceleration. In Veronis cells, the fluid parcel

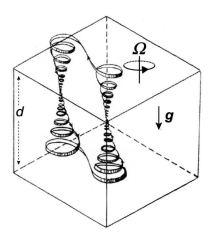

FIG. 4.21 Sketch of streamlines in a Veronis cell, the preferred mode at onset of convection in a rotating plane layer and a model for onset of outer core convection inside the inner core tangent cylinder.

trajectories are helical, and indeed this type of convection has a net helicity, which changes sign from positive to negative on crossing the midplane of the layer.

Veronis cell instability depends on the global Ekman number in the same way as Busse roll instability depends on the annular Ekman number, so in spite of the difference between the Ekman number and Rayleigh number definitions in the two regions, the critical superadiabatic temperature gradient is vanishingly small in the outer core for both types of convection, in comparison with the adiabatic temperature gradient. Because of this, the radially symmetric part of the geothermal gradient in the outer core plays a secondary role in determining which region—inside or outside the inner core tangent cylinder—is more prone to thermal convection. Instead, it is the *lateral variations* in temperature and heat flux that primarily determine which region is more convection prone. Evidence of outer core heterogeneity near the core-mantle boundary and the inner core boundary is described in Chapter 5. Later in this chapter, we examine how this boundary lateral heterogeneity affects the structure of convection in the outer core.

In summary, Busse rolls and Veronis cells provide logical starting points for understanding outer core convection outside and inside the tangent cylinder, respectively. Nevertheless, the character of each type of convection raises some thorny questions. For example, extreme elongation of the convection structures is predicted, both inside and outside the inner core tangent cylinder. According to the formulas in this section, with outer core properties the predicted length-to-width ratio of Busse rolls and Veronis cells are of order $kd \simeq 10^4$. This aspect ratio is geometrically similar to a long strand of spider web. The dynamical stability of such extremely elongated convection cells in the outer core is highly debatable, and suggests that Busse rolls and Veronis cells would readily be replaced by more equidimensional, dynamically stable structures.

Other questions concern the capacity of these flows to induce dynamo action. Near the critical Rayleigh number, the Busse roll structure contains very little helicity, and the Veronis cell helicity, although large, changes sign with depth such that it is zero in volume average. Accordingly, we need to find out if either type of convection contributes to the geodynamo. And lastly, we want to know how efficiently these forms of convection transport heat and composition so as to fit within the energy constraints on the core outlined in Chapter 2. In order to address these issues, it is necessary to examine rotating convection in its fully developed nonlinear form, far beyond the critical Rayleigh number. For that, we turn to laboratory and numerical models.

4.6.2 Fully developed convection in the outer core

Although theory is useful for determining the conditions at convective onset, laboratory and numerical experiments are the primary tools for exploring the properties of fully developed outer core convection, in which the Rayleigh number is far beyond critical and the dynamics are nonlinear. Laboratory experiments naturally include the instabilities and scale separations that rotating fluid systems are famous for, and numerical experiments are agile tools for investigating the effects of different physical properties and boundary conditions. Both approaches, however, suffer from the same limitation, that is, reaching the extreme values of key dimensionless parameters that characterize convection in the outer core, particularly the vanishingly small Ekman number and the astronomically large Rayleigh number. Yet both of these tools have proven to be of enormous value, for multiple reasons. First, they provide model images of what the convection inside the core may look like, images that cannot be obtained other ways. Second, they include strong nonlinear interactions between the various forces, which neither theory nor observations have been able to capture. Third, they generate a statistical database of results that can be extrapolated to outer core conditions. As for the choice of laboratory fluids, experiments using water are easiest for flow visualization, whereas experiments using liquid metals such as gallium alloys and molten sodium are more realistic in terms of the Prandtl numbers. For numerical experimentation, most geophysicists now use community-supported codes. Links to some of these codes are given in the Resource listing at the end of this chapter.

Here, we follow the organization in our discussion of convective onset, initially focusing on the flow structures of fully developed rotating convection inside and outside the inner core tangent cylinder, in that order. Fig. 4.22 shows the flow regimes of rotating thermal convection in a tall cylinder, from experiments by Cheng et al. (2015). A tall, right cylinder represents a geometrical improvement over the basic plane-layer model for convection inside the tangent cylinder. As seen in Fig. 4.22, with increasing Rayleigh number Ra_T the axial structure of the convection evolves from laminar columns (actually, these are very elongated Veronis cells) to a wavy time-dependent form of quasigeostrophic convection in which the time dependence is continuously excited by rising plumes, and finally to what is called *quasigeostrophic turbulence*, highly time-dependent convection consisting of bundles of rising and sinking fluid filaments, all of which are elongated in the direction of the rotation axis.

Which of these flow regimes most closely matches conditions within the inner core tangent cylinder? The answer to this question and others like it are based on extrapolating laboratory and numerical results to the outer core, following some arbitrarily selected path in Fig. 4.3. Clearly, there is an infinity of possible paths, so there is a lot of room for

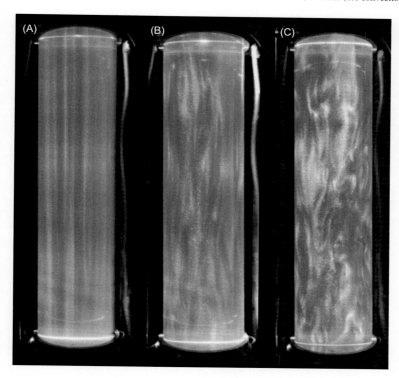

FIG. 4.22 Thermal convection experiments in a rotating right cylinder heated from below and cooled from above by Cheng et al. (2015). Convective regimes illustrated using flake visualization in water. (A) Laminar columns at $Ek = 1 \times 10^{-7}$ and $Ra/Ra_{crit} = 3.8$; (B) quasigeostrophic plume convection at $Ek = 8 \times 10^{-7}$ and $Ra/Ra_{crit} = 23$; and (C) quasigeostrophic turbulence at $Ek = 2 \times 10^{-6}$ and $Ra/Ra_{crit} = 77$. Reproduced from Fig. 7 of Aurnou, J.M., Calkins, M.A., Cheng, J.S., Julien, K., King, E.M., Nieves, D., Soderlund, K.M., Stellmach, S., 2015. Rotating convective turbulence in Earth and planetary cores. Phys. Earth Planet. Inter. 246, 52–71.

uncertainty here. Nevertheless, an extrapolation of laboratory convection experiments similar to Fig. 4.22 to outer core conditions by Aurnou et al. (2015) predicts that if $Ra_T/Ra_{crit} > 30$ inside the inner core tangent cylinder, the convection in that part of the outer core is predicted to be in the regime of quasigeostrophic turbulence. Is it likely that the Rayleigh number inside the tangent cylinder is more than 30 times critical? Assuming $Ek = 3 \times 10^{-15}$, Eq. (4.78) yields $Ra_{crit} \sim 5 \times 10^{20}$, so that Ra_T need only be 1.5×10^{22} or larger. According to Table 4.2, Ra_T may well be several orders of magnitude larger than this. However, this assumes the adiabatic temperature gradient in the outer core is not too large. Indeed, Ra_T might actually be negative if the adiabatic temperature gradient is toward the upper end of the range given in Table S2 from Core Properties and Parameters.

Laboratory experiments on convection outside the tangent cylinder exploit a technique developed by Carrigan and Busse (1983) in which a rapidly rotating spherical shell or hemispherical shell is heated at the outer boundary and cooled at the inner boundary, such that the combination of laboratory gravity, centrifugal acceleration from the rotation, and the applied thermal gradient produce radially outward buoyancy forces in the fluid shell that approximate the orientation of the buoyancy forces at low latitudes in the outer core.

Fig. 4.23 shows experimental results of this technique applied to thermal convection in a rotating water-filled spherical shell with an inner to outer radius ratio r_i/r_o similar to the outer core (Cardin and Olson, 1994). Near Ra_{crit}, the motion consists of a periodic array of Busse rolls arrayed around the inner tangent cylinder and aligned with the rotation axis, as predicted by the theory in the previous section and illustrated schematically in Fig. 4.20. These Busse rolls,

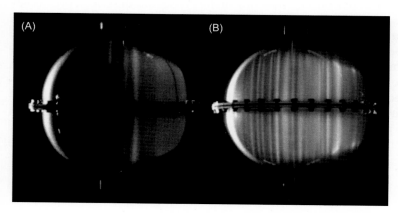

FIG. 4.23 Thermal convection experiments in a water-filled rotating spherical shell cooled at the inner sphere and heated at the outer sphere with centrifugal gravity. (A) Laminar quasigeostrophic convection at $Ek = 2 \times 10^{-6}$ and $Ra \simeq Ra_{crit}$ and (B) turbulent quasigeostrophic convection at $Ek = 2 \times 10^{-6}$ and $Ra/Ra_{crit} \simeq 50$. Reproduced from Fig. 4 of Cardin, P., Olson, P., 1994. Chaotic thermal convection in a rapidly rotating spherical shell: consequences for flow in the outer core. Phys. Earth Planet. Inter. 82, 235–259.

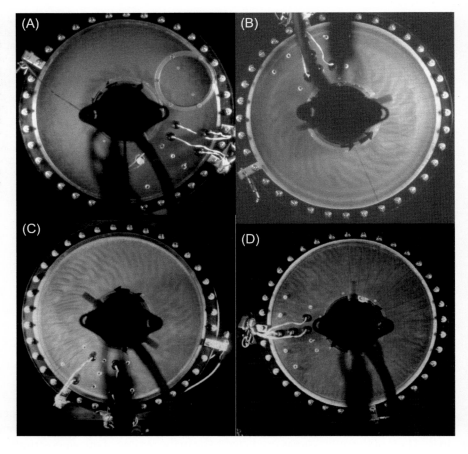

FIG. 4.24 Thermal convection experiments in water in a rotating water-filled hemispherical shell cooled from the inside and heated from the outside with centrifugal gravity at $Ek = 4.7 \times 10^{-6}$. Equatorial planforms illustrated using flake visualization. (A) Laminar columns with spiral cross sections arrayed around the inner sphere tangent cylinder at $Ra \simeq Ra_{crit}$. (B, C) Laminar penetrative convection planforms at $Ra \simeq 6Ra_{crit}$ in growth phase (B) and fully developed in (C). (D) Dual convection at $Ra \simeq 45Ra_{crit}$. Reproduced from Figs. 3, 4, and 6 of Sumita, I., Olson, P., 2000. Laboratory experiments on high Rayleigh number thermal convection in a rapidly rotating hemispherical shell. Phys. Earth Planet. Inter. 117, 153–170.

revealed by light-reflecting flakes in Fig. 4.23A, are narrow columns extending across the entire spherical shell. They propagate in the prograde (eastward) direction, as predicted for thermal Rossby waves. At Rayleigh numbers far beyond Ra_{crit}, the periodic Busse rolls are replaced by time-dependent convection in the form of numerous small-scale quasigeostrophic vortices irregularly distributed throughout the spherical shell, each vortex carrying a column of fluid with it. Although the flow appears turbulent when viewed in planes perpendicular to the rotation axis, it is in fact quasigeostrophic and nearly uniform in the direction of the rotation axis, as revealed by Fig. 4.23B.

Fig. 4.24 shows equatorial plane planforms of fully developed thermal convection in a rotating water-filled hemispherical shell with an outer core radius ratio. With this "lower hemisphere" geometry, a rotation rate can be chosen so that the resultant gravity, the vector sum of laboratory gravity and the centrifugal force due to the apparatus rotation, is nearly perpendicular to the outer hemisphere boundary. By reversing the thermal gradient (heating the outer boundary and cooling the inner boundary), buoyancy forces oriented similar to the outer core are produced and the effects of spurious thermal wind flows are minimized (Sumita and Olson, 2000).

Fig. 4.24A shows the equatorial planform near convective onset, consisting of the Busse rolls described previously and illustrated schematically in Fig. 4.20. In cross section, the Busse rolls in Fig. 4.24A are elongated in the meridional direction and slightly tilted in the prograde azimuthal direction. Fig. 4.24B–D shows two additional flow regimes at Rayleigh numbers beyond critical: the *penetrative convection* regime for $1 < Ra/Ra_{crit} < 8$ and the *dual convective* regime for $Ra/Ra_{crit} > 8$. By virtue of its strong temporal and spatial variability, the dual convective regime is a form of quasigeostrophic turbulence. In the penetrative regime, the convection consists of a nearly periodic array of elongated quasigeostrophic vortices, each vortex tilted in the prograde azimuthal direction so as to form a spiral planform. The vortex array is driven by buoyant plumes originating at the inner boundary with the spacing between plumes equal to one vortex pair. In Fig. 4.24B, the penetrative convection stops short of the outer boundary equator. Between the penetrative convection and the outer boundary is a thin, convectively stable layer containing large-scale

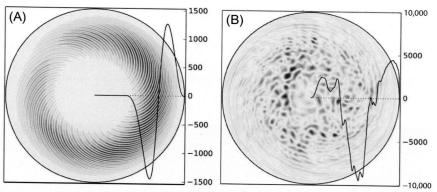

FIG. 4.25 Snapshots of streamfunction with overprinted profile of azimuthal velocity (in Re units) showing spiral cross section structure during development phase (A) and dual convection during fully developed phase (B), from calculations of quasigeostrophic (ζ-integrated) thermal convection in a rotating sphere at $Ek = 1 \times 10^{-8}$ and $Pr = 10^{-2}$ and $Ra/Ra_{crit} = 1.06$. *Reproduced from Fig. 4 of Guervilly, C., Cardin, P., 2016. Subcritical convection of liquid metals in a rotating sphere using a quasigeostrophic model. J. Fluid Mech. 808, 61–89.*

shear flows, probably thermal winds (Section 4.4). In Fig. 4.24C, the penetrative convection extends to the outer boundary equator, eliminating the convectively stable layer. At larger Rayleigh numbers, the penetrative convective regime is replaced by the dual convective regime in Fig. 4.24D, so named because it is driven by plumes from the inner boundary, and in addition, small-scale plumes originating at the outer boundary. The mismatch in scale between these two plume sets destroys the periodicity of each set, producing quasigeostrophic turbulence with an irregular, strongly time-variable planform that nevertheless retains its elongated columnar structure in the ζ-direction.

The transition from a periodic array of spiraling vortices to quasigeostrophic turbulence occurs increasingly close to convective onset as the Prandtl number is decreased from $Pr = 7$ in the water experiments just described to the lower Prandtl number value of the outer core. Fig. 4.25 shows snapshots of the streamfunction color coded by the sign of vorticity with superimposed azimuthal velocity profiles from a high-resolution numerical model of rotating thermal convection in a liquid sphere with $Pr = 0.01$ and $Ra/Ra_{crit} = 1.06$, by Guervilly and Cardin (2016). Their numerical model enforces quasigeostrophy by solving the ζ-integrated equations of motion, rather than the fully 3D equations of motion, in the region outside the inner core tangent cylinder. This partial integration reduces the computational cost and allows access to more extreme values of the dimensionless control parameters. During its development phase, shown in Fig. 4.25A, the convection consists of a periodic prograde spiral of thermal Rossby waves seen in Fig. 4.24A, but over time this planform is replaced by dual planform, quasigeostrophic turbulence similar to Fig. 4.24B. Note that both planforms in Fig. 4.25 are associated with a strong azimuthal shear flows, directed in the prograde (eastward) direction near the outer boundary and in the retrograde (westward) direction closer to the rotation axis. These azimuthal shear flows are driven by Reynolds stresses derived from the tilt of the convective vortices, in the manner illustrated in Fig. 4.10.

Up to this point, we have made emphasized the distinctions between convection inside and outside the inner core tangent cylinder. It has been found that fully 3D numerical models of outer core convection reflect these distinctions to a large extent, while allowing the two regions to interact to some extent. Fig. 4.26 shows two visualizations of temperature perturbations from a high-resolution numerical model of fully 3D thermal convection in a rotating spherical shell by Gastine et al. (2016), where the competing effects of buoyancy and rotation can be seen inside and outside the tangent cylinder. For the case shown in Fig. 4.26B with $Ek = 1 \times 10^{-6}$ and $Ra_T/Ra_{crit} \simeq 16$, the effects of rotation on the flow pattern outweigh the effects of buoyancy, and the temperature structure, although highly irregular in the equatorial plane, remains strongly columnar inside and outside the tangent cylinder. The convection in this case is quasigeostrophic. Note the abrupt change in scale of the temperature fluctuations on the spherical surface as the tangent cylinder boundary is crossed. For the case shown in Fig. 4.26A with $Ek = 1 \times 10^{-5}$ and $Ra/Ra_{crit} \simeq 110$, the effects of buoyancy are substantially stronger than in Fig. 4.26B. In this case, the columnar structure of the temperature perturbations is broken, and although these structures remains highly elongated both inside and outside the tangent cylinder, they are no longer coherent across the entire fluid shell. In this case, the convection is in the geostrophic turbulence regime.

For the purposes of extrapolating their numerical results to outer core conditions, Gastine et al. (2016) found that the r.m.s. convective velocity scales as predicted for a balance between the fluid inertia, the Coriolis acceleration, and the buoyancy force, the so-called *CIA* force balance. In particular, they found

$$v_{rms} \simeq \frac{\nu}{2d} Ra_q^{2/5} Pr^{-4/5} Ek^{1/5}, \tag{4.79}$$

where Ra_q is the Rayleigh number based on the superadiabatic heat flux, defined in Table 4.2. Using $Ra_q = 10^{29}$, $Ek = 3 \times 10^{-15}$, $Pr = 0.2$, $d = 2.26 \times 10^6$, and $\nu = 10^{-6}$ as before, Eq. (4.79) predicts $v_{rms} \simeq 4 \times 10^{-4}$ m/s, comparable to what is inferred in Chapter 3 for outer core fluid velocities from frozen flux analyses of the geomagnetic secular variation.

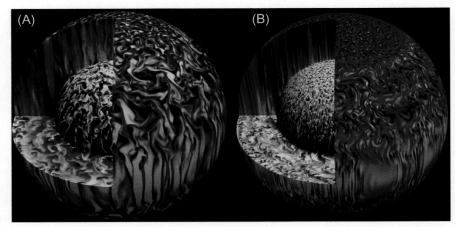

FIG. 4.26 Meridional sections, equatorial cut, and radial surfaces showing snapshots of temperature fluctuations from numerical calculations of rotating convection in a spherical shell (*white = hot; black = cool*). (A) Transitional regime at $Pr = 1$, $Ek = 10^{-5}$; $Ra/Ra_{crit} = 110$; (B) rapidly rotating regime at $Pr = 1$, $Ek = 10^{-6}$, and $Ra/Ra_{crit} = 16$. *Modified from Fig. 2 of Gastine, T., Wicht, J., Aubert, J., 2016. Scaling regimes in spherical shell rotating convection. J. Fluid Mech. 808, 690–732.*

4.6.3 Mantle heterogeneity effects on outer core convection

The so-called D″ region, the several hundred-kilometer-thick region at the base of the lower mantle above the core-mantle boundary, is among the most heterogeneous regions in the Earth, having lateral variations in seismic properties larger than anywhere else in the lower mantle (see Chapter 5). Furthermore, this region is dynamically active. Its long-wavelength lateral structure includes the two *large low seismic velocity provinces* (*LLSVPs*) separated by what appear to be graveyards of subducted slabs, all of which are expression of the large-scale pattern of mantle convection. Closer to the core-mantle boundary are the smaller and more isolated *ultra-low velocity zones* (*ULVZs*) and the possibility of a pressure-induced mineralogical transformation from the perovskite (a.k.a. bridgemanite) structure to a postperovskite structure. When all of these sources of heterogeneity are coupled with the thermal regime of the lower mantle, which includes a basal thermal boundary layer with a superadiabatic temperature variation in excess of 1000 K, the inevitable result is a very large lateral variation in the local heat flux over the core-mantle boundary. Various estimates place the peak-to-peak amplitude of the core-mantle boundary heat flux variation comparable to its spherical mean value, in which case the standard deviation of the local heat flux on the core-mantle boundary is large, in the range of 30%–60% of its spherical mean value.

Because the D″ region is so heterogeneous in terms of both temperature and composition, it is important to delineate the effects of heterogeneous boundary heat flux on outer core convection, to determine whether any of these effects are observable, and how such observations might be used to further constrain the thermal structure of the core-mantle boundary region. Numerical models of rotating convection that include lateral heat flux variations on the outer boundary are natural tools for addressing these issues. But in order to enhance their realism, it is desirable to have a complete mapping of the heat flux over the core-mantle boundary. Unfortunately, no such map exists, although there are a few point-wise estimates of the heat flux based on interpretations of local seismic structure (see Chapter 5).

In place of a global core-mantle boundary heat flux map, the standard modeling approach is to convert a global seismic tomographic image of the shear wave velocity heterogeneity near the base of the mantle into a synthetic heat flux map. The underlying assumption is that reduced seismic shear wave velocity correlates with increased temperature, so that the temperature difference between the lower mantle and the outer core is smaller and the local heat flux is less at such places, compared to other places at the same depth where the lower mantle seismic shear wave velocity is higher.

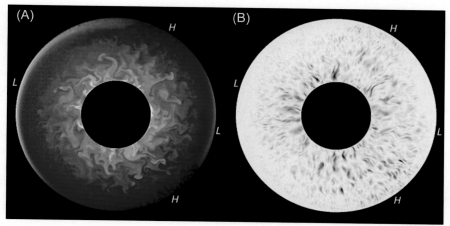

FIG. 4.27 Equatorial plane sections of codensity (A) and radial velocity (B) snapshots from numerical calculations of rotating convection in a spherical shell with heterogeneous (tomographic) outer boundary heat flux, at $Pr = 1$, $Ek = 2 \times 10^{-6}$, $Ra/Ra_{crit} = 440$, and $q^* = 5$. Labels H and L indicate sectors with high and low boundary heat flux, respectively. *Reproduced from Mound, J., Davies, C., Rost, S., Aurnou, J., 2019. Regional stratification at the top of Earth's core due to core-mantle boundary heat flux variations. Nat. Geosci. 12 (7), 575–580.*

Following this recipe, we denote the lateral variations in the core-mantle boundary heat flux by q'_{CMB}, the boundary average superadiabatic heat flux by $\bar{q}_{CMB} - q_{ad}$, and the spherical average and lateral variations in lower mantle seismic shear wave velocity by V_S and ΔV_S, respectively, the notation used in Chapter 5. We then set

$$\frac{q'_{CMB}}{\bar{q}_{CMB} - q_{ad}} \propto \frac{\Delta V_S}{V_S}. \tag{4.80}$$

The coefficient of proportionality needed to convert Eq. (4.80) to an equality is not well constrained, and furthermore it is not certain if the proportionality between core-mantle boundary heat flux heterogeneity and lower mantle shear wave perturbations is actually linear. Modelers usually deal with these uncertainties by examining a range of numerical values for the ratio $q^* = \Delta q'_{CMB}/(\bar{q}_{CMB} - q_{ad})$, in which $\Delta q'_{CMB}$ denotes the peak-to-peak boundary heat flux variation. Note, however, that with this definition q^* is singular if $\bar{q}_{CMB} = q_{ad}$, so the size of q^* may not accurately reflect the amplitude of the boundary heterogeneity. Ignoring this problem for now, the resulting heat flux maps, when combined with the spherical average superadiabatic heat flux, are referred to as the *tomographic boundary condition*. Fig. 4.33A is such a map. It shows higher than average core-mantle boundary heat flux around the rim of the Pacific basin, where sustained deep subduction is predicted to have occurred. Conversely, it shows lower than average core-mantle boundary heat flux beneath the two LLSVP structures.

Fig. 4.27 shows snapshots of codensity and radial fluid velocity in the equatorial plane from a numerical model of rotating spherical shell convection by Mound et al. (2019) that applies a tomographic boundary condition similar to Fig. 4.33A with $q^* = 5$. Apart from boundary heterogeneity, the convection in Fig. 4.27 is roughly comparable to the laboratory experiment in Fig. 4.24D in terms of the other control parameters.

The labels H and L in Fig. 4.27 mark the sectors where the local boundary heat flux is anomalously high and low, respectively. Dual-style convection (defined in Section 4.7.2) extends to the outer boundary in the H-sectors, whereas stable stratification is present near the outer boundary in the L-sectors. Because the L-sector on the left (representing the Pacific Ocean hemisphere) is so large, the stratification there is more extensive and deeper than elsewhere, and as a consequence, the spherical average structure near the outer boundary has an overall stable stratification. This has ramifications for how to interpret anomalous seismic structure in the E' layer. Seismic evidence for stable stratification there could mean $\bar{q}_{CMB} < q_{ad}$. Alternatively, according to the results in Fig. 4.27, it could instead mean that $\bar{q}_{CMB} > q_{ad}$ and that the local heat flux heterogeneity (measured by q^*) is large.

4.7 Numerical dynamos

Our knowledge of convection in the outer core continues to advance, thanks to a combination of theory, laboratory experimentation, and numerical modeling, in conjunction with geomagnetic and other observations. In contrast, the problem of generation of the geomagnetic field by convection in the outer core has proven to be less amenable to attack using theory, geomagnetic observations, or laboratory experiments alone. Although important insight has come from these approaches (see Chapter 3), theoretical progress has slowed because the geodynamo process is strongly nonlinear and magnetic observations are mostly limited to the core-mantle boundary region. In addition, laboratory experimental progress has been slow because electrically conducting fluids are challenging to control and difficult to probe. Accordingly, most of the recent progress on understanding how the geomagnetic field is generated has come through application and interpretation of numerical dynamos.

Fully self-consistent numerical dynamos that directly solve the 3D primitive governing equations listed in Appendices 4.4 and 4.5 without using ad hoc parameterizations are relatively recent accomplishments. Individuals closely associated with their early development include G. Glatzmaier and W. Kuang in the United States, J. Wicht in Germany, E. Dormy in France, and A. Kageyama and M. Sakuraba in Japan. Now in their third decade, numerical dynamos are being applied worldwide to a great variety of phenomena in planetary magnetism.

What do we hope to learn by modeling the geodynamo? Here is a short list of objectives:

- Connect the structure of the core field to core dynamics.
- Determine what controls the intensity of the core field.
- Identify the causes of geodynamo variability over time scales ranging from 1 year to several hundred million years.
- Determine the force balance in the fluid outer core that supports dynamo action.
- Understand why the core field is mainly dipolar, with a preference for axial alignment.
- Identify what roles the complexities in the mantle, inner core, and the E' and F regions of the outer core play.
- Identify the causes and consequences of geomagnetic reversals.

There is a general consensus on the minimum set of attributes for a numerical dynamo to be applicable to the geodynamo. First, the geometry and the forcing must be realistic. Convection in a spherical shell driven by thermochemical buoyancy released at the inner boundary with some additional help from thermal buoyancy released at the outer boundary is generally considered the most realistic modeling choice for the present-day geodynamo. Second, the effects of rotation through the Coriolis acceleration must be a dominant control. Third, the electrical conductivity of the fluid must be large enough to maintain strong electric currents and magnetic field. Many additional complicating effects have been added to this short list, but for our purposes it is best to first examine the behavior of numerical dynamos with only these basics, before turning our attention to more complex models.

The governing equations to be solved are listed in Appendices 4.3–4.5. They consist of the Navier-Stokes equation with Lorentz force added, the continuity equation (usually the Boussinesq form of mass conservation), the codensity transport-diffusion equation, and for the geomagnetic field, Gauss' law plus the induction equation for the magnetic field. In their dimensionless form, four control parameters appear in these equations: the Ekman number Ek, the Prandtl number Pr, magnetic Prandtl number Pm, and one of the various Rayleigh numbers Ra. An additional dimensionless parameter describes the spherical shell geometry of the fluid, but in most studies this parameter is fixed and correctly modeled for the core.

Fig. 4.3 illustrates the major problem we face when applying numerical dynamos to the core and the geodynamo. Of the four control parameters in the outer core, only Pr is accessible to direct numerical simulation. Because of limitations in computer power, it is not possible to reach Earth-like values for the other control parameters in a fully 3D, fully resolved numerical dynamo model. Indeed, now it is not possible to get close. The significance of this fact cannot be overemphasized, because it means that numerical dynamo result cannot be applied directly to the core; all such results must be extrapolated. How that extrapolation is made determines whether the results are valid for the core or not.

Yet in spite of the need to extrapolate, numerical dynamos have proven their worth time and again, and have now become the primary tool for exploring how the geodynamo works. Because of this, much effort has gone into their construction, with improvements in their computational efficiency and flexibility being major goals. It is far beyond the scope of this book to fully describe how numerical dynamos are built; for that the reader is referred to several excellent texts (particularly the 2013 textbook by G. Glatzmaier) and review articles listed as references at the end of this chapter, along with links to community-supported numerical dynamo codes.

4.7.1 A low magnetic Reynolds number dynamo

Fig. 4.28 shows a particularly simple but illustrative numerical dynamo example, one that has been repeatedly used as a benchmark for the validation of numerical codes (Christensen et al., 2001). This dynamo is driven by thermal convection in a rotating spherical shell between isothermal, no-slip boundaries representing the core-mantle and inner core boundaries. The control parameters are $Ek = 10^{-3}$, $Pr = 1$, $Pm = 5$, and Rayleigh number $Ra_T = 10^5$, which corresponds to about $1.8Ra_{crit}$. The resulting flow and magnetic field patterns shown in Fig. 4.28 have fourfold symmetries and drift steadily in the retrograde direction, that is, westward relative to coordinates fixed to the spherical boundaries.

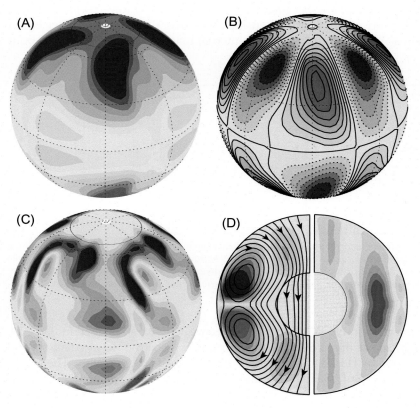

FIG. 4.28 A low magnetic Reynolds number dynamo. Snapshot images of radial magnetic field on the outer boundary (A), streamfunction of the horizontal circulation just below the outer boundary (B), radial helicity at the same depth (C), and azimuthal velocity (D, *right*) and toroidal magnetic intensity with poloidal magnetic field lines (D, *left*) from a numerical dynamo with $Ek = 10^{-3}$, $Pr = 1$, $Pm = 5$, $Ra_T = 10^5$, and $Rm = 39$. *Red (dark gray in print version)* = positive, *blue (darker gray in print version)* = negative values. *Based on Christensen, U.R., Aubert, J., Busse, F.H., et al. 2001. A numerical dynamo benchmark. Phys. Earth Planet. Inter. 128, 25–34.*

The influence of the Coriolis acceleration is evident in Fig. 4.28 in the elongation of the azimuthal component of the flow in the direction of the rotation axis. Four sets of quasigeostrophic vortices define the convection, each set consisting of a cyclonic and anticyclonic pair located outside the inner boundary tangent cylinder. Convection inside the tangent cylinder is weak at this Rayleigh number. In addition to the convective vortices, there are azimuthal thermal winds seen in the right half of Fig. 4.28D. These thermal winds include a strong retrograde (westward) flowing equatorial undercurrent, as described in Section 4.4. The westward drift of the convective vortices and the magnetic field pattern is partly due to this equatorial undercurrent.

The combination of Ekman boundary layer pumping (Section 4.4) plus ageostrophic flow within the convective vortices concentrates the magnetic field on the outer boundary into the cyclonic vortices, as seen by comparing the locations of the intense magnetic field patches in Fig. 4.28A with the circulation pattern in Fig. 4.28B. As shown in the left half of Fig. 4.28D, the poloidal part of the magnetic field is predominantly an axial dipole, while the toroidal part of the magnetic field is concentrated in azimuthal flux bundles, the strongest two bundles located at low latitudes, one such bundle in the northern hemisphere and another at the same latitude in the southern hemisphere and pointing in the opposite direction. Two weaker toroidal magnetic flux bundles are located at higher latitudes in each hemisphere, with the opposite polarity of their low-latitude counterparts. The dominant polarity of the magnetic field, normal polarity in Fig. 4.28, is stable in time for this dynamo and is predetermined by the polarity of the seed magnetic field used to initiate the calculation.

In terms of dimensionless response parameters, the outputs for this dynamo are as follows. The magnetic Reynolds number based on the r.m.s. fluid velocity is $Rm \simeq 39$, which supports the claims made in Chapter 3 that the critical magnetic Reynolds number for rotating convection is low, generally of the order of 40. Most of the kinetic energy in this dynamo is concentrated in the convective vortices, rather than in the axisymmetric azimuthal flows. For example, the maximum velocity in the equatorial undercurrent corresponds to a magnetic Reynolds number $Rm \simeq 38$, but the magnetic Reynolds number based on the r.m.s. velocity of the axisymmetric flows is nearly an order of magnitude less than this. Because their magnetic Reynolds number is relatively low, the axisymmetric flows do not generate very strong toroidal magnetic fields through the ω-effect (Section 3.3 of Chapter 3). Instead, the primary induction mechanism is through an α-effect associated with the helicity of the convection, which Fig. 4.28C reveals is mostly negative in the northern hemisphere and positive in the southern hemisphere. Because both the poloidal and toroidal fields in this dynamo are generated by the same α-effect, they have comparable intensities. The overall Elsasser number based on the r.m.s. magnetic field intensity in the shell is $\Lambda \simeq 6.3$ so that the r.m.s. internal magnetic field intensity is about 2.5 in dimensionless units.

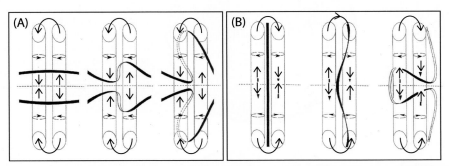

FIG. 4.29 Dynamo mechanisms in columnar convection. (A) Generation of poloidal magnetic field from toroidal magnetic field; (B) generation of toroidal magnetic field from poloidal magnetic field. Magnetic flux depicted by heavy ribbons. *Filled horizontal arrows* indicate primary geostrophic flow around columns; *unfilled vertical arrows* indicate secondary ageostrophic flows. *Dashed line* denotes the equator. *Modified from Olson, P., Christensen, U.R., Glatzmaier, G.A., 1999. Numerical modeling of the geodynamo: mechanism of field generation and equilibration. J. Geophys. Res. 104, 10383–10404.*

The kinematic mechanisms by which the convection in Fig. 4.28 produces dynamo action are sketched in Fig. 4.29. The elongated, quasigeostrophic vortices are depicted as Busse rolls with their primary (geostrophic) circulations and their secondary (ageostrophic) circulations indicated by closed and open sets of arrowheads, respectively. The resulting helicity is negative in the northern hemisphere and positive in the southern hemisphere (Olson et al., 1999), similar to the numerical dynamo helicity in Fig. 4.28C. Fig. 4.29A illustrates production of poloidal magnetic field from toroidal magnetic field. Initially, straight toroidal field lines are simultaneously twisted by the geostrophic flow and transported up and down the rolls by the ageostrophic flow, resulting in the loop-shaped flux bundles shown in the final stage of Fig. 4.29A. When averaged over several vortex pairs, this action is kinematically similar to the α-effect described in Section 3.3 of Chapter 3.

The production of toroidal magnetic field in this dynamo also occurs through the helicity in the vortices. As illustrated in Fig. 4.29B, as the initially straight poloidal field lines are transported around the vortices by the geostrophic flow they are twisted by the ageostrophic flow, resulting in a pair of toroidal magnetic flux bundles positioned in an antisymmetric manner across the equator. Shifting the final image in Fig. 4.29B to the left so that the vortices line up reproduces the final magnetic field line configuration in Fig. 4.29A, confirming the positive nature of the feedback between toroidal and poloidal magnetic field components.

With both toroidal and the poloidal field generation occurring through convective vortex helicity, this dynamo qualifies as being of the α^2 type. Applied to the geodynamo, this α^2 feedback mechanism implies roughly comparable toroidal and poloidal magnetic field intensities inside the outer core. In addition, it implies that sites where poloidal magnetic flux is concentrated on the core-mantle boundary correspond to cyclonic vortices, whereas sites where magnetic flux is dispersed correspond to anticyclonic vortices, in which the sign of the helicity is opposite to that in the cyclones.

However, these interpretations are not as clear-cut as may seem, because of the point emphasized previously in this section—the need to extrapolate numerical dynamos to outer core conditions. For example, in this low magnetic Reynolds number dynamo, there is a nearly one-to-one correspondence between the structure of the convection and the magnetic field structure on the outer boundary, and both are large enough in scale to be seen through the crustal filter. Do these same attributes also apply to numerical dynamos at more realistic magnetic Reynolds numbers? If so, we have reason to think the earlier interpretations are applicable; if not, then alternative interpretations become possible.

4.7.2 A high magnetic Reynolds number dynamo

The restrictions imposed by computational resources not only limit the ranges of Ek, Pr, and Ra accessible to numerical dynamo modeling, but also they impose limitations on the model run times. This means that high-resolution numerical dynamos can only simulate a short-time interval, typically less than one magnetic diffusion time and often not enough to reach equilibrium. Accordingly, in order for these to be applicable to the core, clever schemes are needed to artificially accelerate high-resolution numerical dynamos toward equilibrium.

One such scheme is to initiate a high-resolution dynamo with the output of an equilibrated lower-resolution case. Fig. 4.30 shows the 3D structure of a numerical dynamo at $Ek = 3 \times 10^{-7}$, $Pr = 1$, and $Pm = 0.25$ calculated by Aubert (2019) using this type of accelerated convergence. The modified Rayleigh number in this case, $Ra_m = 2.75 \times 10^{-5}$, is based on codensity uniformly released from the inner boundary and uniformly absorbed in the fluid, as defined in Table 4.2. Fig. 4.30 shows the axial (rotationally aligned) component of vorticity, the azimuthal velocity, codensity perturbations, and the radial component of the magnetic field on spherical surfaces and planar cuts through the fluid shell.

With a magnetic Reynolds number $Rm \simeq 1100$, the dynamics of this dynamo are considerably more complex and irregular than the low Rm dynamo just described, yet many of the attributes seen at low Rm carry over to this case. In particular, the elongation of convective structures in the direction of the rotation axis is also seen in this case, although there is distinctly less structural continuity in the ζ-direction than at low Rm. In addition, the inner core tangent cylinder plays a similar role here as in the low Rm dynamo, with an array of high-intensity magnetic flux bundles on the outer boundary located just outside the tangent cylinder, and low outer boundary magnetic intensity inside the tangent cylinder. One difference with the previously described low Rm dynamo is that in this dynamo the convection is entirely driven by codensity plumes originating at the inner boundary, with no buoyancy produced at the outer boundary. Outside the tangent cylinder, the plumes curl to form intermediate-scale vortices; inside the tangent cylinder they merge with increasing height to form a large axial upwelling. The geostrophic response to this axial upwelling is a strong thermal wind, with generally retrograde (westward) flow near the outer boundary—a polar vortex (Section 4.5)—and generally prograde (eastward) flow close to the inner boundary.

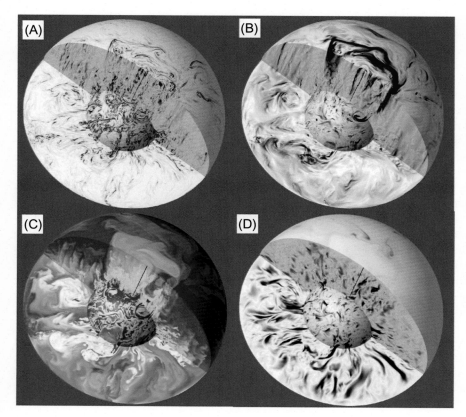

FIG. 4.30 A high magnetic Reynolds number dynamo. Snapshots of 3D structure of a numerical dynamo with $Ek = 3 \times 10^{-7}$, $Pr = 1$ and $Pm = 0.25$, and $Rm = 1100$. Images show the axial (rotationally aligned) vorticity (A), the azimuthal velocity (B), codensity perturbations (C), and the radial component of the magnetic field (D). *Reproduced from Figs. 5–7 of Aubert, J., 2019. Approaching Earth's core conditions in high-resolution geodynamo simulations. Geophys. J. Int. 219, 137–151.*

A part of the reason for the increased complexity in this dynamo is the presence of multiple length scales. There is a different characteristic length scale for each variable in Fig. 4.30, with vorticity residing on smaller length scales than the other three variables. Because of this scale separation, several small vortices are included within, and contribute to, each magnetic flux concentration. This contrasts with the low Rm dynamo in Fig. 4.28, in which there is a one-to-one correspondence between individual vortices and individual magnetic flux concentrations.

The balance of forces in this dynamo is also scale dependent. At large length scales, that is, for spherical harmonic degrees $n \leq 15$, the primary balance is geostrophic, with the pressure gradient force largely balanced by the Coriolis acceleration. The second-order balance in this spherical harmonic range is between the buoyancy force and the part of the Coriolis acceleration left over after the first-order balance is removed. The Lorentz force only becomes important at intermediate length scales, in the range of spherical harmonic degrees $15 < n < 100$. And even in this range the Lorentz force mostly contributes through the Maxwell pressure associated with axially aligned magnetic flux concentrations, so it does not destroy the quasigeostrophic structure of the vortices. Finally, viscous effects only become significant at very high spherical harmonic degrees, above $n = 100$. Accordingly, viscosity plays almost no role in this dynamo at large and intermediate length scales, the length scales that are actually observable in the core field and its secular variation. This last point is important because it suggests there exists a practical lower limit for the Ekman number in numerical simulations of the geodynamo, beyond which little insight is to be gained by further decreases of this parameter.

An important similarity between these high and low Rm dynamos is the dominance of the Coriolis acceleration in each case. The Lorentz force tends to damp the fluid motions, especially the azimuthal wind flow, and is critical for equilibrating the magnetic field intensity. But it plays a relatively small role in structuring the convection, compared to the Coriolis acceleration. We can quantify this relationship by comparing the relative magnitudes of the Lorentz force to the Coriolis acceleration in both dynamos. Using the definitions in Appendices 4.1 and 4.4, the ratio of the Lorentz force to the Coriolis acceleration is given by Λ/Rm, the ratio of our two primary dynamo response parameters. Dominance of the Lorentz force, which implies magnetostrophic flow, corresponds to $\Lambda > Rm$, whereas dominance of the Coriolis acceleration, which implies quasigeostrophic flow, corresponds to $\Lambda < Rm$. Both of the numerical dynamos described here satisfy this second inequality, particularly the high Rm dynamo, and this accounts for why they are more quasigeostrophic than magnetostrophic. In addition, according to the parameter values in Table 4.1, $\Lambda \ll Rm$ in the outer core, similar to the high Rm numerical dynamo. This is another argument in favor of quasigeostrophy (and against magnetostrophy) as the basic force balance in the outer core that maintains the geodynamo.

4.7.3 Crustal filtering effects

Like many numerical dynamos, the examples just described reproduce some of the main structural characteristics of the observed core field. Among these are a dominant and persistent axial dipole, tendency for westward drift of nondipole magnetic structures, high-latitude, high-intensity magnetic flux concentrations, and expressions of the inner core tangent cylinder. However, in order to make direct comparisons with the observed core field, it is necessary to filter the magnetic field of the numerical dynamos in the same way that crustal magnetization filters the core field.

Fig. 4.31 shows the result of applying a low-pass filter to a snapshot of the radial magnetic field on the outer boundary of a high-resolution numerical dynamo. The low-pass filter designed by Christensen and Wicht (2015) attenuates spherical harmonic degrees beyond $n = 12$ in the magnetic field on the outer boundary, such that all spherical harmonic degrees greater than $n = 14$ are essentially removed, similar to the way crustal magnetization filters the core field. Application of this filter transforms the morphology of the boundary magnetic field, from small-scale structures elongated in the rotation direction to larger, more equidimensional structures. Comparison with the present-day core field (Fig. 3.3 of Chapter 3) reveals additional points of similarity that only become evident after the dynamo magnetic field is properly filtered. These include reduced magnetic field intensity inside the tangent cylinder, high-intensity magnetic flux concentrations just outside the tangent cylinder, and pairs of reversed polarity magnetic flux concentrations near the equator.

Crustal filtering applied to high-resolution numerical dynamos demonstrates that some of the large and intermediate-scale structures observed in the core field may actually consist of concentrations of small-scale magnetic field structures with mixed polarity, in which one polarity is dominant. Consider, for example, the origin of the high-intensity magnetic flux concentrations located just outside the tangent cylinder in the core field in Fig. 3.3 of Chapter 3. The numerical dynamos described here suggest two distinct interpretations of these flux concentrations, illustrated schematically in Fig. 4.32.

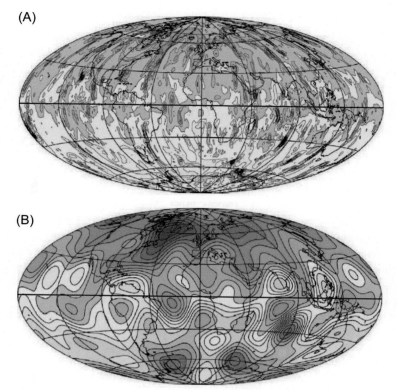

FIG. 4.31 Effects of crustal filtering. (A) Unfiltered radial magnetic field on the outer boundary from a numerical dynamo with $Ek = 3 \times 10^{-5}$, $Pr = Pm = 1$, $Ra/Ra_{crit} = 42$. (B) Same magnetic field with crustal filtering applied. *Modified from Fig. 12 of Christensen, U.R., Wicht, J., 2015. Numerical dynamo simulations. In: Schubert, G. (editor-in-chief), Treatise on Geophysics, vol. 8, second ed. Elsevier, Oxford, pp. 245–277.*

According to the interpretation illustrated in left-hand side of Fig. 4.32, magnetic flux concentrations are produced by, and located inside, the large-scale cyclonic vortices. This interpretation is consistent with the low Rm, low-resolution numerical dynamo in Fig. 4.28. According to the other interpretation, illustrated in right-hand side of Fig. 4.32, geomagnetic flux concentrations actually consist of many small-scale magnetic flux bundles with a statistically preferred polarity, each small flux bundle associated with a small-scale cyclonic vortex. This interpretation is more consistent with the high Rm, high-resolution numerical dynamo in Fig. 4.30. Unfortunately, crustal filtering makes it difficult to distinguish between these alternatives, particularly if only time snapshots of the core field are used. However, numerical dynamos may reveal differences in their time dependence, so it may be possible to distinguish between them based on the geomagnetic secular variation.

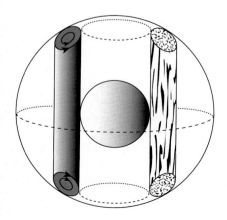

FIG. 4.32 Sketch illustrating two possible interpretations of the high-latitude, high-intensity magnetic flux concentrations on the core-mantle boundary. *Left side*: large-scale laminar cyclonic vortex; *right side*: concentrations of small-scale quasigeostrophic turbulent vortices. *Dashed lines* mark the outer core equator and inner core tangent cylinder rim.

4.7.4 Scaling laws for outer core convection and the geodynamo

The search for a universal scaling law for planetary dynamos is a venerable quest. In 1947, W.M. Elsasser, atomic physicist and dynamo theory pioneer, proposed a scaling for dynamo magnetic intensity based on a balance between Coriolis and Lorentz forces (i.e., $\Lambda \simeq 1$), while in the same year Nobel Prize winner P.M.S. Blackett argued that planetary magnetic field intensity depends only on the planet core size, electrical conductivity, and angular momentum of rotation. Since then, many scaling laws for planetary dynamos have been advanced, usually founded on some combination of theory and simulation. These laws have had considerable success in systematizing numerical dynamo results, as well as some limited success in rationalizing the magnetic fields of Jupiter and Saturn with the geomagnetic field. So far, however, no single scaling law approach has been shown to account for the dynamo properties of all the planets.

It is fair to ask if we actually need dynamo scaling laws. Why not skip the scaling, and compare dynamo models directly with the observations? There are several answers to this question. First, scaling laws are useful for purely theoretical reasons. Because the geodynamo involves so many interacting parts, in order to achieve full understanding of how the parts interact we need to put observations, numerical models, and laboratory experiments into a common framework. Second, as emphasized earlier in this chapter, no numerical simulation or experiment applies directly to outer core convection or to the geodynamo. Fig. 4.3 shows why this is so: all numerical and laboratory experiments are far from the core and the geodynamo in terms of one or more of their dimensionless control parameters. And a third reason is that dynamo scaling law construction is good for geophysics. Comparisons between the various proposed scaling laws have generated many lively and informative discussions.

As with other systems in which complex fluid dynamics are at work, scaling laws are vehicles for extrapolating computational and experimental results to various parts of the Earth system—the ocean, atmosphere, mantle, and core. Although they may be theoretically self-consistent, dynamo scaling relationships are nevertheless nonunique, a consequence of the fact that outer and inner core dynamics involve so many physical and chemical processes that are poorly constrained. Obviously, the most useful scaling laws are consistent with theory, numerical models, and lab experiments, and are also testable against observations.

4.7.4.1 Power-law scaling

Most of the scaling relationships that have been applied to the outer core can be derived by combining dimensional analysis with similarity assumptions, without having to make direct reference to the governing equations or to particular force balances. Recall that there are four basic nondimensional control parameters for the geodynamo, plus at least one geometrical parameter. For the latter, we choose the outer core thickness normalized by its radius, so this parameter is fixed and does not enter our scaling explicitly. The independent control (input) parameters include the Ekman number Ek, the Prandtl number Pr, the magnetic Prandtl number Pm, and one Rayleigh number. Here, we choose Ra_F, based on the average buoyancy flux F. Independent response (output) parameters include the magnetic Reynolds number Rm for the outer core fluid velocity and the Elsasser number Λ, which measures the r.m.s. magnetic field intensity internal to the core. These dimensionless parameters are derived in Section 4.3 and defined in Tables 4.1 and 4.2.

To begin, we write the response parameters, the Elsasser and magnetic Reynolds numbers, as products of the control parameters raised to powers:

$$(Rm, \Lambda) \propto Ra_F^\alpha Ek^\beta Pr^\gamma Pm^\delta. \tag{4.81}$$

Throughout this section, α, β, γ, δ denote the power-law exponents to be determined for each response parameter. Substituting the definitions of the four control parameters into Eq. (4.81) yields

$$(Rm, \Lambda) \propto \nu^{(\beta - 2\alpha + \gamma + \delta)} \kappa_\chi^{(-\alpha - \gamma)} \kappa_B^{(-\delta)} F^{(\alpha)} \Omega^{(-\beta)} d^{(4\alpha - 2\beta)}, \tag{4.82}$$

where ν, κ_χ, κ_B, and Ω are kinematic viscosity, codensity diffusivity, magnetic diffusivity, and rotation rate, respectively. Here, we are mostly concerned with determining the exponents in Eq. (4.82); the coefficients needed to convert these proportionalities to equalities usually come from numerical or laboratory experimentation.

In order to determine the power-law exponents, we need four independent constraints involving the fluid velocity (for Rm) or the magnetic field intensity (for Λ). Physical intuition, geophysical observations, along with the results of laboratory and numerical experimentation are all legitimate sources for these constraints. Here are some that are frequently applied to the outer core and the geodynamo.

Diffusion-free velocity scaling

Suppose the fluid velocity in the outer core is independent of the three diffusivities κ_T, ν, and κ_B, and also independent of the outer core thickness d. Applying these constraints to Eq. (4.82) implies for Rm

$$\alpha = 1/2 \quad \beta = 1/2 \quad \gamma = -1/2 \quad \delta = 1, \tag{4.83}$$

and for the outer core fluid velocity

$$v \propto \left(\frac{F}{\Omega}\right)^{1/2}, \tag{4.84}$$

which is a standard scaling relationship for the fluid velocity in rotating convection, applicable to thermal as well as compositional buoyancy. Some additional generalization of Eq. (4.84) appears to be warranted, because as discussed earlier in this chapter, convection in the form of quasigeostrophic, columnar vortices is likely to be sensitive to the thickness of the outer core. Allowing for dependence on the outer core thickness d in the previous analysis leads to a more general relation,

$$v \propto \Omega^{1-3\alpha} d^{1-2\alpha} F^{\alpha}, \tag{4.85}$$

which reduces to Eq. (4.84) when $\alpha = 1/2$.

There are differing opinions as to the best value of α in Eq. (4.85) for outer core convection. Based on numerical dynamo results, Christensen and Aubert (2006) argue for $\alpha = 2/5$, whereas Davidson (2013) advocates $\alpha = 4/9$, and Starchenko and Jones (2002) propose a velocity scaling law like Eq. (4.84) with a prefactor $(d/l)^{1/2}$, where l is the width (cross-sectional dimension) of the convective vortices.

Diffusion-free magnetic field scaling

Now suppose the magnetic field intensity in the outer core is independent of all three diffusivities and also independent of the rotation rate Ω. Applying these constraints to Eq. (4.82) implies for Λ

$$\alpha = 2/3 \quad \beta = 1 \quad \gamma = -2/3 \quad \delta = 1, \tag{4.86}$$

and for the magnetic field intensity in the outer core

$$B \propto (\rho \mu_0)^{1/2} (Fd)^{1/3}, \tag{4.87}$$

which is a commonly used scaling relationship for the magnetic intensity in dynamos driven by rotating convection. Similar to the diffusion-free fluid the velocity scaling, some further generalization of Eq. (4.87) appears to be warranted, in order that the magnetic field intensity also depends on the rotation rate. Allowing for dependence on the rotation rate in the previous analysis leads to a more general relation,

$$B \propto (\rho \mu_0)^{1/2} \Omega^{(1-\beta)/2} d^{(2-\beta)/3} F^{(1+\beta)/6}, \tag{4.88}$$

which reduces to Eq. (4.87) when $\beta = 1$.

As with diffusion-free velocity scaling, different numerical data or theoretical considerations yield somewhat different exponents for diffusion-free magnetic field scaling applied to the outer core. Christensen and Aubert (2006) find $\beta = 4/5$ in Eq. (4.88), Davidson (2013) prefers $\beta = 1$ but replaces d with the actual length of the convective vortices, while Starchenko and Jones (2002) advocate for $\beta = 1/2$.

Diffusive magnetic field scaling

Here, we suppose the magnetic field intensity in the outer core is independent of the small diffusivities κ_χ and ν and also independent of the outer core thickness and the rotation rate, but does depend on the larger magnetic diffusivity κ_B. Applying these constraints to Eq. (4.82) implies for Λ

$$\alpha = 1/2 \quad \beta = 1 \quad \gamma = -1/2 \quad \delta = 1/2, \tag{4.89}$$

and for the magnetic field intensity

$$B \propto (\rho\mu_0)^{1/2}(\kappa_B F)^{1/4}. \tag{4.90}$$

Lastly, suppose the magnetic field intensity in the outer core is independent of both of the small diffusivities, the outer core thickness, and also independent of the buoyancy flux. This set of constraints implies that Λ in Eq. (4.82) is independent of all the listed properties except rotation rate and electrical conductivity, so that

$$B \propto (\rho\mu_0\kappa_B\Omega)^{1/2}, \tag{4.91}$$

which is called *Elsasser number scaling*.

4.7.4.2 Scaling law applications

The next step is to compare the fluid velocities and magnetic field intensities predicted by these scaling laws with observationally based estimates for the outer core, from this chapter and from Chapter 3. Using $\mu_0 = 4\pi \times 10^{-7}$, $\rho = 10^4$ kg/m^3, $d = 2.26 \times 10^6$ m, $\Omega = 7.29 \times 10^{-5}$ s^{-1}, and $F = 2 \times 10^{-12}$ m^2/s^3 as in Tables S1 and S3 from Core Properties and Parameters, substitution into Eq. (4.85) and assuming the missing proportionality coefficients equal one, the predicted outer core r.m.s. fluid velocities are $v = 1.7 \times 10^{-4}$, 7.7×10^{-4}, and 2.6×10^{-3} m/s, respectively, for the exponent choices $\alpha = 1/2$, $4/9$, and $2/5$, respectively. Note that the fluid velocity prediction for $\alpha = 1/2$ would be a factor of about 10 higher when the $(d/l)^{1/2}$ prefactor is included, raising it to about 1.7×10^{-3} m/s. For comparison, outer core fluid velocities obtained from frozen flux analyses (Chapter 3) are also in the range of $0.5-1 \times 10^{-3}$ m/s. In short, all of these scaling laws predict reasonable outer core fluid velocities.

Repeating this analysis for the magnetic field intensity in the outer core using the same core property values in Eq. (4.88) and again assuming the missing proportionality coefficients equal 1, the predicted r.m.s. magnetic intensities are $B = 2.3, 4.7$, and 18 mT, respectively, for the exponent choices $\beta = 1, 4/5$, and $1/2$, respectively. All of these predictions are based on diffusion-free scaling, and they bracket the outer core r.m.s. magnetic field intensity of $B \simeq 5$ mT inferred from observations of torsional oscillations (Section 4.5). For the diffusive scaling, using $\kappa_B = 0.8$ along with the other parameters, Eqs. (4.90), (4.91) yield 0.13 and 0.86 mT, respectively, rather too weak compared to what is inferred for the outer core magnetic intensity. However, it is important to note that we have assumed the missing proportionality coefficient in Eq. (4.91) equals one, which is equivalent to assuming $\Lambda = 1$ a priori. But as demonstrated in Section 4.7.4.1, numerical dynamos commonly produce internal magnetic field intensities corresponding to $\Lambda \simeq 5-10$, for which Eq. (4.91) predicts $B \simeq 2-3$ mT, closer to the magnetic intensity in the outer core inferred from torsional oscillations.

4.7.5 Effects of mantle heterogeneity and E′ layer stratification

Because the solid mantle has a larger thermal mass than the core, and because convective overturn in the mantle is much slower than in the outer core, heat transfer from the core to the mantle is the ultimate rate limiter for convection in the outer core as well as for inner core growth. Although estimates of the total heat flow from the core to the mantle remain uncertain, these have recently been revised upward and are now in the range of 8–16 TW (Chapter 2). At the same time, there has been some upward revision of the adiabatic thermal gradient in the outer core, which may be in the range 0.8–1 K/km close to the core-mantle boundary. Multiplying this adiabatic gradient by a prospective higher thermal conductivity and the core surface area implies the total amount of heat conducted down the core adiabat at the core-mantle boundary may approach or possibly exceed the total heat loss from the core. In the latter situation, thermal stratification would develop just below the core-mantle boundary, contributing to the seismic stratification proposed for the E′ layer (Chapter 5). In addition, large variations are expected in the local heat flux on the core-mantle boundary. As shown in Section 4.6, the combination of nearly adiabatic but highly variable boundary heat flux has major effects on the structure of outer core convection, especially in the region below the core-mantle boundary and possibly throughout the core.

Other possible contributors to E′ layer dynamics come from light element enrichment. As discussed in Chapters 5 and 8, excess light elements in the E′ layer might be a relic left over from the formation of the core during Earth's accretion, a result of chemical reactions with the mantle, or an accumulation of compositionally buoyant material released from the growing inner core. Regardless of its origin, compositional stratification due to light element excess has the potential to create very strong stratification in the E′ layer—far stronger than thermal stratification.

Fig. 4.33 illustrates the combined effects of a compositionally stratified layer and the tomographic heat flux boundary condition on numerical dynamos (Christensen, 2018). Fig. 4.33A is a map of the heat flux heterogeneity applied to the outer boundary. The images in Fig. 4.33B and C are snapshots of the radial magnetic field on the outer boundary from two dynamos driven by codensity released uniformly at the inner boundary and absorbed uniformly in the fluid shell. A tomographic outer boundary condition has been applied to each case. The dynamo shown in Fig. 4.33C includes a 750 km thick, chemically stratified outer layer (one-third of the fluid shell thickness), whereas the dynamo shown in Fig. 4.33B has no chemical stratification imposed and the amplitude of the boundary heat flux heterogeneity is somewhat reduced compared to the dynamo in Fig. 4.33C. Otherwise, both dynamos have similar control parameters, $Ek = 3 \times 10^{-5}$, $Pr = 1$, $Pm = 3$, and $Ra/Ra_{crit} \simeq 100$, and both yield similar magnetic Reynolds numbers, $Rm \simeq 345$. In addition, a crustal filter has been applied to the magnetic field of each, as described in Section 4.7.3.

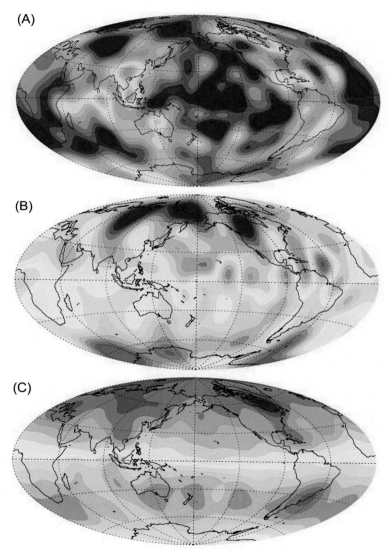

FIG. 4.33 Numerical dynamo results illustrating the combined effects of core-mantle boundary heat flux heterogeneity and stable stratification in the E′ layer of the outer core. (A) Outer boundary heat flux pattern (*red* = high, *blue* = low); (B) snapshot of outer boundary radial magnetic field without stratification; (C) same dynamo with 750 km of stable stratification below the outer boundary. $Ek = 3 \times 10^{-5}$, $Pr = 1$, $Pm = 3$, and $Ra/Ra_{crit} \simeq 100$ in each case. Continental outlines are shown for reference. *Modified from Figs. 2 and 8 of Christensen, U.R., 2018. Geodynamo models with a stable layer and heterogeneous heat flow at the top of the core. Geophys. J. Int. 215, 1338–1351.*

The major difference between the dynamos in Fig. 4.33B and C can be seen in their outer boundary magnetic field structures. The relatively thick stratified layer in Fig. 4.33C dynamo attenuates the nonaxisymmetric components of the internal magnetic field, so that the radial magnetic field on its outer boundary is dominated by axisymmetric components. It also lacks patches of reversed magnetic flux on its outer boundary. In contrast, the absence of a global stratified layer in the dynamo in Fig. 4.33B preserves the nonaxially symmetric magnetic field components as well as reversed magnetic flux on the outer boundary, allowing the effects of the tomographic boundary condition to show clearly. In particular, comparison with the tomographic heat flux pattern in Fig. 4.33A reveals that high-intensity magnetic field concentrations in this dynamo correlate with high boundary heat flux and low-intensity magnetic field generally correlates with low boundary heat flux. Furthermore, the size and shape of the high-latitude high-intensity magnetic field concentrations in Fig. 4.33B dynamo are comparable to those in the core field shown in Fig. 3.3 of Chapter 3. In contrast, the magnetic field concentrations in the strongly stratified dynamo in Fig. 4.33C are far more axisymmetric.

In summary, stratification and lateral heterogeneity in the core-mantle boundary region have competing effects on the geomagnetic field, according to numerical dynamo results. Deep stratification makes the core field more axisymmetric, while lateral heterogeneity in the heat flux over the core-mantle boundary has the opposite effect, enhancing the nonaxisymmetric part of the core field. Because these controls have opposing effects, there is likely a trade-off between the amplitude of heat flux heterogeneity over the core-mantle boundary and the amount of stratification below the core-mantle boundary, as far as the core field structure is concerned. Nevertheless, we can say that too much stratification, more than a few hundred kilometers in depth extent, would make the core field far more axisymmetric than what is observed. Another important finding is that numerical dynamos show a positive correlation between boundary magnetic field strength and boundary heat flux, a useful guide for interpreting the core field in terms of mantle-core thermal interaction.

4.7.6 Effects of heterogeneous inner core growth

As described in Chapters 5 and 6, there is growing consensus on the existence of a globally anomalous F layer above the inner core boundary, characterized by anomalously low seismic P-wave velocity and a positive density anomaly with respect to the adiabatic density profile in the outer core.

A variety of mechanisms have been advanced to explain the origin of the F layer, each with implications for outer core dynamics. Possible explanations include the F layer as a relic of inhomogeneous core formation or a product of iron snowfall. But in terms of broad dynamical implications, the possible origin that has received the most scrutiny is the proposal by Alboussiere et al. (2010) that the inner core is translating sideways, and experiences faster growth in one hemisphere and slower growth with possible melting in the other hemisphere. Chapter 7 gives a detailed analysis of the mechanics of inner core translation within the broader context of convection in the inner core, and Chapter 6 reviews the seismic evidence for and against this interpretation. In this chapter, we focus on the implications of heterogeneous inner core growth for outer core convection and for the geodynamo.

Fig. 4.34 shows results from numerical models of thermal convection in a rotating spherical shell by Deguen et al. (2014) that illustrate how the combination of heterogeneous inner core growth and inner core translation would affect flow in the outer core. The images in Fig. 4.34 are snapshots of codensity, axial vorticity, azimuthal fluid velocity, and the time-averaged streamfunction of the flow from two numerical models of rotating convection with differing amplitudes of codensity heterogeneity applied over the inner boundary. In these models, codensity heterogeneity on the inner boundary represents the uneven release of light elements at the inner core boundary. In Fig. 4.34, the inner boundary codensity is hemispherical and consists of a spherical harmonic degree and order one pattern, with a maximum on the right, the *solidification pole*, and a minimum on the left, the *melting pole*. The implied inner core translation direction is, therefore, right to left in these images, and the gray shading indicates the implied age anomaly of the inner core solids, increasing from light to dark. In addition to the global Rayleigh number Ra_χ defined in Table 4.2, this system is characterized by an additional Rayleigh number for the inner boundary heterogeneity, $Ra_{\Delta\chi} = \alpha_C g_o d^3 \Delta C / \kappa_C \nu$, where ΔC denotes the amplitude of the inner boundary light element heterogeneity and the subscript o denotes the outer boundary of the fluid shell.

The top row in Fig. 4.34 shows the effects of rather weak inner core boundary heterogeneity, with $Ra_{\Delta\chi}/Ra_\chi = 0.125$. The inner boundary heterogeneity in this case is strong enough to produce a hemispherical difference in the convective vigor, with the convectively active hemisphere above the solidification pole, as expected on thermodynamic grounds. But in this case, the codensity is positive over the whole inner boundary, implying heterogeneous solidification with no melting.

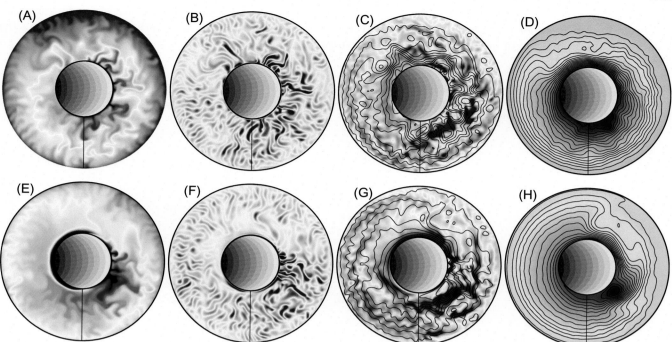

FIG. 4.34 Convection with a heterogeneous inner core boundary. Equatorial plane views of rotating convection in a spherical shell with hemispherical (spherical harmonic degree and order one) codensity variations applied at the inner boundary. Images show snapshots of codensity (A, E) axial vorticity (B, F), azimuthal velocity (C, G), and time-average streamfunction (D, H). Melting pole is on the *left*, solidification pole on the *right*, *solid line* marks the melting equator, *gray shading* shows qualitatively the implied inner core age anomaly, increasing from light to dark. Control parameters $Ek = 3 \times 10^{-5}$, $Pr = 1$, $Pm = 1$. Top row: $Ra_\chi = 6 \times 10^7$; $Ra_{\Delta\chi} = 7.5 \times 10^6$. Bottom row: $Ra_\chi = 6 \times 10^7$; $Ra_{\Delta\chi} = 3 \times 10^7$. Red (dark gray in print version) = positive, *blue* (darker gray in print version) = negative in all images. *Modified from Figs. 2 and 4 of Deguen, R., Olson, P., Reynolds, E., 2014. F-layer formation in the outer core with asymmetric inner core growth. C. R. Geosci. 346, 101–109.*

In contrast, the bottom row of Fig. 4.34 shows the effects of stronger inner core boundary heterogeneity implying faster inner core translation, with $Ra_{\Delta\chi}/Ra_\chi = 0.5$. In this case, the codensity is negative over a large portion of the left hemisphere of the inner boundary, signifying that melting is occurring there. The difference in convective vigor between the active and quiet hemispheres is even more pronounced in this case, but the longitudes of these two hemispheres are now shifted in the retrograde direction with respect to the melting and solidification poles. In addition, melting produces a new structure in the flow pattern, evident in the azimuthal velocity pattern in Fig. 4.34G. The new flow structure consists of a thin gravity current, circling the inner spherical boundary and transporting dense fluid from the melting pole toward the solidification pole. This gravity current concentrates the convective plumes in the active hemisphere, producing a tight plume cluster that further enhances the dichotomy between the two hemispheres.

In Fig. 4.34, the streamfunction of the generally retrograde (westward) motion in the equatorial plane also shows a hemispherical dichotomy, but near the outer boundary this dichotomy is shifted in longitude with respect to the inner core dichotomy, by nearly 90 degrees in both cases. Accordingly, the weakest and strongest westward flows are nearly 180 degrees in longitude apart, and rotated nearly 90 degrees from the axis of inner core translation. For purposes of interpretation, if we were to identify the sector with slow azimuthal flow in Fig. 4.34D with the Pacific hemisphere and the sector with fast azimuthal flow with the Atlantic hemisphere, then the implied direction of inner core translation is from east to west. This arrangement is broadly consistent with the interpretation from a numerical dynamo study by Aubert et al. (2008). However, the streamfunction pattern in Fig. 4.34D implies no inner core melting, even at the "melting pole," which would then be at odds with the F layer being a product of inner core melting.

In addition to hemispheric dichotomies in outer core convection and azimuthal flow, hemispheric inner core growth is expected to generate an observable signature in the time-average geomagnetic field. Fig. 4.35 shows the time-average radial magnetic field over the northern hemisphere of the outer boundary from a numerical dynamo with control parameters similar to those in Fig. 4.34E–H. The effect of hemispheric dichotomies in outer core convection and large azimuthal flow is to displace the time-average location of the dipole axis away from the rotation axis, creating an *eccentric dipole*. In principle then, eccentricity of the time-average geomagnetic dipole is an observable that constraints the amplitude and orientation of heterogeneous inner core growth.

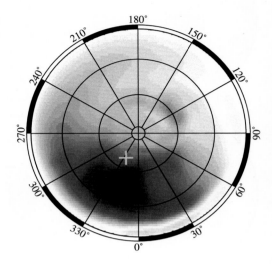

FIG. 4.35 North polar view of time-average radial magnetic field intensity on the outer boundary from a numerical dynamo with inner core heterogeneity similar to Fig. 4.34E–H. *Plus* sign marks the dipole axis location.

4.7.7 Dynamo model reversals

Numerical dynamos provide unique insights into what is probably the most widely known yet poorly understood geomagnetic field behavior—polarity reversals. Chapter 3 summarizes the main characteristics of polarity reversals as revealed by the paleomagnetic record, and includes the claim that polarity reversals are an intrinsic property of the geodynamo. In this section, we show how numerical dynamos support that claim. As with geomagnetic reversals, there is a wide variety of numerical dynamo reversals, ranging from abrupt, rather simple axial dipole flips to drawn-out, very complex transitions involving false starts and recoveries. We are now able to identify some of the factors that promote dynamo model reversals, but like geomagnetic reversals, a full understanding of the underlying mechanisms remains elusive.

To begin with, most numerical dynamos that have a strong dipole do not reverse polarity. This simple fact explains why, as first-principles numerical dynamos were being developed, the ability to produce spontaneous and realistic-looking reversals was given so much weight, and why the early numerical dynamo reversals reported by Glatzmaier and Roberts (1995) were deemed so consequential.

As mentioned previously, numerical dynamo reversals come in many forms, but broadly speaking, there are three main types. Each type offers clues to the underlying mechanism. First, there is periodic reversal behavior, in which reversals are equally or nearly equally spaced in time and the lengths of the stable polarity chrons is, therefore, almost the same. Reversals of this type are most often found in numerical dynamos with strong azimuthal shear flows such as thermal wind flows (Section 4.3), signifying those dynamos are of the $\alpha\omega$ variety. In these dynamos, the toroidal and poloidal magnetic fields oscillate out of phase in space and time, and the polarity reversals an expression of this oscillation. Dynamos that reverse periodically because of intrinsic oscillatory behavior are sometimes referred to as *a.c. dynamos* because they feature alternating electric current systems.

Second, there are reversals that appear to be triggered by major fluctuations in the fluid velocity, meaning that the magnetic field reversal is a direct expression of a global change in the structure of the flow. Numerical dynamos typically do not reverse very many times by this mechanism, since it requires the flow to undergo major structural changes repeatedly. Indeed, one or two reversals of this type often occur during the transient spin-up of a numerical dynamo, but in these cases the reversals reflect the choice of initial conditions, and they cease to occur once the effects of the initial conditions fade.

Third, and likely the most relevant for the outer core, are the dynamo model reversals that occur at seemingly random times, through the combined effects of many small fluctuations in the flow. The global structure of the flow does not appreciably change during this type of reversal, although the total kinetic energy often does change. Because only minor changes in the flow statistics are required to precipitate them, this type of reversal can occur repeatedly, even

4.7.7.1 Simple and complex reversals

Fig. 4.36 shows an example of what could be called a *simple reversal*, from a study by Glatzmaier et al. (1999) of the external controls on dynamo reversal frequency. In this case, the dynamo is driven by thermochemical convection, including simulated light element release at the inner core boundary and uniform heat flux at the core-mantle boundary, and the results are given in physical (dimensional) units. Two reversals and parts of three polarity chrons are shown, the reversals spaced about 170 kyr apart. Both reversals are preceded by a dipole moment collapse, in which the dipole moment decreases by more than one order of magnitude over about 20 kyr. Dipole latitude fluctuations increase in amplitude during the dipole moment collapse but remain limited until the actual directional transition occurs, which lasts only a few thousand years. At that point, the dipole moment recovery begins, and extends over the same 20 kyr time scale on which it collapsed. Virtual geomagnetic pole paths calculated for this model reversal show a simple rotation from polar latitudes in one hemisphere to polar latitudes in the other. Although it occasionally reverses, this is an example of *d.c. dynamo* behavior, because apart from the brief polarity transitions its largest-scale electric current system has a steady direction.

In contrast to the dynamo in Fig. 4.36, Fig. 4.37 shows an example of a more *complex reversal*. In this case, only about 20 kyr of simulated time around the actual reversal is shown. As seen in Fig. 4.37A, prior to the directional transition the energy in the dipole field fluctuates, while the energy in the nondipole field and the tilt angle of the dipole axis gradually increase with time. In addition, the symmetry of the internal field changes, as shown by the *e/o ratio*, the ratio of even (quadrupolar-type) to odd (octupolar-type) nondipolar magnetic fields. Once the e/o ratio becomes large, the dipole field collapses and a directional instability occurs that include a precursory dipole excursion, a transit of the dipole axis across the equator, and a rebound excursion of the dipole, followed by a gradual settling of the dipole into the new polarity. The images at the bottom of Fig. 4.37 show snapshots of the internal magnetic field structure at four times, with the field lines color coded to highlight their individual polarities. The mixed polarity corresponds to times with high e/o ratios in the nondipolar field, whereas uniform polarity corresponds to times with low e/o ratio, suggesting a connection between axial dipole instability and the production of quadrupolar-type nondipole fields by the convection.

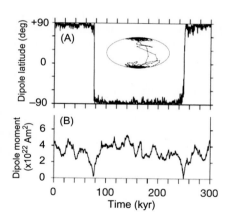

FIG. 4.36 Simple numerical dynamo polarity reversals. (A) Time series of dipole latitude including two reversals. *Inset* shows the path of the dipole axis. (B) Time series of dipole moment showing dipole collapse events synchronous with the reversals. *Modified from Fig. 2 of Glatzmaier, G.A., Coe, R.S., Hongre, L., Roberts, P.H., 1999. The role of the Earth's mantle in controlling the frequency of geomagnetic reversals. Nature 401, 885–890.*

4.7.7.2 Reversal onset

The onset of reversal behavior in most numerical dynamos is dictated by the amount of time variability, primarily in the magnetic field but also in the fluid velocity. As mentioned earlier, dynamo model reversals that occur repeatedly but at random times are typically a result of time variability in the statistical properties of flow, rather than a transition in the overall structure of the flow.

The most common expression of the statistical variability that leads to polarity reversals is localized production of reverse magnetic flux. Fig. 4.38 illustrates how a fluctuation in the intensity of a single convective eddy can produce reverse magnetic flux. First, the convective upwelling lifts a bundle of toroidal magnetic flux, generating a poloidal magnetic flux loop. Meanwhile, the quasigeostrophic circulation around the upwelling twists the poloidal flux loop

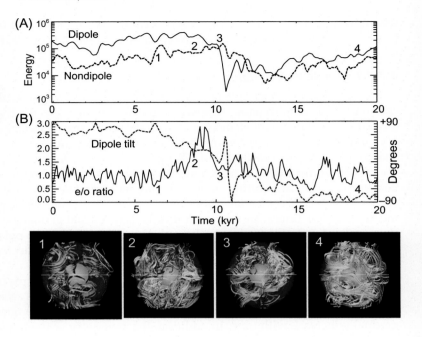

FIG. 4.37 A complex numerical dynamo reversal. (A) Time series of dipole and nondipole magnetic energies. (B) Time series of dipole axis tilt angle and e/o, the ratio of even (quadrupole family) to odd (octupole family) nondipole components. *Bottom images*: Snapshots of internal magnetic flux patterns colored by equatorial polarity at the labeled times. *White line* marks the core equator. *Modified from Figs. 5 and 6 of Olson, P.L., Glatzmaier, G.A., Coe, R.S., 2011. Complex polarity reversals in a geodynamo model. Earth Planet. Sci. Lett. 304, 168–179.*

in the direction indicated in Fig. 4.38A. If the poloidal flux loop is twisted by less than 180 degrees, it will have "Normal" polarity, as shown on the left in Fig. 4.38B. If, however, the flow is so vigorous that the poloidal flux loop is twisted more than 180 degrees, or alternatively, if the twisting is in the opposite direction, the poloidal loop will acquire "Reverse" polarity, as shown on the right in Fig. 4.38B. Both of these mechanisms for producing reverse magnetic flux are possible if the convection is sufficiently irregular and time dependent.

For example, the production of reverse poloidal magnetic flux by excessive twisting can occur simply by increasing the vigor of the convection for a while, so that the velocity of the quasigeostrophic flow is increased. By increasing the velocity and hence the magnetic Reynolds number of the convection, over twisted, reverse polarity poloidal flux loops become more numerous. Similarly, reverse polarity poloidal flux loops can be produced where the helicity has the "wrong sign," that is, where reverse twisting occurs. As the ratio of reverse to normal polarity flux loops increases by these mechanisms, both the geomagnetic equator and the helicity equator deviate from the geographic equator, the axial dipole moment decreases, and the dipole axis tends to become destabilized. Continuation of this process leads to collapse of the axial dipole and increases the probability of a polarity excursion or polarity reversal.

Numerical dynamos offer multiple ways to test this concept. First, we can determine if there is a threshold convective velocity, below which there are no dynamo reversals but above which reversals occur. Second, we can determine if the frequency of reversals increases as convective velocities pass this threshold. A number of studies have

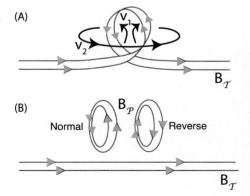

FIG. 4.38 Reverse magnetic flux generation. (A) Twisting of a poloidal magnetic flux loop by a helical convective upwelling. *Black arrows*, v_1 and v_2 denote fluid velocity; *gray arrows*, B_T and B_P denote magnetic field. (B) Final poloidal magnetic flux loops B_P in two polarities. Normal polarity loop produced by right-hand twisting through 180 degrees or less; reverse polarity loop produced by right-hand twisting between 180 and 360 degrees or by left-hand twisting.

systematically addressed the first test by identifying regimes of reversing and nonreversing dynamos in terms of the dimensionless control parameters. In convective dynamos, it is found that reversals are suppressed by faster rotation and enhanced by the buoyancy force, qualitatively consistent with the conceptual model in Fig. 4.38.

Fig. 4.39 illustrates the transition from nonreversing to reversing dynamo behavior as a function of the Rayleigh number, assuming the other control parameters are fixed. As the Rayleigh number Ra increases, the strength of the buoyancy forces driving the convection increase, along with the convective velocities and the production of reverse magnetic flux. Three distinct magnetic regimes are typically found. At low Ra, the convection is either subcritical or not strong enough for the magnetic Reynolds number Rm to be supercritical, and no dynamo action occurs. Dynamo onset occurs where Ra is large enough such that the magnetic Reynolds number satisfies $Rm > Rm_{crit}$. Dynamos in the slightly supercritical, power-law regime is dominated by a strong, stable axial dipole field and typically do not reverse. The magnetic field intensity, represented in Fig. 4.39 by the dipole moment, increases in this regime with the Rayleigh number as Ra^α, where $\alpha \simeq 1/3$, as described in Section 4.7.4.

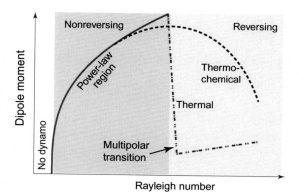

FIG. 4.39 Dynamo regime schematic, plotted as dipole moment versus Rayleigh number. *Solid curve* shows power-law region. *Dash and dash-dot curves* show postpower-law behaviors of numerical dynamos driven by thermochemical and thermal buoyancy, respectively. Reversing and nonreversing regimes indicated by *shadings*.

However, the power-law increase in dipole moment does not continue indefinitely as the dynamo forcing is increased. In fact, further increase in Ra ultimately leads to a decrease in dipole moment. In many numerical dynamos, this dipole moment decrease corresponds roughly to the onset of polarity reversals. The precise Ra-value where reversal onset occurs depends on the other control parameters, and beyond reversal onset, the structure of the dynamo magnetic field also depends on the type of buoyancy forcing used. As indicated in Fig. 4.39, dynamos with thermal buoyancy forcing defined by Ra_T (Table 4.2) typically experience a precipitous drop in dipole moment, called the *multipolar transition*. Reversals occur in these dynamos in the multipolar regime, but because the dipole is no stronger or more stable than the higher multipoles, these reversals are poor analogs for geomagnetic reversals.

In contrast, thermochemical dynamos driven by a codensity flux from the inner boundary with the codensity absorbed in the fluid interior behave more Earth like in terms of polarity reversals. As depicted in Fig. 4.39, these dynamos remain dipole-dominant well into the reversing regime, so that the reversing magnetic field in these dynamos is controlled by the dipole component. In this regard, these are better analogs for geomagnetic reversals. There are two primary reasons why thermochemical dynamos reverse differently: first, they use a flux condition at the outer boundary, rather than fixed temperature, and second, the codensity sink ϵ fosters "deep-dynamo" behavior. Both of these attributes favor a strong dipole and tend to suppress the abrupt multipolar transition found in thermal convection dynamos.

The Rayleigh number is not the only parameter that measures reversal onset, and indeed it is probably not the best. A better choice would be a parameter that explicitly includes the convective velocity v and the smaller length scale l over which the convective velocity varies. Two candidates that satisfy these requirements are the *local magnetic Reynolds number* $Rm_l = vl/\kappa_B$ and the *local Rossby number* $Ro_l = v/l\Omega$. The local magnetic Reynolds number is implicit in the reverse flux generation mechanism depicted in Fig. 4.38. The local Rossby number differs from the global Rossby number (4.2) by the ratio d/l. This ratio is probably quite large in the outer core, which opens the possibility that Ro_l may be as large as 0.1 in the outer core (Christensen and Aubert, 2006), implying that the inertial acceleration of outer core fluid parcels may play a role in geomagnetic reversals. Independent support for this latter possibility comes from a numerical investigation of dynamo behavior with and without fluid inertia by Sreenivasan and Jones (2006). These authors find that when inertial effects become significant, the convection becomes more irregular and less symmetric about the

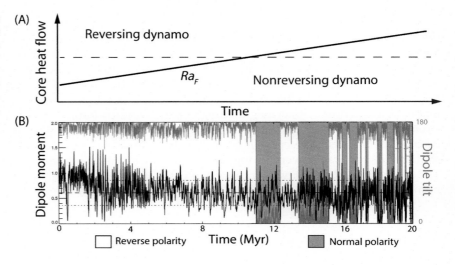

FIG. 4.40 Polarity reversal sensitivity to core heat flow. Time series of dimensionless dipole moment and dipole axis tilt angle (B) from a numerical dynamo driven by an inner boundary codensity flux and an outer boundary heat flow increasing linearly with time as in (A). Normal and reverse polarity chrons indicated by *shadings*. Modified from Fig. 1 of Driscoll, P., Olson, P., 2011. Superchron cycles driven by variable core heat flow. Geophys. Res. Lett. 38, L09304.

equator. Such flows produce weaker, more unstable dipole magnetic fields along with a proliferation of reversed magnetic flux—precisely the conditions that are thought to initiate polarity reversals.

4.7.7.3 Reversal frequency

Another test that numerical dynamos can offer is to determine the *reversal sensitivity*, that is, the frequency of reversals as a function of one or more of the control parameters. If the frequency of otherwise random reversals increases with the convective forcing and/or decreases with the rate of rotation, for example, this should be evident in changes in dynamo model reversal frequency as these control parameters change.

Fig. 4.40 shows a numerical dynamo calculation by Driscoll and Olson (2011) that demonstrates the reversal sensitivity to the strength of the convective forcing. In this numerical experiment, the buoyancy flux Rayleigh number Ra_F of a thermochemical dynamo is gradually increased with time as shown in Fig. 4.40A, simulating a gradual increase in both the total core-mantle boundary heat flow and the rate of light element release from the inner core growth. Fig. 4.40B shows the numerical dynamo response in terms of the history of the dipole moment and dipole tilt angle, with time expressed in dimensional units. As the convective forcing increases, the time-average dipole moment decreases while the fluctuations in the dipole moment remain nearly stationary in terms of their amplitude. Approximately halfway through the calculation, the time-average dipole moment approaches the amplitude of the largest dipole moment fluctuations, and at that point polarity reversals and excursions commence. As the convective forcing increases farther, the frequencies of the polarity excursions and reversals increase, as shown in Fig. 4.40B. In short, this dynamo calculation demonstrates (1) reversal onset at a particular Ra_F, and (2) reversal sensitivity, with reversal frequency increasing with the strength of the convective forcing.

The simulations used in polarity reversal studies require exceptionally long run times and are, therefore, made using low-resolution numerical dynamos. This means they require extrapolations that are even larger than the dynamos described earlier in this chapter, in order to reach outer core conditions. Nevertheless, the qualitative implication of these low-resolution numerical experiments is clear enough: dynamo polarity reversal frequency is sensitive to the strength of the convective forcing. Furthermore, reversal sensitivity extends beyond the overall convective forcing; other numerical dynamo studies (Glatzmaier et al., 1999; Kutzner and Christensen, 2004) have demonstrated there is some reversal sensitivity to the *pattern* of the convective forcing, in particular the pattern of heat flux at the core-mantle boundary. In short, numerical dynamos point to a connection between the thermal and dynamical state of the core and the polarity reversal history of the geodynamo. Efforts to interpret the paleomagnetic reversal record in light of this connection are now underway, and will no doubt continue to be refined as new paleomagnetic data and new insight into polarity reversal mechanics come available.

4.8 Summary

- Outer core dynamics are dominated by Earth's rotation. Large-scale flows are partly geostrophic, reflecting a balance between Coriolis accelerations and pressure gradient forces. Buoyancy forces and Lorentz forces from magnetic fields are also important.

- Outer core dynamics is best described by dimensionless parameters, including the Ekman number (ratio of viscous to Coriolis effects), the Prandtl and magnetic Prandtl numbers (ratios of viscous to thermal and magnetic diffusion), and the Rayleigh number (ratio of buoyancy to diffusion effects). Buoyancy forces in the outer core are thermochemical, due to unstable gradients in both temperature and light element concentration, the latter being somewhat larger.

- Laminar flow in the outer core includes geostrophic flows, shear flows driven by thermal and magnetic field gradients, and turbulent Reynolds stresses. Waves in the outer core include Alfvén waves (possible cause of geomagnetic jerks), torsional oscillations (responsible for length-of-day variations), inertial and Rossby waves, and in stably stratified regions, internal gravity and MAC waves.

- Numerical and laboratory models predict that convection in the outer core is quasigeostrophic, with motion elongated in the direction of the rotation axis. Convection inside the inner core tangent cylinder contrasts with convection outside the tangent cylinder because of differences in the orientation of gravity. Outer core convection is also affected by heterogeneity at both the core-mantle and the inner core boundaries.

- Numerical dynamos driven by thermochemical convection in a rotating spherical shell reproduce many observed properties of the geomagnetic field, including a dominant axial dipole, secular variation, torsional oscillations, and polarity reversals. Although far from realistic in terms of input properties, numerical dynamos are scalable to core conditions.

- Polarity reversals in numerical dynamos occur spontaneously and are associated with large amplitude fluctuations in the dipole moment.

Appendix

Appendix 4.1 Accelerations in Earth's rotating coordinates

Here, we derive the acceleration terms in the conservation of momentum equation that is due to Earth's rotation. In this derivation, we ignore the accelerations due to Earth's orbital motions, including revolution about the Sun and revolutions in the Earth-Moon system, focusing instead on the effects of Earth's rotation alone.

The basic geometry is shown in Fig. 4.41. Due to Earth's rotation, the particle labeled P moves in space by a small amount indicated by the displacement vector $\Delta \mathbf{r}$ in a short time Δt. Therefore, its velocity in space (i.e., its velocity relative to a fixed coordinate system) is given by

$$\lim_{\Delta t \to 0} \left(\frac{\Delta \mathbf{r}}{\Delta t} \right) \equiv \frac{d\mathbf{r}}{dt}. \tag{4.92}$$

Note that the vector $d\mathbf{r}/dt$ in Eq. (4.92) is perpendicular to both the position vector $\mathbf{r}$ and Earth's rotation vector $\mathbf{\Omega}$. These three vectors are related by

$$\frac{d\mathbf{r}}{dt} = \mathbf{\Omega} \times \mathbf{r}. \tag{4.93}$$

Now suppose that the particle P also moves relative to Earth's geographical coordinates, that is, moves relative to coordinates that are fixed in the solid Earth. These "earth" coordinates rotate with angular velocity $\mathbf{\Omega}$ and have the same origin as the "space" coordinates. Then, according to Eq. (4.93)

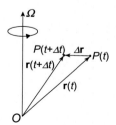

FIG. 4.41 Geometry for accelerations in Earth's rotating coordinate system. Symbols defined in the text.

$$\left.\frac{d\mathbf{r}}{dt}\right)_{space} = \left.\frac{d\mathbf{r}}{dt}\right)_{earth} + (\mathbf{\Omega} \times \mathbf{r})_{earth}, \tag{4.94}$$

where the subscripts *space* and *earth* denote the two coordinate systems. We use Eq. (4.94) to transform the acceleration $\mathbf{a} = d^2\mathbf{r}/dt^2$ from the fixed in space coordinate system to the rotating earth coordinate system, as follows:

$$\left.\frac{d^2\mathbf{r}}{dt^2}\right)_{space} = \left(\frac{d}{dt} + \mathbf{\Omega}\times\right)\left(\frac{d\mathbf{r}}{dt} + \mathbf{\Omega}\times\mathbf{r}\right)_{earth}. \tag{4.95}$$

Expanding Eq. (4.95), performing the derivatives, and collecting terms yields

$$\left.\frac{d^2\mathbf{r}}{dt^2}\right)_{space} = \left(\frac{d^2\mathbf{r}}{dt^2} + 2\mathbf{\Omega}\times\mathbf{v} + \mathbf{\Omega}\times\mathbf{\Omega}\times\mathbf{r} + \frac{d\mathbf{\Omega}}{dt}\times\mathbf{r}\right)_{earth}, \tag{4.96}$$

where $\mathbf{v} = d\mathbf{r}/dt$ is the velocity of particle P relative to the rotating earth coordinate system.

Each of the terms on the r.h.s. of Eq. (4.96) has generic names. Omitting their subscripts for now, $d^2\mathbf{r}/dt^2 = \mathbf{a}$ is called the *inertial acceleration*, $2\mathbf{\Omega}\times\mathbf{v}$ is the *Coriolis acceleration*, $\mathbf{\Omega}\times\mathbf{\Omega}\times\mathbf{r}$ is the *centrifugal acceleration*, and $(d\mathbf{\Omega}/dt)\times\mathbf{r}$ is the *transverse* or *Poincaré acceleration*.

Two of the previous terms, the Coriolis and inertial accelerations, play major roles in outer core dynamics and feature prominently in this chapter. In the conservation of momentum equation, these two accelerations are multiplied by density. Sometimes the Coriolis term is moved to the r.h.s. of the momentum equation, in which case that term is called the *Coriolis force*.

The roles of the other two accelerations are more specialized. Transverse accelerations are present whenever the vector angular velocity of Earth's rotation is unsteady. Phenomena in this category include changes in the speed of rotation, called *length-of-day* variations, or LOD for short. Also in this category are changes in the direction of rotation, such as those associated with precession, nutation, and wobble. Except when considering these specialized phenomena, it is justifiable to assume Earth's rotation is steady and to neglect the transverse acceleration.

As for the centrifugal acceleration, it can be written as the gradient of the *centrifugal potential* ψ_{cent}

$$\mathbf{\Omega}\times\mathbf{\Omega}\times\mathbf{r} = -\nabla\psi_{cent}, \tag{4.97}$$

where

$$\psi_{cent} = \frac{(\Omega r \sin\theta)^2}{2}. \tag{4.98}$$

Note that ψ_{cent} varies only with $r\sin\theta$, the perpendicular distance from the rotation axis. Accordingly, in the conservation of momentum equation, the centrifugal potential can be moved to the r.h.s. and combined with the gravitational potential, their sum being termed the *geopotential*. The effect of the centrifugal potential is to give Earth's equipotential surfaces, such as mean sea level, a dominant oblate spheroidal shape as described in Chapter 1, including the equatorial bulge and polar flattening. However, because it only contributes to the small ellipticity of equipotential surfaces in the core (less than one part in 300), it is customary to ignore its contribution to outer core dynamics. Exceptions to this rule include the dynamics related to precession and nutation.

Lastly, we can write the inertial acceleration term in Eq. (4.96) in terms of displacements from a reference state $\mathbf{u}$, or in terms of velocity $\mathbf{v}$. For displacements, $\mathbf{u} = \mathbf{r} - \mathbf{r}_0$, where $\mathbf{r}_0$ is the reference position of the particle, and the inertial acceleration on the r.h.s. of Eq. (4.96) becomes

$$\left.\frac{d^2\mathbf{r}}{dt^2}\right)_{earth} = \left.\frac{d^2\mathbf{u}}{dt^2}\right)_{earth}, \tag{4.99}$$

which is the form used for seismic waves. For fluid flow and other continuous motions, the inertial acceleration is expressed in terms of velocity, and takes one of the following two forms:

$$\left.\frac{d^2\mathbf{r}}{dt^2}\right)_{earth} = \left.\frac{d\mathbf{v}}{dt}\right)_{earth} \equiv \left.\frac{\partial\mathbf{v}}{\partial t} + \mathbf{v}\cdot\nabla\mathbf{v}\right)_{earth} \tag{4.100}$$

the first form on the r.h.s. involving just the material time derivative d/dt, the second form involving the local time derivative $\partial/\partial t$.

Appendix 4.2 Equations of motion

We start with a description of the equations of motion in their dimensional forms, and then consider various ways they can be nondimensionalized. The *conservation of momentum* (a.k.a. the Navier-Stokes equation) for outer core flows can be written as

$$\rho_0 \left(\frac{d\mathbf{v}}{dt} + 2\mathbf{\Omega} \times \mathbf{v} \right) = -\nabla P' + \rho' \mathbf{g} + \eta \nabla^2 \mathbf{v} + \mathbf{F}_L, \qquad (4.101)$$

where $\mathbf{v}$ is the fluid velocity, t is the time, ρ_0 is a reference density, typically equated to the mean outer core density or the mean density of some part of the outer core, $\mathbf{\Omega} = \Omega \hat{\zeta}$ is the angular velocity vector of Earth's rotation, $\hat{\zeta}$ is the unit axial vector, P' is the reduced pressure, $\mathbf{g}$ is the gravity vector, ρ' is the density perturbation, and η is the dynamic viscosity. Here, "reduced" means a deviation from the hydrostatic state usually connected to the dynamics. The terms on the l.h.s. of Eq. (4.101) are the inertial and Coriolis accelerations. The terms on the r.h.s. are the pressure, buoyancy, viscous, and Lorentz forces, respectively. In Eq. (4.101), all material properties are represented by constant values except the density perturbation. In addition, gravity is allowed to vary with radius. The Lorentz force $\mathbf{F}_L$ is specified below. A hydrostatic balance of the form

$$\nabla \overline{P} - \overline{\rho} \mathbf{g} = 0 \qquad (4.102)$$

has been subtracted from Eq. (4.101), the overlines denoting spherically averaged but radially variable pressure and density.

The validity of Eq. (4.101) in the outer core depends on a number of conditions. First, it assumes uniform dynamic viscosity η. Second, it assumes that we can replace the variable density of the outer core with a constant reference value ρ_0 in every term except the buoyancy force term. This is called the *Boussinesq approximation*. In applying the Boussinesq approximation to Eq. (4.101), we are basically ignoring effects of compressibility (through the elastic bulk modulus) on the flow. Compressibility effects need to be included in Eq. (4.101) if the variation in hydrostatic density across the fluid $\Delta \overline{\rho}$ in Eq. (4.102) is comparable to the reference density ρ_0. From the seismic models in Chapter 1, $\Delta \overline{\rho} \simeq 2.3 \times 10^3$ kg/m^3, whereas $\rho_0 \simeq 10.9 \times 10^3$ kg/m^3, nearly five times larger. Another condition derives from the codensity equation, which requires the so-called dissipation number $Di = \alpha_T g d / c_p$ be small. The property Tables S1 and S2 from Core Properties and Parameter yield $Di \simeq 0.3$. Although hardly "slam dunks," these comparisons validate the Boussinesq approximation in Eq. (4.101) to first order in the outer core.

The *conservation of mass* with compressibility effects included is given by

$$\frac{\partial \rho}{\partial t} + \nabla \cdot (\rho \mathbf{v}) = 0. \qquad (4.103)$$

Under the assumption that $\rho' \ll \overline{\rho}$ (a very safe assumption for the outer core) along with the conditions validating the Boussinesq approximation, this can be broken into the *continuity equation* for the flow

$$\nabla \cdot \mathbf{v} = 0 \qquad (4.104)$$

plus an equation for adiabatic density perturbations that takes into account compressibility and stratification. Using thermodynamic relationships from Chapter 2, this can be written

$$\frac{\partial \rho'}{\partial t} = v_r \left(\frac{d\rho_{ad}}{dr} - \frac{d\overline{\rho}}{dr} \right), \qquad (4.105)$$

where v_r is the radial component of $\mathbf{v}$ and $d\rho_{ad}/dr$ is the adiabatic density gradient. Eq. (4.105) is used for internal gravity wave motion, for example.

The *heat transport equation* for the outer core can be written in terms of temperature T as

$$\frac{dT}{dt} = \kappa_T \nabla^2 T + \frac{h}{c_P}, \qquad (4.106)$$

where κ_T is thermal diffusivity (assumed uniform), h is the radioactive heat production per unit of mass, and c_P is the specific heat. As with density, we express the temperature as the sum of a spherical average (but radially varying) part plus a much smaller perturbation related to the dynamics: $T = \overline{T} + T'$. Substituting into Eq. (4.106) and collecting terms of similar size yields

$$\frac{dT'}{dt} - \kappa_T \nabla^2 T' = -\left(\frac{\partial \overline{T}}{\partial t} - \kappa_T \nabla^2 \overline{T} - \frac{h}{c_P}\right), \qquad (4.107)$$

with the terms on l.h.s. being functions of radius and time only. To the extent that the spherical mean temperature is (approximately) adiabatic, we can replace $\overline{T}$ in Eq. (4.107) with T_{ad}.

The *light elements transport equation* for the outer core can be written in terms of the light element concentration

$$\frac{dC}{dt} = \kappa_C \nabla^2 C, \qquad (4.108)$$

where κ_C is the light element diffusivity in the fluid and C is its concentration in kg/kg units. We then express the concentration as the sum of a uniform mean part C_0 and a much smaller perturbation C'. Substitution into Eq. (4.108) yields

$$\frac{dC'}{dt} - \kappa_C \nabla^2 C' = -\frac{dC_0}{dt}. \qquad (4.109)$$

The *codensity transport equation* combines the equations for heat and mass (light element) transport. We define codensity as the sum of a mean value

$$\chi_0 = \rho_0(\alpha_T T_0 + \alpha_C C_0) \qquad (4.110)$$

plus a perturbation.

$$\chi' = \rho_0(\alpha_T T' + \alpha_C C') \qquad (4.111)$$

with $\alpha_T = -1/\rho(\partial \rho/\partial T) > 0$ and $\alpha_C = -1/\rho(\partial \rho/\partial C) > 0$.

On combining Eqs. (4.107)–(4.111) and assuming $\kappa_C = \kappa_T = \kappa_\chi$, we obtain

$$\frac{d\chi'}{dt} - \kappa_\chi \nabla^2 \chi' = \rho_0 \alpha_T \left(\frac{h}{c_P} - \frac{dT_0}{dt}\right) - \rho_0 \alpha_C \frac{dC_0}{dt}, \qquad (4.112)$$

where dT_0/dt is defined as

$$\frac{dT_0}{dt} = \frac{\partial T_{ad}}{\partial t} - \kappa_T \nabla^2 T_{ad}. \qquad (4.113)$$

Appendix 4.3 Nondimensional equations

To nondimensionalize these equations, we scale lengths by d, time by the viscous diffusion time d^2/ν, velocity by ν/d, and reduced pressure by $\kappa_B \Omega$. Different scalings, based on different choices of the convective forcing, are given here. These scalings involve dimensionless control parameters from Section 4.2 and Table 4.1, namely the Prandtl number Pr, the Ekman number Ek, and several versions of the Rayleigh number Ra. In what follows, these Rayleigh number versions are identified by subscripts, and their definitions and relationships to the others are listed in Table 4.2.

Appendix 4.3.1 Temperature scaling

This is a traditional scaling, used in the initial studies of rotating thermal convection in the outer core. It is based on fixing ΔT, the temperature at the inner core boundary minus that at the core-mantle boundary, with the adiabatic temperature change across the outer core subtracted out. A fixed temperature at the core-mantle boundary is not a particularly realistic choice for that boundary condition, and more importantly, it is difficult to estimate ΔT in the outer core because it is the small difference between the vastly larger radial variations in T and T_{ad}. However, being the traditional choice, it makes for easier comparison with thermal convection in other fluid systems, laboratory experiments and theoretical treatments in particular. With the codensity perturbation consisting only of a thermal part, the perturbation density becomes $\rho' = -\rho_0 \alpha_T T'$ and the dimensionless form of the Navier-Stokes equation (4.101) without the Lorentz force becomes

$$Ek\left(\frac{d\mathbf{v}}{dt} - \nabla^2 \mathbf{v}\right) + 2\hat{\zeta} \times \mathbf{v} = -\nabla P' + (Ek\, Pr^{-1}\, Ra_T)\mathbf{r}^* T', \qquad (4.114)$$

where $\hat{\zeta}$ is the unit vector parallel to Ω, $\mathbf{r}^* = \mathbf{r}/r_{CMB}$, and we have assumed that gravity varies linearly with radius, $\mathbf{g} = -g_{CMB}\mathbf{r}^*$. The dimensionless heat transport equation, with the various heat sources included, becomes

$$\frac{dT'}{dt} = Pr^{-1}\nabla^2 T' + \epsilon, \tag{4.115}$$

where $\epsilon = Ra_h/Ra_T$ and Ra_h is the Rayleigh number based on radioactive heat production, defined in Table 4.2.

The continuity equation (4.104) completes this set. The usual boundary conditions for temperature are given in non-dimensional form by $T' = 0$ and $T' = 1$ at the core-mantle and inner core boundaries. Usual boundary conditions on the fluid velocity are no-slip $\mathbf{v} = 0$ at the core-mantle and inner core boundaries.

Appendix 4.3.2 Codensity scaling

Here, we substitute the perturbation codensity χ' for the perturbation density ρ' and define a scale factor for χ' based on the codensity buoyancy flux F, given by $\rho_0 F d/\nu g_{CMB}$. With these, the dimensionless form of the Navier-Stokes equation (4.101) becomes

$$Ek\left(\frac{d\mathbf{v}}{dt} - \nabla^2 \mathbf{v}\right) + 2\hat{\zeta} \times \mathbf{v} = -\nabla P' + (Ek\, Pr^{-1}\, Ra_F)\mathbf{r}^*\chi', \tag{4.116}$$

the continuity equation is Eq. (4.104), and the dimensionless version of the codensity equation (4.112) becomes

$$\frac{d\chi'}{dt} = Pr^{-1}\nabla^2 \chi' + \epsilon, \tag{4.117}$$

where

$$\epsilon = -\frac{Ra_{\dot\chi}}{Ra_F} + \frac{Ra_h}{Ra_F}, \tag{4.118}$$

with the Rayleigh numbers Ra_F, $Ra_{\dot\chi}$, and Ra_h defined in Table 4.2. Typical boundary conditions include a specified codensity χ' on the inner core boundary and a specified codensity flux $\partial \chi'/\partial r$ at the core-mantle boundary.

Appendix 4.4 Lorentz force and Maxwell stress

The *Lorentz force* arises from the interaction between an electrically conducting fluid and a magnetic field, when there is relative motion between the two. In terms of the magnetic field intensity $\mathbf{B}$ and the electric current density $\mathbf{J}$, the Lorentz force per unit volume $\mathbf{F}_L$ can be written alternatively as

$$\mathbf{F}_L = \mathbf{J} \times \mathbf{B} = \frac{1}{\mu_0}(\nabla \times \mathbf{B}) \times \mathbf{B} = -\nabla\left(\frac{\mathbf{B}\cdot\mathbf{B}}{2\mu_0}\right) + \nabla\cdot\left(\frac{\mathbf{BB}}{\mu_0}\right), \tag{4.119}$$

or using indicial notation

$$F_i = \frac{\partial}{\partial x_j} M_{ij}, \tag{4.120}$$

where M is the *Maxwell stress tensor*, with components

$$M_{ij} = -\frac{B_k B_k}{2\mu_0}\delta_{ij} + \frac{B_i B_j}{\mu_0}. \tag{4.121}$$

The first term in Eq. (4.121) represents *magnetic pressure*, the second term represents *magnetic tension*. In vector notation, the magnetic pressure force is

$$\mathbf{F}_L^P = -\nabla\left(\frac{\mathbf{B}\cdot\mathbf{B}}{2\mu_0}\right) \tag{4.122}$$

and the magnetic tension force is

$$\mathbf{F}_L^T = (\mathbf{B}\cdot\nabla)\frac{\mathbf{B}}{\mu_0}. \tag{4.123}$$

Appendix 4.5 Convective dynamos

To adapt the rotating convection equations of motion for convective dynamos, we first add the Lorentz force F_L (derived above) to the r.h.s. of the Navier-Stokes equation (4.101). We also include Gauss' law for the magnetic intensity **B**

$$\nabla \cdot \mathbf{B} = 0 \qquad (4.124)$$

and the hydromagnetic induction equation from the Appendix of Chapter 3. For incompressible flow (i.e., invoking the Boussinesq approximation) and for constant magnetic diffusivity κ_B, the induction equation is

$$\frac{d\mathbf{B}}{dt} = (\mathbf{B} \cdot \nabla)\mathbf{v} + \kappa_B \nabla^2 \mathbf{B}, \qquad (4.125)$$

where **v** is the fluid velocity.

Typical boundary and continuity conditions on the magnetic field are as follows. At material interfaces, we require continuity of the normal and tangential magnetic field components, continuity of the normal component of the electrical current density and the tangential component of the electric field. The way these continuity conditions express themselves in terms of the magnetic field alone depends on the electrical properties of the materials on both sides of the interface.

The most commonly used condition at the core-mantle boundary assumes the electrical conductivity of the mantle is negligible, in which case the normal component of the current density vanishes there, which implies that the toroidal component of the magnetic field vanishes there,

$$\mathbf{B}_T = 0 \quad \text{at} \quad r = r_{CMB}. \qquad (4.126)$$

In addition, the absence of electric currents means that the poloidal magnetic field emerging from the outer core is a potential field in the mantle. Continuity between the poloidal magnetic field inside the core and the potential field in the mantle cannot be expressed in terms of a local condition in physical space, but it can be expressed as a local condition in the space of the spherical harmonics of the two fields. In terms of the amplitude of the poloidal magnetic potential at spherical harmonic degree n, continuity requires

$$\frac{d\mathcal{P}_n}{dr} + \frac{n+1}{r}\mathcal{P}_n = 0 \quad \text{at} \quad r = r_{CMB}. \qquad (4.127)$$

To model electrical conductivity in the inner core, Eq. (4.125) is solved with the velocity **v** constrained to solid-body rotation determined by a suitable torque balance. In addition, continuity of the toroidal magnetic potential, continuity of the poloidal magnetic potential and its radial derivative, and continuity of the tangential electric field are needed.

The magnetic boundary conditions at the core-mantle boundary listed earlier are homogeneous and do not introduce new dimensionless parameters. The new dimensionless parameters come into the Lorentz force and the induction equation. Using $(\rho_0 \Omega/\sigma)^{1/2}$ to scale the magnetic field, the dimensionless forms of Eqs. (4.120), (4.125) are, for the Lorentz force

$$\mathbf{F}_L = \frac{1}{Pm}(\nabla \times \mathbf{B}) \times \mathbf{B} \qquad (4.128)$$

and for the induction equation

$$\frac{d\mathbf{B}}{dt} = (\mathbf{B} \cdot \nabla)\mathbf{v} + \frac{1}{Pm}\nabla^2 \mathbf{B}, \qquad (4.129)$$

where $Pm = \nu/\kappa_B$ is the magnetic Prandtl number. The other new dimensionless parameter is the square of the magnetic field intensity scale, the Elsasser number $\Lambda = (\sigma B^2/\rho_0 \Omega)$.

Appendix 4.6 Global kinetic energy balance

The global kinetic energy balance for outer core fluid motions is obtained by taking the inner product of the fluid velocity v and the Navier-Stokes equation (4.101) with the Lorentz force (4.119) included, and integrating the result over the outer core volume. The first term on the left-hand side of the resulting equation (the full equation not shown here) is the temporal change in kinetic energy:

$$\int \rho_0 \left(\mathbf{v} \cdot \frac{\partial \mathbf{v}}{\partial t}\right) dV = \int \rho_0 \frac{\partial}{\partial t}\left(\frac{v^2}{2}\right) dV = \frac{\partial}{\partial t} E_{kin}, \tag{4.130}$$

where E_{kin} denotes the total kinetic energy of the fluid motion. The second term on the left-hand side of that equation vanishes on volume integration. The third term on the left-hand side of that equation vanishes because $\mathbf{v}$ is perpendicular to the Coriolis acceleration. The pressure force term on the right-hand side becomes, using Eq. (4.104) and Stokes' theorem,

$$\int_V (\mathbf{v} \cdot \nabla P) dV = \int_{CMB} (P v_r) dA. \tag{4.131}$$

The area integral in this term is zero if the radial component of the fluid velocity v_r vanishes at the CMB, which is a standard assumption.

The second term on the right-hand side, involving the buoyancy force, becomes the production of kinetic energy by the buoyancy force, called *buoyancy production* for short, here denoted by $\mathcal{B}$. Only the radial component of the velocity appears in this term:

$$\mathcal{B} = \int (\rho_0 g' v_r) dV. \tag{4.132}$$

The third term on the right-hand side, involving the viscous force, becomes the viscous heating:

$$Q_\nu = \int \eta (\nabla \times \mathbf{v})^2 dV. \tag{4.133}$$

The last term on the right-hand side includes the Lorentz force and involves the rate of work done against the Lorentz force, denoted by W_L:

$$W_L = -\int_V (\mathbf{v} \cdot \mathbf{F_L}) dV. \tag{4.134}$$

In shorthand form, the resulting kinetic energy balance for outer core fluid motions is then

$$\frac{\partial}{\partial t} E_{kin} = \mathcal{B} - (Q_\nu + W_L). \tag{4.135}$$

References

Acheson, D.J., 1978. Magnetohydrodynamic waves and instabilities in rotating fluids. In: Roberts, P.H., Soward, A.M. (Eds.), Rotating Fluids in Geophysics. Academic Press, New York, pp. 315–349.

Alboussiere, T., Deguen, R., Melzani, M., 2010. Melting-induced stratification above the Earth's inner core due to convective translation. Nature 466, 744–747.

Aubert, J., 2019. Approaching Earth's core conditions in high-resolution geodynamo simulations. Geophys. J. Int. 219, 137–151.

Aubert, J., Amit, H., Hulot, G., Olson, P., 2008. Thermo-chemical wind flows couple Earth's inner core growth to mantle heterogeneity. Nature 454, 758–762.

Aurnou, J.M., Calkins, M.A., Cheng, J.S., Julien, K., King, E.M., Nieves, D., Soderlund, K.M., Stellmach, S., 2015. Rotating convective turbulence in Earth and planetary cores. Phys. Earth Planet. Inter. 246, 52–71.

Braginsky, S.I., 1993. MAC-oscillations of the hidden ocean of the core. J. Geomagn. Geoelectr. 45, 1517–1538.

Brito, D., Alboussiere, T., Cardin, P., Jault, D., et al., 2011. Zonal shear and super-rotation in a magnetized spherical Couette-flow experiment. Phys. Rev. E 83, 066310.

Buffett, B.A., 2010. Tidal dissipation and the strength of the Earth's internal magnetic field. Nature 468, 952–955.

Buffett, B.A., 2014. Geomagnetic fluctuations reveal stable stratification at the top of the Earth's core. Nature 507, 484–487.

Buffett, B.A., Knezek, N., Holme, R., 2016. Evidence for MAC waves at the top of Earth's core and implications for variations in length of day. Geophys. J. Int. 204 (3), 1789–1800.

Busse, F.H., 1970. Thermal instabilities in rapidly rotating systems. J. Fluid Mech. 44, 441–460.

Cardin, P., Olson, P., 1994. Chaotic thermal convection in a rapidly rotating spherical shell: consequences for flow in the outer core. Phys. Earth Planet. Inter. 82, 235–259.

Carrigan, C.R., Busse, F.H., 1983. An experimental and theoretical investigation of the onset of convection in rotating spherical shells. J. Fluid Mech. 126, 287–305.

Chandrasekhar, S., 1961. Hydrodynamic and Hydromagnetic Stability. Clarendon Press, Oxford.

Cheng, J.S., Stellmach, S., Ribeiro, A., Grannan, A., King, E.M., Aurnou, J.M., 2015. Laboratory-numerical models of small-scale flows in planetary cores. Geophys. J. Int. 201, 1–17.

Christensen, U.R., 2018. Geodynamo models with a stable layer and heterogeneous heat flow at the top of the core. Geophys. J. Int. 215, 1338–1351.
Christensen, U.R., Aubert, J., 2006. Scaling properties of convection driven dynamos in rotating spherical shells and application to planetary magnetic fields. Geophys. J. Int. 166, 97–114.
Christensen, U.R., Wicht, J., 2015. Numerical dynamo simulations. In: Schubert, G. (Ed.), second ed. Treatise on Geophysics, vol. 8. Elsevier, Oxford.
Christensen, U.R., Aubert, J., Busse, F.H., et al., 2001. A numerical dynamo benchmark. Phys. Earth Planet. Inter. 128, 25–34.
Deguen, R., Olson, P., Reynolds, E., 2014. F-layer formation in the outer core with asymmetric inner core growth. C. R. Geosci. 346, 101–109.
Driscoll, P., Olson, P., 2011. Superchron cycles driven by variable core heat flow. Geophys. Res. Lett. 38, L09304.
Finlay, C.C., Jackson, A., 2003. Equatorially dominated magnetic field change at the surface of Earth's core. Science 300, 2084–2086.
Gastine, T., Wicht, J., Aubert, J., 2016. Scaling regimes in spherical shell rotating convection. J. Fluid Mech. 808, 690–732.
Gillet, N., Jault, D., Canet, E., Fournier, A., 2010. Fast torsional waves and strong magnetic field within the Earth's core. Nature 465, 74–77.
Glatzmaier, G.A., Roberts, P.H., 1995. A three-dimensional self-consistent computer simulation of a geomagnetic field reversal. Nature 337, 203–209.
Glatzmaier, G.A., Coe, R.S., Hongre, L., Roberts, P.H., 1999. The role of the Earth's mantle in controlling the frequency of geomagnetic reversals. Nature 401, 885–890.
Guervilly, C., Cardin, P., 2016. Subcritical convection of liquid metals in a rotating sphere using a quasigeostrophic model. J. Fluid Mech. 808, 61–89.
Kutzner, C., Christensen, U.R., 2004. Simulated geomagnetic reversals and preferred virtual geomagnetic pole paths. Geophys. J. Int. 157, 1105–1118.
Mound, J., Davies, C., Rost, S., Aurnou, J., 2019. Regional stratification at the top of Earth's core due to core-mantle boundary heat flux variations. Nat. Geosci. 12 (7), 575–580.
Olson, P., Christensen, U.R., Glatzmaier, G.A., 1999. Numerical modeling of the geodynamo: mechanism of field generation and equilibration. J. Geophys. Res. 104, 10383–10404.
Sreenivasan, B., Jones, C.A., 2006. The role of inertia in the evolution of spherical dynamos. Geophys. J. Int. 164, 467–476.
Starchenko, S., Jones, C.A., 2002. Typical velocities and magnetic field strengths in planetary interiors. Icarus 157, 426–435.
Sumita, I., Olson, P., 2000. Laboratory experiments on high Rayleigh number thermal convection in a rapidly rotating hemispherical shell. Phys. Earth Planet. Inter. 117, 153–170.
Wicht, J., Christensen, U.R., 2010. Torsional oscillations in dynamo simulations. Geophys. J. Int. 181, 1367–1380.

Further readings and resources

Numerical and laboratory modeling resources

MagIC is a numerical code developed by Johannes Wicht and Thomas Gastine that simulates fluid dynamics in a spherical shell, including dynamo action. MagIC solves for the Navier-Stokes equation including Coriolis force, optionally coupled with an induction equation for magnetohydrodynamics (MHD), a temperature (or entropy) equation and an equation for chemical composition under both the anelastic and the Boussinesq approximations (https://magic-sph.github.io/).

Rayleigh is a 3D convection code developed by Nick Featherstone for the study of dynamo behavior in spherical shell geometries. It solves the incompressible and anelastic MHD equations of motion in spherical shell geometry using advanced parallelization methods for improved efficiency. Hosted by the Computational Infrastructure for Geodynamics (CIG) organization (https://geodynamics.org/cig/software/rayleigh/).

SPINLAB is a laboratory in the Earth and Space Sciences Department at UCLA devoted to geophysical fluid dynamics experimentation. In addition to research projects on the dynamics of planetary cores and atmospheres, this site provides teachers at all levels with access to and instructions for classroom demonstrations. Contact: Jonathan Aurnou (https://spinlab.ess.ucla.edu/)

Maryland Sodium Dynamo Project. An enormous 3 m diameter rotating spherical shell filled with liquid sodium, used for a variety of scale model experiments on outer core dynamics and magnetic field generation processes. Contact: Daniel Lathrop (https://complex.umd.edu/research/MHDdynamos/MHDdynamos.php/).

In addition to the references cited in the text of this chapter, here are some additional readings, organized by subject matter. These include general texts and monographs, review papers, a selection of older classic studies, plus some papers that offer different perspectives on the topics presented in this chapter.

Geophysical fluid dynamics (GFD) and magnetohydrodynamics (MHD)

Davidson, P.A., 2001. An Introduction to Magnetohydrodynamics. Cambridge University Press, Cambridge.
Davidson, P.A., 2013. Turbulence in Rotating, Stratified and Electrically Conducting Fluids. Cambridge University Press.
Gillet, N., Schaeffer, N., Jault, D., 2011. Rationale and geophysical evidence for quasigeostrophic rapid dynamics within the Earth's outer core. Phys. Earth Planet. Inter. 187, 78–88.
Glatzmaier, G.A., 2013. Introduction to Modeling Convection in Planets and Stars: Magnetic Field, Density Stratification, Rotation. Princeton Series in Astrophysics, vol. 24 Princeton University Press, Princeton, NJ.
Greenspan, H.P., 1968. The Theory of Rotating Fluids. Cambridge University Press, Cambridge.

Gubbins, D., Roberts, P.H., 1987. Magnetohydrodynamics of the Earth's core. In: Jacobs, J.A. (Ed.), Geomagnetism. vol. 2. Academic Press, New York, pp. 1–183.
Kundu, P.K., Cohen, I.M., Dowling, D., 2018. Fluid Mechanics, sixth ed. Academic Press.
Schubert, G., Olson, P. (Eds.), 2015. Core dynamics. In: The Treatise on Geophysics. vol. 8. Elsevier, Oxford.
Smylie, D.E., 2013. Earth's Dynamics, Deformations and Oscillations of the Rotating Earth. Cambridge University Press, New York.
Soward, A.M., Roberts, P.H., 2007. Taylor's constraint. In: Dormy, E., Soward, A.M. (Eds.), Mathematical Aspects of Natural Dynamos. CRC Press, Boca Raton, FL, pp. 419–424.
Sreenivasan, B., Jones, C.A., 2005. Structure and dynamics of the polar vortex in the Earth's core. Geophys. Res. Lett. 32, L20301.

Perspectives on convection inside and outside the tangent cylinder

Aubert, J., Amit, H., Hulot, G., Olson, P., 2008. Thermo-chemical wind flows couple Earth's inner core growth to mantle heterogeneity. Nature 454, 758–762.
Aurnou, J., Andreadis, S., Zhu, L., Olson, P., 2003. Experiments on convection in the Earth's core tangent cylinder. Earth Planet. Sci. Lett. 21, 119–134.
Zhang, K., 1995. Spherical shell rotating convection in the presence of a toroidal magnetic field. Proc. R. Soc. Lond. A 448, 245–268.

Numerical dynamos

Aubert, J., Labrosse, S., Poitou, C., 2009. Modelling the palaeo-evolution of the geodynamo. Geophys. J. Int. 179, 1414–1428.
Christensen, U.R., 2011. Geodynamo models: tools for understanding properties of Earth's magnetic field. Phys. Earth Planet. Inter. 187, 157–169.
Christensen, U.R., Aubert, J., Hulot, G., 2010. Conditions for Earth-like geodynamo models. Earth Planet. Sci. Lett. 296, 487–496.
Davidson, P.A., 2016. Dynamos driven by helical waves: scaling laws for numerical dynamos and for the planets. Geophys. J. Int. 207, 680–690.
Dormy, E., Valet, J.P., Courtillot, V., 2000. Numerical models of the geodynamo and observational constraints. Geochem. Geophys. Geosyst. 1. 2000GFC000062.
Glatzmaier, G.A., 2002. Geodynamo simulations—how realistic are they? Ann. Rev. Earth Planet. Sci. 30, 237–257.
Glatzmaier, G.A., Roberts, P.H., 2002. Simulating the geodynamo. Contemp. Phys. 38, 269–288.
Olson, P., Christensen, U.R., 2002. The time-averaged magnetic field in numerical dynamos with nonuniform boundary heat flow. Geophys. J. Int. 151, 809–823.
Sakuraba, M., Roberts, P.H., 2009. Generation of a strong magnetic field using uniform heat flux at the surface of the core. Nat. Geosci. 2, 802–805.
Tilgner, A., 2005. Precession driven dynamos. Phys. Fluids 17, 034104.

CHAPTER
5
Boundary regions

5.1 D″ (lowermost mantle region)

The lowermost mantle is a thermal boundary layer through which heat from the core is conducted into the mantle across a steepened temperature gradient. This temperature gradient is estimated from connecting a temperature extrapolated from the adiabatic temperature gradient in the mantle to the temperature estimated for the liquid iron alloyed core. Lateral variations in this temperature profile induce lateral variations in both heat transport out of the core and fluid flow of the outer core near the core-mantle boundary. These flow variations in turn control spatial and secular variations of the magnetic field. Important goals of imaging the seismic structure of D″ are to improve estimates of its vertical temperature profile and its lateral variation and to quantify the separate contributions of conductive and convective heat transport across the CMB.

Complicating these goals is the lowermost 200–600 km of the mantle is perhaps the most structurally complex region of Earth outside its near surface. It has consequently proved to be the most difficult to spatially resolve and separate thermal effects from competing compositional effects on physical properties inferred from the seismic wavefields sampling it. Among the laterally varying structures, characterized by a broad spatial spectrum, are dense piles of chemically distinct and primitive mantle, fragments of subducted slabs and phase changes within the slabs, silicate-iron reaction products at and above the CMB entrained by mantle flow, partial melts, ascending plume structures, and regions of elastic anisotropy due to intrinsic lattice structure and flow-oriented heterogeneities. (Fig. 5.1).

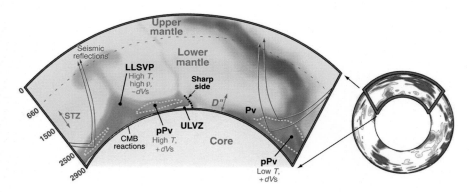

FIG. 5.1 Interpretations of structural complexity in the lowermost mantle. *From Garnero, E.J., McNamara, A.K., 2008. Structure and dynamics of Earth's lower mantle, Science, 320 (626), https://doi.org/10.1126/science.1148028.*

5.1.1 Travel-time tomography

Global tomographic images (Lekić et al., 2012) robustly agree on the form of large-scale seismic structure: two antipodal regions of low P and S velocity centered beneath South Africa and the South Pacific, ringed by high velocity (Fig. 5.2). When spherical harmonic analysis is applied to these images, the spatial spectrum of heterogeneity of the lowermost mantle is slightly reddened compared to the upper mantle. Regional imaging studies, however, suggest that this reddening does not signify any fundamental change in the scales of heterogeneity, but rather is due to a diminished capability of many global studies to resolve shorter wavelength structures in the lowermost mantle. Simplified hypotheses of the two large-scale, low-velocity structures are that they are dense, chemically distinct, piles of primitive mantle, large-scale hot regions, regional averages of concentrated thinner hot plumes, or a combination of these. A simple interpretation of the fast regions ringing the two slow patches is that they are regions of cold, dense, sinking lithospheric slabs.

FIG. 5.2 A comparison of S wave anomalies at 2800 km depth near the CMB from five global tomographic studies (A–E) and a summary of S wave travel time anomalies sensitive to lowermost mantle structure (F) where blue is fast and red slow. *Modified from Lekić, V., Cottar, S., Dziewonski, A., and Romanowicz, B., 2012. Cluster analysis of global lower mantle tomography: a new class of structure and implications for chemical heterogeneity, Earth Planet. Int., 357–358, 68-77.*

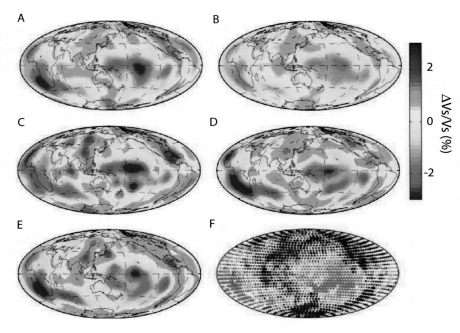

5.1.2 Slabs and plumes

Dense receiver arrays, such as Earthscope in North America, and dense seismicity, such as that along the subduction zones of South America, have made possible the imaging of smaller scale structures along the core-mantle boundary in several regions. Combined P and S velocity images can assist in interpreting whether a velocity anomaly is thermal or chemical in nature. The CMB region beneath Central America is a particularly well-studied area (Fig. 5.3). Full-waveform inversions of S, ScS, SKS, and SKKS have illuminated a complex structure of high-velocity, slab-like anomalies in this region, interpreted to be related to subducted and deformed remnants of the Farallon plate (Wang et al., 2008; Shang et al., 2014; Borgeaud et al., 2017). The high velocities are presumed to signify a *post-perovskite* phase transition (Murakami et al., 2004; Oganov and Ono, 2004), which has been predicted to occur in the MORB (mid-ocean ridge basalt) veneer of a subducted lithospheric slab.

FIG. 5.3 Image of the structure near the CMB beneath Central America from Borgeaud et al. (2017).

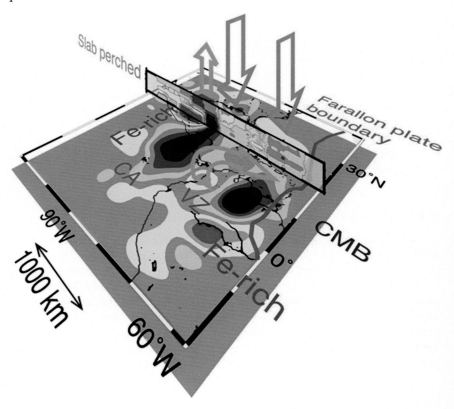

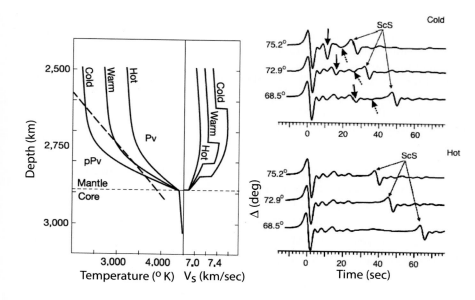

FIG. 5.4 Left: Geotherms and Clapeyron slope predicted for the perovskite (Pv) to post-perovskite phase transition in the lowermost mantle, and predicted V_S velocity profiles for the geotherms. Right: Observed SH component seismograms showing the reflection from the predicted V_S discontinuities. The reflections from the upper and lower surfaces of the Pv to pPv and pPv to PV transitions are marked by arrows. Adapted from Hernlund, J.W., Thomas, C., Tackley, P.J., 2005. A doubling of the post-perovskite phase boundary and structure of the Earth's lowermost mantle. Nature 434(7034), 882–886, https://doi.org/10.1038/nature03472.

5.1.3 The post-perovskite phase transition

The earliest seismic identification of the post-perovskite phase transition occurred in the work of Lay and Helmberger (1983), who interpreted a body wave arriving between S and ScS as a total reflection, part of a travel-time triplication associated with a rapid or discontinuous velocity increase of 200–400 km above the CMB. Subsequent searches for this lower mantle discontinuity have found both its expressions and depth to be laterally varying, absent beneath lower shear velocity (presumed higher temperature) regions of the lower mantle. Depending on the shape of the geotherm, a double-crossing of the post-perovskite phase transition has been predicted (Hernlund et al., 2005). This double-crossing can be observed in seismic body waves sampling D″ by the effects on body waves of paired discontinuities in S velocity having opposite signs (Fig. 5.4). Paired discontinuities of lower magnitude are also predicted for P velocity, anticorrelated with the signs of the jumps in S velocity. From the seismically determined depths of these discontinuities, which determine the pressures of the phase changes, both the absolute temperature and the temperature gradient above the core-mantle boundary can be estimated. The product of this estimated temperature gradient with an estimate of the thermal conductivity of the lowermost mantle allows an estimate of the heat transported out of the outer core into the mantle across the CMB, which must be satisfied by any source of energy proposed for driving the convective motion in the outer core. A recent ab initio calculation of the elastic moduli for the post-perovskite transition predicts a frequency dependence of the reflection coefficients of the top and bottom of the zones in which the phase transition occurs. This frequency dependence is due to the possibility that the two silicate phases may coexist over a range of depths and that the cycle stress of the incident elastic wave may locally induce the phase change (Langrand et al., 2019). The induced phase change can change the reflection coefficients by up to 2 orders of magnitude if the kinetics of the phase change is such that it can be induced during a cycle of time on the order of the dominant period of the incident wave (order of 1–10 s for low-frequency S waves). Unfortunately, the uncertainties of this complex behavior add to the challenge in estimating the temperature profile near the CMB. These challenges are already daunting due to the strong lateral variations of chemistry, temperature, and seismic attenuation in D″.

5.1.4 Ultralow velocity zones and LLVPs

While patches of high seismic velocity anomalies near the CMB have been largely interpreted as signatures of either subducted slab remnants or phase changes within subducted slabs, larger uncertainties remain in the interpretation of low-velocity anomalies near the CMB. These include the two large low velocity provinces (LLVPs) imaged by global tomography and smaller, locally thinner, ultralow velocity zones (ULVZs) along the core-mantle boundary. Debates include the geometry and association of deep mantle plumes with the LLVPs and ULVZs and their associated surface volcanism and chemistry. The ULVZs are thin zones (10s of km) of extremely low S velocity and P velocity (reductions up to 50% and 25%, respectively) extending laterally along the core-mantle boundary for several 100–1000 km or more (Yu and Garnero, 2018). Both the distribution of surface hotspots presumed to originate from the conduits of thin upwelling plumes and the distribution of imaged ULVZs are more concentrated near the edges of the two LLVPs (Thorne et al., 2004). Exceptions and complexities added to these apparent correlations may be explained by patterns of mantle convection (Hassan et al., 2015). Long-term persistence of the LLVPs and ULVZs in the midst of mantle convection requires a positive density difference between their composition and that of the rest of the mantle, estimated to be up to +6% for LLVPs and up to +20% for ULVZs. Suggestive of a chemical anomaly in LLVPs is an anticorrelation between bulk sound velocity, $V_K = \sqrt{\frac{V_P^2 - 4/3 V_S^2}{\rho}}$, and shear velocity V_S (Masters et al., 2000). The correlation of plume and hotspot locations with the edges of LLVPs and ULVZs may be related to the absorption of core heat beneath subducted slabs, allowing the formation of dense partial melt expressed as ULVZs (Fig. 5.5). The formation of positively buoyant plumes would require a sufficiently high temperature due to core heating within the overlying LLVP slab-derived material to overcome its intrinsically higher density. A competing hypothesis, factoring in the differing chemistry of subducted slabs (MORBs) versus that measured from ocean island basalts (OIBs) in surface hotspots, is that the plumes are derived from mantle material outside of the LLVPs along their edge (Niu, 2018).

Lateral gradients in seismic velocities have also been proposed as an additional discriminant between a chemical versus a thermal origin for LLVPs. A chemical origin for the LLVPs is expected to have a stronger lateral gradient, or near discontinuous change, in P and S velocity. With accuracies achieved thus far, however, this discriminant has proved inconclusive. An emerging consensus is that the structure of the two antipodally located LLVPs are that they are lower resolution images of structures characterized by mixed chemistry and weaker lateral gradients near their boundaries and unresolved plumes near their centers.

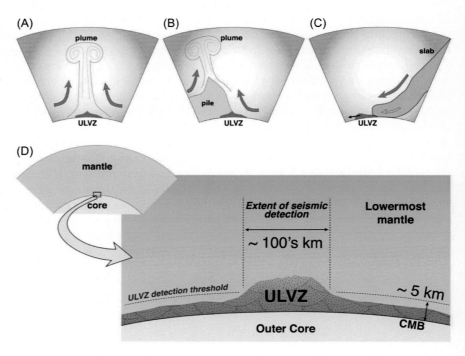

FIG. 5.5 Suggested relations between ULVZs plumes (A), LLVPs and *thermochemical piles* (B), slabs (C), and core–mantle reactions (D). *From Yu, S., Garnero, E.J., 2018. Ultralow velocity zone locations: a global assessment. Geochem. Geophys. Geosyst. 19(2), 396–414. https://doi.org/10.1002/2017GC007281.*

5.1.5 Constraints from scattering

Except for the thickness of laterally coherent ULVZs, heterogeneity having scale lengths less than 100 km cannot be imaged from travel-time or full-waveform tomography. The scattered wavefield generated by these smaller scale heterogeneities, however, can be observed in body waves and their coda interacting with the CMB (Chapter 1, Fig. 1.5). Forward modeling of coda shapes is feasible with methods of radiative transport, in which discrete packets of elastic body wave energy at specific frequencies are tracked through multiple scattering events that occur between mean free paths governed by models of a mantle heterogeneity spectrum. Increasingly it will become possible to model the effects of small-scale heterogeneity with fully numerical methods of solving the elastic wave equation at frequencies 2 Hz and higher. The observations sensitive to small-scale structures in the CMB region having the longest history of observation and modeling are the high-frequency precursory energy to the PKIKP wave in the 110–140° distance range.

The results of radiative transport modeling of 150,000 precursor observations (Fig. 5.6) by Mancinelli et al. (2016) support the persistent existence of small-scale perturbations in P velocity on the order of 0.1% at spatial scales of 10 km in a relatively thick region of the lower mantle. The inferred thickness of this region of scattering is far above the 200–400 km thickness commonly defined as the D″ region. Thermal heterogeneities at this small scale have short lifetimes and hence would quickly diffuse away. They are thus more likely small-scale compositional heterogeneity in the mantle. The shape of PKIKP precursor coda summarized in Fig. 5.6 is consistent with such small-scale compositional heterogeneity distributed throughout the entire mantle. In the stacked global averages obtained in the Mancinelli et al. study, enriched heterogeneity power at small scales concentrated in a thinner 200–400 km D″ is thus not evident, but an examination of the spatial coherence of the precursor amplitudes finds that amplitudes can vary by up to a factor of 5 when grouped in regional bins of 2000 × 2000 km. It is possible, however, to reconcile both the large-scale seismic images that show an increased concentration of heterogeneity in a thinner region of the lowermost mantle with

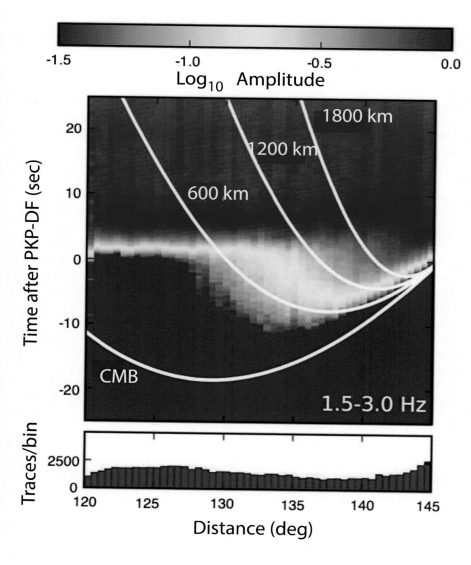

FIG. 5.6 PKIKP precursor amplitudes as a function of distance observed in stacked envelopes of seismogram coda. White curves denote the minimum arrival time of waves scattered from heterogeneities at the CMB, 600, 1200, and 1800 km above the CMB. *From Mancinelli, N., Shearer, P., Thomas, C., 2016. On the frequency dependence and spatial coherence of precursor amplitudes, J. Geophys. Res. 121, https://doi.org/10.1002/2015JB012768.*

the existence of more spatially uniform, smaller scale heterogeneity by a multiscale statistical image of the lower mantle heterogeneity (Fig. 5.7). In this model the coda of high-frequency (>1 Hz) body waves are primarily affected by smaller scale heterogeneity distributed throughout the entire lower mantle.

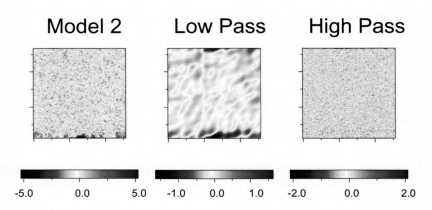

FIG. 5.7 Images a 2000 × 2000 km patch of a possible lower mantle having a multi-scale structure. The color scale denotes per cent perturbations in S velocity (associated P velocity perturbations will be a factor 2–3 smaller). The low-passed structure, resolvable by tomographic imaging, capable of inducing waveform perturbations in low-frequency (<0.2 Hz) body waves, removes all heterogeneities having scale lengths shorter than 200 km. The high-passed structure, observable in scattered PKIKP precursors at frequencies higher than 1 Hz, removes all heterogeneities having scale lengths longer than 20 km. D″ disappears as a distinct feature in the high-passed image. *From Cormier, V.F., 2000. D as a transition in the heterogeneity spectrum of the lowermost mantle. J. Geophys. Res. 105, 16193–16205.*

5.1.6 Constraints from elastic anisotropy

Observations of S waveforms interacting with D″ have found evidence of both *transverse isotropy* and more general anisotropy at the edges of the two LLVPs in regions suspected of having remnants of plate subduction (Fig. 5.8). Two possible interpretations of this anisotropy are shape-preferred orientation (SPO) and lattice preferred orientation (LPO). SPO anisotropy can originate from a preferred orientation of heterogeneity shapes that are elongated along one axis. LPO anisotropy occurs from plastic deformation along planes defined by the intrinsic lattice structure of minerals, in which atomic bonding strengths differ along the axes (Burger vectors) of the unit cell of a mineral crystal. Both SPO and LPO orientations can be affected by the plastic deformations induced by convective flow.

Numerical simulations of S waves in 2- and 3D models having the spectrum of D″ heterogeneity imaged by tomography demonstrate that an apparent transverse isotropy with a vertical axis of symmetry can be induced by even an isotropic distribution of heterogeneity scale lengths (Parisi et al., 2018). Fig. 5.9, for example, demonstrates that plausible heterogeneous models of D″, can produce an apparent transverse isotropy in S waves sampling D″. Hence, any estimate of increased transverse isotropy with a vertical axis of symmetry in the lowermost mantle may simply be the effect of increased heterogeneity over a broad spectrum of wavelengths in D″. More general forms of anisotropy, particularly TTI (transverse isotropy with a tilted axis of symmetry), have been proposed from the inversion of S, ScS, SKS, and SKKS waveforms, in which the orientation of the tilted symmetry axis is determined from the interference of two quasi S waves observed on the SH component of motion using methods first developed by Silver and Chan (1991). TTI has been observed along the edges of the South African LLVP (Cottar and Romanowicz, 2013; Lynner and Long, 2014) and a spatially smaller zone of D″ low velocity beneath Perm, western Russia (Long and Lynner, 2015). Nowacki et al., 2012 have found TTI associated with regions of paleo-subduction beneath North and Central America. The preferred interpretation of TTI observed along both the edges of LLVPs and in regions of paleo-subduction is that it is the effect of more vertical flow associated with either plume formation or localized deflection of mantle flow by slabs enriched in the post-perovskite phase near the CMB.

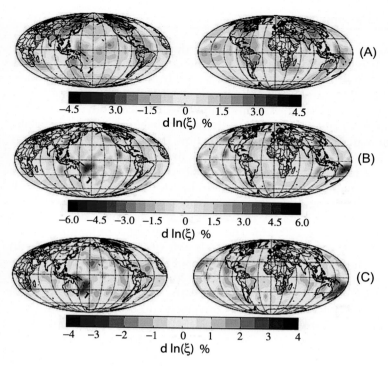

FIG. 5.8 A comparison of three global models of anisotropy in D″ (75 km above the CMB). (A) SAW642AN (Panning and Romanowicz, 2006), (B) S362WMANI (Kustowski et al., 2008), (C) SAW642ANb (Panning et al., 2010). A positive parameter d ln(ξ) % implies a more horizontal symmetry axis of TTI and vertical convective flow or vertically inclined layers, and negative a more vertical symmetry axis and either horizontal flow or horizontal layering. *From Walker, A.M., Forte, A.M., Wookey, J., Nowacki, A., Kendall, J.M., 2011. Elastic anisotropy of D predicted from global models of mantle flow. Geochem. Geophys. Geosyst. 12(10), https://doi.org/10.1029/2011GC003732.*

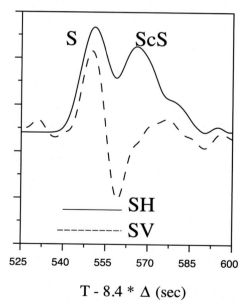

FIG. 5.9 Displacement S seismograms, synthesized by a pseudospectral modeling method, interacting with an elastically isotropic but heterogeneous 2D model of the lowermost D″ region of the mantle. The Earth model assumed a gradient increase in RMS perturbation strength in S velocity from 1% to 3% between 700 km and 50 km above the CMB. To simulate the existence of laterally intermittent ultralow velocity zones in the lowermost 50 km a random negative S velocity perturbation is assumed with a gradient increase in perturbation strength increasing from 3% to 30% in the lowermost 50 km. An exponential autocorrelation and isotropic distribution of scale lengths between 1000 and 30 km is assumed at all depths and a scaling between P velocity and S velocity fluctuations such that $\frac{\Delta V_P}{V_P} = 0.3 \frac{\Delta V_S}{V_S}$.

5.2 CMB topography

The discontinuous increase in density at the core-mantle boundary is the largest density contrast across any of Earth's other discontinuities in phase or chemistry. It is not surprising then that the effects of larger lateral scales of CMB topography are observable in the gravity and gravity potential (*geoid*) measured from low-orbiting satellites. The signal of CMB topography observed in the gravity field can be reconciled with topography estimates obtained from core-reflected seismic body waves and normal mode splitting. Assuming thermo-chemical convection in the mantle, geodynamic modeling studies that seek this reconciliation assume a scaling between density and seismic velocities together with a depth-dependent model of mantle viscosity. The viscous flow that dynamically maintains the topography critically depends on the seismic velocity/density scalings as well as the depth dependence for viscosity near the core-mantle boundary. Key observations to fit include an *excess ellipticity* of the geoid expressed in the degree 2 harmonic and an anticorrelation in topography height and seismic velocities up to at least harmonic degree 4. Joint inversions of the low-order free-air gravity anomaly, tomographic images, normal mode splitting, and observations of Stoneley modes (Koelemeijer et al., 2017) sensitive to the CMB have converged on accepted limits of 1.5–4 km to the height of CMB topography at lateral scale lengths exceeding 2000 km. The sense of the anticorrelation between CMB height and seismic velocities is such that low seismic velocities on the mantle side of the CMB correlate with an elevated boundary, and high with a depressed boundary. Assuming only the effect of thermal expansion and negative temperature derivatives of the seismic velocities in a chemically homogeneous lower mantle, observations of large low-velocity provinces (LLVPs) at and near the CMB beneath southern Africa and the South Pacific are consistent with an observation of an elevated CMB in these regions (Koelemeijer, 2020). Conversely, descending slabs are predicted to be associated with depressed CMB topography at shorter lateral scale and CMB height approaching 10 km (Fig. 5.10). Departures from the assumption of a homogeneous chemistry and phase in the lower mantle, including the effects of dense piles and deformed slabs and a post-perovskite phase transition have been explored by Deschamps et al. (2018) and Yoshida (2008).

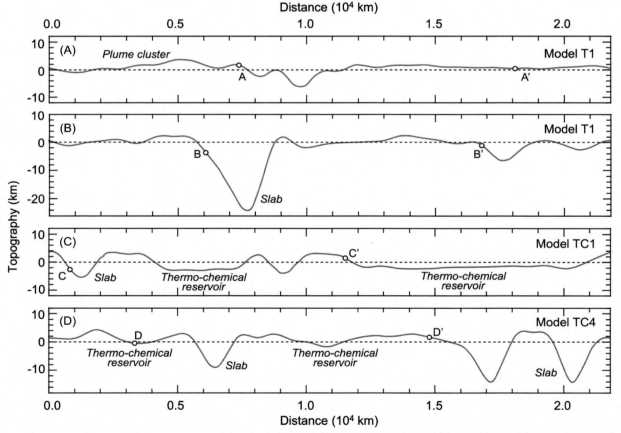

FIG. 5.10 Predicted CMB topography from geodynamic simulations of Deschamps et al. (2018), designed to be in agreement with the height and lateral scales inferred from seismology and gravity field. Model details and keys to four great circle cross-sections are available in Deschamps et al., 2018.

Motions of the fluid outer core that sustain the geodynamo can apply torques to the topographic irregularities of the CMB, altering its rotation rate (Hide et al., 1996). These are observed in measurements of length-of-day fluctuations (LODs) on the order of milliseconds in magnitude with periods of fluctuation on the order of 10 years (Chapter 3). The predicted LODs are consistent with the maximum heights and horizontal scales of CMB topography predicted from seismology and geodynamic simulations, and offer an alternative to the torsional oscillation interpretation discussed in Chapter 4.

In 2D simulations of outer core flow, Calkins et al. (2012) have shown that CMB topography on these scales can excite stationary Rossby waves and cyclonic motions along the inner core boundary. These stationary cyclonic motions can in turn affect the heat transport across the inner core boundary and the rates of its solidification and melting. These lateral variations in heat flow across the CMB can in turn affect outer core convection, e.g., (Sumita and Olson, 2002). They occur in a turbulent flow regime and strongly depend on assumptions for the Ekman number and viscosity of the outer core.

Differential travel times of seismic body waves, e.g., P4KP-PcP (Tanaka, 2010), PKP-PcP (Sze and van der Hilst, 2003), can remove the contaminating effects of mantle heterogeneity on the topography signal if the rays of the two waves have nearly equal takeoff angles. Depending on the density of seismic sources and stations, differential travel times can potentially resolve smaller and regionally varying structures on the core-mantle boundary that may have higher amplitude and steeper slopes than are resolvable from global tomography or normal mode inversions. Nonetheless, the tendency of most body wave studies has been to confirm the range of heights 1.5–4 km and at the lateral scales greater than 2000 km. The agreement between the estimates of topography constructed from differencing the travel times of seismic waves reflected by versus those transmitted through the CMB confirms that there is no detectable lateral heterogeneity in the liquid core. In higher seismic frequency bands (>1 Hz), constraints on much smaller scale CMB topography have been estimated by comparing observed and predicted waveforms of PcP (Menke, 1986) and from CMB scattered precursors to PKIKP (Doornbos and Vlaar, 1973). These studies estimate a maximum height of 100–200 m at lateral scales of 10–100 km.

5.3 E′ region (uppermost outer core)

5.3.1 Seismic velocity anomalies

If the outer core is a vigorously convecting, chemically homogeneous, liquid metal alloy from its top to its bottom, it will be unlikely to find any detectable changes in its physical properties near the CMB and ICB. Evidence against this are studies suggesting a reduced P velocity and an apparent steepening of the P wave velocity gradient in a 100–700 km thick E′ region below the CMB (Fig. 5.11). Its signal, however, is small and must be separated from competing effects of D″ heterogeneity and anisotropy on the differential travel times of SKS and SmKS waves (e.g., Fig. 5.12). At least one study (Irving et al., 2018) has suggested an equally good fit to seismic observations can be obtained with an equation of state consistent with a chemically homogeneous outer core, without the need for a region of altered velocity gradients or reduced P velocity either at the top or at the bottom of the outer core. Instead, slightly lower velocity and higher

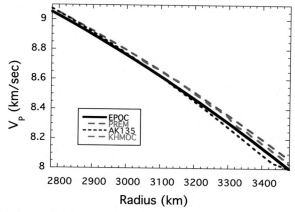

FIG. 5.11 V_p in the upper outer core from several studies of differential travel times of SmKS waves, including KHOMC by Tanaka and Tkalčić (2015) and EPOC (Irving et al., 2018). The EPOC V_p model is derived from the parameterization of an equation of state that best satisfies combined differential travel times of SmKS waves and free oscillation splitting measurements.

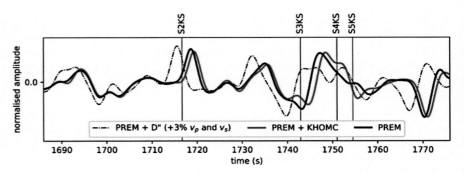

FIG. 5.12 Radial-component reflectivity seismograms for two of E′ models for outer core P velocity shown in Fig. 5.11, including the effect of a D″ perturbation to PREM (280 km thick, +3 per cent for v_p and v_s). Predicted SmKS arrivals for PREM are also shown. The distance for the synthetic seismograms is 139.0° for an earthquake at a depth of 150 km. *From Van Tent, R., Deuss, A., Kaneshima, S., Thomas, C., 2020. The signal of outermost-core stratification in body-wave and normal-mode data. Geophys. J. Int. 223(2), 1338–1354. https://doi.org/10.1093/gji/ggaa368.*

density relative to PREM are distributed throughout the outer core. Stoneley modes, which are sensitive to both velocity and density in the outermost core, suggest that this lower velocity and higher density are concentrated in the E′ region (Van Tent et al., 2020). A higher density perturbation at the top of the outer core in E′ region does not favor a simple intuitive process for forming a stably stratified region at the top of the outer core.

5.3.2 Constraints from chemistry and MAC waves

From the small size of seismically predicted perturbations of standard Earth models, it is not surprising that the earliest hint of a chemically distinct layer in the E′ region did originate from the interpretation of seismic observations. Laboratory studies have suggested the possibility of chemical species transfer across the CMB and the existence of a chemically distinct, stably stratified layer in E′. Additional evidence is provided from length of day variations and magnetic secular variations (Chapters 3 and 4). In heated perovskite samples placed in diamond anvils to simulate the P–T conditions of the CMB, Knittle and Jeanloz (1991) found that perovskite reacts with liquid iron, forming FeSi, FeO, and a high-pressure form of quartz (*stishovite*). Depending on the relative buoyancies and their chemical diffusivity, these CMB reaction products may enrich the iron content of the lowermost mantle and the light element components (e.g., Si and O) of the uppermost core. Buffett and Seagle (2010) predicted that the adding of Si and O to core from CMB reactions will create a stably stratified layer at the top of the outer core, isolated from the deeper convection in the outer core. Over time the layer evolves according to a process in which its composition and buoyancy depend on differing vertical gradients in composition and temperature. Thinner layers of varying composition may form within the layer with narrow fingers of material at the bottom of the layer interacting with and penetrating into the deeper core circulation. This process has been termed double diffusion, and was first developed to describe the formation of a stable layer at the surface of the ocean created by differing gradients in temperature and salinity (Huppert and Turner, 1981). Braginsky (1993) demonstrated that convective core motion interacting with the bottom of such a layer, having an order of 100 km in thickness, may excite so-called MAC waves of fluid motion having periods near 65 years, detectable in the LOD signal. Core surface flows inferred from the downward continuation of magnetic secular variations having periods from 10s to 100 years have provided observational constraints on MAC waves.

5.3.3 Alternative theories for the formation of E′

A problem in the formation of a stably stratified layer in E′ in models proposed by Buffett and Seagle is that the sign of their predicted P velocity anomaly is opposite to that observed from the travel times of seismic body waves. Successful models must find mechanisms that both populate the E′ layer with lighter, lower density elements, but at the same time achieve an average composition with alloying iron to have a sufficiently low bulk modulus to overcome the lower density of the composition. Two alternative models for the formation of the stratified layer in E′ have attempted to create it by adjusting the specific composition of light elements in different ways. Brodholdt and Badro (2017) propose a model in which the stably stratified layer is enriched in C, O, or S, but depleted in Si, inherited from the composition of a giant impactor. Bouffard et al. (2020) examined models in which a stably stratified layer of differing composition is created by sequences of planetary accretion in which the P–T conditions of the early core vary together with the composition of accreting objects impacting early Earth.

5.4 F region

5.4.1 Seismic velocity anomalies

Early models of Earth's core by H. Jeffreys and K. Bullen show a low-velocity zone existing above the inner core at the base of the mantle. This F structure was originally proposed to explain precursors to the PKIKP wave observed in the 110–140° range, which later were instead demonstrated to be a combination of diffraction from the PKP-B caustic in the low-frequency band (<0.2 Hz) and scattering from the CMB and D″ region in the high-frequency band (Chapter 1, Fig. 1.5).

More contemporary Earth models obtained solely from fits of global travel times, e.g., AK135, IASP91, predict a region of flattened P velocity gradient in the lower 200–400 km, which is our current understanding of F region properties. The P velocity value and its gradient near the ICB can constrain possible chemical differences in F and determine whether F is another stably stratified region of the outer core. The P waves that nearly graze the ICB used in constraining F structure, however, are nearly insensitive to the density and shear wave jumps at the ICB. These jumps are of more primary interest to the understanding of inner core solidification.

The F region remains an active area of research, characterized by uncertainties due to trade-offs in inner core versus outer core side structure on the fitting of differential travel times of PKiKP-PKIKP in the range 110–140° and PKP-C/PKP-Cdiff-PKIKP in the range 152–180°. Further complicating the analyses are lateral variations in the inner core structure described in Chapter 6, aggravated by weakened signals of the PKIKP wave due to the high intrinsic attenuation of the inner core. Additional constraints on the P velocity gradient in the F region are the decay and change in frequency content with increasing distance of the inner core diffraction (PKP-Cdiff). Nearly all studies agree that the P wave velocity in the lowermost outer core is less than that of PREM in the lowermost 250–300 km, although both its gradient structure and absolute values near the ICB differ to the extent that different models neither confirm nor negate the requirement that F be a stably stratified layer. For example, model FFV (Fig. 5.13) having a lower P velocity than PREM in the F region beneath Japan, which smoothly joins the AK135 profile above F, does not definitively support the existence of a thick stably stratified layer enriched in Fe. In the study that lead to the FFV velocity profile, Ohtaki et al. (2018) pointed out that the Bullen parameter computed for PREM's F region slightly exceeds neutral buoyancy to be consistent with a homogeneous outer core and that a revision similar to either the milder P velocity gradient of FFV or the EPOC model might better characterize the state of the F beneath Japan. Using the same analysis, however, Ohtaki et al. find a quite different F structure (FVA) beneath Australia, consistent with 0.8% O enrichment. Since there is no evidence for time-dependent changes in F structure, any lateral variation in F, if confirmed in more studies, would be evidence of stably stratified structures in at least some areas above the ICB.

An even more complicated seismic structure has been proposed by Adam and Romanowicz (2015) to explain a coherent-scattered, high-frequency, arrival-observed following PKP-BC and PKP-Cdiff. This structure consists of a high-velocity perturbation 400–500 km above the ICB, followed by a low-velocity perturbation in a 50 km thick zone immediately above the ICB. Such a structure may be consistent with the structure predicted by some *double-diffusive* models of F chemistry. Sens-Schonfelder et al. (2021), however, interpret the arrival observed by Adam and Romanowicz to be P waves scattered from the lowermost D″ region of the mantle, preceding PKP-AB, rather than the effect of inner core scattering affected by multipathing in a complex F structure (see dotted travel time curve after cusp C in Fig. 1.5, Chapter 1).

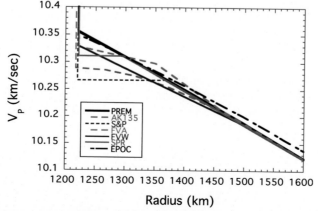

FIG. 5.13 P velocity models proposed for the lowermost outer core (F region), determined from differential travel times of P waves at near grazing incidence to the ICB. *Adapted from Ohtaki, T., Kaneshima, S., Ishikawa, H., Tsuchiya, T., 2018. Seismological evidence for a laterally heterogeneous lowermost outer core of the earth, J. Geophys. Res. 123, 10903–10917, https://doi.org/10.1029/2018JB01587.*

5.4.2 Theories for formation and stability

Similar to the E' region, seismic modeling has yet to provide a clear answer to whether F is a stably stratified layer that does not participate in the convective motion that sustains the geodynamo. Nonseismic constraints can be invoked to determine whether it is even feasible to create and maintain a stably stratified F region (Chapter 7). Models to explain chemical differences in F include (1) a model in which iron freezes near the top of F and melts as it descends toward the ICB, (2) a model in which the chemical differences in F are products of an evolving chemistry of accreted material during core formation, (3) a *slurry* model in which iron particles crystallize at the *liquidus* temperature throughout the F layer, and (4) a model in which lateral variations in F are related to variations in freezing and melting of the inner core.

Wong et al. (2021) investigated the stability range of parameters for stable stratification of an F region slurry, finding that stability requires an upper bound on the density jump of the inner core such that $\Delta \rho = 534 \, \text{kg/m}^3$ and a high value of $100 \, \text{W} \, \text{m}^{-1} \, \text{K}^{-1}$ for the thermal conductivity of the outer core. In the F region their predicted P velocity profile falls between PREM and FVW in Fig. 5.13.

Freezing/melting models invoke the existence of a double-diffusive F layer consisting of an enriched iron melt blanketing the ICB with inner core melting in one hemisphere and freezing in the other. This type of model involves a mode of slow, dominantly horizontal, convective motion in the inner core. Its behavior is dependent on the viscosity of the uppermost inner core. It is sometimes termed the translating/convecting model. Since this model has implications for other inner core observables, including elastic anisotropy, hemispherical differences, and super-rotation of the inner core, its details are discussed further in Chapters 6 and 7.

Unlike the E region, there are fewer observable constraints to include in relevant geodynamic and geochemical modeling of the F region. The ICB density jump and thermal conductivity are important to the slurry model, but the seismically inferred density jump can have large error bars (Section 6.6.6, Chapter 6) and the thermal conductivity of the outer core cannot be directly observed. In some instances, however, there are constraints determined from modeling the convection in the outer core required by the paleo and current magnetic field. Modeling by Deguen et al. (2014) explores a translating/convecting model of asymmetric growth and melting of the inner core, and concludes that a laterally homogeneous layer of negatively buoyant, iron-enriched melt above the ICB would inhibit the release of light elements from the solidifying inner core, and be inconsistent with the observed strength and behavior of the magnetic field unless it was laterally varying with separate regions of light element enrichment and iron enrichment.

5.5 ICB topography

The density contrast of the inner core boundary is much smaller than that of the core–mantle boundary, reflecting a simple change in solid to liquid phase of an iron alloy with smaller chemical differences between the light element enrichment of the liquid outer core and light element depleted, nearly pure iron/nickel alloy of the inner core. Combining observations of the LOD, geoid, and seismic wavefields, the behavior of the spectrum of lateral variations in ICB topography differs from that of the CMB. CMB topography is on the order of several km at lateral scales greater than 1000s of km decreasing to 100s of meters at lateral scales of 10–100 km. In contrast, the spatial power spectrum of ICB topography appears to be whiter, slowly increasing from an order of 100 m in its ellipticity to an upper bound of 1–2 km at lateral scales of 10–100 km. To account for LOD observations due to torques from fluid motions in the outer core applied to the ellipticity of the inner core, Chao (2016) estimates an upper bound of 140 m in geoid height due to the ellipticity of the inner core.

Observations of the waveforms and reflectivity of high-frequency (1 Hz and greater) P waves interacting with the ICB provide some upper bounds on topography at shorter lateral scales that may be important to understanding the processes of solidification of the inner core (Fig. 5.14). P waves reflected by the ICB can be observed at pre-critical

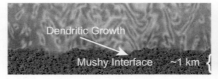 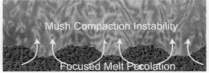

FIG. 5.14 Focused flow of outer core fluid, enriched in elements lighter than iron, buoyantly exits from an uppermost mushy inner core in a mechanism that can locally increase the height of ICB topography. *From Hernlund (personal communication).*

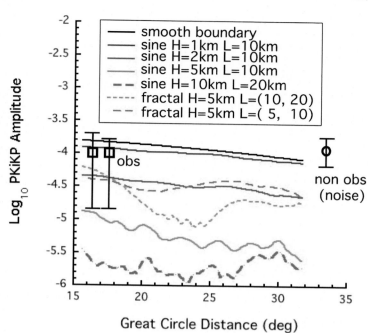

FIG. 5.15 Comparison of simulated PKiKP amplitudes for different ICB topographies for a range of pre-critical distances (de Silva et al., 2018). Boxes with error bars show PKiKP amplitudes from Tkalčić et al. (2009) observed from PKiKP/PcP ratios corrected for PcP amplitude and normalized with respect to PKiKP amplitude of PREM having a smooth ICB. With similar corrections is the amplitude of a noise estimate in the time window of PKiKP determined from nonobservation of PKiKP by Waszek and Deuss (2015).

incidence, where the ICB reflection coefficient is small, and a post-critical incidence, where the reflection coefficient rapidly approaches total reflection between great circle distance 110° and 135° (Chapter 1, Fig. 1.6).

In the pre-critical range, Tanaka and Tkalčić (2015) estimate an upper bound of 1–1.5 km for height and lateral scale lengths of ICB topography in numerical modeling of the complex spectra of PKiKP waveforms observed at the high frequency, HiNet array in Japan. From numerical modeling across pre-critical, post-critical, and diffraction ranges, de Silva et al. (2018) suggested a similar upper bound to a height of 2 km at lateral scales between 1 and 10 km (Fig. 5.15). Their study found that larger topography heights tend to destroy the PKiKP waveform at pre-critical range. Numerical modeling demonstrates that post-critical and diffraction (PKP-Cdiff) are relatively insensitive to higher ICB topography approaching 10 km in the same lateral scale band. If a global upper height bound of 1–2 km is accepted, higher ICB topography heights suggested in studies by Dai et al. (2012) (4–14 km) and Cao et al. (2007) (3–5 km) may be explained by localized anomalies, possibly induced by narrow upwelling plumes emanating from the ICB.

5.6 Summary

Although the structure of the lowermost mantle abounds in complexity, a growing framework of interpretations begins with the identification of two antipodal separated regions of lower seismic velocity (LLVPs), with their edges separating narrower corridors of high velocity in which reside remnants of descending and deformed slabs or narrow corridors of low-velocity ascending plumes. Superposed on this larger scale heterogeneity is smaller scale (<10 km) heterogeneity, indicative of pervasive small-scale chemical heterogeneity throughout the mantle. The effects of lower mantle heterogeneity for the operation of the core dynamo manifest themselves through the conductive heat transport across the CMB. Although the origin and persistence of the LLVPs are still in debate, regions of high velocity and presumably colder D″ often show enough sharpness in depth and lateral coherence along the CMB to induce triplicated body waves, allowing the depth of a post-perovskite phase change to be detected. The joint action of increasing pressure and temperature with depth as the CMB is approached from above is predicted to induce a double-crossing of the post-perovskite P–T phase boundary, which can be used to significantly reduce the uncertainties in the temperature profile in the lowermost mantle. Improved estimates of the thermal conductivity of this region, together with an improved temperature profile obtained from modeling the post-perovskite phase transition, will drive future efforts of geodynamic modeling to examine the effects on the geodynamo of laterally varying D″ structure.

Evidence for stable boundary layers E′ and F at both the top and bottom, respectively, of the outer core have been found from seismic modeling. The perturbations of seismic velocities that require these layers are quite small, amounting to 1% and less, requiring novel techniques and error analyses to have confidence in their results. The consequences, however, of these small perturbations in E' and F can be large in their effects on the more vigorous motion of the core fluid between them. At least one or more studies have demonstrated that it is possible to fit seismic data without E' and F layers, although in most cases the fits to travel-time data sample laterally limited regions of the CMB and ICB. If stable E' and F structures do exist, chemistry consistent with their stable stratification is not yet understood. Current geodynamic modeling of core motions requires them to be laterally varying, in agreement with at least some of the lateral variations seen in body wave studies.

References

Adam, J.-C., Romanowicz, B., 2015. Global scale observations of energy scattered near the inner core boundary: seismic constraints on the base of the outer-core. Phys. Earth Planet. Inter. 245, 103–116.

Borgeaud, A.F.E., Kawai, K., Konishi, K., Geller, R.J., 2017. Imaging paleoslabs in the D layer beneath Central America and the Carribean using seismic waveform inversion. Sci. Adv. 3 (11). https://doi.org/10.1126/sciadv.1602700, e1602700.

Bouffard, M., Landeau, M., Goument, A., 2020. Convective erosion of a primordial stratification atop Earth's core. Geophys. Res. Lett. 47 (14). https://doi.org/10.1029/2020GL087109.

Braginsky, S.I., 1993. MAC-Oscillations of the hidden ocean of the core. J. Geomag. Gelectr. 45, 1517–1538.

Brodholdt, J., Badro, J., 2017. Composition of the low seismic velocity E' layer at the top of Earth's core. Geophys. Res. Lett. 44 (16), 8303–8310. https://doi.org/10.1002/2017GL074261.

Buffett, B.A., Seagle, C.T., 2010. Stratification due to chemical interactions with the mantle. J. Geophys. Res. 115 (B4). https://doi.org/10.1029/2009JB006751.

Calkins, M.A., Noir, J., Eldreddge, J.D., Aurnou, J.M., 2012. The effects of boundary topography on convection in Earth's core. Geophys. J. Int. 189, 798–814.

Cao, A., Masson, Y., Romanowicz, B., 2007. Short wavelength topography on the inner core boundary. Proc. Natl. Acad. Sci. 104 (1), 31–35. https://doi.org/10.1073/pnas.0609810104.

Chao, B.F., 2016. Dynamics of axial torsional libration under the mantle-inner core gravitational interaction. J. Geophys. Res. 122 (1). https://doi.org/10.1002/2016JB013515.

Cottar, S., Romanowicz, 2013. Observations of chaing anisotropy across the southern margin of the African LLSVP. Geophys. Res. Lett. 195 (2), 1184–1195.

Dai, Z., Wang, W., Wen, L., 2012. Irregular topography at the Earth's inner core boundary. Proc. Natl. Acad. Sci. 109 (20), 7654–7658. https://doi.org/10.1073/pnas.1116342109.

de Silva, S., Cormier, V.F., Zheng, Y., 2018. Inner core boundary topography explored with reflected and diffracted P waves. Phys. Earth Planet. Inter. 276, 202–214. https://doi.org/10.1016/j.pepi.2017.04.008.

Deguen, R., Olson, P., Reynolds, E., 2014. F-layer formation in the outer core with assymetric inner core growth. Compt. Rendus Geosci. 346 (5–6), 101–109. https://doi.org/10.1016/j.crte.2014.04.003.

Deschamps, F., Rogister, Y., Tackley, P.J., 2018. Constraints on core-mantle topography from thermal and thermochemical convections. Geophys. J. Int. 212, 164–188.

Doornbos, D., Vlaar, N., 1973. Regions of seismic wave scattering in the Earth's mantle and precursors to PKP. Nature Phys. Sci. 243 (126), 58.

Hassan, R., Flament, N., Gurnis, M., Bower, D.J., Muller, D., 2015. Provenance of plumes in global convection models. Geochem. Geophys. Geosyst. 16 (5), 1465–1489. https://doi.org/10.1002/2015GC005751.

Hernlund, J.W., Thomas, C., Tackley, P.J., 2005. A doubling of the post-perovskite phase boundary and structure of the Earth's lowermost mantle. Nature 434 (7034), 882–886. https://doi.org/10.1038/nature03472.

Hide, R., Boggs, D.H., Dickey, J.O., Dong, D., Gross, R.S., Jackson, A., 1996. Topographic core-mantle coupling and polar motions on decadal timescales. Geophys. J. Int. 125, 599–607.

Huppert, H.E., Turner, J.S., 1981. Double-diffusive convection. J. Fluid Mech. 106, 290–329.

Irving, J.C.E., Cottaar, Lekić, V., 2018. Seismically determined elastic parameters for Earth's outer core. Sci. Adv. 4 (6). https://doi.org/10.1126/sciadv.aar2538, eaar2538.

Knittle, R., Jeanloz, R., 1991. Earth's core-mantle boundary: results of experiments at high pressures and temperatures. Science 251 (5000), 1438–1443. https://doi.org/10.1126/science.251.5000.1438.

Koelemeijer, P., 2020. Towards consistent seismological models of the core-mantle boundary landscape. In: Marquardt, Cottaar, Ballmer, Konter (Eds.), Book Chapter in Revision for AGU Monograph "Mantle Upwellings and Their Surface Expressions".

Koelemeijer, P., Deuss, A., Ritsema, J., 2017. Density structure of Earth's lowermost mantle from Stoneley mode splitting observations. Nat. Commun. 8. https://doi.org/10.1038/ncomms15241.

Kustowski, B., Ekstrom, G., Dziewonski, A.M., 2008. Anisotropic shear-wave velocity structure of Earth's mantle: a global model. J. Geophys. Res. 113 (B6). https://doi.org/10.1029/2008JB005169.

Langrand, C., Adrault, D., Durand, S., Konopkova, Z., Hilairet, N., Thomas, C., Merkel, S., 2019. Kinetics and detectability of the bridgmanite to post-perovskite transformation in the Earth's D layer. Nat. Commun. 10, 5690. https://doi.org/10.1038/s41467-019-13482-x.

Lay, T., Helmberger, D.V., 1983. A lower mantle S-wave triplication and the shear velocity structure of D". Geophys. J. R. Astron. Soc. 75, 799–837.

Lekić, V., Cottar, S., Dziewonski, A., Romanowicz, B., 2012. Cluster analysis of global lower mantle tomography: a new class of structure and implications for chemical heterogeneity. Earth Planet. Int. 357–358, 68–77.

Long, M., Lynner, C., 2015. Seismic anisotropy in the lowermost mantle near the Perm anomaly. Geophys. Res. Lett. 42 (17), 7073–7080. https://doi.org/10.1002/2015GL065506.

Lynner, C., Long, 2014. Lowermost mantle anisotropy and deformation along the boundary of the African LLSP. Geophys. Res. Lett. 41. https://doi.org/10.1002/214GL059875.

Mancinelli, N., Shearer, P., Thomas, C., 2016. On the frequency dependence and spatial coherence of precursor amplitudes. J. Geophys. Res. 121. https://doi.org/10.1002/2015JB012768.

Masters, G., Laske, G., Bolton, H., Dziewonski, A., 2000. The relative behavior of shear velocity, bulk sound speed and compressional velocity in the mantle: implications for chemical and thermal structure. In: Karato, S., Forte, A., Liebermann, R., Masters, G., Stixrude, L. (Eds.), Earth's Deep Interior: Mineral Physics and Tomography From the Atomic to the Global Scale. Geophysical Monograph 117. American Geophysical Union, Washington, pp. 63–87.

Menke, W., 1986. Few 2–50 km corrugations on the core-mantle boundary. Geophys. Res. Lett. 13 (13), 1501–1504.

Murakami, M., Hirose, K., Kawamura, K., Sata, N., Ohishi, Y., 2004. Post-perovskite phase transition in MgSiO3. Science 304 (5672), 855–858.

Niu, Y., 2018. Origin of the LLSVPs at the base of the mantle is a consequence of plate tectonics—a petrological and geochemical perspective. Geosci. Front. 9 (5), 1265–1278.

Nowacki, A., Walker, A.M., Wookey, J., Kendall, J.M., 2012. Evaluating post-perovskie as a cuase of D anisotropy in regions of paleo-suduction. Geophys. J. Int. 192, 1085–1090.

Oganov, A.R., Ono, S., 2004. Theoretical and experimental evidence for a post-perovskite phase of MgSiO3 in Earth's D layer. Nature 430 (6998), 445–448.

Ohtaki, T., Kaneshima, S., Ishikawa, H., Tsuchiya, T., 2018. Seismological evidence for a laterally heterogeneous lowermost outer core of the earth. J. Geophys. Res. 123, 10903–10917. https://doi.org/10.1029/2018JB01587.

Panning, M., Romanowicz, B., 2006. A three-dimensional radially anisotropic model of shear velocity in the whole mantle. Geophys. J. Int. 167, 361–379. https://doi.org/10.1111/j.1365-246X.2006.03100.x.

Panning, M.P., Lekić, V., Romanowicz, B.A., 2010. Importance of crustal corrections in the development of a new global model of radial anisotropy. J. Geophys. Res. 115. https://doi.org/10.1029/2010JB007520, B12325.

Parisi, L., Ferreira, A.M.G., Ritsema, J., 2018. Apparent splitting of S waves propagating through an isotropic lowermost mantle. J. Geophys. Res. 123 (5), 3909–3922. https://doi.org/10.1002/2017JB014394.

Sens-Schonfelder, C., Bataille, K., Bianchi, M., 2021. High frequency (6 Hz) PKP-ab precursors and their sensitivity to deep Earth heterogeneity. Geophys. Res. Lett. 48 (2). https://doi.org/10.1029/GL089203.

Shang, X., Shim, S.-H., deHoop, M., van der Hilst, 2014. Multiple seismic reflectors in Earth's lowermost mantle. Proc. Natl. Acad. Sci. 111 (7), 2242–2446. https://doi.org/10.1073/pnas.1312647111.

Silver, P., Chan, W., 1991. Shear wave splitting and subcontinental mantle deformation. J. Geophys. Res. 96 (B10), 16429–16454. https://doi.org/10.1029/91JB00899.

Sumita, I., Olson, P., 2002. Rotating thermal convection experiments in a hemispherical shell with hetergeneous boundary heat flux: implications for the Earth's core. J. Geophys. Res. 107 (B8). https://doi.org/10.1029/2001JB000548.

Sze, E., van der Hilst, R., 2003. Core mantle boundary topography fromshort period PcP, PKP, and PKKP data. Phys. Earth Planet. Inter. 135 (1), 27–46.

Tanaka, S., 2010. Constraints on the core-mantle boundary topography from P4KP-PcP differential travel times. J. Geophys. Res. 115 (B4), B04310.

Tanaka, S., Tkalčić, H., 2015. Complex inner core boundary from frequency characteristics of the reflection coefficients of PKiKP waves observed by Hi-net. Prog. Earth Planet Sci. 2 (34). https://doi.org/10.1186/s40645-015-0064-3.

Thorne, M., Garnero, E.J., Grand, S.P., 2004. Geographic correlation between hot spots and deep mantle lateral shear-wave velocity gradients. Phys. Earth Planet. Inter. 146 (1–2), 47–63. https://doi.org/10.1016/j.pepi.2003.09.026.

Tkalčić, H., Kennett, B.L.N., Cormier, V.F., 2009. On the outer-inner core density contrast from PKiKP/PcP amplitude ratios and uncertainties caused by seismic noise. Geophys. J. Int. 179, 425–443. https://doi.org/10.1111/j.1365-246X.2009.04294.x.

Van Tent, R., Deuss, A., Kaneshima, S., Thomas, C., 2020. The signal of outermost-core stratification in body-wave and normal-mode data. Geophys. J. Int. 223 (2), 1338–1354. https://doi.org/10.1093/gji/ggaa368.

Wang, P., de Hoop, M., van der Hilst, R.D., 2008. Imaging the lowermost mantle (D″) andthe core-mantle boundary with SKKS coda waves. Geophys. J. Int. 75, 103–115.

Waszek, L., Deuss, A., 2015. Anomalously strong observations of PKiKP/PcP amplitude ratios on a global scale. J. Geophys. Res. 120, 5175–5190.

Wong, J., Davies, C.J., Jones, C.A., 2021. A regime diagram for the slurry F-layer at the base of Earth's outer core. Earth Planet. Sci. Lett. 560. https://doi.org/10.1016/j.epsl.2021.116791.

Yoshida, M., 2008. Core-mantle boundary topography estimated from numerical simulations of instantaneous mantle flow. Geochem. Geophys. Geosyst. 9 (7). https://doi.org/10.1029/2008GC002008.

Yu, S., Garnero, E.J., 2018. Ultralow velocity zone locations: a global assessment. Geochem. Geophys. Geosyst. 19 (2), 396–414. https://doi.org/10.1002/2017GC007281.

CHAPTER 6

Inner core explored with seismology

6.1 Elastic anisotropy

After the discovery of the inner core and a determination of its average P velocity profile, the next most significant refinement of its structure has been the quantification of its elastic anisotropy. The addition of anisotropy addresses the fact that travel times of P waves transiting the inner core are faster when propagating more nearly parallel to its rotational axis than close to its equatorial plane. This has led to investigations in mineral physics to determine the stable lattice structure and elastic behavior of iron alloys at inner core pressures and temperatures. Explanations for observed anisotropy typically seek mechanisms of inner core crystallization and deformation related to heat extraction at the ICB coupled to outer core convection. A depth-dependence of anisotropy has also been confirmed from seismic observations. It has been a continuing challenge to determine the mechanisms creating elastic anisotropy and its depth dependence, including the nature of crystallization and plastic deformation beneath the ICB. There is also yet to be a consensus on the regional dependencies seen in observations.

6.1.1 Differential travel times

To remove the effects of heterogeneity and anisotropy in the mantle and isolate the effects of inner core structure, body wave studies have concentrated on measuring the time differences between PKiKP and PKIKP, PKP-BC and PKIKP, and PKP-AB and PKIKP (Fig. 6.1). Except for the PKIKP and PKP-AB pair near the CMB, the rays of each pair are typically separated by no more than several 10s of kilometers throughout the mantle. Incorporating all possible pairs of measurements where one member of the pair does not penetrate the inner core permits P velocities to be estimated from 50 km below the ICB to the inner core's center. The accuracy of the time difference between PKIKP and PKP-AB, whose rays have larger separation near the CMB, will be strongly governed by corrections applied for the effects of 3D velocity structure in the heterogeneous lowermost mantle. A more rarely reported differential time is that of PKIKP-PKIIKP. This can potentially sense velocity changes all the way to the center of the inner core, but can be difficult to observe because PKIIKP rarely exceeds background noise. From PKIKP-PKIIKP times, Wang and Song (2018) have suggested a significant change in anisotropy near 600 km below the ICB depth, but their identification of PKIIKP has been questioned by Wang and Tkalčić (2020).

Progress has been limited by the locations of earthquakes and stations, particularly the sparseness of polar paths. Most studies fit the observed signal in differential travel time with a transverse isotropy having an axis of symmetry close to Earth's rotational axis, fitting the $\sin^2(2\theta)$ and $\cos^4(\theta)$ terms in the perturbation approach detailed in Appendix 1.7 of Chapter 1, where θ is the angle between a P wave ray and Earth's rotational axis. This type of transverse isotropy is commonly termed cylindrical anisotropy.

6.1.2 Depth dependence

Depth dependence of inner core anisotropy has been found in both differential travel times and free oscillations. The differential times sensitive to the uppermost 50–100 km are consistent with isotropy, with anisotropy only emerging at greater depth (Fig. 6.1). Free oscillation studies, however, have suggested that this region may not be truly isotropic, but rather radially anisotropic. In this hypothesis the differential times corresponding to P wave rays sampling

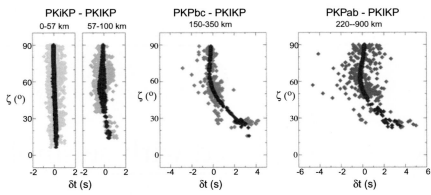

FIG. 6.1 Observed and predicted differential travel times sensitive to the elastic anisotropy of the inner core for different ranges of depth below the ICB (Deuss, 2014). Vertical axis is the angle with respect to Earth's rotational axis. Strong onset of anisotropy does not begin until at least 100 km depth.

the uppermost inner core are presumed to be perpendicular to a radially directed axis of symmetry. Radial anisotropy can be detected by the center frequency of spheroidal radial modes, but it is a nonunique solution with an isotropic uppermost layer still permitted by the relevant modes. In addition, it is also possible for a vertical gradient in the distribution of isotropic heterogeneity near a boundary to behave as if it is radially anisotropic for wavelengths larger than the characteristic scale length (Chapter 5). Inversion and forward modeling experiments over a broad frequency band are thus consistent with at least three possible hypotheses for the elastic structure of the uppermost inner core. These are a layer that is either (1) homogeneous and uniformly isotropic; (2) heterogeneous and isotropic, consisting of randomly oriented small scale (1–10 km) heterogeneities, but with each crystal having a similar LPO in the radial direction; or (3) radially anisotropic, consisting of either LPO or SPO crystals with their fast axes in the radial direction. If (2) or (3), the elasticity of the uppermost 50–100 km is consistent with a mechanism in which heat is transported out of the inner core in the radial direction, with iron dendrites growing perpendicular to the ICB. Such a radial texture places constraints on either the shape of patches of crystals and or on the crystalline phase of iron near the ICB. Radially growing solid dendrites would be required to be fast in the radial direction. The dendrites themselves may consist of patches of crystals having an isotropic distribution of scale lengths with similar LPO in the radial direction. They may also exist as columnar structures separated by thin lenses of interstitial melt or solute-enriched crystallites (Section 7.1.1 in Chapter 7).

In any of the possible interpretations of isotropy or radial anisotropy in the uppermost inner core, a mechanism must exist to change the character of elastic anisotropy such that the fast direction of P waves aligns close to Earth's rotational axis deeper in the inner core. This is required to satisfy combined observations of body wave differential travel times and free oscillation splitting functions sensitive to deeper inner core structure, e.g., Fig. 1.15 in Chapter 1. This might take the form of a change in the character of a convective flow or plastic deformation or changes in the shapes of oriented patches of crystals, or the vanishing or changing of the orientation of any included lenses of melt. In the case of convective flow or plastic deformation, estimates of the thickness of this uppermost isotropic, or radially anisotropic, region can place constraints on the inner core's viscosity and the relaxation time of any topography on the ICB.

Measurements of differential travel times attempt to incorporate the perturbation of rays by descending slabs as well as the effects of more broadly resolved 3D mantle structure. The majority of studies suggest some smaller changes in the anisotropy of the lower half of the inner core as well as a more complex lateral variation of anisotropy in the upper half of the inner core. Mattesini et al. (2013) favor the existence of three or more regions of differing intensity of cylindrical anisotropy, consistent with varying mixtures of hcp and bcc Fe, rather than a quasi-hemispherical structure. Analyses that find evidence for a change in anisotropy in the deeper inner core include one by Stephenson and Tkalčić (2021), who invert ISC bulletin reports of PKIKP travel times, assuming corrections for well-resolved anisotropic structure in the upper 400 km of the inner core. They favor a gradual change in the slow direction of P waves at about 650 km depth with no significant change in the overall strength of anisotropy. Applying mantle and ray path corrections to differential travel times and including new polar data, Brett and Deuss (2020), in contrast, favor a strengthening of cylindrical anisotropy at 530 km depth. These changes with depth in the strength or direction of anisotropy may be related to changes with time in the processes orienting iron crystals, a change with pressure in the stable phase of Fe, or a change in the light-alloying component(s) with depth.

6.2 Attenuation and scattering

6.2.1 Frequency dependence and parameter trade-offs

Unlike the liquid outer core, the inner core significantly attenuates seismic body waves and free oscillations. Until recently, significant differences were found between attenuation measured from either the spectra or waveforms of body waves and from splitting functions of normal modes measured in the free-oscillation band. When corrections for cross-coupling of modes (see Chapter 1) are applied, however, both body wave and free oscillation measurements of attenuation now agree that the inner core is highly attenuating with a shear Q as low as 80 in the uppermost 100–500 km. In the body wave band, this attenuation may be a combination of viscoelastic attenuation and scattering attenuation induced by small-scale (wavelength and less) heterogeneity, which removes energy from the front of a pulse to its later coda.

Although intrinsic viscoelasticity occurs with velocity dispersion as well as attenuation, the dispersion is a difficult quantity to accurately measure, requiring an accurate estimate of the source-time function. Fig. 6.2 illustrates how a PKIKP waveform has been fit with parameters describing the viscoelasticity of the inner core, assuming an average mantle attenuation operator and a far-field source-time function determined from P waves confined to the mantle. Attenuation in the outer core is assumed to be zero. This modeling predicts a weaker velocity dispersion than would be predicted by a viscoelastic Q constant in frequency. Diminished velocity dispersion, compared to that expected for a near-constant attenuation with frequency, suggests that either the low- or high-frequency corner of the inner core's relaxation spectrum lies in the 0.2–2 Hz frequency band (Fig. 6.3).

Although Doornbos (1983) first suggested that it is the low-frequency corner that lies in the body band, recent work by Mäkinen and Deuss (2013) finds agreement between free-oscillation and body wave measurements. This removes the ambiguity of choosing which of the two peaks provides a better waveform fit in the histogram of Fig. 6.3. When the high-frequency corner lies in the body wave band, attenuation is nearly constant from frequencies near 1 mHz until the beginning of the body wave band near 0.2 Hz. Stronger frequency dependence begins when the frequency equals or

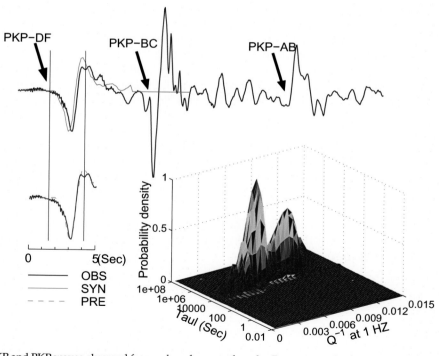

FIG. 6.2 Top left: PKIKP and PKP waves observed from a deep focus earthquake. Parameters of an inner core attenuation operator are varied to obtain the best fit to the PKIKP waveform by convolving the attenuation operator with a far-field source-time function and mantle attenuation operator. Bottom right: A probability density for a two-parameter model of the inner core attenuation operator is constructed from an L_2 norm of the observed minus predicted waveform, normalized by a covariance related to a signal to noise ratio (from Li and Cormier, 2002). The existence of two peaks in the probability density demonstrates an important nonlinearity in the inverse problem.

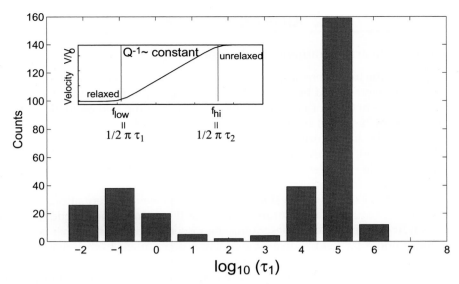

FIG. 6.3 The dispersion of an elastic wave velocity (upper left inset) is maximum between low- and high-frequency corners of a viscoelastic relaxation spectrum and minimum in the vicinity of the corners, decreasing to zero in the frequency bands corresponding to relaxed and unrelaxed states. A histogram of estimates of the low-frequency corner of the relaxation spectrum inverted from PKIKP waveforms sampling the inner core (Li and Cormier, 2002) exhibits two peaks. Assuming a five-decade width for the frequency band of near constant high attenuation, the highest peak at $\log_{10}(\tau_1)=5$ implies that the relaxation time τ_2 is on the order of 1 s, which is near the middle of the passband of high-frequency body waves.

exceeds 1 Hz. If the relaxation spectrum were instead such that free oscillations suffered weaker attenuation than body waves, it might be possible to explain an observed correlation between attenuation and velocity. (Fast PKIKP travel times correlate high PKIKP attenuation.) Since measured attenuation is relatively constant from 1 mHz to 0.2 Hz, other mechanisms are needed to explain the sense of this correlation. One explanation may be related to the behavior of fluids in a transversely isotropic, porous, medium (e.g., Carcione and Cavalini, 1994). In the case of the inner core, this fluid could be melt. Any significant trapped melt in the deep inner core, however, would likely be buoyantly expelled into the outer core, except in a thin region near the ICB. Another possibility to explain weaker pulse dispersion of body waves would be a significant contribution from scattering to the total attenuation.

Waveform broadening and dispersion of PKIKP can be fit equally well by a pure viscoelastic or a pure scattering attenuation. Current measurements of inner core scattering from the shape of the coda envelopes of PKiKP waves, however, limit the maximum scattering contribution to about 50% of the total apparent attenuation in the upper 300 km of the inner core (Leyton and Koper, 2007a,b). Inversion for the parameters describing a heterogeneity spectrum capable of inducing pulse broadening and coda of body waves suffers problems similar to that encountered in inversions for viscoelastic parameters. Well-separated domains of P velocity fluctuation and corner scale length produce equally good fits to PKIKP waveforms (Fig. 6.4). Similarly, a range of parameters describing the spatial spectrum of inner core heterogeneity can fit the scattered coda of PKiKP waves at pre-critical range. Peng et al. (2008) found trade-offs in the parameters describing the heterogeneity spectrum such that good fits to coda envelopes can be found by assuming either an RMS P velocity fluctuation of 0.8% with a corner scale length of 1 km or by an RMS P velocity fluctuation of 10% at a corner scale length of 16 km. The lack of significant high-frequency scattered coda in PKIKP waveforms at distances between 160° and 180°, however, excludes models having RMS P velocity fluctuations exceeding 2% and scale lengths larger than several kilometers.

6.2.2 Depth dependence

Although trade-offs exist in parameters describing both viscoelastic and scattering attenuation, estimates of attenuation at 1 Hz, either dominantly viscoelastic or dominantly scattering, exhibit strong dependence both on depth and on path with respect to the axis of rotation (Fig. 6.5). Although relatively complex models of the depth dependence have been proposed in the uppermost 200 km of the inner core, including lateral variations, the simplest, globally consistent models agree that attenuation is highest in the upper half of the inner core, averaging $Q_P=200$ and $Q_S=80$, with Q_P increasing up to 1000 in the lower half. The measurements shown in Fig. 6.5 are consistent with a slow decrease in

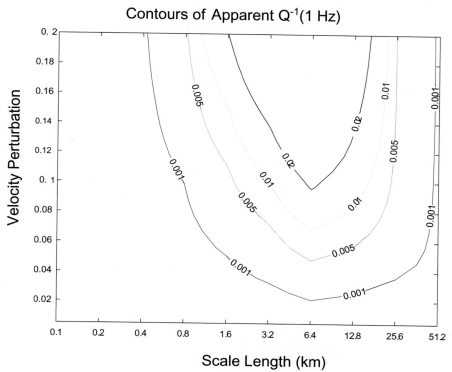

FIG. 6.4 Pulse attenuation in a pure scattering model of inner core attenuation as a function of per-cent P velocity perturbation and scale length calculated by the DYCEM technique of Kaelin and Johnson (1998). The nonlinear behavior in the inverse problem is evident in the existence of two possible scale lengths at the same velocity fluctuation required to achieve an apparent Q^{-1} to match a PKIKP waveform. *From Cormier, V.F., Li, X., 2002. Frequency dependent attenuation in the inner core: Part II. A scattering and fabric interpretation. J. Geophys. Res. 107(B12), https://doi.org/10.1029/2002JB1796.*

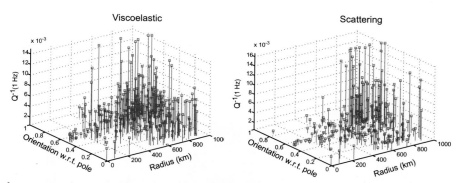

FIG. 6.5 Left: Inverted attenuation from PKIKP waveforms, assuming attenuation is dominantly shear and viscoelastic. Right: Inverted attenuation assuming attenuation is purely from scattering. *From Li, X., Cormier, V.F., 2002. Frequency-dependent seismic attenuation in the inner core 1. A viscoelastic interpretation. J. Geophys. Res. 107 (B12), https://doi.org/10.1029/2002JB0011795 and Cormier, V.F., Li, X., 2002. Frequency dependent attenuation in the inner core: Part II. A scattering and fabric interpretation. J. Geophys. Res. 107(B12), https://doi.org/10.1029/2002JB1796.*

P attenuation in the upper half of the inner core. In pure scattering models, a more abrupt decrease occurs halfway to the center, suggestive of sharp change in crystalline texture at that depth that may be indicative of the existence of an inner-inner core. Global inversions for inner core attenuation exhibit slight increases in attenuation as ray paths become more parallel to the axis of rotation. A similar anisotropy of attenuation is found in the inversion of normal modes. Some regionally localized experiments with body waves exhibit a much stronger anisotropy of attenuation (e.g., Souriau and Romanowicz, 1996). A majority of studies find the previously cited correlation of velocity with attenuation, in which fast velocity correlates with high attenuation and slow velocity with low attenuation (a reversed correlation from that commonly observed in the mantle).

6.2.3 Lateral variations in scattering

Estimates of the spectrum of small-scale heterogeneities in the inner core can be determined from the coda envelope of PKiKP at narrow angle incidence on the inner core (Fig. 6.6). The inverted spatial spectrum is often described by the parameters that define either an exponential or von Kármán spectrum in wavenumber (Section 1.5, Chapter 1). These are the maximum percent RMS fluctuation in P velocity and the corner scale-length parameter of the heterogeneity spectrum. Leyton and Koper (2007a,b) examined the effects of both ICB topography and volumetric heterogeneity in the inner core assuming single scattering and concluded that there is evidence for volumetric heterogeneity in the inner core with an *RMS* fluctuation on the order of 1% with a corner parameter on the order of several km. Wu and Irving (2017), modeling with a radiative transport theory for multiple scattering estimated an RMS fluctuation of 0.5%–1% and a corner parameter of 2 km. Similar studies by Vidale and Earle (2000) and Peng et al. (2008) are mutually consistent when trade-offs between the RMS fluctuation and corner parameter are considered. The stronger lateral variations in the Leyton and Koper study seem to coincide with a strong gradient change in buoyancy flux at the ICB predicted from compositional convection in the outer core driven by heat flow variations at the CMB (compare Fig. 6.7 with Fig. 6.10C).

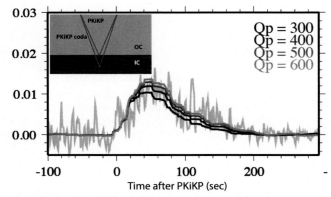

FIG. 6.6 Scattered coda envelope of PKiKP waves and smoothed fits assuming parameters describing an exponential autocorrelation of fluctuations in P velocity with different assumed values for intrinsic viscoelastic Q_P (Peng et al., 2008). Note the relatively weak sensitivity of pre-critical PKiKP coda envelopes to Q_P. This is due to the short path length in the inner core of the single-scattered waves (upper left inset) that contribute to the coda shown in time window following PKiKP.

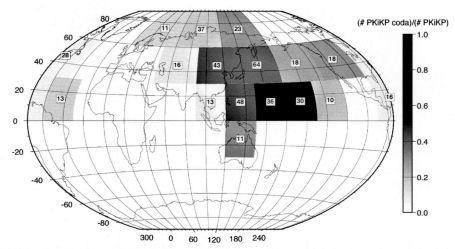

FIG. 6.7 Regional variations in PKiKP coda amplitudes due to scattering in the uppermost inner core (Leyton and Koper, 2007b).

6.2.4 Anisotropy of the heterogeneity spectrum

Cormier et al. (2011) examined the effect of anisotropy of scale lengths at the top of the inner core on PKiKP coda and concluded that the volumetric scattering is consistent with a relatively more isotropic or uniform distribution of scale lengths (Fig. 6.8), a result which is important for discriminating between the existence LPO or SPO anisotropy in the uppermost inner core. An isotropic distribution of scale-lengths of heterogeneity, having a peak in power around 2 km scale, explains PKiKP coda up to 200 s after its first break. This suggests that the onset of elastic anisotropy in the inner core beneath 100 km depth is best-explained by LPO rather than SPO anisotropy. It is important to note, however, that an isotropic distribution of heterogeneity scale lengths does not always result in an equivalent elastically isotropic medium at wavelengths greater than the scale lengths. In the presence of vertical gradients in average velocity such a medium may exhibit some weak radial anisotropy (vertical axis of symmetry in an equivalent TI medium), as shown by computational experiments that assume small-scale isotropic heterogeneity in D″ of the lowermost mantle. For example, see Fig. 5.8 in Chapter 5 and Parisi et al. (2018).

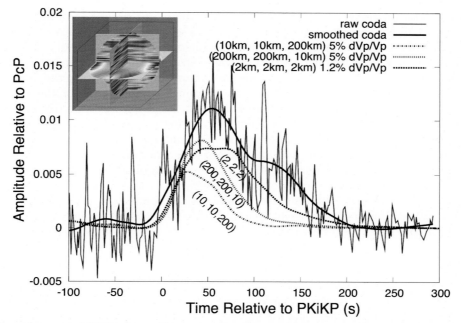

FIG. 6.8 Coda envelope of pre-critical PKiKP observed by Peng et al. (2008) at LASA compared with predictions by Cormier et al. (2011) from *radiative transport* modeling of volumetric scattering in the inner core for distributions of small-scale inner core heterogeneity that have isotropic, vertical, and horizontally stretched anisotropic distributions of scale lengths. Inset shows a possible inner core texture proposed by work of Bergman (1997) to explain cylindrical anisotropy and differences in the attenuation of seismic body waves traveling polar versus equatorial paths. Note that the observed coda shape, however, is best-explained by a 3D isotropic distribution of scale lengths (2 km, 2 km, 2 km) rather than by scattering from shapes that are stretched as thin sheets, oriented either radially perpendicular to the ICB, (10 km, 10 km, 200 km), or horizontally, parallel to the ICB (200 km, 200 km, 10 km).

6.3 Hemispherical differences

6.3.1 Observations

Although body wave sampling of the polar regions is still relatively poor, sampling of the equatorial and mid-latitude regions is sufficient to show that there are approximately hemispherical or degree-one differences in inner core structure. This global behavior has been confirmed from studies of the splitting functions of cross-coupled normal modes (Irving et al., 2009). Both body wave and free-oscillation studies require relatively sharp longitudinal transitions. Less attenuation and lower isotropic velocity are generally observed in a western quasi-hemisphere compared to a quasi-eastern hemisphere. The implications of this degree one structure are significant to the understanding of the dynamics of the outer core and behavior of the geodynamo. This is strongly suggested by the fact that the seismic hemispherical transitions (Fig. 6.9) correlate with features of the magnetic field. The seismic transitions separate

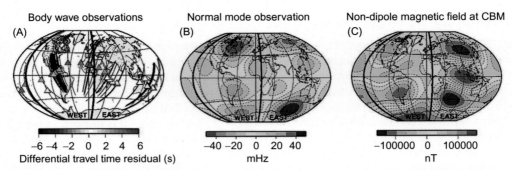

FIG. 6.9 Hemispherical differences in inner core properties: (A) Differential travel time residuals from PKP-DF-PKP-BC and PKP-DF-PKP-AB (Deuss et al., 2010); (B) cross-coupled splitting function for normal mode pair $_{16}S^5$ and $_{17}S^4$; and (C) Non-dipole magnetic field downward extrapolated to the core–mantle boundary (Kelly and Gubbins, 1997).

regions of different intensities of the non-dipole field when the field is downward continued to the CMB. The centers of the four blue-colored patches of intense flux patches in Fig. 6.9C intersect a projection of the boundaries of a cylinder tangent to the inner core at its equator. The transition of inner core seismic properties near 0° longitude correlates with patches of high secular variation of the magnetic field concentrated at low and mid-latitude (Section 3.4, Chapter 3). The correlation of these features of the magnetic field with elastic structure, and their persistence in time in the midst of convection in the outer core, points to a central role for the geodynamo in understanding the structure of the inner core.

When elastic anisotropy has been incorporated into tomographic imaging, however, the regional variation of the intensity of anisotropy is seen to be more complex than a simple hemispherical pattern (Irving and Deuss, 2015). The general pattern of PKIKP travel times suggests stronger anisotropy in the western hemisphere and a weaker and more complex pattern in the eastern hemisphere. Significant latitudinal and depth variations in anisotropy have also been identified. This regional complexity may be the natural result of either solidification or deformation mechanisms being affected by variations of heat transport across the CMB, which are subsequently transmitted via outer core convection to the ICB. Lateral variations in CMB heat transport are likely correlated with the structural complexities of the D″ region (Chapter 5) and their evolution with time. Understanding the complexity of anisotropy in the inner core will require increased body wave sampling, additional advances in the treatment of mode cross-coupling, and confidence in estimating how outer core convection near the ICB is impacted by heat transport across the CMB.

6.3.2 Theories

Two theories have been proposed for the degree-one structure, having opposite predictions for regions of crystallization and melting and different implications for the rheology of the uppermost inner core, the development of anisotropy, and the age of the inner core. One theory formulated by Aubert et al. (2008) depends on lateral variations in heat flow across the core-mantle boundary to predict fluid core motions from geodynamo simulations. The lateral variations in CMB heat flow are inferred from the global tomographic images, which resolve two antipodal located large low-velocity provinces above the CMB (Chapter 5). Although this large-scale structure near the CMB is dominantly degree-two, it predicts a thermochemical flow of light element released from the solidifying ICB in which a cyclonic circulation below Asia affects the intensity of magnetic field at the CMB. The CMB magnetic field and core flow along the CMB determined from numerical dynamo modeling with a laterally varying CMB heat flux agree with that predicted from the downward continuation of the magnetic secular variations. In this model the eastern hemisphere of the inner core is dominantly freezing and a source of maximal upward flux of light elements. The faster freezing in the eastern hemisphere results in a textured solid with more random crystal orientations (weaker anisotropy). Conversely, lower freezing in the western hemisphere allows time for the formation of larger crystals, preferentially oriented in the predominantly radial direction of heat transport. Longitudinal changes in predicted ICB heat flux from this model (Fig. 6.10), which are concentrated in the equatorial region, are close to the boundaries of quasi-hemispherical structure determined from global body wave studies. Some interesting correlations, however, exist with the position of non-dipole field components in the eastern hemisphere, with the position of either lateral changes in inner core scattering (Fig. 6.7) or lateral gradients in a possible thin (<40 km) low velocity or a more intensely scattering layer at the top of the inner core (Fig. 6.10B) and with the eastern edge of a predicted ICB heat flux anomaly (Fig. 6.10C).

The competing theory to explain degree-one structure is translating convection, in which the inner core crystallizes in the western hemisphere and melts in the eastern hemisphere (Fig. 6.11). This has been termed convective translation

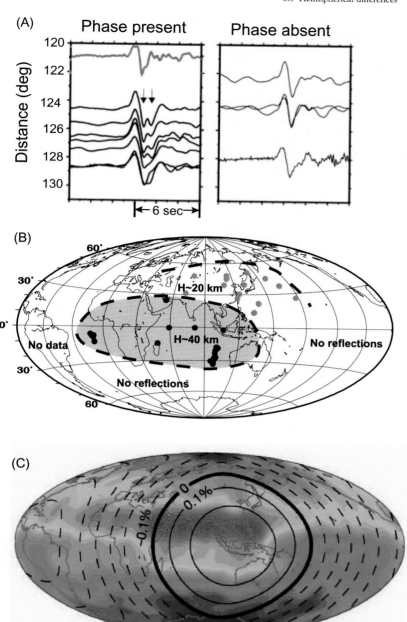

FIG. 6.10 (A) Observations of an anomalous PKIKP+PKiKP waveform; (B) surface projected iso-contours of the thickness of either a low-velocity layer (Stroujkova and Cormier, 2004) or a region of more intense scattering producing a multipathed arrival between PKIKP and PKiKP; (C) predicted variations in heat extraction at the inner core boundary from numerical dynamo modeling (Aubert et al., 2008). High inner core heat extraction is denoted by red and low by blue. The P velocity perturbation at top of inner core is shown by closed solid contours indicating high P velocity and negative dashed contours indicating low P velocity.

because it involves a particular mode of inner core convection that can exist if the inner core possesses sufficiently high viscosity. In this regime, a super-adiabatic temperature gradient induces differences between the density gradients in the two hemispheres of the inner core and a consequent translation of the inner core toward the higher temperature, lower density, hemisphere to maintain bouyant equilibrium. The inner core surface melts in the higher temperature hemisphere and crystallizes in the opposite hemisphere as the inner core position shifts. Geodynamic modeling of convective translation predicts that the entire inner core would be renewed by this process every 100 m.y.

Measurements of lateral variations in elastic scattering in the inner core have been invoked as one of the tests of translating convection, in which smaller sizes of patches of commonly oriented crystals exist in a crystallizing western hemisphere, inducing stronger scattering in PKiKP coda. Larger patches of commonly oriented crystals exist in a melting, older, eastern hemisphere. Scattering effects have also been suggested to explain the anomalously low shear modulus of the inner core as the effect of a long-wavelength equivalent medium of randomly oriented anisotropic iron

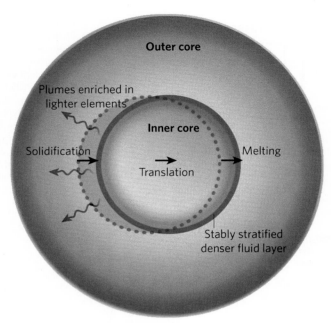

FIG. 6.11 Translating convection model from Alboussière et al. (2010).

crystals having scale lengths between 300 and 700 m and 7 to 15 km (Calvet and Margerin, 2008). Unfortunately, estimates of small-scale inner core heterogeneity from the scattered coda of PKiKP do not yet have the global coverage to test such models. Strong nonlinearities and trade-offs in the inverse problem for scattering parameters, demonstrated by PKIKP waveforms (Fig. 6.4) and coda envelopes of PKiKP (Fig. 6.6), suggest that similar nonlinearities might need to be considered in inversions for a long-wavelength equivalent medium.

Other tests of the translating convection model are any observations that constrain inner core viscosity. Models of the *free inner core nutation*, in which the axis of the inner core rotation wobbles about the axis of rotation of the mantle, constrain the inner core viscosity to 2 to 7×10^{14} Pa s (Koot and Dumberry, 2011). This value is beneath the $>10^{18}$ Pa s regime predicted for translating convection (Deguen et al., 2013). Unless scattering is the dominant mechanism to explain the anomalously low shear modulus and high seismic attenuation of the inner core, its relatively low shear modulus and high viscoelastic relaxation times do not support the high viscosities required for translating convection.

6.4 Differential rotation

6.4.1 Observations

Since an intervening liquid outer core separates the solid inner core from the solid mantle, it is possible, in principle, to have the inner core rotate at a different rate about a different axis from the mantle even if the solid mantle and inner core initially form having a common rotational rate and axis. This situation can occur if there exist unique torques acting on the inner core. If the elastic structure of the inner core is either elastically anisotropic, possesses 3D heterogeneity, or surface topography then any difference in its rotation with respect to the mantle may be detected from variations in the travel time of P waves transmitted through it (Fig. 6.12).

Experiments for detecting this inner core differential rotation have primarily sought to measure the variations in the travel time difference between the P wave transmitted just above the inner core (PKP-BC) or reflected by it (PKiKP or PKP-CD) and the wave transmitted through it (PKIKP or PKP-DF). The observed size of these time variations suggests that differential inner core rotation may be as large as 0.1–0.5 deg./year.

To achieve the required accuracy in travel time measurement, it is necessary to use a method that is insensitive to small variations in the shape and frequency of earthquake source-time functions that control the shape of the waves and their apparent time differences. This is usually accomplished by comparing the differential travel times measured from two earthquakes at different times, termed doublets, which are determined to have nearly identical source-time functions of P waves and epicenters. In addition to network refined relocations, the identical location, focal mechanism, and source-time functions of the event pairs are typically confirmed by examining the correlation of the body

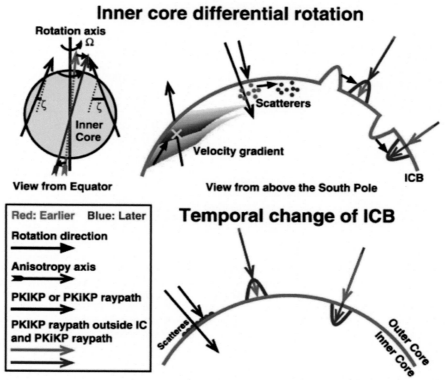

FIG. 6.12 Possible mechanisms that can produce temporal changes in the travel times of PKIKP. Red (light gray in print version) and blue (dark gray in print version) denote states at different times. Upper left: A difference in the symmetry axis of inner core transverse isotropy compared to its rotation axis results in differences in elastically anisotropic structure sampled at two different times. Upper right: Differential rotation of the inner core results in observations of PKIKP from an *earthquake doublet* sampling different topography of the inner core boundary or different portions of laterally varying structure at two different times. Lower right: No differential rotation of the inner core but a change in its surface at two different times. *From Yao, J., Tian, D., Sun, L., Wen, L., 2019. Temporal change of seismic Earth's inner core phases: Inner core differential rotation or temporal change of inner core surface?, J. Geophys. Res. 124(7), 6720–6736, https://doi.org/10.1029/2019JB017532.*

waveforms and their codas that follow the PKIKP wave. The quality of the designation of a true identical doublet is usually judged by the quality of the waveform cross-correlation following PKIKP. A measurement from a less common type of source doublet is to retrieve time shifts from the cross-correlation of the scattered coda of PKiKP observed by receiver arrays from the seismograms of nuclear tests within a common test site having a common size and emplacement. Lythgoe et al. (2020) demonstrate that in principle it is possible to achieve a precision of 0.02 s to differential travel times measured by cross-correlation of signals that are digitized at 0.05 s. A 0.02 s precision, however, is a lower bound likely only achievable with near perfectly matched earthquake doublets. An estimated 0.4 deg./year rotational difference, which is at the high end of obtaining from waveform cross-correlations of doublets one year apart, corresponds to a lateral difference of less than 10 km for the separation between the doublet's PKIKP rays incident on the ICB. Considering imperfections in the quality of doublet pairs and the width of the sensitivity kernel of high-frequency PKIKP waves, accurate measurements will typically require observations of a decade or more in time. An example of the challenges in measuring the small changes in differential travel times in earthquake doublets is shown in Fig. 6.13.

6.4.2 Theories

Early body wave experiments predicted differential rotation rates as high as 1 deg./year, inconsistent with nondetectability from free oscillations. Subsequent experiments, which have included corrections for known laterally varying structure in the uppermost inner core, have generally found lower average rates on the order of 0.1–0.5 deg./year. Tkalčić et al. (2013) found strong evidence for temporal irregularity in the differential rotation, consistent with an average mean in this range over a 40-year period, but with decadal fluctuations as high as 1 deg./year in both prograde and retrograde directions. They measured an exceptionally average low rate in the time period in which

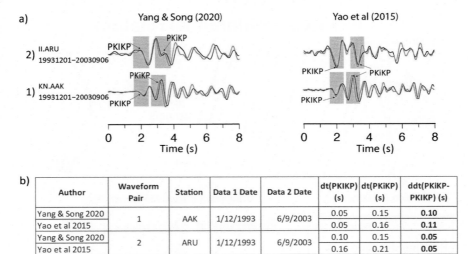

FIG. 6.13 (A) Comparison of two earthquake doublet pairs cross-correlated by two separate groups to measure the change in PKiKP–PKIKP differential travel time over a nearly 10 year time period. Each doublet pair is recorded at the same seismic station, with each member presumed to have occurred at nearly the same location with nearly the same source mechanism and source-time function. Waveforms are aligned on the PKIKP arrival. Differences in waveform pairs between the left and right columns are due to differences in filtering between the two groups. (B) Table of measured differential travel times for each doublet by each group for PKIKP, PKiKP, and PKiKP–PKIKP. *From Lythgoe, K.H., Inggrid, M.I., Yao, J., 2020. On waveform correlation measurement uncertainty with implications for temporal changes in inner core seismic waves. Phys. Earth Planet. Inter. 309, 105506, https://doi.org/10.1016/j.pepi.2020.106606.*

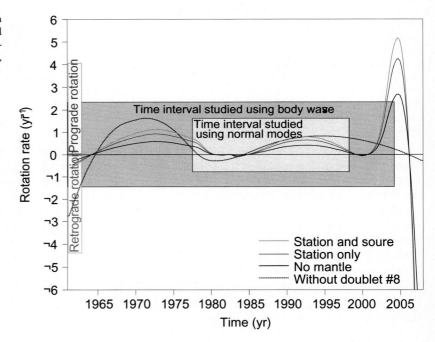

FIG. 6.14 Estimated differential rotation rate in deg./yr of the inner core from the differential travel time of PKP-DF–PKP-BC measured from PKP waveforms recorded in earthquake doublets (Tkalčić et al., 2013).

free oscillations could not detect a differential rotation and also found agreement in the time period in which a low rotation rate was measured in a study of scattered PKiKP coda (Vidale, 2019).

Times when the acceleration of the differential rotation rate is zero (time derivative of the differential travel times in Fig. 6.14), correlate well with times of geomagnetic jerks observed in 1970, 1981, 1992, and 2003 (see Fig. 3.8 in Chapter 3). Geomagnetic jerks are characterized by a sudden large change in the second time derivative of the magnetic field. A possible mechanism for exciting a geomagnetic jerk, which may also induce electromagnetic torques

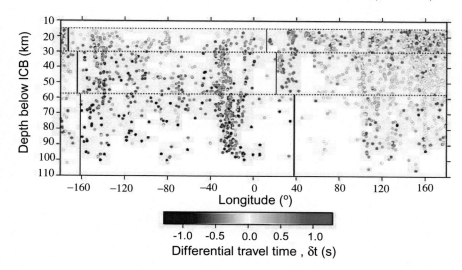

FIG. 6.15 Differential PKIKP-PKP travel times as a function of depth of PKIKP ray turning points in the inner core. Shifts to the east in differential travel times can be measured to estimate a very slow net average inner core super-rotation with respect to the mantle on the order of 0.1 deg./m.y., assuming an inner core growth rate of 1 mm/year (Waszek et al., 2011).

exciting a wobbling motion of the inner core with respect to the mantle, has been proposed from numerical dynamo modeling by Aubert and Finlay (2019). In this model, narrow plumes rapidly ascend toward the CMB from reservoirs of positively buoyant material enriched from light elements released from the solidification of the inner core.

A confirmation of either the inner core's net zero position or a much slower average *super-rotation* on the order of 0.1 deg./m.y. is a study by Waszek et al. (2011). From differential travel times, they demonstrated that hemispherical differences in the inner core persist with depth with very small lateral drift in their boundaries (Fig. 6.15). The very small depth independence of the hemispherical boundaries is not consistent with larger magnitudes of super-rotation of the inner core as it solidifies.

From the time differences of features in the coda of PKiKP of doublet pairs, Yao et al. (2019) suggest the alternative possibility of fast temporal changes in the inner core radius, e.g., up to 1 km in 80 days. Yang and Song (2020), however, argue that errors in station timing and the interference by scattered PKIKP coda overlying the coda of PKiKP make it difficult to achieve the timing accuracy claimed in the Yao et al. studies. Instead they conclude that the best current estimates of inner core differential rotation are about 0.05–0.1° per year. Larger decadal fluctuations about that value, such as those identified by Tkalčić et al. (2013) cannot yet be excluded. Several studies have pointed out that the inner core ray paths in many of these studies are dominated by a single corridor of ray paths from the South Sandwich Islands to Alaska, and require careful estimation of the contaminating effects of lateral variations in small-scale structure in the uppermost inner core as well as by the larger scale structure of the descending slab beneath receivers in Alaska.

At least two conclusions can confidently be inferred from seismic studies of inner core differential rotation: (1) the motion of the inner core on average is small and irregular and (2) its near-zero time average in position is consistent with the existence of restoring forces due to gravitational locking between density anomalies in the inner core and mantle. In seeking to retrieve the small time signals of its differential rotation such studies have also confirmed the existence of small-scale (order of 10–20 km) heterogeneity in the uppermost inner core important for understanding its processes of solidification.

6.5 Shear modulus, density, and viscosity

6.5.1 Shear modulus from seismology

Once the discontinuous increase in P velocity at the ICB was discovered, Bullen proposed that the simplest explanation for that increase would be the appearance of a nonzero shear modulus beneath the ICB due to the solidification of the liquid outer core. Surprisingly the average shear velocity of the inner core of 3.5 km/s estimated from Bullen's initial hypothesis remains close to all recent estimates obtained from the studies of normal modes and body waves. In contrast, assuming compositions of iron/nickel alloys at inner core conditions, mineral physics estimates of the V_S of the inner core are significantly higher than 3.5 km/s, unless the temperature of the inner core approaches its melting temperature. Together with uncertainty in estimates of inner core density, the difference between the seismically measured and theoretical or laboratory observed shear velocity of iron alloys remains one of the major unsolved problems in core structure.

The first reliable estimates of the shear modulus were obtained from the observations of spheroidal normal modes $_nS_l$ with low order number l and high overtone number n, which are most sensitive to inner core structure. These observations resulted in the inner core shear model of PREM, and recent updates by Deuss have not deviated much from original PREM values, with Vs=3.5 to 3.67 km/s. Analysis of recent mode observations, which have included the effects of mode cross-coupling, have enabled the identification of hemispherical differences and improved estimates of viscoelastic attenuation. These have established error bounds on the equivalent isotropic shear modulus between 149.0 ± 1.6 GPa and 167.4 ± 1.6 GPa.

Observations in the body wave band (0.01–1 Hz) have concentrated on the identification of the PKJKP wave, in which a compressional wave in the liquid outer core propagates as a shear (J) wave in the inner core. The PKJKP wave is predicted to be a very weak body wave band with a significantly lower frequency content due to the small conversion coefficient of longitudinally polarized P waves to transversely polarized S waves at the ICB and the high shear attenuation of the inner core. Consequently, many body wave studies have encountered problems in separating the weak PKJKP wave from other weak interfering body waves and scattered coda. This has been overcome by new techniques of seismic interferometry, including correlation wavefields measured across seismic arrays. In this technique, seismograms from pairs of earthquakes are cross-correlated, which helps in identifying waves of weak amplitude, normally buried in interfering coda. The correlation wavefield takes advantage of the fact that the pairs of waves will be highly correlated when they share the same source-time function and similarly sample the heterogeneous structure in the mantle beneath receivers when their wavefronts arrive at the same angle. From *correlegrams* between pairs of waves sampling the inner core, including those shown in Fig. 6.16, Tkalčić and Pham (2018) estimate a Vs=3.42 ± 0.2 km/s at the ICB and

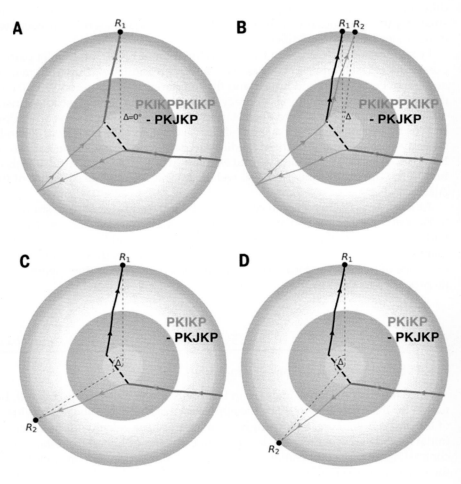

FIG. 6.16 Ray paths of body waves that are cross-correlated to identify the PKJKP wave from Tkalčić and Pham (2018).

3.58 ± 0.02 km/s at the center of the inner core. These values are in agreement with the most recent estimates of Vs from mode cross-coupling, which continue to agree with PREM if its radius is reduced by 0.05%. Applying this technique to estimate inner core attenuation from the correlation of PKIIKPPKIIKP and PKJKP predicts a 50% lower average Q_P for the inner core than PREM. The effects, however, of correlegram processing, depth dependence of attenuation, and shear wave splitting by IC anisotropy, on this estimate have not yet been fully investigated.

6.5.2 Transition defining the ICB

In either a formation of the ICB as growing solid dendrites from below or solid particles precipitating from above, there will be a transition region in which the shear modulus gradually rises from zero over some depth range corresponding to processes that extract melt and solids compact. If the ICB is growing from below, this transition could be the dendritic mushy zone described in Chapter 7. PKiKP waves reflected from the transition when their wavelengths are much less than their thickness cannot easily detect the thickness of such a transition zone. The amplitude of the PKIIKP wave at near antipodal distances is strongly sensitive to the shear modulus change and transition at the ICB. This is because, in this distance range, the angles of incidence of a compressional K wave on the ICB are such that the conversion to a shear J wave in the inner core is maximum. Hence, the excitation of the PKJKP wave reduces the amplitude of the PKIIKP wave. PKIIKP will be larger for a lower inner core shear modulus. If the shear modulus grows from zero to a value near its inner core average over a transition zone, the PKIIKP wave will be larger for a thicker transition and smaller for a thinner transition. These relations have been used in a study by Attanayake et al. (2018) to infer a maximum 2–4 km thickness for the thickness of an inner core transition zone from liquid to solid (Fig. 6.17). This thickness agrees with estimates for the upper bound to inner core topography over a broad spectrum of scales. It also allows the ICB to be accurately treated as a first-order discontinuity in physical properties in the majority of seismic wavefield simulations at frequencies up to 2 Hz.

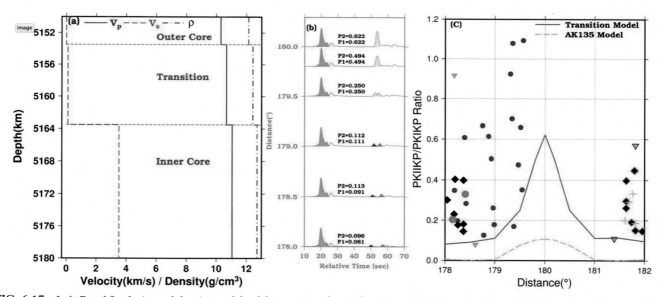

FIG. 6.17 Left: P and S velocity and density models of the transition from a liquid to solid at the ICB; middle: PKIKP and PKIIKP displacement waveforms at antipodal distances. Right: Modeled amplitude ratio of PKIIKP/PKIKP obtained from integrated displacements. *From Attanayake, J., Thomas, C., Cormier, V.F., Miller, M.S., Koper, K.D., 2018. Irregular transition layer beneath Earth's inner core boundary from observations of antipodal PKIKP and PKIIKP waves. Geochem. Geophys. Geosyst. 19(10), 3607–3622, https://doi.org/10.1029/2018GC007562.*

6.5.3 Predictions from mineral physics

The seismically estimated P velocity is 4%–10% lower and the S velocity 30% lower than those estimated from computational and experimental predictions for the properties of hcp and bcc iron. These observations suggest that lighter elements are incompletely expelled into the liquid outer core as the inner core solidifies as an Fe—Ni alloy. Since iron meteorites, which typically contain 5%–25% Ni, are presumed to be the remnants of the cores of differentiated *planetesimals*, the effects of Ni in iron alloys have been explored as an important contributor to the lower seismic velocities.

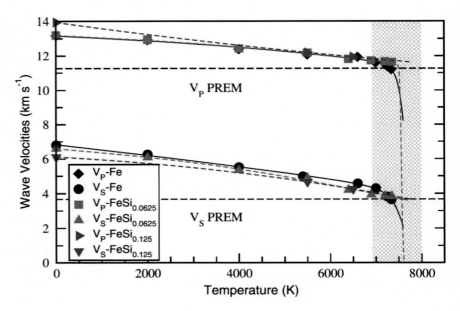

FIG. 6.18 A comparison of ab initio estimated seismic velocities in the inner core for various Fe—Si alloys. Dashed lines indicate PREM seismic velocities and the shaded region are estimated upper bounds to the temperature ranges of the inner core required to match the PREM velocities. *From Martorell, B., Vocadlo, L., Brodholt, J., Wood, I.G., 2013. Strong premelting effect in the elastic properties of hcp-Fe under inner core conditions. Science 342 (6157) 466–468, https://doi.org/10.1126/science.1243651.*

The predicted effects of Ni in plausible concentrations are generally found not to be large enough to explain the seismic velocities of the inner core without also strong assumptions on the temperature of the inner core. This has prompted a search for candidate lighter elements other than Ni alloyed with iron to satisfy both the seismically estimated velocities and the densities of the inner core.

In addition to examining the effect of varying the alloy composition, the temperature may be varied to match the observed elastic velocities as long as it does not excessively exceed the predicted solidus temperature. Fig. 6.18 illustrates how a possible range of inner core temperatures has been estimated from Fe—Si alloys having an hcp lattice structure. Studies by this group and others have examined the effects of Si, S, O, H, and C alloyed with Fe in both hcp and bcc lattice structures, but have yet to converge on an ideal structure. Thus far, it appears that at least a ternary alloy is required in a solid solution form.

The results illustrated in Fig. 6.18 match the observed V_s only if the entire inner core is close to its predicted melting temperature ($T/T_m \sim 0.99$), which they termed pre-melting. Allowing concentrations of Ni and an expanded group of light elements, more recent estimates by this group can match the seismically measured V_s without as strong a pre-melting effect.

The stability of a bcc Fe phase in the inner core has been questioned, but using a larger number of atoms in the cell of an ab initio calculation, Belonoshko et al. (2019) has found that bcc Fe can indeed be a stable phase at inner core P–T conditions. Their predicted bcc properties for both V_s and the elastic anisotropy of the inner core can match seismic observations. Using ab initio predictions, Mattesini et al. (2013) have developed schemes to fit observations of more complex spatial patterns of anisotropy from combinations of hcp and bcc Fe.

6.5.4 Viscosity

In addition to constraining differential rotation of the inner core, its viscosity and its effect on the modes of plastic deformation of crystalline lattices are essential to understanding the mechanisms responsible for its elastic and anelastic anisotropy. Unfortunately, the scatter in viscosity estimates for the inner core is larger than that of the liquid outer core, ranging from 10^{-2} to 10^{27} Pa s These are discussed in greater detail in Chapter 7, but for now, it is useful to summarize three major groupings of these estimates. The highest range 10^{19}–10^{27} Pa s presumes either diffusion or dislocation creep. The low end, 10^{-2} to 10^{2} Pa s is obtained from the predicted self-diffusion of bcc iron (Belonoshko et al., 2019). This exceptionally low value is in the range of outer core viscosities, but with ab initio estimates of other rheologic properties it is still possible for the inner core to be sufficiently solid-like. The range 10^{10}–10^{18} Pa s is obtained from free inner core nutation and estimates of deformation texturing to match its elastic anisotropy. This range also roughly agrees with what might be expected from the seismically measured shear attenuation. Lasbleis and Deguen (2015) have built a regime diagram to isolate the dominant factors controlling the strain rate responsible for observed elastic anisotropy. At the high end of estimated viscosities, they conclude that the timescale for texture development to match the observed anisotropy would be similar to the age of the inner core, but the low end ($\leq 10^{12}$ Pa s) would be effective in allowing the Lorentz force of the magnetic field to impose cylindrical anisotropy.

6.5.5 Density discontinuity at the ICB

Experiments with both free oscillations and body waves are useful for placing bounds on the density jump at the ICB. With gravity becoming more important as a restoring force for elastic motions at lower frequencies, free oscillations are intrinsically more sensitive to density structure than body waves. Exceptions are the reflection coefficients of body waves at near vertical incidence on radial structural discontinuities. Reference Earth models PREM and AK135 report a density jump of 500–600 kg/m^3 at the ICB. Body wave experiments, however, report a higher upper bound of 1000 kg/m^3. Since acceptance of this higher upper bound allows a greater energy input to models of compositional convection from a solidifying inner core, it is important to critically understand the origin and background of body wave experiments that have estimated the ICB density jump. The ICB density jump has also been shown to be a key parameter controlling the stability of the F region in the lowermost 150–400 km of the outer core (Chapter 5) when its reduced P velocity gradient is explained by the properties of an Fe slurry layer (Wong et al., 2021).

At vertical incidence the amplitude of the P wave reflected from a first-order discontinuity in P velocity, S velocity, and density does not depend on the S velocity discontinuity. It is well approximated by the acoustic approximation for the ICB reflection coefficient, which depends only on P velocity and density. If the P velocity jump at the ICB is well determined from the location of the D cusp of the triplicated travel times curves of the PKIKP and PKiKP, the density jump of the inner core boundary can be determined from the amplitude of near vertical PKiKP. This amplitude depends on an estimate of both the source strength and the effects of propagation along the paths in the core and mantle. The amplitude ratio PKiKP/PcP is often measured to cancel these effects (Fig. 6.19). Within error estimates of noise and uncertainties of CMB discontinuities, an upper bound of for $\Delta\rho = 1000$ kg/m^3 is obtained in studies by Shearer and Masters (1990) and 1100 kg/m^3 by Tkalčić et al. (2009), but median values within the error estimates of these studies are all consistent with the free oscillation estimates of $\Delta\rho = 500$–600 kg/m^3. These and similar experiments are usually highly scattered, with the detection of either PKiKP or PcP not coincident with the detection of the other wave at the same receiver.

Numerical tests with heterogeneous Earth structures have shown that the largest effects on the PcP/PKiKP amplitude ratio are due to heterogeneous structure in the crust and upper mantle beneath the receiver rather than to

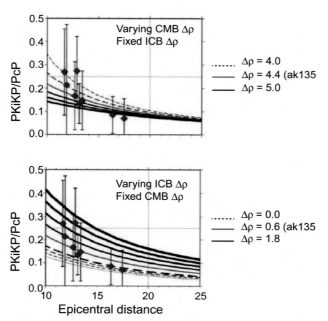

FIG. 6.19 Measured (diamonds) PKiKP/PKP amplitude ratios compared with predicted values for varying densities jumps at the CMB, holding the ICB jump equal to its AK135 (Kennett et al., 1995) value, and ICB, holding the CMB jump equal to its AK135 value.

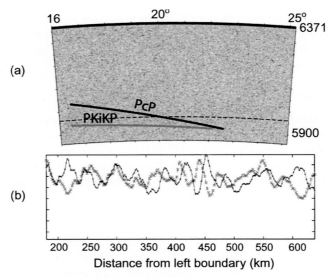

FIG. 6.20 Top: Wavefronts of PcP (blue [dark gray in print version]) and PKiKP (red [light gray in print version]) in the range 16–25° incident on an Earth model perturbed by statistically described heterogeneity in the crust and upper mantle beneath a receiver array. A 5% RMS perturbation in P velocity is assumed in the crust and upper mantle with an exponential autocorrelation having a corner scale length of 2 km. Bottom: Predicted amplitude fluctuations of PKiKP and PcP from the incident wavefronts shown at the top at an array of surface receivers measured from bandpassed seismograms of particle velocity synthesized by a numerical pseudospectral algorithm. *Adapted from Tkalčić, H., Cormier, V.F., Kennett, B.L.N., He, K., 2010. Steep reflections from the Earth's core reveal small-scale heterogeneity in the upper mantle. Phys. Earth Planet. Inter. 178, 80–91, https://doi.org/10.1016/j.pepi.2009.08.004.*

heterogeneous structure near either the CMB or the ICB. For example, a plausible isotropic distribution of heterogeneity scale lengths in the upper mantle can produce a factor of 1.4 or more variation in the PKiKP/PcP amplitude ratio in the high-frequency body wave band (Fig. 6.20). This is entirely consistent with the scatter in measured ratios seen in Fig. 6.19. In the case of distributions having scale lengths stretched in the vertical direction the predicted variations in the PKiKP/PcP ratio can exceed a factor of 4. Since pre-critical PKiKP is already a weak phase, often barely above background noise, there will be a tendency for body waves studies to overestimate the amplitude of the ICB reflection coefficient because many of the PKiKP identifications will occur only when upper mantle effects have enhanced its amplitude above the background noise.

Investigations with free oscillations (Masters and Gubbins, 2003), in which the biasing effects of mode coupling and splitting have been addressed, raise the estimate of the inner core density jump to 0.83 ± 0.18 kg/m^3. The difference between this value and the value of 0.55 ± 0.05 kg/m^3 inferred from reflected body waves may be due to the width of the sensitivity kernel in the free oscillation band, which is on the order of 200 km, versus that in the 1–2 Hz band of reflected body waves, which is on the order of 2–4 km. Hence, the higher free oscillation density jump may be indicative of a compositional gradient in the uppermost inner core. If such a compositional gradient does exist, it is difficult to constrain from the differential travel time of PKIKP-PKiKP because of its trade-off with structure in the upper, F region, side of the ICB.

6.6 Summary

Recent improvements in the measurements of mode splitting functions in the free-oscillation band and correlation wavefields in the body wave band have allowed inference of the average properties of Earth's inner core with greater confidence. Its average shear velocity is 3.5–3.6 km/s with a weak radial gradient. Density increases sharply over a 2–4 km zone straddling the identified ICB by 500–600 kg/m^3, with possibly an additional 200 kg/m^3 density increase occurring in its uppermost 100–200 km. At the largest scale, the inner core is elastically anisotropic with an axis of symmetry close to the axis of Earth's rotation. It strongly attenuates elastic energy with shear Q's in at least its upper half as low as 80 across a broad frequency band. Its low shear velocity, high attenuation, and comparisons with estimates of the properties of likely iron alloys suggest that its temperature is relatively close to its melting temperature from the ICB to its center.

It exhibits a first-order, degree-one structure in angularly averaged seismic velocities, anisotropy, attenuation, and possibly scattering. Two competing hypotheses have been proposed to explain this quasi-hemispherical structure: (1) translating convection and (2) lateral variation in CMB heat flux, which controls inner core solidification and outer core convection. The bulk of observational evidence at this point seems to favor CMB heat flux as the more probable mechanism to explain this structure. Evidence in its favor includes a correlation of differing intensities of the non-dipole magnetic field with the seismically resolved hemispheres, correlation of a strong gradient in elastic scattering with a predicted strong gradient in predicted heat flux at the ICB from dynamo modeling, and the correlation of a region of P wave multipathing in the uppermost inner core with a predicted region of maximum heat flux out of the inner core in the equatorial region of the eastern hemisphere.

With less confidence in resolution, both anisotropy and attenuation exhibit depth dependence. A thin zone of solidification, between 2 and 4 km in thickness, defines the inner core boundary in which the shear modulus rapidly rises to its average value. This may represent a compaction zone of solid iron mush or a region in which the last outer core fluid or inner core melt is expelled from a dendritic matrix. The uppermost 50–100 km is either isotropic or radially anisotropic, with cylindrical anisotropy developing below this region with an axis of symmetry close to the Earth's rotational axis, extending at least through the upper half of the inner core. The thickness of this outermost isotropic or radially anisotropic region may be useful for placing constraints on the viscosity of the inner core. A change in texture or relaxation spectrum and viscosity is suggested by a rapid decrease in body wave attenuation near 600 km depth, near the depth in which a transition in the form or intensity of anisotropy has been suggested in some studies. The deeper half of the inner core has not been as well examined in body wave studies due to sparser sampling of its small volume and lack of strong reference waves to measure differential travel times. The evidence for possibly up to three depth zones of differing elastic attenuation, scattering, and anisotropy may record a history of the inner core's changing composition and aid in identifying changes in convection or stable stratification with depth, e.g., Deguen and Cardin (2011) and Cottar and Buffett (2012).

Both free-oscillation studies and body wave studies can be affected by strong nonlinear behavior and parameter trade-offs in the inversion of cross-coupled modes and waveform shapes. Examples of these problems are trade-offs in the depth dependence of types of anisotropy in the case of modes and strong bimodal minima of L_2 norms in parameter fits to inner core attenuation.

Differential rotation of the inner core with respect to the mantle has been confirmed from variations in travel times of inner core–outer core body waves and the cross-correlations of scattered PKiKP coda observed from seismic doublets. Over a 40+ year period of observations, which have progressively improved in the identification of *earthquake doublets* and the treatment of laterally varying inner core structure, an evolving consensus is that the differential rotation of the inner core is irregular and its net average motion over long periods of time may be near zero with a very small *superrotation*. Its net zero position is consistent with both exciting and restoring torques, which have contributions from both internal (electromagnetic and gravitational coupling) and external (tidal) sources. Its motion places an upper limit to dynamic viscosity of $2-7 \times 10^{14}$ Pa s. Elastic anisotropy and attenuation suggest a viscosity range $10^{10}-10^{18}$ Pa s. These values are below the domain in which convective translation is predicted to exist.

References

Alboussière, T., Deguen, R., Melzani, M., 2010. Melting-induced stratification above the Earth's inner core due to convective translation. Nature 466 (7307), 744–747. https://doi.org/10.1038/nature09257.

Attanayake, J., Thomas, C., Cormier, V.F., Miller, M.S., Koper, K.D., 2018. Irregular transition layer beneath Earth's inner core boundary from observations of antipodal PKIKP and PKIIKP waves. Geochem. Geophys. Geosyst. 19 (10), 3607–3622. https://doi.org/10.1029/2018GC007562.

Aubert, J., Finlay, C.C., 2019. Geomagnetic jerks and rapid hydrodynamic waves focusing at Earth's core surface. Nat. Geosci. 12, 393–398.

Aubert, J., Amit, H., Hulot, G., Olson, P., 2008. Thermochemical flows couple the Earth's inner core growth to mantle heterogeneity. Nature 454 (7205), 758–761. https://doi.org/10.1038/natre07109.

Belonoshko, A.B., Fu, J., Bryk, T., Simak, S.I., Mattesini, M., 2019. Low viscosity of Earth's inner core. Nat. Commun. 10, 2483. https://doi.org/10.1038/s41467-019-10346-2.

Bergman, M., 1997. Measurements of elastic anisotropy due to solidification texturing and the implications for the Earth's inner core. Nature 389 (6646), 60–63.

Brett, H., Deuss, A., 2020. Inner core anisotropy measured new-ultra-polar PKIKP paths. Geophys. J. Int. 223 (2), 1230–1246. https://doi.org/10.1093/gji/ggaa358.

Calvet, M., Margerin, L., 2008. Constraints on grain size and stable ironphases in the uppermost inner core from multiple scattering modeling ofseismic velocity and attenuation. Earth Planet. Sci. Lett. 267, 200–212. https://doi.org/10.1016/j.epsl.2007.11.048.

Carcione, J.M., Cavalini, F., 1994. A rheological model for anelastic propagation with applications to seismic wave propagation. Geophys. J. Int. 119 (1), 338–348. https://doi.org/10.1111/j.1365-246X.1994.tb00931.x.

Cormier, V.F., Attanayake, J., He, K., 2011. Inner core freezing and melting: Constraints from seismic body waves. Phys. Earth Planet. Inter. 188, 163–172.

Cottar, S., Buffett, B., 2012. Convection in the Earth's inner core. Phys. Earth Planet. Inter. 188–189, 67–68. doi: 10.1016.j.pepi.2012.03.008.

Deguen, R., Alboussière, T., Cardin, P., 2013. Thermal convection in Earth's inner core with phase change at its boundary. Geophys. J. Int. 194 (3), 1310–1334. https://doi.org/10.1093/gji/ggt202.

Deguen, R., Cardin, P., 2011. Thermochemical convection in Earth's inner core. Geophys. J. Int. 187 (3), 1101–1118.

Deuss, A., 2014. Heterogeneity and anisotropy of Earth's inner core. Annu. Rev. Earth Planet. Sci. 42. https://doi.org/10.1146/annurev-earth-060313-054658.

Deuss, A., Irving, J.C.E., Woodhouse, J.H., 2010. Regional variation of inner core anisotropy from seismic normal mode observations. Science 328, 1018–1020.

Doornbos, J.D., 1983. Observable effects of the seismic absorption band in the Earth. Geophys. J. R. Astron. Soc. 75, 693–711.

Irving, J.C.E., Deuss, A., 2015. Regional seismic variations in the inner core under the North Pacific. Geophys. J. Int. 203, 2189–2199. https://doi.org/10.1093/gji/ggv435.

Irving, J.C.E., Deuss, A., Woodhouse, J.H., 2009. Normal mode coupling due to hemispherical anisotropic structure in Earth's inner core. Geophys. J. Int. 178 (2), 962–975. https://doi.org/10.1111/j.1365-246.2009.04211.x.

Kaelin, B., Johnson, L.R., 1998. Dynamic composite elastic medium theory. Part II. Three-dimensional media. J. Appl. Phys. 84, 5458–5468.

Kelly, P., Gubbins, D., 1997. The geomagnetic field over the past 5 million years. Geophys. J. Int. 128, 315–330.

Kennett, B.L.N., Engdahl, E.R., Buland, R., 1995. Constraints on seismic velocities in the earth from travel times. Geophys. J. Int. 122, 108–124.

Koot, L., Dumberry, M., 2011. Viscosity of the Earth's inner core: constrinats from nutation observations. Earth Planet. Sci. Lett. 308 (3), 343–349. https://doi.org/10.1016/j.epsl.2011.06.004.

Lasbleis, M., Deguen, R., 2015. Building a regime diagram for Earth's inner core. Phys. Earth Planet. Inter. 247, 80–93. https://doi.org/10.1016/j.pepi.2015.02.001.

Leyton, F., Koper, K.D., 2007a. Using PKiKP coda to determine inner core structure: 1. Synthesis of coda envelopes using single-scattering theories. J. Geophys. Res. 112 (B5). https://doi.org/10.1029/2006JB004369.

Leyton, F., Koper, K.D., 2007b. Using PKiKP coda to determine inner core structure: 2. Determination of Q_C. J. Geophys. Res. 112 (B5). https://doi.org/10.1029/2006JB004370.

Li, X., Cormier, V.F., 2002. Frequency-dependent seismic attenuation in the inner core 1. A viscoelastic interpretation. J. Geophys. Res. 107 (B12). https://doi.org/10.1029/2002JB001795.

Lythgoe, K.H., Inggrid, M.I., Yao, J., 2020. On waveform correlation measurement uncertainty with implications for temporal changes in inner core seismic waves. Phys. Earth Planet. Inter. 309, 105506. https://doi.org/10.1016/j.pepi.2020.106606.

Mäkinen, A., Deuss, A., 2013. Normal mode splitting function measurements of anelasticity and attenuation in the Earth's inner core. Geophys. J. Int. 194 (1), 401–416.

Masters, G., Gubbins, D., 2003. On the resolution of density within the Earth. Phys. Earth Planet. Inter. 140, 159–167.

Mattesini, M., Belonoshko, A.B., Tkalčić, H., Buforn, A., Udias, A., Ahuja, R., 2013. Candy wrapper for the Earth's inner core. Sci. Rep. 3 (1). https://doi.org/10.1038/srep02096.

Parisi, L., Ferreira, A.M.G., Ritsema, J., 2018. Apparent splitting of S waves propagating through an isotropic lowermost mantle. J. Geophys. Res. 123 (5), 3902–3922. https://doi.org/10.1002/2017JB014394.

Peng, Z., Koper, K.D., Vidale, J.E., Leyton, F., Shearer, P., 2008. Inner-core fine-scale structure from scattered waves recorded by LASA. J. Geophys. Res. 113. https://doi.org/10.1029/2007JB005412.

Shearer, P., Masters, G., 1990. The density and shear velocity contrast at the inner core boundary. Geophys. J. Int. 102, 491–498.

Souriau, A., Romanowicz, B., 1996. Anisotropy in inner core attenuation: A new type of data to constrain the nature of the solid core. Geophys. Res. Lett. 23 (1), 1–4. https://doi.org/10.1029/95GL03583.

Stephenson, J., Tkalčić, H., 2021. Evidence for the innermost inner core: Robust search for radially varying anisotropy, using the Neighborhood Algorithm. J. Geophys. Res. 126 (1). https://doi.org/10.1029/2020JB020545.

Stroujkova, A., Cormier, V.F., 2004. Regional variations in the uppermost 100 km of the Earth's inner core. J. Geophys. Res. 109 (B10). doi: 10.129/2004JB002976.

Tkalčić, H., Pham, T.-S., 2018. Shear properties of Earth's inner core constrained by a detection of J waves in global correlation wavefield. Science 362, 329–332.

Tkalčić, H., Kennett, B.L.N., Cormier, V.F., 2009. On the inner-outer core density contrast from PKiKP/PcP amplitude ratios and uncertainties caused by seismic noise. Geophys. J. Int.. https://doi.org/10.1111/j.1365-246X.2009.04294.x.

Tkalčić, H., Young, M.K., Bodin, T., Sambridge, M., 2013. The shuffling rotation of the Earth's inner core revealed by earthquake doublets. Nat. Geosci. 6 (6), 497–502. https://doi.org/10.1038/ngeo1813.

Vidale, J.E., 2019. Very slow rotation rate of Earth's inner core 1971 to 1974. Geophys. Res. Lett. 46 (16), 9483–9488. https://doi.org/10.1029/2019GL083774.

Vidale, J.E., Earle, P.S., 2000. Fine-scale heterogeneity in the Earth'sinner core. Nature 404, 273–275. https://doi.org/10.1038/35005059.

Wang, T., Song, X., 2018. Support for equatorial anisotropy of Earth's inner-inner core from seismic interferometry at low latitudes. Phys. Earth Planet. Inter. 276, 247–257.

Wang, S., Tkalčić, H., 2020. Seismic event coda-correlation: toward global coda-correlation tomography. J. Geophys. Res. 125. https://doi.org/10.1029/2018/jb018848.

Waszek, L., Irving, J., Deuss, A., 2011. Reconciling the hemispherical structure of Earth's inner core with its super-rotation. Nat. Geosci. 4 (4), 265–267. https://doi.org/10.1038/ngeo1083.

Wong, J., Davies, C.J., Jones, C.A., 2021. A regime diagram for the slurry F-layer at the base of Earth's outer core. Earth Planet. Sci. Lett. 560, 116791.

Wu, W., Irving, J.C.E., 2017. Using PKiKP coda to study heterogeneity in the top layer of the inner core's western hemisphere. Geophys. J. Int. 209 (2), 672–687. https://doi.org/10.1093/gji/ggx047.

Yang, Y., Song, X., 2020. Origin of temporal changes of inner-core seismic waves. Earth Planet. Sci. Lett. 541, 11267.

Yao, J., Sun, L., Wen, L., 2015. Two decades of temporal change of Earth's inner core boundary. J. Geophys. Res. 120, 6263–6283. https://doi.org/10.1002/2015JB012339.

Yao, J., Tian, D., Sun, L., Wen, L., 2019. Temporal change of seismic Earth's inner core phases: Inner core differential rotation or temporal change of inner core surface? J. Geophys. Res. 124 (7), 6720–6736. https://doi.org/10.1029/2019JB017532.

CHAPTER

7

Inner core dynamics

The materials processes of solidification, deformation, and annealing, along with the geodynamics and material properties of the core Fe alloy, determine the inner core microstructure (including grain size) and texture. We first review solidification, including nucleation, translation, and the F layer. We then examine geodynamic sources of stress/strain, and review dislocation and diffusion creep. At core temperatures, deformed materials can easily anneal, which includes the competing processes of recovery, and recrystallization and grain growth. The stress and grain size, along with the stable phase, self-diffusivity, shear modulus, and available slip systems of Fe, determine the deformation mechanism and hence the solid-state viscosity. The viscosity in part controls the geodynamics, however, and material properties are still uncertain. Modeling that includes the relevant materials processes, correct geodynamics, and accurate material properties offers the best prospect for reproducing the seismically inferred inner core elastic and anelastic properties.

7.1 Solidification of the inner core

Although the inner core is hotter than the outer core, the liquidus slope as a function of pressure (i.e., depth) is steeper than the geotherm (which approximates the adiabat for a well-mixed outer core), so that as the core cools, the geotherm first falls beneath the liquidus at the center (Fig. 2.8). The inner core is essentially thus directionally solidifying outward. Metallurgical studies have shown that directional solidification of alloys typically takes the form of *columnar dendritic* crystals (crystal = grain) with a solidification texture that may play a role in causing the seismic elastic anisotropy. In an effort to understand the F layer at the base of the outer core some have suggested, on the other hand, a slurry model that yields heavy, Fe-rich "snow" from this layer to build up the inner core. Another topic that has garnered recent attention is inner core nucleation, because in the absence of nucleation sites metals can often undercool by hundreds of degrees centigrade. If you are rusty you might first want to review phase diagrams, Section 2.1 and Appendix 2.1. Incidentally, in the geophysical literature crystallization is sometimes used interchangeably with solidification, though the former term can encompass means to form a solid other than freezing of a liquid.

7.1.1 Morphological instability and dendritic growth

In a pure melt where heat is extracted from the liquid through the growing solid and into a sidewall, the solid–liquid boundary is macroscopically flat, because any solid protrusion has a smaller radius of curvature than a flat interface, thereby increasing the heat flow into the solid protrusion and melting it back (Fig. 7.1, left). On the other hand, when nucleation occurs at sites within the melt it happens at a temperature beneath the equilibrium melting temperature, as a result of the surface energy associated with the solid/liquid interface. This is known as thermal undercooling or *supercooling*. In this case the heat flow is from the solid into the liquid, and if a protrusion occurs on a solid droplet, the protrusion's smaller radius of curvature can result in more conductive heat flow into the supercooled liquid, which leads to further growth of the protrusion away from the solid droplet (Fig. 7.1, right). This is known as a *morphological instability* (Mullins and Sekerka, 1963), leading to thermal dendrites, the best-known example, all too familiar to New Englanders, are snowflakes (growing from supercooled water droplets on dust particles).

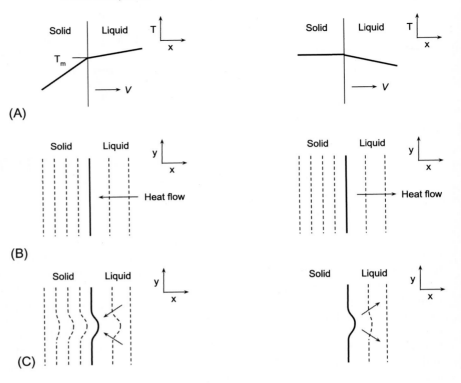

FIG. 7.1 (Left) (A) Temperature T versus position x as the solidification front moves rightward into a pure melt at growth speed V, due to a chilled sidewall to the left, (B) The direction of heat flow, and (C) a solid protrusion increases the heat flow, melting the protrusion back, keeping the solidification front flat; (Right) (A) and (B) As for left panel, but for supercooled (or thermally undercooled) liquid. A solid protrusion again increases the heat flow, which in this case leads to continued growth of the protrusion. This morphological instability of the flat interface leads to thermal dendrites. *Adapted from Porter, D.A., Easterling, K.E., 1992. Phase Transformations in Metals and Alloys, second ed. Chapman and Hall, London.*

In a typical laboratory alloy solidification experiment, heat is extracted from the liquid through the growing solid and into a chilled sidewall, but in the case of an alloy the solid–liquid boundary can become morphologically unstable due to *constitutional supercooling*. This occurs because rejection of solute during solidification of an alloy leads to a solute boundary layer whose concentration decreases with distance from the solid, and hence a liquidus temperature T_L that increases with distance from the solid. Although the temperature T increases with distance from the solid, it may nevertheless lie beneath T_L. A solid perturbation into the liquid may thus continue to grow until it reaches a height where $T = T_L$. The solute boundary layer has a scale thickness κ_C/V, where κ_C is the liquid solute (compositional) diffusivity and V is the solid growth velocity. If the gradient of the liquidus,

$$dT_L/dz = \Delta T/(\kappa_C/V), \tag{7.1}$$

where ΔT is the depression of T_L due to the excess solute in the boundary layer, exceeds the actual temperature gradient in the fluid, dT/dz, then freezing is predicted ahead of the flat solid/liquid interface (Fig. 7.2). An interesting issue arises in the core, however, because alloy melts can also thermally supercool due to the excess surface energy between the solid and liquid, and the lack of a sidewall or solid impurities where solidification is likely to commence. This is the cause of the so-called inner core "nucleation paradox" (Section 7.1.3).

The solid protrusions are known as dendrites, and the forest of these dendrites is known as a *mushy zone*, where in equilibrium the composition and temperature are tied to the liquidus. At lower temperatures the system is fully solid, at higher, fully liquid. As a result of solute rejection the solid dendrites are relatively solute-poor and the interdendritic liquid pockets solute-rich. Primary dendrites oriented by chance close to the direction of heat flow grow most rapidly. Due to anisotropic growth kinetics, dendrites also grow in particular crystallographic directions, so that the direction of heat flow can become a lattice preferred direction; this is the origin of solidification texturing. Secondary and tertiary side-branching dendrites also occur due to solute depression of the melting temperature in boundary layers along dendrites, particularly in high-symmetry cubic materials. Directionally solidified alloys thus typically exhibit columnar dendritic crystals, elongated in the direction of growth (Fig. 7.3). Mushy zones are ubiquitous features in directionally solidifying metallic alloys, organic systems, and aqueous salt solutions on both sides of the eutectic, though they are less common in silicates because of the low entropy (and hence high Gibbs free energy, Eq. 2.2) associated with their faceted dendrites.

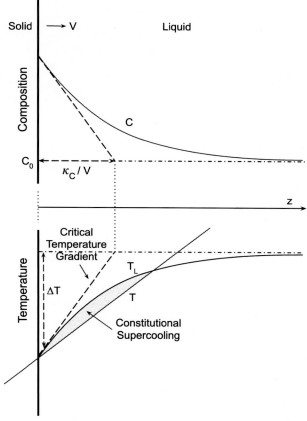

FIG. 7.2 Dendritic growth in directionally solidifying alloys can occur when the liquid is constitutionally supercooled. Adjacent to the solid at $z=0$ is a solute boundary layer of scale thickness κ_C/V, where κ_C is the liquid solute diffusivity and V is the growth rate of the solid. The solute concentration C in the boundary layer is enriched relative to its value C_0 far from the solid. This enrichment depresses the equilibrium freezing (liquidus) temperature T_L by an amount ΔT from its value far from the solid. If the temperature T in the liquid is less than T_L, then solid is predicted ahead of the solid/liquid interface, a condition known as constitutional supercooling. The criterion for constitutional supercooling is that the temperature gradient in the liquid $dT/dz < \Delta T/(\kappa_C/V)$. Adapted from Porter, D.A., Easterling, K.E., 1992. Phase Transformations in Metals and Alloys, second ed. Chapman and Hall, London.

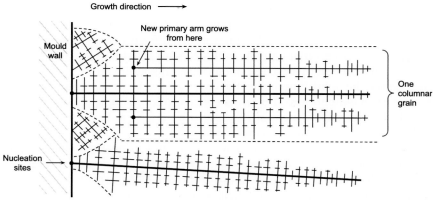

FIG. 7.3 Typical structure of a directionally solidified (cast) alloy. Away from the chill zone adjacent to the mold wall, columnar grains encompass many parallel primary dendrites, all with the same crystallographic orientation (this is what defines a single crystal!). The primary dendrites in two different columnar grains might be at a small angle from each other in the growth direction, and there is typically no preferred orientation *transverse* to the growth direction. Primary dendrites grow close to the direction of heat flow, and in a particular crystallographic direction; this is the origin of solidification texturing. Adapted from Porter, D.A., Easterling, K.E., 1992. Phase Transformations in Metals and Alloys, second ed. Chapman and Hall, London.

FIG. 7.4 Primary dendrites growing upward in a (cubic) NH$_4$Cl-rich aqueous solution. One can see significant side-branching off the primary dendrites. Courtesy of G. Worster.

The solidification microstructure and textures of castings are well-established. Outside the thin chill zone the solid is composed of columnar dendritic crystals, each comprising many parallel primary dendrites with the same crystallographic orientation. Within a crystal, between dendrites, are interdendritic, solute-enriched crystallites that typically have some other crystallographic orientation. Although the physical angle of primary dendrites between different columnar crystals can differ by a small amount (Fig. 7.3), they are close enough that a solidification texture in the growth direction results.

Primary dendrites in cubic crystals grow in $\langle 100 \rangle$ directions, with secondary and tertiary dendrites in equivalent $\langle 010 \rangle$ and $\langle 001 \rangle$ directions (Hellawell and Herbert, 1962). Fig. 7.4 shows typical dendritic growth in a cubic crystal. In hcp crystals, primary dendrites grow in $<10\bar{1}0>$ directions, i.e., in the basal plane (this may seem surprising because the c-axis is a more obvious axis of symmetry). Because of the slower growth kinetics out of the basal plane, side-branching is less common than in cubic crystals, though it does occur. Dendrites in hcp crystals look less like trees, and instead take on the morphology of *platelets* parallel to the basal plane, much like dishes in a dish rack (Fig. 7.5). In the absence of fluid flow in the melt, the crystallographic direction transverse to the growth direction is arbitrary; a crystal can have any transverse orientation. There are field observations in sea ice (Weeks and Gow, 1980) and experiments on sea ice and hcp Zn alloys (Bergman et al., 2003), however, which show that a shear flow can transversely align hcp crystals, likely having to do with the platelet nature of dendrites in hcp crystals.

7.1.2 The solidification microstructure and texture of the inner core

In the Earth's core, one must of course also consider the effect of the pressure axis on the phase diagram. Essentially, the direction of heat flow is reversed from a typical laboratory solidification experiment, i.e., heat is flowing from the hotter solid inner core into the cooler liquid outer core, with pressure providing the driving force for solidification. As in the laboratory the expulsion of the alloying component(s) upon solidification results in the solute concentration decreasing with distance above the solid inner core (the core is Fe-rich, so the inner core is enriched in Fe relative to the outer core). The decrease in temperature and solute concentration with height above the inner core favors solidification, countering the effect of decreasing pressure. The net effect is that, as in the laboratory, the geotherm can cross the liquidus twice, resulting in a mixed-phase mushy zone (Fig. 7.2).

Several studies have estimated that the inner core growth velocity V is nearly 500 times that required for morphological instability (Loper and Roberts, 1981; Deguen et al., 2007; Eq. (7.1)), because although V is likely very small (1200 km over a fraction of a billion years), the temperature gradient dT/dz near the inner–outer core boundary is also very small (roughly 0.5 K/km). These studies assume that heat and solute are removed from the mushy zone only by diffusion in spite of convective instability in the outer core, because one expects advection very near the boundary to be weak. The metallurgical argument for columnar dendritic growth of the inner core is suggestive, but by no means certain. Using the outer core compositional diffusivity $\kappa_C = 10^{-9}$ m^2/s and $V = \dot{r}_{ICB} = 0.63 \times 10^{-3}$ m/yr (for $Q_{CMB} = 10$ TW) from Chapter 2, one can estimate the solute boundary layer thickness in the core, κ_C/V, to be about 50 m. Although the mushy zone can be thicker, whether it is resolvable seismically is debatable.

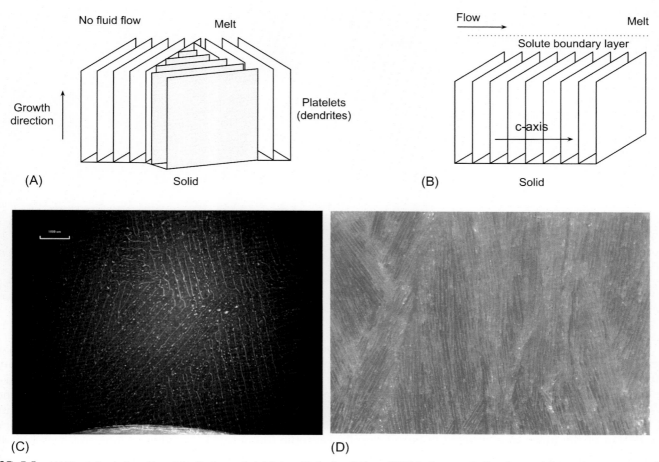

FIG. 7.5 (A) Dendrites in hcp alloys take the form of platelets, with the basal plane (0001) in the growth direction, and the c-axis transverse to the growth direction. Within the same crystal c-axes are parallel, between different crystals the c-axes are randomly oriented, in the absence of fluid flow. (B) In the presence of an externally forced fluid flow such as current, c-axes tend to align parallel to the flow, which is known as transverse solidification texturing. (C) Micrograph of directionally solidified hcp Zn-3 wt-pct Sn, transverse to the growth direction. The primary Zn-rich phase is dark, the Sn-rich phase is light. Platelets within a single crystal are parallel, with grain boundaries between regions of differently trending platelets. Limited side-branching from platelet boundaries is present. The scale represents 1mm. (D) An example of transverse solidification texturing: platelets growing horizontally (out of the page) in an (hcp) aqueous-rich NaCl solution, with a vertical flow as a result of convection driven by solidification. The platelets all trend vertically, parallel to the flow (which is opposite that sketched in (B), likely due to the different nature of the forcing fluid flow), but grain boundaries can be seen where the trending of the platelets changes slightly. Photograph size 5.5 × 4.0 cm. *(C) From Bergman, M.I., Yu, J., Lewis, D.J., Parker, G.K., 2017. Grain boundary sliding in high-temperature deformation of directionally solidified hcp Zn alloys, and implications for the deformation mechanism of Earth's inner core. J. Geophys. Res. Solid Earth 123, 189–203. (D) From Bergman, M.I., Agrawal, S., Carter, M., Macleod-Silberstein, M., 2003. Transverse solidification textures in hexagonal close-packed alloys. J. Cryst. Growth 255, 204–211.*

Iron meteorites, thought to be remnants from *planetesimal* cores, can provide some clues about the existence of dendritic growth. Pieces from the Cape York meteorite shower exhibit compositional gradients that are too large to arise from the general fractionation of a planetesimal core, but instead are more likely due to microsegregation between secondary and tertiary dendrite arms. Moreover, the FeS (troilite) nodules in meteorites are elongated and oriented (Fig. 7.6, left), suggesting interdendritic pockets of melt during directional solidification. Given the uncertainties connecting iron meteorites to planetesimal cores, these conclusions are tenuous, and more meteoritic studies may prove insightful.

Columnar crystals in a directionally solidifying inner core may be seismically observable due to the resulting elastic anisotropy, scattering attenuation, and/or scattering attenuative anisotropy. Theory and experiments have shown that under certain assumptions the primary dendrite spacing scales as $V^{-1/4}$ and $(dT/dz)^{-1/2}$, with the cooling rate $dT/dt = V \cdot (dT/dz)$ (Kurz and Fisher, 1992). Scaling to values in the range expected for the inner core suggests inner core primary dendritic spacing on the order of tens of meters, and grain widths on the order of hundreds of meters (Bergman, 1998). Columnar dendritic crystals have a shape anisotropy, so that the grains might be much longer in the growth direction (Figs. 7.3–7.5).

FIG. 7.6 (Left) Section from an iron meteorite from the Cape York shower, showing elongated troilite nodules, suggestive of dendritic growth. (Right) Section from the Carbo iron meteorite showing Widmanstätten pattern with the same orientation across its surface, indicating it cooled from a single crystal. The entire meteorite is about a meter across. *Left: From Esbensen, K.H., Buchwald, V.F., 1982. Planet(oid) core crystallization and fractionation-evidence from the Agpalilik mass of the Cape York iron meteorite shower. Phys. Earth Planet. Inter. 29, 218–232. Right: Photograph by M. Bergman (Harvard Museum of Natural History).*

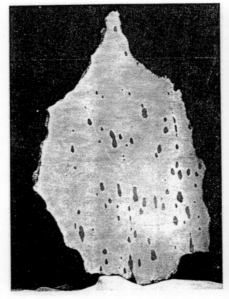

Although the elongated growth direction is likely to be primarily radial, due to the effects of rapid rotation on heat transfer from outer core convection (see Figs. 4.20 and 4.23 and discussion in Chapter 4) or east–west translation (Sections 7.1.3 and 7.2.1.1), one might expect some tendency toward cylindrical symmetry in the pattern of columnar dendritic growth, and hence in the polycrystalline elastic anisotropy (Fig. 7.7). This is in accordance with seismic inferences of inner core elastic anisotropy with a dominant cylindrically symmetric signature (Chapter 6). Although a smooth increase in the anisotropy with depth does result from this model, the upper 50–100 km of the inner core appears to be seismically isotropic (though it is possible to interpret it as radially

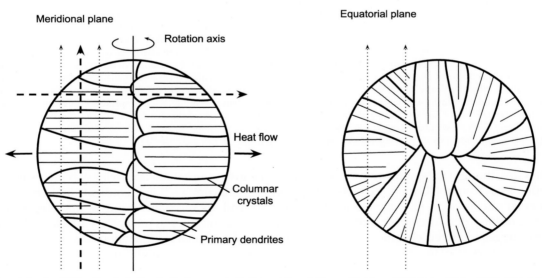

FIG. 7.7 The origin of solidification texturing in the inner core. Left: meridional plane, showing columnar crystals oriented east–west due to the pattern of heat flow in the outer core. Right: equatorial plane, showing columnar crystals oriented radially. The dashed and dotted arrows represent seismic rays. The dashed arrow parallel to the rotation axis in the meridional plane tends to be perpendicular to dendrites, that perpendicular to the rotation axis tends to be parallel to dendrites; together with dendrites growing in particular crystallographic directions and single crystal elastic anisotropy this is the origin of cylindrical anisotropy. The depth dependence of the anisotropy arises from the dotted arrows nearer the inner core boundary having similar orientation (perpendicular) relative to the dendrites in both planes, whereas with depth into the inner core the dotted arrows in the equatorial plane become increasingly parallel to dendrites. *Adapted from Bergman, M.I., 1997. Measurements of elastic anisotropy due to solidification texturing and the implications for the Earth's inner core. Nature 389, 60–63.*

anisotropic). It is thus not clear that a simple model of solidification texturing can by itself explain the more complex pattern of inner core elastic anisotropy that seismologists have begun revealing. There have also been suggestions that mantle-driven fluid flow in the outer core may affect transverse solidification textures in the inner core (Fig. 7.5; Aubert et al., 2008), but the detailed connections with seismic observations have not yet been fully deciphered and need further exploration.

Scattering attenuation is a maximum when ka, where k is the wavenumber and a is the lengthscale of the scatterer, is of order one. For a P-wave of 1 Hz, k is approximately 10^{-3} m^{-1} in the inner core. For elastically anisotropic crystals, grain boundaries can serve as scatterers because the wavespeed V and hence the acoustical impedance $Z = \rho V$, where ρ is the density, can change at a boundary, but the inner core grain size a remains hidden from us (Section 7.4). Large stresses typically produce fine grains, so that $ka \ll 1$, rendering scattering attenuation unimportant. On the other hand, the columnar dendritic crystals produced by solidification of the inner core may be hundreds of meters across, so that ka is of order one, increasing attenuation by scattering. Although recent seismic studies suggest intrinsic attenuation may dominate in the inner core, particularly in the deep inner core, elongated columnar crystals can exhibit an anisotropy in scattering attenuation (because in the elongated direction $ka \gg 1$, and/or because the impedance variation is less in the elongated growth direction, due to texturing), which may contribute to inferences of inner core attenuative anisotropy. Ultrasonic attenuative anisotropy due to columnar crystals has been observed in the laboratory.

It is observed that when a directionally solidifying alloy ceases to actively solidify and the temperature is above the eutectic, the solid dendritic structure can begin to slump and compact if the solid phase is denser than the liquid. Although the inner core is likely actively solidifying (although perhaps not all over the ICB if the inner core is translating), there may nevertheless be a tendency for the liquid to be expelled (Sumita et al., 1996), particularly if the melt is connected. It is unknown by how much the dendritic structure and texture change upon compaction or slumping, given the uncertainty about inner core viscosity (Section 7.5). The solidification microstructure and texture may also be modified by deformation resulting from shear stress that could arise from a variety of sources (Sections 7.2).

7.1.3 The origin of the F layer, inner core translation, the "snowing" core model, and inner core nucleation

Chapter 5 discusses the seismological evidence for a layer at the base of the core, perhaps some few hundred kilometers thick, where the P-wave speed increases with depth less slowly than PREM. It also explores briefly its possible formation, which we pick up on further here. One explanation for this so-called F layer could be liquid enriched in Fe relative to the overlying outer core, but this is somewhat difficult to reconcile with an outer core that is compositionally convecting as a result of light elements being released at the ICB as a result of inner core solidification. However, it may be possible to maintain both compositional convection driven by solidification at the ICB and a dense Fe-rich layer directly above the ICB by means of a translating inner core (Alboussiere et al., 2010).

In this scenario melting of the Fe-rich inner core occurs on one side of the core, producing the dense Fe-rich liquid layer, while solidification occurs on the other, releasing the light elements that stir up the outer core above (Fig. 7.8). Translation is in essence a degree one mode of inner core convection (Section 7.2.1.1), occurring because the solidifying side of the inner core is denser than the melting side, thereby sinking the solidifying side toward the center, and hence translating the inner core. It is an unusual mode of convection, made possible because the ICB is a phase change boundary, unlike, for example, the CMB. The translation may also explain hemispherical variations of inner core seismic properties (Chapter 6). Interestingly, to produce the necessary volume of Fe melt in the F layer requires a translation rate that greatly exceeds the inner core's overall growth rate, so that the inner core has regenerated new solid many times over its existence. Rigid translation requires a high dynamic viscosity ($\eta > 10^{18}$ Pa s; Deguen, 2012) inner core, else the degree one convection breaks down into higher modes of convection, at the expense of translation. Moreover, more work needs to be done to understand the details of penetrative convection through a stably stratified layer.

A second scenario that could produce an Fe-rich, seismically slower than expected F layer involves an inhomogeneous mantle heat flow, discussed in Chapter 4. In this scenario, forced outer core convection could bring heat down to some regions of the inner core boundary in spite of an overall heat flow that is obviously up from the inner core. This could result in melting parts of the inner core surface despite its net solidification (Gubbins et al., 2011). Like translation, it could potentially result in hemispherical and also regional variations in inner core seismic properties.

A third idea to interpret the F layer involves the material properties of Fe at F layer temperatures and pressures. Some first principles calculations have found that under these conditions liquid Fe exhibits an anomalously high viscosity (Section 2.3.1), sometimes described as a glassy state, but a physical understanding is unclear, as the slow

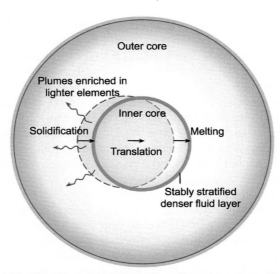

FIG. 7.8 Translation of the inner core, resulting in an Fe-enriched F layer that still provides compositional convection resulting from solidification. Translation results from solidification on one side of the inner core, and melting on the other, with the degree one convection driven by the denser solidifying side sinking. *From Bergman, M.I., 2010. An inner core slip-sliding away. Nature 466, 697–698. Adapted from Alboussiere, T., Deguen, R., Melzani, M., 2010. Melting-induced stratification above the Earth's inner core due to convective translation. Nature 466, 744–747.*

cooling conditions of the core are exactly the opposite of those for which glasses typically form. Yet another idea is that the F layer, like the E' layer at the top of the core, involves changes in chemical solubilities as the core has evolved after its formation (see Chapter 8).

Last, it has been suggested that the F layer might be a slurry layer in which Fe snow forms and settles to form the inner core. A snowing core model has been proposed for smaller planets such as Mercury. The argument is that in a smaller planet such as Mercury, the increase of the liquidus with pressure (depth) is less than the increase in temperature, so that solidification commences at the top of the core, either at the core–mantle boundary, or as unattached nuclei at the top of the core. For the Earth, it has been proposed that nucleation of solid Fe particles occurs throughout the F layer (Gubbins et al., 2008). The Fe-enriched solid particles are denser than the liquid, and sink. Moreover, they may not completely melt as they sink because although the deeper core is warmer, it may also have a lower concentration of light elements. The settling solid particles can then form a solid inner core. Although this scenario has its proponents as a means to understand the Earth's F layer, it is somewhat difficult to understand because any slurry that occurs above the ICB would be at a height below which solid dendrites from the inner core would have already grown (Shimizu et al., 2005), i.e., the morphological instability results in dendrites to the height where the liquidus crosses the geotherm (Fig. 7.2). In summary, an understanding of the seismically inferred F layer remains elusive.

There is a further problem with the slurry model, which makes even the existence of the inner core puzzling: nucleation. Because of the interfacial surface energy associated with solid droplets in a melt, metals and their alloys can remain liquid far beneath their melting temperatures. The Gibbs free energy of a solid droplet of volume V in a melt differs from that of the melt without the droplet by an amount.

$$\Delta G = -V(G^L - G^S) + A\gamma, \tag{7.2}$$

where G^L is the Gibbs free energy of the droplet if it were in the liquid phase, G^S is that if it were in the solid phase, A is the surface area of the droplet, and γ is the surface energy between the solid and liquid, assumed isotropic for simplicity. Notice that because we are interested in nucleation of the solid, the temperature is beneath the melting temperature, so that $G^L > G^S$, and the first term is negative. The second term is positive, the surface energy cost of forming the solid droplet. If one assumes a spherical droplet of radius r, Eq. (7.2) becomes

$$\Delta G_r = -4/3\pi r^3 (G^L - G^S) + 4\pi r^2 \gamma. \tag{7.3}$$

We plot these energy terms in Fig. 7.9. ΔG_r exhibits a maximum at a critical radius

$$r_{crit} = 2\gamma/(G^L - G^S), \text{with a value} \tag{7.4}$$

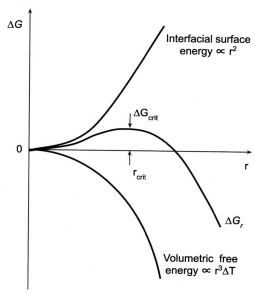

FIG. 7.9 Plot of the terms in Eq. (7.3). *Adapted from Porter, D.A., Easterling, K.E., 1992. Phase Transformations in Metals and Alloys, second ed. Chapman and Hall, London.*

$$\Delta G_{crit} = 16\pi\gamma^3 / \left(3\left(G^L - G^S\right)^2\right). \tag{7.5}$$

For solid clusters with $r < r_{crit}$ the system decreases its free energy if the droplet melts, but for those with $r > r_{crit}$ the system decreases its free energy if the droplet grows, and solidification commences. The existence of a nonzero r_{crit} is the origin of undercooling. For undercooling ΔT that is not too large compared with the melting temperature T_M, $(G^L - G^S) = L\Delta T / T_M$ is a good approximation, where L is the latent heat of solidification (per volume). Eq. (7.4) thus becomes

$$r_{crit} = (2\gamma T_M)/(L\Delta T). \tag{7.6}$$

The significance of this result is that r_{crit} decreases with increasing undercooling, i.e., as the undercooling increases random solid clusters are more likely to exceed the critical radius.

In the laboratory crucible walls and solid impurities typically serve as nucleation sites, preventing large undercooling. However, neither of these are likely for helping to nucleate the inner core, let alone Fe snowflakes, and the amount of undercooling in the core has been estimated to be as large as 1000 K, begging the question how the inner core ever formed (Huguet et al., 2018). This is the core nucleation paradox.

7.2 Deformation in the inner core

In addition to solidification, deformation has been identified as the possible cause for the lattice preferred orientation that might cause the elastic anisotropy of the inner core. Deformation occurs when solids are subject to stress or strain, and in this section, we look at possible geodynamic causes for stress or strain in the inner core. We will then turn to the inner core *deformation mechanism*, i.e., its rheology. In this context, we are interested in plastic flow, also known as creep, where the deformation is due to a steady stress and is typically nonreversible. The subject is immense, as there are many ways in which solids can flow, depending on the temperature, pressure, crystal structure, purity, stress, grain size, etc. We are interested here in high-temperature (i.e., close to the melting temperature) deformation, likely at low stress (as compared with the shear modulus), but the inner core deformation mechanism remains unknown, in part due to uncertainty about the grain size. The stress/strain and deformation mechanism determine the deformation texture. In the following section, we consider *annealing* and *recrystallization*, which refer to competing processes whereby a solid lowers its free energy associated with microstructures and defects introduced during solidification and/or deformation. During annealing the original grains remain but with some high-energy features removed; in recrystallization new grains nucleate with fewer of these features. These processes are typically thermally activated and hence likely relevant to Earth's inner core.

7.2.1 Stress/strain

A variety of driving stresses and/or strains have been suggested for the inner core, often with an eye toward explaining inner core elastic anisotropy, as an alternative to or in combination with solidification texturing. We first discuss inner core convection and translation, before turning to other possible driving stresses.

7.2.1.1 Convection

As in much of the Earth thermal convection is a possible culprit for driving motion in the inner core (Jeanloz and Wenk, 1988). Unlike in the outer core however, rotational effects are not dominant, though their influence is likely felt through coupling with the outer core. Although long-lived radioisotopes of K, U, and Th have been suggested as an energy source for inner core convection, the role of radiogenic heating is now thought to be small compared with secular cooling because of the unlikelihood that these elements would partition into the inner core, especially as their concentration in the outer core may also be low. The basic requirement for thermal instability, cooling from above, is satisfied, but unlike the outer core with its low viscosity, or the mantle with its low thermal conductivity, it is unclear whether the inner core can overcome the retarding effects of viscosity and thermal diffusivity and is thermally convecting at present or was for how long after its nucleation.

As discussed in Chapter 4 the Rayleigh number Ra is a measure of the vigor of thermal convection. There are different ways to define Ra, but for studying inner core convection we will adopt (Deguen, 2012)

$$Ra = \alpha_T \rho g r_{ICB}^5 S / (6\kappa_T^2 \eta), \tag{7.7}$$

where α_T is the thermal expansivity of Fe under inner core conditions, ρ is its density, g is the gravitational acceleration at r_{ICB}, where $r_{ICB} = r_{ICB}(t)$ is the inner core radius that increases with time, κ_T is the inner core thermal diffusivity, and η is its dynamic viscosity. S is defined as

$$S = S(t) = \kappa_T \nabla^2 T_{ad} - \dot{T}_{ad} \tag{7.8}$$

is a measure of the inner core thermal instability, where T_{ad} is the inner core adiabat. The first term in $S(t)$ is the cooling rate due to conductive heat flow, which obviates the need for thermal convection, and is negative (Fig. 2.8). The second term is the secular cooling, which promotes convection, and is also negative. For thermal convection to occur S must be sufficiently positive, so the latter term must exceed the former. In other words, S is positive if the rate at which the inner core is removing heat is greater than what can be conducted up its adiabat. Note that in Eq. (7.8) the inner core geotherm is assumed to be at the adiabat, a good approximation if the inner core is convecting. Also note that the first term on the right-hand side of Eq. (7.8) could have a smaller magnitude if the inner core is subadiabatic, but by definition it is then not convecting.

Eqs. (7.7) and (7.8) confirm that a small thermal conductivity k $(= \kappa_T \rho c_P$, where c_P is the specific heat at constant pressure) and inner core viscosity η favor thermal instability and convection, as does a rapidly cooling core, i.e., a young inner core or equivalently a high CMB heat flux Q_{CMB} (Section 2.4.3). If the inner core has been thermally unstable, it was most likely when it was young and more rapidly cooling, in spite of the smaller $r_{ICB}(t)$, because of the cooling rate's effect on $S(t)$.

The measure for thermal instability, Eq. (7.8), can be nondimensionalized in terms of a normalized inner core age,

$$\mathcal{T}_{IC} = (\tau_{IC}/\tau_\kappa)/[(dT_M/dr)/(dT_{ad}/dr) - 1], \tag{7.9}$$

where τ_{IC} is the inner core age, $\tau_\kappa = r_{ICB}^2/\kappa_T$ is its thermal diffusion timescale, and dT_M/dr and dT_{ad}/dr are the melting temperatures and adiabatic gradients at the ICB (Fig. 2.17). One finds that for $\mathcal{T}_{IC} > 1.6$ the inner core was never thermally unstable, for $0.9 < \mathcal{T}_{IC} < 1.6$ the inner core was initially unstable but is now stable, and for $\mathcal{T}_{IC} < 0.9$ it is still unstable. Using $\tau_{IC} = 500 \, Ma$ from Section 2.4.3, and values for dT_M/dr and dT_{ad}/dr from the Table on Core Thermodynamic and Transport Properties at the end of the book, $\mathcal{T}_{IC} = 0.6$ for $k = 20 \, W/m/K$ and $\mathcal{T}_{IC} = 3.0$ for $k = 100 \, W/m/K$. For not unrealistically low values of k the inner core is still thermally unstable, but for not unrealistically high values of k the inner core was never thermally unstable. There is enough slop in these numbers that it is difficult to be certain of the present or past thermal instability of the inner core.

In most convective scenarios the boundaries are rigid or free slip, but in either case they are usually impermeable. Such is not the case for the ICB, since it represents a phase transition. Translation is a degree one mode of convection, so

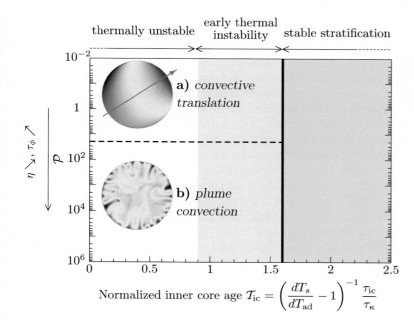

FIG. 7.10 Diagram showing regimes for inner core convective translation, more "typical" plume style convection, and stable stratification, as a function of the ratio of the phase change timescale to the viscous relaxation timescale, $\mathcal{P}$, and the normalized inner core age $\mathcal{T}_{IC}$. A young, rigid inner core favors translation. τ_{IC} is the age of the inner core, $\tau_\kappa = r_{ICB}^2/\kappa_T$ is its thermal diffusion timescale. *From Deguen, R., 2012. Structure and dynamics of Earth's inner core. Earth Planet. Sci. Lett. 333–334, 211–225, where further details can be found.*

that in order for it to occur the inner core must be convectively unstable, but in addition the ratio $\mathcal{P}$ of the phase change timescale to the viscous relaxation timescale must be sufficiently small. If that is the case Fe at the ICB can melt/solidify faster than it can viscously flow, and translation can occur. If $\mathcal{P}$ is large the inner core can more easily flow, and convection takes on the more typical plume style associated with impermeable boundaries. By far the largest unknown in determining $\mathcal{P}$ is η, with values exceeding 10^{18} Pa s required for translation. We will return to inner core viscosity in Section 7.5, but while $\eta = 10^{18}$ Pa s is a large value, it is not outside the range that has been reasonably suggested. Fig. 7.10 summarizes the convective regimes in terms of the nondimensional numbers $\mathcal{T}_{IC}$ and $\mathcal{P}$.

The role of composition for inner core convection is ambiguous. If the solid-state compositional diffusivity κ_C of less dense components were large, then the compositional profile in the inner core would look like that sketched in Fig. 7.11A (see Appendix 2.1 and Fig. 2.30B), with the uniform composition of the solid C_S evolving along the solidus as the inner core continues to solidify. However, because solid-state diffusion is so slow ($r_{ICB}^2/\kappa_C = (10^6 \, \text{m})^2/(10^{-12} \, \text{m}^2/\text{s}) = 10^{10} \, Ma$), the composition of a nonconvecting inner core is likely not in thermodynamic equilibrium, and the inner core does not have a single composition that follows the solidus. Instead, as the outer core becomes increasingly enriched in less dense components, then so too does the alloy solidifying at the top of the inner core, resulting in a compositionally stably stratified inner core (Fig. 7.11B, also see Fig. 2.31B)). On the other hand, the partition coefficients $k = C_S/C_L$ of light elements (O, S, Si, etc.) may not be constant (i.e., a nonconstant solidus slope k). If as the core cools those components' ability to crystallize into the solid decreases (Fig. 7.11C), then the inner core could become

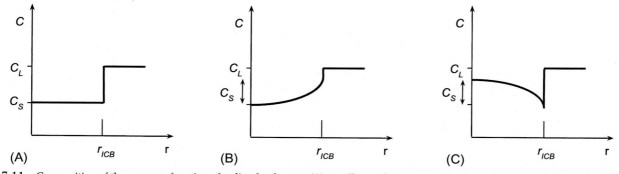

FIG. 7.11 Composition of the core as a function of radius for the case (A) a well-mixed inner core due to rapid solid-state diffusion, (B) an inner core whose light component composition increases with radius due to slow solid-state diffusion, and (C) an inner core whose light component composition decreases with radius due to a partition coefficient that decreases with decreasing temperature. All cases assume a well-mixed outer core.

compositionally unstable (Gubbins et al., 2013). Moreover, because κ_C is much smaller than the thermal diffusivity κ_T, a compositionally unstably stratified inner core is more likely to result in convection. And unlike the likelihood for inner core thermal convection, which has decreased as the core has cooled, the likelihood for compositional convection may be increasing due to the r_{ICB}^5 dependence in Eq. (7.7). It remains uncertain whether a decrease in the partition coefficients with decreasing temperature would outweigh the increase in C_L, and C_S at the top of the inner core, as the core solidifies.

In addition to its role in core evolution, the occurrence of inner core convection may help interpret inner core elastic anisotropy (Chapter 6). Through *lattice rotations* the strain associated with inner core convection can texture the inner core, if *dislocation creep* is the operative deformation mechanism (Section 7.2.2.1). However, aside from uncertainty concerning the deformation mechanism, questions remain about inner core convection as the cause for stress/strain. First, as discussed in this section, is the inner core convecting? Second, the basic seismic elastic anisotropy is aligned with the spin axis. This suggests convection must be low order and aligned with the spin axis (Fig. 7.12A). Why? Even small centrifugal effects may explain the latter, and the former is most likely explained by the inner core being only slightly supercritical (Buffett, 2009), but it is unclear for how long convection can remain in this low-order state, and whether sufficient strain can develop while in this state to explain the seismic elastic anisotropy. It has also been proposed that the depth dependence of the elastic anisotropy might be caused by changes in or cessation of inner core convection, but clearly this remains speculative. No deformation results from pure translation.

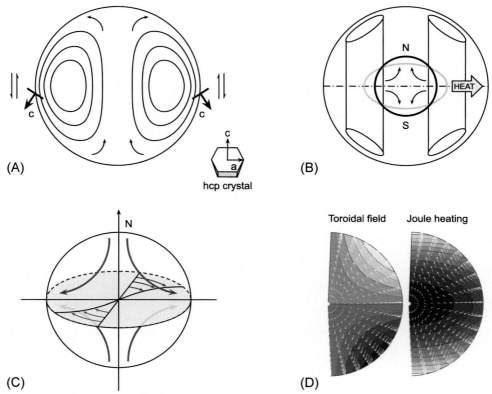

FIG. 7.12 Geodynamic sources of stress/strain that might result in inner core deformation. (A) Low-order convection aligned with the spin axis, showing strain and proposed orientation of hcp Fe crystals for an assumed slip system. (B) Anisotropic outer core convection and heat flow driving excess equatorial solidification (gray ellipsoid), resulting in misalignment with the gravitational equipotential (black sphere), resulting in inner core poloidal flow (arrows), i.e., topographical relaxation. (C) Magnetic stresses due to latitudinal variations in the toroidal magnetic field can drive a poloidal flow (red arrows, long gray arrows in print version, notice that for this particular toroidal magnetic field the poloidal flow is in the opposite direction to that in (B)); magnetic stresses arising from a combination of toroidal and poloidal magnetic fields can drive a toroidal flow (blue arrows, short gray arrows in print version). (D) A heterogeneous distribution of Joule heating in the inner core can generate a flow. The left figure shows the toroidal magnetic field (light: into page, dark: out of page) with arrows indicating the electric current field. The right figure shows the Joule heating (lighter areas with larger values) with arrows indicating the flow field. *(A) Adapted from Jeanloz, R., Wenk, H.-R., 1988. Convection and anisotropy of the inner core. Geophys. Res. Lett. 15, 72–75. (B) From Yoshida, S., Sumita, I., Kumazawa, M., 1996. Growth model of the inner core coupled with outer core dynamics and the resulting elastic anisotropy. J. Geophys. Res. 101, 28085–28103. (C) Adapted from Deguen, R., 2012. Structure and dynamics of Earth's inner core. Earth Planet. Sci. Lett. 333–334, 211–225, after Karato, S.-I., 1999. Seismic anisotropy of the Earth's inner core resulting from flow induced by Maxwell stresses. Nature 402, 871–873 and Buffett, B.A., Wenk, H.-R., 2001. Texturing of the Earth's inner core by Maxwell stresses. Nature 413, 60–63. (D) Adapted from Takehiro, S., 2011. Fluid motions induced by horizontally heterogeneous joule heating in the Earth's inner core. Phys. Earth Planet. Inter. 184, 134–142.*

7.2.1.2 Other causes for stress/strain

Other geodynamic causes for inner core deformation have also been suggested, generally in an effort to explain the seismic elastic anisotropy. One that inherently gives the required large-scale pattern of elastic anisotropy that is aligned with the spin axis involves inner core forced convection driven by inhomogeneous outer core heat flow (Yoshida et al., 1996). As has been suggested for inner core solidification texturing, the effects of the Coriolis force may lead to an outer core convective heat flow that is larger perpendicular to the spin axis. This in turn leads to preferential equatorial growth of the inner core, which has been seen in laboratory model experiments. Because of the density difference between the inner and outer cores, the inner core must deform isostatically to maintain its spherical shape (aside from the asphericity due to centripetal effects), with a resulting overall stress field with uniaxial tension in the polar direction (Fig. 7.12B). This is known as *topographical relaxation*.

A problem with topographical relaxation is that the strain rate resulting from anisotropic growth is quite small due to the slow solidification rate of the inner core. As a result, a lattice preferred orientation could take a geologically long time (on the order of 10^9 years) to develop. If the inner core is compositionally stratified the forced convection becomes confined to the uppermost inner core, which could speed up the texture development there, though that is not where the inner core texture is largest. Deeper in the inner core the texture would then need to be a remnant of earlier dynamics or solidification.

Another possible source of stress involves the magnetic field, i.e., the Lorentz force $J \times B$, where B is the magnetic field and the current $J = (1/\mu_o)\, \nabla \times B$, where μ_o is the permeability of free space. The geomagnetic field is generated in the outer core, with the larger scale components diffusing into the inner core. This is attractive for generating large-scale, inner core flow that could yield seismic elastic anisotropy. Latitudinal variations in the strength of the toroidal field result in latitudinal variations in the Maxwell (magnetic) stresses that could drive poloidal (i.e., radial) flow (Karato, 1999; Fig. 7.12C). Unfortunately, the toroidal component of the magnetic field is unseen at the surface, so its strength and morphology are uncertain. More problematic, if the inner core is stably stratified against thermal and compositional convection, then poloidal flow may be inhibited (Buffett and Bloxham, 2000), and the speed at which the inner core can melt/solidify in order to supply a poloidal flow may be limited, as for topographical relaxation. Toroidal (i.e., horizontal) flow (Buffett and Wenk, 2001; Fig. 7.12C) driven by Maxwell stresses that result from a combination of toroidal and poloidal components of the magnetic field in the inner core avoids these issues, but it is not clear that the resulting sense of the elastic anisotropy is correct, or that it can give the observed depth dependence.

If the inner core is thermally or compositionally unstable the magnetic field can also influence the style of convection, including aligning large-scale flow with the spin axis. A measure of the relative importance of magnetic to convective (viscous) stresses is given by the Hartmann number,

$$Ha = \sigma B^2 r_{ICB}^2 / \eta, \tag{7.10}$$

where σ is the electrical conductivity, B is the strength of the relevant magnetic field component(s), and η is the dynamic viscosity. Once again, unfortunately, we have wide bounds on B and η, but inner core magnetoconvection remains an interesting avenue to explore. The Table on Core Nondimensional Parameters gives estimates for Ha.

Heterogeneous Joule heating due to a heterogeneous magnetic field in the inner core can also force inner core flow, and hence deformation (Takehiro, 2011; Fig. 7.12D). In particular, Joule heating due to a large-scale, latitudinally varying toroidal magnetic field that diffuses into the inner core will drive a poloidal flow even against stable stratification. On the other hand, limits on the speed of the phase change at the ICB may again weaken the flow. Outer and inner core convection, forced inner core flows, and magnetic stresses in the inner core are all coupled through the requirement that the ICB be in thermodynamic equilibrium, so a self-consistent model of inner core deformation needs to properly incorporate all of these effects.

Table 7.1 gives stress/strain estimates, along with comments, for the sources of stress/strain that have been suggested for the inner core. We'll return to causes for the lattice preferred orientation and seismic elastic anisotropy of the inner core in Section 7.6.1. We should perhaps say here that there may well be geodynamical processes that result in inner core stress/strain that have yet to be realized!

7.2.2 Deformation mechanisms

The stresses/strains discussed in Section 7.2.1 can result in deformation, annealing, and/or recrystallization, which can create new or modify existing microstructures and textures. In this section, we discuss inner core rheology. The step between stress/strain and texture development requires knowledge of the operative deformation mechanism(s), which requires knowledge of material properties such as atomic self-diffusivities and active slip systems, as well as

TABLE 7.1 Sources of inner core stress/strain.

Source of stress/strain	Size of stress/strain rate (values depend on assumed deformation mechanism, Eq. (7.11))	Comments
Convection	$\dot{\varepsilon} = 10^{-15}\,\text{s}^{-1}$ corresponding to $\sigma = 10^2\,\text{Pa}$	Required low order convection exists for a limited time window, but perhaps explains depth dependence?
Topographical relaxation due to equatorial growth	$\dot{\varepsilon} = 3 \times 10^{-18}\,\text{s}^{-1}$ corresponding to $\sigma = 10^4\,\text{Pa}$ (due to high η)	Radial flow inhibited by stable stratification; crystallographic alignment slow to develop as compared with IC age
Magnetic stresses: Poloidal flow due to toroidal magnetic field	$\dot{\varepsilon} = 10^{-15} - 10^{-18}\,\text{s}^{-1}$ corresponding to $\sigma = 10^2\,\text{Pa}$	Radial flow inhibited by stable stratification; finite time for ICB phase change lessens strain rate
Magnetic stresses: Toroidal flow due to poloidal/toroidal magnetic field	$\dot{\varepsilon} = 10^{-16}\,\text{s}^{-1}$ corresponding to $\sigma = 10^0\,\text{Pa}$	Likely very small stress; sense of anisotropy may be opposite to observed
Heterogeneous Joule heating	$\sigma = 10^2\,\text{Pa}$ No deformation mechanism specified	Finite time for ICB phase change lessens strain rate

geodynamically determined factors such as the stress level and the grain size. For obvious reasons, we will be concerned here with high-temperature deformation, typically greater than 0.9 times the *homologous temperature* T/T_M, where T_M is the melting temperature. Although practical knowledge of the mechanical properties of metals has been accumulating since the dawn of the Bronze Age over 5000 years ago, a theoretical understanding of plastic flow didn't really get going until the concept of dislocations was introduced independently by E. Orowan, M. Polanyi, and G.I. Taylor in 1934. Interestingly, this lack of understanding of plastic flow delayed the paradigm of plate tectonics.

In a defect-free crystal, the shear strength of a material is of the order of $\mu/2\pi$, where μ is the shear modulus, which is at least two orders of magnitude greater than what is typically observed, even at low and moderate homologous temperatures. Dislocations are linear defects that allow plastic flow at much lower stresses than predicted for perfect crystals, but they are not the only mechanism to facilitate creep. *Diffusion creep* occurs via the diffusion of point defects, i.e., vacancies, either through the lattice or along grain boundaries. *Grain boundary sliding*, also known as superplasticity, can also accompany diffusion creep, and twinning is another deformation mechanism.

High-temperature creep is governed by the empirical relation between strain rate $\dot{\varepsilon} = d\varepsilon/dt$ and stress σ,

$$\dot{\varepsilon} = A\,(D\mu b/k_B T)\,(b/d)^p\,(\sigma/\mu)^n, \tag{7.11}$$

where D is the atomic self-diffusivity, b is the magnitude of the *Burgers vector*, k_B is Boltzmann's constant, T is temperature, d is the grain size, and p, n, and A are dimensionless constants. If $n=1$, there is a linear relationship between strain rate and stress, and the viscosity is said to be linear, or *Newtonian*, with $\eta = \sigma/2\dot{\varepsilon}$.

For diffusion creep the stress exponent $n=1$, whereas for dislocation creep n is typically in the range 4–6. Higher stress levels thus favor dislocation creep as it yields the larger strain rate. There is no grain size dependence for dislocation creep, i.e., $p=0$, whereas for diffusion of atoms through the lattice (Nabarro–Herring creep), $p=2$, and for diffusion along grain boundaries (Coble creep) the grain size dependence is even stronger, $p=3$. Smaller grain sizes thus favor diffusion creep. Diffusion creep is thermally activated, so higher temperatures also generally favor diffusion creep, though certain dislocation processes are also diffusive. Deformation mechanisms can act in parallel, with the strain rate being additive. On the other hand, the mechanism that gives the highest strain rate for a given stress (or the lowest stress for a given strain rate) and grain size is most efficient and considered most operative. We now discuss each deformation mechanism.

7.2.2.1 Dislocation creep

Dislocation creep involves several processes, including *glide*, which accommodates much of the strain, but also *climb* and *cross-slip*, which enable dislocations to avoid obstacles that can otherwise impede glide. There are two basic types of dislocations, edge and screw. An edge dislocation involves an extra partial plane of atoms (Fig. 7.13). The actual edge dislocation is the line running along the edge of the extra partial plane of atoms. The Burgers vector $\boldsymbol{b}$ gives the magnitude and direction of slip due to dislocation glide. For an edge dislocation, $\boldsymbol{b}$ is perpendicular to the lattice plane containing the dislocation line, with a magnitude equal to the distance between that plane and one adjacent. Edge dislocation glide thus occurs as the extra partial plane of atoms moves past successive partial planes of atoms on the other side of the slip plane. In practice, certain crystallographic planes tend to serve as active slip planes.

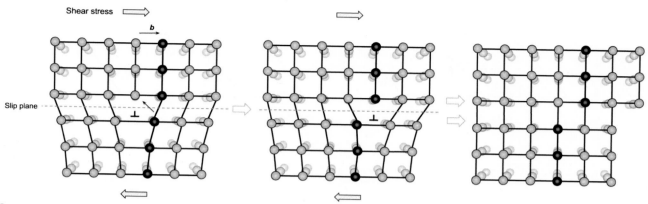

FIG. 7.13 Edge dislocation glide, showing the shear stress (large arrows), the dislocation line (into the page, marked with a dislocation core "nail"), the slip plane (perpendicular to the page, marked by the dotted line), and the Burgers vector b. For an edge dislocation, b is perpendicular to the dislocation line, and both lie in the slip plane. Looking from left to right one can see the glide of the two black half planes of atoms.

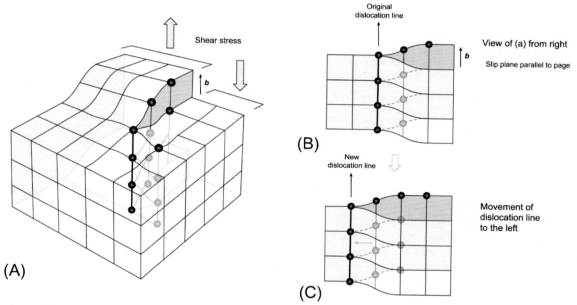

FIG. 7.14 Screw dislocation glide, (A) showing the stress (large arrows) and the Burgers vector b. (B) shows just the slip plane, which like for an edge dislocation contains both b and the dislocation line, but for a screw dislocation b and the dislocation line are parallel. (C) shows that as the stress causes glide on the slip plane, the dislocation line moves in that plane, perpendicularly to b, similar to the point of tearing moving perpendicularly to the direction in which one shears in order to rip a piece of paper. Any plane containing the screw dislocation can potentially serve as the slip plane.

For a screw dislocation, b is parallel to the dislocation line (Fig. 7.14). As atoms move in the direction of b, parallel to the dislocation, the dislocation line itself moves in a perpendicular direction, similar to the point of tearing moving perpendicularly to the direction in which one shears in order to rip a piece of paper. Any plane containing the screw dislocation can potentially serve as the slip plane, though as for edge dislocation glide only certain crystallographic planes are active. The arrangement of atoms around a screw dislocation is helical, hence the name. The magnitude of b is again equal to the distance between atomic planes in the direction of b.

In practice, dislocations need not be straight lines, and b can be neither parallel nor perpendicular to the dislocation, resulting in a mixed dislocation. Unlike in a perfect crystal where the bonds on an entire plane of atoms must be broken in order for deformation to occur, dislocation creep requires breaking of bonds only along a line of atoms, thus providing an important mechanism for the ductility of metals.

Dislocation creep is sometimes known as power law creep, with $n = 4$–6 in Eq. (7.11). This is a result of stress introducing dislocations, further enabling dislocation creep. As the dislocation density increases yet more dislocations eventually begin to interfere with each other's ability to move, eventually impeding dislocation creep. This is known

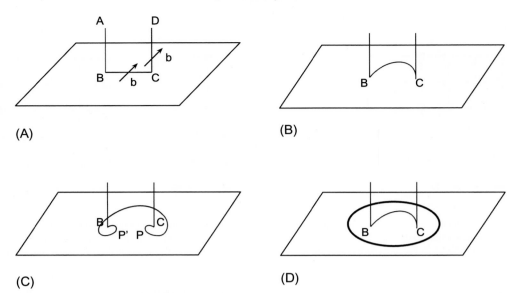

FIG. 7.15 The Frank-Read source of dislocations, an example of how stress can introduce dislocations. (A) Only the segment BC of the dislocation ABCD is in the slip plane, which contains the Burgers vector *b*. Segments AB and CD are immobile in the absence of climb. (B) The applied stress causes the segment BC to bow out since the points B and C are pinned. (C) As the force arising from the curvature of the dislocation exceeds the applied stress, the configuration becomes unstable, and with the points P and P′ having Burgers vectors of opposite sign, they annihilate. (D) A new (in this case, circular) dislocation is formed, and the process can repeat. *Adapted from Weertman, J., Weertman, J.R., 1992. Elementary Dislocation Theory. Oxford Press, Oxford.*

as strain or work hardening, and is particularly relevant at low and moderate temperatures, $T < .4T_M$, for which *recovery* and recrystallization are slower (Section 7.3). Stress typically introduces dislocations at grain boundaries, because homogeneous nucleation of a dislocation in the interior of a crystal likewise requires breaking of a semi-plane of atoms, which is energetically expensive. A well-known example of how stress can introduce dislocations is the *Frank-Read source* (Fig. 7.15).

An ill-understood deformation mechanism is *Harper–Dorn creep*, a form of dislocation creep in which $n = 1$ and $p = 0$ in Eq. (7.11). It has been reported in the deformation of Al, Pb, and Sn, among other materials, at low stress levels and large grain sizes, leading to these power dependencies. However, it has not been universally observed, with some arguing it is only operative in very pure materials with low initial dislocation densities. It has been suggested as a possible deformation mechanism in the inner core because of the inner core's likely low stress levels and large grain sizes, but questions remain about its existence and the resulting predicted viscosity.

In general, dislocation glide is not strongly temperature-dependent, so that dislocation creep is a primary creep mechanism for low and moderate temperatures, as well as higher stresses, due to the $n = 4$–6 power law dependence. However, other dislocations or point defects such as solute atoms can impede dislocation glide. Partial dislocations (see next paragraph) can sometimes facilitate slip past defects, as can two other phenomena, dislocation climb and cross-slip. Dislocation climb involves the movement of an edge dislocation in the direction perpendicular to both *b* and the dislocation (Fig. 7.16). It involves the diffusion of atoms off or on to the partial plane, allowing for a new slip plane so as to avoid the obstacle. Unlike glide, climb is thermally activated because it is a diffusive process. Whereas glide is associated with shear stress, climb is associated with compressive or tensile stress, as a comparison of Figs. 7.13 and 7.16 suggests. Glide along a screw dislocation can avoid a defect or obstacle by changing its slip plane, a process known as cross-slip (Fig. 7.17). Unlike climb it does not require diffusion of atoms. Climb and cross-slip allow for greater ductility during dislocation creep. Glide, climb, and cross-slip work together, with one of the processes often being rate-limiting.

Stacking faults are imperfections in crystals in which the normal order of atomic planes in close-packed structures is disrupted (in this and following paragraphs one might refer back to the basic crystallography discussed in Section 2.5). The classic examples of stacking faults occur in fcc and hcp materials. In the former the usual sequence of planes is *ABCABCA* and in the latter *ABABABA*, but a stacking fault results in a local change to this sequence. They are created by dislocation glide due to two partial (Shockley) dislocations shifting part of a close-packed plane (Fig. 7.18). The sum of the Burgers vectors of two partial dislocations must equal that of a full dislocation, but their strain energy is lower than that of the full dislocation. For instance, in fcc systems,

7.2 Deformation in the inner core

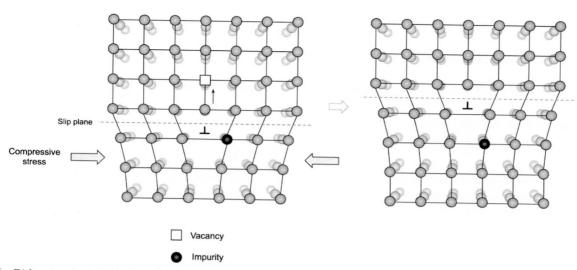

FIG. 7.16 Dislocation climb. Glide of an edge dislocation in a slip plane can be impeded by an impurity. Diffusion of an atom into a vacancy, aided by compressive stress, allows the slip plane to move upward and hence avoid the obstacle; this is known as dislocation climb. In this example, tensile stress would allow the slip plane to move downward. *Adapted from Weertman, J., Weertman, J.R., 1992. Elementary Dislocation Theory. Oxford Press, Oxford.*

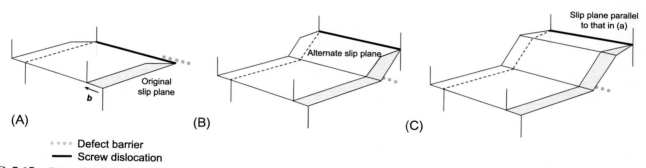

FIG. 7.17 Cross-slip allows for a change in the slip plane during glide along a screw dislocation, for instance, when the dislocation line encounters a defect barrier. As the dislocation approaches the barrier in (A), it can begin to glide along an alternate plane (B) in order to avoid the barrier (assuming there is another available crystallographic easy slip plane), and then (C) may revert to a slip plane parallel to the original one. The shaded part of the slip plane here corresponds to that in Fig. 7.14. The vertical lines and dashed line represent that this is, of course, a section of a 3D structure. Cross-slip is not a diffusive process.

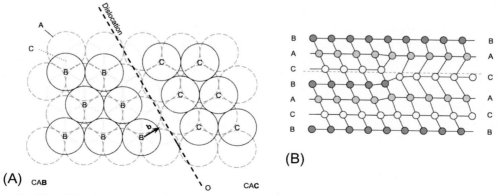

FIG. 7.18 An example of a stacking fault due to partial dislocations shifting part of a close-packed plane. (A) Top view showing the position of atoms beneath the plane denoted by the dotted line in (B), which is a side view. To the left of the partial dislocation, the stacking order beneath this plane is CAB, to the right CAC. *Adapted from Weertman, J., Weertman, J.R., 1992. Elementary Dislocation Theory. Oxford Press, Oxford.*

$$\tfrac{1}{2}\left[10\bar{1}\right] = 1/6\left[2\bar{1}\bar{1}\right] + 1/6\left[11\bar{2}\right],\tag{7.12}$$

with the right-hand side being the two partial dislocations, but

$$\left|\tfrac{1}{2}\left\{1^2 + 0^2 + (-1)^2\right\}^{1/2}\right|^2 > \left|1/6\left\{2^2 + (-1)^2 + (-1)^2\right\}^{1/2}\right|^2 + \left|1/6\left\{1^2 + 1^2 + (-2)^2\right\}^{1/2}\right|^2$$

(the dislocation strain energy $E = \tfrac{1}{2}\mu b^2$, where μ = shear modulus). An example of a Shockley partial dislocation in hcp systems is

$$\left[11\bar{2}0\right] = \left[10\bar{1}0\right] + \left[01\bar{1}0\right].\tag{7.13}$$

However, stacking faults introduce extra stacking fault energy into a crystal. A large stacking fault energy thus inhibits partial dislocations from forming. In this case, cross-slip of screw dislocations may be more energetically favorable for facilitating dislocation creep. This has implications for textural development during dislocation creep, as discussed in the next section, and for the occurrence of dynamic recrystallization (Section 7.3).

7.2.2.2 Slip systems

One can show that in order for dislocation creep to result in homogeneous deformation of a polycrystalline material for an arbitrary strain while preserving volume, there must be five independent slip systems, which is known as the von Mises condition (Karato, 2008). A slip system consists of a family of slip planes {} and slip directions <> in those planes. If this condition is not satisfied, the material may fracture (at least at low pressure) or deform by a mechanism other than dislocation creep, typically by diffusion creep (Section 7.2.2.3). Inhomogeneous deformation or cross-slip can also sometimes reduce the required number of slip systems. One form of inhomogeneous deformation is twinning, which occurs by the motion of partial dislocations and hence requires a small stacking fault energy. Twinning by itself is likely unimportant for inner core deformation, however, because it typically can accommodate only a small fraction of the total strain, and is primarily operative at lower homologous temperatures.

Texture development arising from dislocation creep is primarily associated with the lattice rotation that occurs during dislocation glide and/or twinning (but not climb, which does not involve shear stress) in each of the grains. In order to understand the texture that develops during dislocation creep it is therefore necessary to know the active slip systems. In general, due to their lower symmetry hcp polycrystals often have fewer available slip systems than cubic materials. Knowing the available slip systems in hcp Fe under inner core conditions would be most helpful for understanding the deformation mechanism and texture development in the inner core, but they are currently unknown.

Experiments may someday determine the active slip systems in hcp Fe under inner core conditions, but dislocations are notoriously difficult to study experimentally. For now we must rely on predictive schemes, analogous materials whose slip systems are easier to study, or modeling observed textures from the applied stresses in laboratory experiments. Such modeling has also been tried using seismically "known" inner core textures, and assumed stresses and hcp Fe elastic moduli (Section 7.6.1).

As a predictive scheme, some have used c/a ratio of hcp crystals (Wenk et al., 1988). For example, Ti has a c/a ratio less than the ideal, so that prismatic planes $\{10\bar{1}0\}$ are closer packed (with atoms), but further apart from each other than are basal planes $\{0001\}$. One might thus expect easier slip along these prismatic planes and in the closest-packed $<1\bar{2}10>$ directions (Fig. 7.19, left). Others have argued, however, that stacking fault energy is a better predictor of the active slip planes than is the c/a ratio, because of stacking fault energy's role in inhibiting the partial dislocations that can facilitate slip, as discussed earlier. Stacking fault energy can be computed quantum mechanically, and from this Poirier and Price (1999) suggested that primary slip in Fe under inner core conditions is likely to be basal $\{0001\}$ $<1\bar{2}10>$ (Fig. 7.19, right). Basal slip in hcp Fe was indeed found at high pressures but room temperature by Wenk et al. (2000) and also at 30 GPa and 1000 K (refer to Fig. 2.25 for the stability field of hcp ε-Fe) by Merkel et al. (2004), who also found some prismatic slip $\{10\bar{1}0\}<1\bar{2}10>$ associated with twinning. In any case, neither the c/a ratio nor the stacking fault energy of Fe under inner core conditions is yet known with any confidence.

The hcp ε-martensite phase of Cr—Ni stainless steel may serve as a suitable analog for the deformation of Fe under inner core conditions, and has the advantage of being metastable at atmospheric pressure, unlike hcp ε-Fe, so that it is easier to study dislocations (Poirier and Langenhorst, 2002). Experiments on this stainless steel at 1000 K find some evidence for secondary pyramidal slip $\{11\bar{2}2\}<11\bar{2}3>$, in addition to basal slip. Along with basal and prismatic slips, some pyramidal slip is required by the von Mises condition to accommodate a general strain. However, pyramidal slip in hcp ε-Fe has never been directly observed (Merkel et al., 2004), and in several hcp metals the critical resolved shear stress for pyramidal slip is known to actually increase with temperature (Poirier and Langenhorst, 2002). The temperature dependence for pyramidal slip in hcp ε-Fe is unknown.

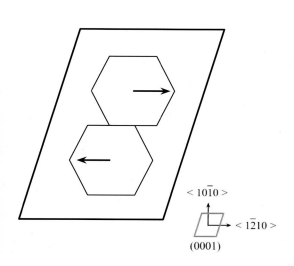
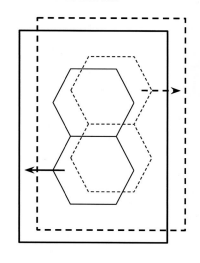

FIG. 7.19 (Left) hcp prismatic slip {10$\bar{1}$0}<1$\bar{2}$10>. In this sketch, prismatic planes are perpendicular to the page. (Right) hcp basal slip {0001}<1$\bar{2}$10>. *Adapted from Bergman, M.I., Yu, J., Lewis, D.J., Parker, G.K., 2017. Grain boundary sliding in high-temperature deformation of directionally solidified hcp Zn alloys, and implications for the deformation mechanism of Earth's inner core. J. Geophys. Res. Solid Earth 123, 189–203.*

Modeling of observed textures from deformation experiments on ε-Fe at up to 30 GPa and a range of temperatures up to 1950 K suggest that in addition to basal and prismatic slip, some secondary pyramidal slip is occurring, especially for large strains (Miyagi et al., 2008; Merkel et al., 2012). Similar modeling of inner core texture inferred seismically, using assumed elastic moduli of hcp ε-Fe and geodynamically derived stresses, also requires secondary pyramidal slip <c+a> such as {$\bar{1}$011}<11$\bar{2}$3> or {11$\bar{2}$2}<11$\bar{2}$3> (Lincot et al., 2016). Clearly the additional pyramidal slip affords more degrees of freedom, which are necessary because these models assume that dislocation creep is the operative deformation mechanism. Modeling by Brennan et al. (*2021*) found that hcp Fe-Ni-Si alloys deformed at 60 GPa and 1650 K exhibit greater pyramidal slip activity than does Fe without Si, but that the alloy is nevertheless mechanically ten times more resistant to dislocation creep.

7.2.2.3 Diffusion creep

Diffusion creep results from a stress gradient causing a flow of atoms/vacancies. In particular, atoms flow toward regions of high tension, vacancies toward regions of large compression (Fig. 7.20). These point defects can flow primarily through the grains (volumetric diffusion), known as Nabarro–Herring creep, for which the grain size exponent $p=2$ in Eq. (7.11), or they can flow primarily along grain boundaries (grain boundary diffusion), known as Coble creep,

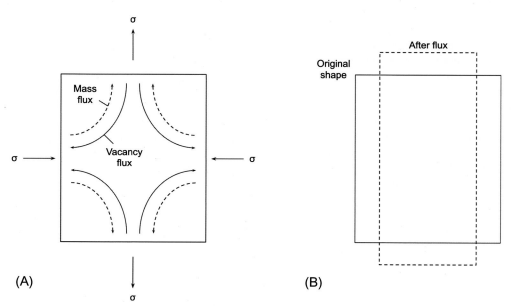

FIG. 7.20 (A) Diffusion creep due to a stress (σ) gradient results in (B) a change in grain shape. *Adapted from Courtney, T.H., 1990. Mechanical Behavior of Materials. McGraw-Hill, New York.*

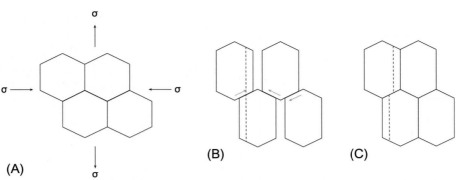

FIG. 7.21 Diffusion creep, (A), results in grain shape change, which can yield voids, (B), which grain boundary sliding can ameliorate, (C). In (B) the arrows show the grain boundary sliding. The vertical dashed line shows the result of the grain boundary sliding. There is no change in texture. *Adapted from Courtney, T.H., 1990. Mechanical Behavior of Materials. McGraw-Hill, New York.*

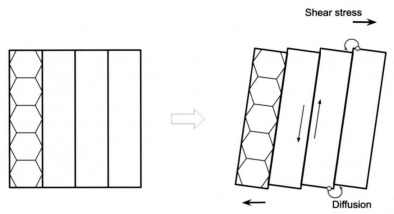

FIG. 7.22 An example of grain boundary sliding that can result in a change in texture. (Left) columnar crystals due to solidification, showing the orientation of the hcp lattice in one crystal. (Right) shear stress results in tilting of the columnar crystals and the original solidification texture, by means of grain boundary sliding (thin arrows). In order to maintain macroscopic shape diffusion of atoms at the top and bottom accompanies the grain boundary sliding. *Adapted from Bergman, M.I., Yu, J., Lewis, D.J., Parker, G.K., 2017. Grain boundary sliding in high-temperature deformation of directionally solidified hcp Zn alloys, and implications for the deformation mechanism of Earth's inner core. J. Geophys. Res. Solid Earth 123, 189–203.*

for which $p=3$. Because the atomic self-diffusivity D ($=D_{vol}$ for Nabarro–Herring creep, $=D_{gb}$ for Coble creep) has the usual Arrhenius temperature dependence, $e^{-Q/k_B T}$, where Q is the activation energy and k_B is Boltzmann's constant, diffusion creep is most active at high temperatures. Similarly, because of the grain size dependence it is typically most active for small grain sizes.

In order to maintain macroscopic shape and avoid voids or cracks, diffusion creep is typically accompanied by grain boundary sliding (Fig. 7.21). Incidentally, the term superplasticity refers to large strain deformation, often in tension, without a change in shape or voids or cracks forming. It typically involves grain boundary sliding. Dislocation creep typically results in a texture due to the lattice rotation that accompanies glide along particular slip planes in each of the grains. Diffusion creep, on the other hand, does not typically result in a completely new texture as there is no lattice rotation when individual atoms and vacancies move about. However, diffusion creep may preserve an existing texture (Fig. 7.21) or, under some conditions, modify one (Fig. 7.22).

7.3 Annealing: Recovery, recrystallization, grain growth, and coarsening

High-energy microstructural features such as dislocations, dendrites, and grain boundaries form as a result of dynamical processes, but are usually thermodynamically unstable. However, at low homologous temperatures they can be metastable so that, for instance, the tangle of dislocations that are introduced during cold working of a metal cannot work themselves out, which eventually leads to loss of ductility. At higher temperatures crystals can more easily evolve to lower the energy associated with these features, this is the process of annealing. Annealing can take competing paths: grains can recover by decreasing the density of high-energy features, while preserving the original grains (Fig. 7.23A–C); or recrystallization can occur, whereby new grains that are relatively free of high-energy features nucleate, and then grow at the expense of the original grains (Fig. 7.23A, D–F). *Coarsening* is a high-temperature process common in two-phase alloys in which high-energy features such as highly curved phase boundaries smooth out and smaller inclusions of the secondary phase coalesce but, like recovery, the original grains remain. This is common, for instance, in alloys with a directional solidification microstructure.

If recovery and recrystallization occur after deformation at low temperature, they are known as static recovery and static recrystallization; if they occur during high-temperature deformation, without an in-between low temperature stage, the processes are known as dynamic recovery and dynamic recrystallization. Dynamic recovery and recrystallization are therefore the relevant processes during dislocation creep of the inner core (they are essentially irrelevant for diffusion creep). The high-energy features that result from solidification can be worked out of the microstructure by the competing processes of coarsening, and static (because the process occurs after solidification) recrystallization and *grain growth*.

We outline the process of recovery in Fig. 7.23A–C. During dislocation creep the very dislocations that enable plastic flow can eventually become so numerous that they impede each other's ability to glide (Fig. 7.23A). Dislocation climb and cross-slip can, however, help crystals remove tangles of dislocations by *annihilation* (Fig. 7.24). Annihilation is generally not complete, and if dislocations of one sign are in excess, sub-grain boundaries can result (Figs. 7.23B and 7.25). Sub-grains then grow (Fig. 7.23C) in a process similar to grain growth following recrystallization (see following paragraphs), with the difference that a growing sub-grain has the same lattice orientation as that which existed previously, since it is not the result of new nucleation.

During recrystallization nucleation typically occurs along high-energy, high-crystallographic angle grain boundaries (Fig. 7.23D), because the new crystals may have relative crystallographic orientations that lower the overall surface energy (this can lead to a recrystallization texture). Because of their lower dislocation density, newly nucleated grains then grow by *grain boundary migration*, in spite of the higher energy associated with greater boundary curvature due to their initial smaller size. For low crystallographic angle grain boundaries, grain boundary migration occurs via the motion of the dislocations that form the grain boundary (Fig. 7.26). For higher angle boundaries, it occurs by grain boundary sliding, which is diffusion-limited (Section 7.2.2.3). The growing grains often have an equiaxed, polygonal

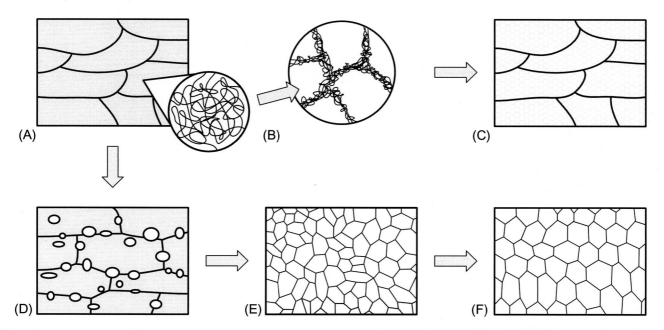

FIG. 7.23 (A) Grains with high energy due to tangles of dislocations can lower their energy through dislocation annihilation, resulting in (B) sub-grain formation, and (C) recovery of the original grain structure but with a lower dislocation density. Alternatively, high-energy grains may recrystallize, beginning with (D) nucleation of relatively dislocation-free grains along grain boundaries, and (E) and (F) grain growth of these new grains. Shading in these figures indicates the dislocation density. *Adapted from Humphreys, F.J., Hatherly, M., 1996. Recrystallization and Related Annealing Phenomena. Pergamon, New York.*

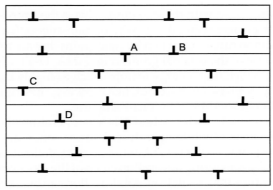

FIG. 7.24 Annihilation of dislocations during recovery occurs when two dislocations on the same slip plane and of opposite signs, such as A and B, merge. Climb and cross-slip can facilitate annihilation, for instance, by allowing C and D to move onto the same slip plane. *Adapted from Humphreys, F.J., Hatherly, M., 1996. Recrystallization and Related Annealing Phenomena. Pergamon, New York.*

FIG. 7.25 Sub-grain formation due to an excess of one sign of dislocation (A), accompanied by annihilation (B). Sub-grain boundaries are formed by a series of dislocations, which yield crystals with slight misorientations (C). *Adapted from Humphreys, F.J., Hatherly, M., 1996. Recrystallization and Related Annealing Phenomena. Pergamon, New York.*

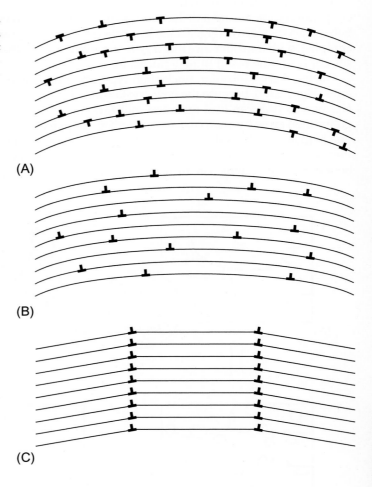

morphology (Fig. 7.23E and F). During dynamic recrystallization the dislocation density of the growing grains increases as deformation continues, ultimately limiting the grain size, as yet new grains nucleate and grow. A steady-state recrystallization grain size d is rapidly reached, given empirically for a surprisingly wide range of materials and conditions by

$$(\sigma/\mu)(d/b)^q = K, \tag{7.14}$$

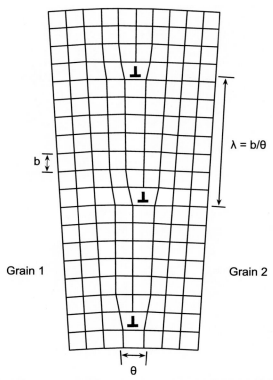

FIG. 7.26 A low angle tilt grain boundary formed by dislocations. λ is the boundary dislocation spacing, and θ is the crystallographic angle between the grains. High-angle grain boundaries have a higher energy, and a more open and disordered structure.

where q and K are empirical dimensionless constants, approximately 0.8 and 15, respectively (Humphreys and Hatherly, 1996).

Because stress introduces dislocations (Fig. 7.15) and hence excess elastic energy that is the driving force for grain growth of newly nucleated grains, dynamic recrystallization typically results at higher stress levels. On the other hand, metals with high stacking fault energies tend to have dislocations that climb and cross-slip, which allows them to more easily annihilate dislocations and break-up tangles, and hence recover. Critical conditions have been developed to predict whether dynamic recrystallization during deformation occurs, but the values of relevant parameters are very much unknown for the inner core, though the low stress levels might suggest recovery. Whereas recrystallization can introduce a new texture, though one often related to the deformation texture, recovery does not change the deformation texture. Interpretations of seismic inner core elastic anisotropy data may perhaps have to consider recrystallization textures.

Coarsening is a form of the colorfully named Ostwald ripening, whereby smaller particles coalesce into lower energy larger ones and highly curved surfaces increase their radii of curvature, by means of diffusion. Coarsening is akin to recovery, for the thermodynamically unstable solidification microstructure of two-phase alloys. During coarsening dendrite side branches tend to dissolve as primary dendrite arms thicken (Fig. 7.27). Some experiments on annealing of directionally solidified hcp Zn alloys also show coarsening, with no loss of solidification texture (Fig. 7.28A), but others instead show static recrystallization and grain growth (Fig. 7.28B), driven by the high-energy phase boundaries. In the latter case there is a loss of solidification texture, which has been suggested as a possible cause for hemispherical variations in elastic and attenuative anisotropies if the inner core is translating, but the conditions for coarsening versus recrystallization in the experiments and in the inner core remain unclear.

Unlike dynamic recrystallization for which ongoing deformation introduces elastic energy in new grains and hence results in a steady-state grain size, during static recrystallization grains may continue to grow since there is no ongoing deformation or solidification (in the solid!). Under the assumptions that (1) the driving pressure P for grain growth is grain boundary curvature, and hence is inversely proportional to the mean grain size d, (2) the boundary velocity is proportional to P, and (3) the surface energy between grains is the same for all boundaries, one can show that d increases with time as $t^{1/s}$, where the grain growth exponent $s=2$ in this simple analysis (Humphreys and Hatherly, 1996). In practice, s tends to be larger, particularly in alloys, where secondary phases can pin boundaries and reduce grain boundary mobility.

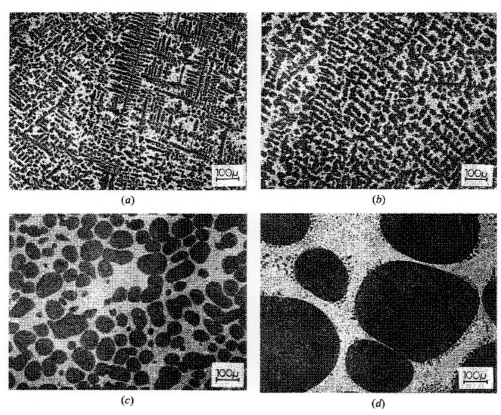

FIG. 7.27 Quenched microstructures of directionally solidified Sn-40 wt pct Bi, annealed just above the eutectic temperature. Micrographs are transverse to the original growth direction, showing the coarsening of the dark Sn (cubic) dendrites as time progresses. (A) As cast, (B) after 10 min, (C) after 2.5 h, and (D) after 10 days. The scale represents 100 microns. *Courtesy of Marsh, S.P., Glicksman, M.E., 1996. Overview of geometric effects on coarsening of mushy zones. Metall. Mater. Trans. A 27, 557–567.*

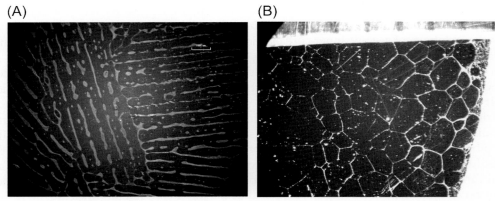

FIG. 7.28 (A) Micrograph of directionally solidified and then annealed hcp Zn-3 wt pct Sn, transverse to the original growth direction. The primary Zn-rich phase is dark, the Sn-rich phase is light. The original, large grains comprised of parallel platelets (dendrites) remain, but platelet boundaries of secondary Sn-rich phase have begun to coalesce, resulting in coarsening. The original solidification texture remains. The scale represents 1 mm. Compare with Fig. 7.5C. (B) As for (A), but instead of coarsening the original, large grains comprised of parallel platelets have statically recrystallized and exhibited polygonal grain growth, resulting in destruction of the solidification texture. Ruled markings at top are mm. *(A) From Al-Khatatbeh, Y., Bergman, M.I., Lewis, D. J., Mason, Z., Zhu, L., Rosenstock, S., 2013. Annealing of directionally solidified alloys revisited: no loss of solidification texture in Earth's inner core. Phys. Earth Planet. Inter. 223, 32–39, https://doi.org/10.1016/j.pepi.2013.04.003. (B) From Bergman, M.I., Lewis, D.J., Myint, I.H., Slivka, L., Karato, S.-I., Abreu, A., 2010. Grain growth and loss of texture during annealing of alloys, and the translation of Earth's inner core. Geophys. Res. Lett. 37, L22313, https://doi.org/10.1029/2010GL045103.*

7.4 Grain size and the deformation mechanism map of the inner core

In Section 7.2.1, we discussed various causes for inner core deformation, in Section 7.2.2 ways in which the inner core might deform, and in Section 7.3 how the inner core might anneal after deformation (or solidification). It is now time to put it together, which is often done in the form of a *deformation mechanism map*. For a given material and grain size, a deformation mechanism map presents the most active deformation mechanism as a function of normalized stress, σ/μ, and homologous temperature, T/T_M. Usually the data is for experiments at atmospheric pressure, though obviously for the inner core we are interested in hcp Fe at high pressure, and pressure normalized to the shear modulus μ could be treated as another parameter. A deformation mechanism map also shows regions of parameter space where recrystallization is likely. Maps may also give contours of isostrain rates, which can be calculated from Eq. (7.11). Fig. 7.29 displays a couple of examples of deformation mechanism maps. Such maps usually assume equiaxed, untextured grains.

Unfortunately, we do not have the data to construct a deformation mechanism map for the inner core. There is some uncertainty in T/T_M because we do not know the inner core geotherm precisely, and because we do not know the exact melting temperature of Fe and its alloys at core pressures. T/T_M obviously varies across the inner core, but even close to the ICB it is likely less than roughly 0.9 because the solid that has already frozen out will have a solute concentration that is less than that of the solidus and because diffusion in the solid state is very slow (Figs. 2.31 and 7.11B). The stress or strain rate resulting from any of the possible causes for inner core deformation discussed in Section 7.2.1 has greater uncertainty, but even with μ as low as 1.5×10^{11} Pa, σ/μ is quite likely $<10^{-6}$ for any of the suggested driving stresses.

By far the greatest uncertainty in the parameters that determine the deformation mechanism is the grain size, along with the microstructure and texture. Estimates for the inner core grain size range from a few mm to the entire inner core, some 1200 km is radius. The small grain size estimate comes from geodynamics, as an inner core viscosity $\eta < 10^{16}$ Pa s is one way to reconcile inner core *super-rotation* of 1 deg./yr (Chapter 6) with *gravitational locking* to the mantle (Buffett, 1997). This is discussed further in the next section. Such a low viscosity implies a small grain size if, as was done, one assumes diffusion creep to be operative. The large grain size estimate comes from a comparison of the large-scale seismic elastic anisotropy of the inner core, 3%, with the quantum mechanically calculated elastic anisotropy of hcp Fe at 0 K, also 3%, suggesting the entire inner core is a single crystal (Stixrude and Cohen, 1995). During an after-dinner talk at the American Association for Crystal Growth, whose members are primarily concerned with Si boules and such, at the mention of the possibility that the entire inner core could be a single crystal, someone from the back of the function room was heard to exclaim "it's the mother of all crystals!". While neither of these extremes are likely correct, they do show the incredible uncertainty in this key parameter.

It is likely that the single crystal elastic anisotropy of Fe under (high temperature) inner core conditions is larger than 3%, so that complete alignment of crystals (i.e., the inner core is a single crystal) is not required. Moreover, the rate of super-rotation may be less than originally suggested, and the assumptions that led to the small grain size may not be warranted. Several lines of evidence suggest intermediate grain sizes: a few meters, based on recrystallization, Eq.

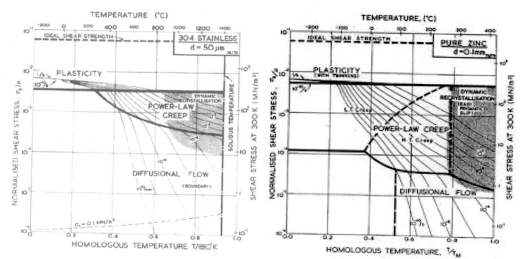

FIG. 7.29 Examples of deformation mechanism maps for 304 stainless steel (left) and Zn (right). *From Frost, H.H., Ashby, M.F., 1982. Deformation Mechanism Maps. Pergamon, Tarrytown, NY.*

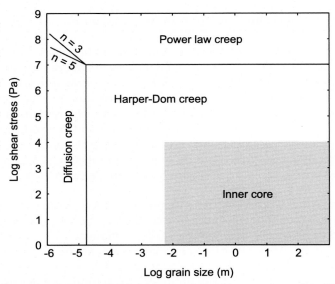

FIG. 7.30 An example of a deformation mechanism map for the inner core. *Adapted from van Orman, J.A., 2004. On the viscosity and creep mechanism of Earth's inner core. Geophys. Res. Lett. 31, L20606. https://doi.org/10.1029/2004GL021209.*

(7.14), with stress levels from Section 7.2.1 (Yoshida et al., 1996); meteorites that might have been parts of planetesimal cores, at least as large as a few meters (Fig. 7.6, right); columnar crystals hundreds of meters across, and longer in the growth direction, from extrapolation of solidification experiments; estimates based on assuming that inner core attenuation is due to scattering, also hundreds of meters (Bergman, 1998); and extrapolation of grain growth experiments on ε-Fe with a grain growth exponent $s = 2$, hundreds of meters to a few kilometers (Yamazaki et al., 2017). None of these estimates are, of course, without assumptions. Moreover, microstructure such as dendrites (Figs. 7.3–7.5) or sub-grains (Fig. 7.23B), which can serve as sources of atoms/vacancies and hence be relevant for the efficacy of diffusion creep, can decrease the effective grain size. This has perhaps not been properly recognized. For instance, slightly misoriented dendrites can be modeled as low angle tilt grain boundaries (Fig. 7.26).

As an example of how one might use a deformation mechanism map for the inner core, the likely low stress and large grain size suggest that the inner core might be in the regime of Harper–Dorn creep (Fig. 7.30) though its existence, especially in materials that are not ultra-pure, remains controversial. It is clear we need more experimental data and theoretical calculations to make a deformation mechanism map for the inner core. One difficulty, however, is that it is not easy to perform high-pressure deformation experiments or numerical simulations on large-grained samples. Surprises could be in store; for example, experiments on the high-temperature deformation of hcp Zn alloys with a directional solidification microstructure and texture at atmospheric pressure show grain boundary sliding (Fig. 7.22) to be operative, in spite of the large grain size. The operating deformation mechanism, i.e., the rheology, may also change with depth in the inner core, as a result of different stresses, grain sizes, or other parameters, which could result in a change of texture with depth.

7.5 Inner core viscosity

Unlike the fluid outer core viscosity that is essentially a material property, the solid inner core dynamic viscosity η depends not only on microphysics, but also on geodynamics, which controls the stress level and microstructure, i.e., grain size, shape, and texture. But the geodynamics in turn depends on the viscosity, a bit of a Catch-22! If asked today on one's qualifying exams what unknown might be most helpful in understanding Earth's deep interior, inner core viscosity would be a good answer.

From a microphysical standpoint, solid-state viscosity in general depends on the crystal structure, shear modulus, atomic self-diffusivity, and available slip systems, which factor into the operative deformation mechanism. Nabarro–Herring diffusion creep, with $n = 1$ in Eq. (7.11), has a Newtonian viscosity $\eta = \sigma/2\dot{\varepsilon} = 10^{21} d^2$ Pa s (Vocadlo, 2007). As discussed in the previous section, the grain size d depends on the geodynamics and varies from 10^{-3} m to 10^3 m or even larger, yielding $\eta = 10^{15}$–10^{27} Pa s for diffusion creep (if sub-grain structure is important, however, estimates toward the smaller end of the range might be more relevant, while smaller values of the self-diffusivity D would increase η).

For high-temperature power law dislocation creep, η depends inversely on the dislocation density ρ because, of course, it is the dislocations that facilitate slip (we have seen that at high temperatures tangles of dislocations can work themselves out). Because ρ increases with the stress σ for power law creep, Eq. (7.11) predicts a nonlinear relationship between stress and strain rate. For power law dislocation creep it has been estimated that $\eta = 10^{23}/\rho$ Pa s, where the dislocation density ρ could be between 10^6 and 10^{13} m^{-2} (Vocadlo, 2007), yielding $\eta = 10^{10}$–10^{17} Pa s, with a recent ab initio calculation finding η for dislocation creep in hcp Fe toward the upper end of this range (Ritterbex and Tsuchiya, 2020). For Harper–Dorn creep (Fig. 7.30), for which the stress exponent $n = 1$, a very low dislocation density $\rho < 10^7$ m^{-2} is required, yielding a Newtonian viscosity $\eta > 10^{16}$ Pa s. Viscosities add inversely because the mechanisms operate in parallel. These simple microphysical models do not take into account certain microstructural effects such as sub-grains and textures, which could change the given dynamic viscosity ranges. A recent experimental estimate for η of hcp Fe-Ni-Si alloys also finds an increase by a factor of ten over that of pure Fe, if dislocation creep is operative (Brennan et al., 2021).

The passage of shear body (J) waves in the inner core on 1 Hz timescales suggests η is likely greater than 10^{10} Pa s, though under unusual conditions it could perhaps be lower. Although the low shear modulus μ and high attenuation Q^{-1} indicate large seismic wave dissipation in the anelastic regime, this does not necessarily imply a low viscosity in the plastic regime. For instance, Eq. (7.11) shows that diffusion creep, with $n = 1$, is independent of the shear modulus μ, presumably because it involves only the movement of atoms and vacancies. On the other hand, for power law dislocation creep, with $n > 1$, η increases with μ. The relationship between μ and η thus depends in part on the timescale of the deformation and the deformation mechanism, i.e., on the geodynamical process being considered.

As an example of a purely geodynamical estimate of inner core viscosity, if the inner core's figure and spin axes are aligned, but not with those of the outer core and mantle (Fig. 7.31), then the inner core will precess and nutate on diurnal timescales about the mantle's figure and spin axis. If the ability of the inner core to deform helps control the frequency and attenuation of the nutation, then under certain assumptions, including a Newtonian inner core viscosity, one can deduce η in the range of 2–7×10^{14} Pa s.

Another geodynamical argument comes from inner core super-rotation. In order to prevent gravitational locking of the inner core to the mantle, $\phi'\tau \ll 1$ in order to allow the inner core to flow (Fig. 7.32), or $\phi'\tau \gg 1$ in order to prevent locking topography from growing. ϕ' is the super-rotation rate, and τ is the topographical viscous relaxation timescale, which is proportional to η. With $\phi' = 1$ deg./yr, the first limit requires $\eta < 10^{16}$ Pa s, the second limit $\eta > 10^{20}$ Pa s. However, as estimates of super-rotation have decreased by perhaps two orders of magnitude, then the upper bound on the low viscosity estimate has increased to $\eta < 10^{18}$ Pa s, a much weaker constraint. The lower bound on the high-viscosity estimate also increases, to $\eta > 10^{22}$ Pa s, which is unreasonably high, likely ruling out this limit.

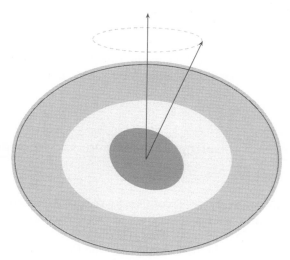

FIG. 7.31 Misalignment between the inner core figure and rotation axes and those of the outer core and mantle results in inner core precession and nutation. *Adapted from Koot, L., Dumberry, M., 2011. Viscosity of the Earth's inner core: constraints from nutation observations. Earth Planet. Sci. Lett. 308, 343–349.*

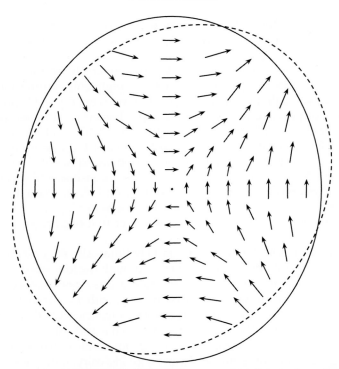

FIG. 7.32 As (if) the inner core super-rotates, it can respond plastically in an attempt to keep its shape in alignment with the gravitational equipotential. *Adapted from Buffett, B.A., 1997. Geodynamic estimates of the viscosity of the Earth's inner core. Nature 388, 571–573.*

The deformation models of Section 7.2.1 and Fig. 7.12 all generate a stress or strain rate, and if one assumes a deformation mechanism, one can estimate the viscosity. For example, if nonequipotential growth of the inner core (Fig. 7.12B) is driving a flow with an estimated strain rate of $3 \times 10^{-18}\,\text{s}^{-1}$, then a dynamically recrystallized grain size of the order of meters results if one assumes diffusion creep is operative. This assumption is justified by diffusion creep yielding lower stress than power law dislocation creep (and not considering Harper–Dorn creep) for that strain rate and grain size. This leads to a rather large Newtonian viscosity of $10^{21}\,\text{Pa s}$. However, unless one is certain one has both the correct stress or strain rate for a deformation model, and deformation mechanism, one cannot be sure one is providing the proper constraint on inner core viscosity.

With uncertainties in the stress/strain, grain size and microstructure, active slip systems, atomic self-diffusivities, and hence operative deformation mechanism, and assumptions required for geodynamical estimates, it is very difficult to even bracket a value for inner core viscosity. This is a case where the authors, with different expertise and trust in certain data, had difficulty agreeing even on a range. V.C. prefers a generous range for η, 10^{10}–$10^{18}\,\text{Pa s}$, while M.B. prefers 10^{14}–$10^{20}\,\text{Pa s}$, but with more probable values in the range 10^{15}–$10^{18}\,\text{Pa s}$. P.O. is agnostic, though he does believe the inner core has a viscosity. Inner core translation, for instance, may thus lie at the upper end of what it plausible for η.

7.6 Inner core elastic anisotropy, attenuation, and isotropic heterogeneity

7.6.1 Elastic anisotropy

Because of its remoteness and comparatively small volume, for many years there was relatively little research on the inner core. Even the realization that compositional convection in the outer core driven by inner core solidification plays a major role in the generation of the Earth's magnetic field did not drive significant research on the inner core, aside from work on the nature of solidification at the ICB. As late as 1988 Paul Roberts quipped, perhaps correctly at that time, that including the inner core in geodynamo models is like "worrying about the size of the tip when one can't pay the bill!" (Roberts, 1988). Interest in the inner core changed about that time, with seismic inferences of inner core elastic anisotropy. The simplicity of the initial parameterization, with large-scale anisotropy aligned with the spin axis, was intriguing, and spurred research in seismology, mineral physics, and geodynamics toward an understanding of the state and evolution of the inner core. As much more as we now understand, the origin of the seismic elastic anisotropy remains largely unsolved, and the inner core appears increasingly complex. The seismic inferences to be explained (Chapter 6), in rough order of confidence, include: 1) large-scale elastic anisotropy aligned with the spin axis (cylindrical anisotropy), 2) a depth dependence where the upper 50–100 km are weakly anisotropic or isotropic, or possibly radially anisotropic, 3) hemispherical variations, where the eastern quasi-hemisphere is less anisotropic and exhibits a more complex pattern, 4) a possibly sharp hemispherical boundary, 5) a deep inner core, some 500 km in radius, with an anisotropy that is not aligned with the spin axis, and 6) regional variations in anisotropy.

The most widely suggested explanation for inner core elastic anisotropy is a lattice preferred orientation (LPO) of the Fe crystals that comprise the inner core. LPO is observed and of much interest in the mantle, where it is associated with convection. LPO requires single crystal elastic anisotropy, and a means to texture, i.e., align, at least one crystallographic axis of the crystals. The single crystal elastic anisotropy depends, of course, on the stable phase of the inner core Fe alloy. These questions were addressed in Sections 2.5.2 and 2.5.3. We found there that the hcp phase is the most stable phase of Fe under inner core conditions, but alloying could arguably make a cubic phase stable. The hcp phase has cylindrical symmetry, but the magnitude, and possibly even the sense, of single crystal anisotropy remains the subject of research. Cubic phases may have a larger single crystal elastic anisotropy, but their symmetry makes it more difficult to explain cylindrical anisotropy (Lincot et al., 2015).

Two broad causes for LPO have been proposed, that due to solidification, and that due to deformation. The former was discussed in Sections 7.1.1 and 7.1.2. Deformation texturing requires a source of stress/strain, discussed in Section 7.2.1, and a deformation mechanism, discussed in Section 7.2.2. Table 7.1 summarizes the proposed deformation texturing models, along with comments on their shortcomings. As we have seen in these sections, no single proposed solidification or deformation texturing model can by itself explain a more complex pattern of inner core seismic elastic anisotropy. Several have built-in a means to explain the depth dependence, though none completely convincing. Long-term mantle control or translation, if it is occurring, could plausibly explain hemispherical or regional lateral variations. A deep inner core with a different orientation of anisotropy could perhaps be understood if a phase change from hcp to cubic Fe alloy occurs at some depth.

For continued progress, seismologists will of course need more data and better data coverage, while mineral physicists improve experimental and computational techniques to understand Fe alloys under inner core conditions. Geophysicists will have to work together on how to best parameterize seismic inversions, for instance, how to capture the sharpness of depth and lateral variations. With seismic inferences and mineral physics inputs still evolving, geodynamicists are faced with trying to hit a moving target. Nevertheless, Fig. 7.33 shows an example of an attempt to combine texturing mechanisms, with inputs from mineral physics, to model seismic elastic anisotropy. In this example, topographical relaxation due to preferential equatorial growth (Fig. 7.12B) is considered with and without a solidification pretexture (Fig. 7.7), and with and without stable stratification (Fig. 7.10). The study finds the best fit to seismic data when there is a solidification pretexture that is modified by topographical relaxation, and when stable stratification is present, because the stable stratification confines the topographical relaxation to modify and weaken the solidification pretexture in the upper inner core, while burying the solidification pretexture deeper in the inner core. Such models could be expanded to include other sources of stress/strain, different deformation mechanisms, annealing, or translation or mantle-driven inner core flow that might explain hemispherical variations.

We would be remiss if we did not mention another means of texturing, known as a shape preferred orientation (SPO). SPO arises from a medium that is layered or composed of structures elongated in 1D or 2D. Wave theory shows that because of the anisotropic shapes' influence on the stress and strain fields, such a medium can result in a large-scale elastic anisotropy even if the material making up the medium is elastically isotropic. In geology, shales often exhibit an SPO. SPO has been suggested for the inner core (Singh et al., 2000), where the anisotropic structures could be elongated liquid pockets such as one finds between dendrites. The required 3%–10% liquid fraction, however, is unrealistically high, since the liquid is likely to be buoyant and squeezed out of the inner core. Nevertheless, SPO remains a perhaps underexplored possible cause for the large-scale inner core elastic anisotropy.

7.6.2 Attenuation and isotropic heterogeneity

Seismic attenuation can be intrinsic, i.e., due to anelasticity, which results in conversion of elastic energy to heat, or due to scattering, in which elastic energy is conserved, but in which seismic waves are reflected off interfaces with an impedance contrast, resulting in a coda. The two types of attenuation can act in parallel. The low values for the shear modulus μ and high values for attenuation Q^{-1} now found for both long wavelength normal modes as well as for body waves, even at great depth into the inner core, suggest that a substantial fraction of the attenuation is intrinsic (see Chapter 6). Nevertheless there could still be considerable attenuation due to 1–10 km size scatterers in the upper inner core, which could perhaps be consistent with smaller, unannealed crystals in that region.

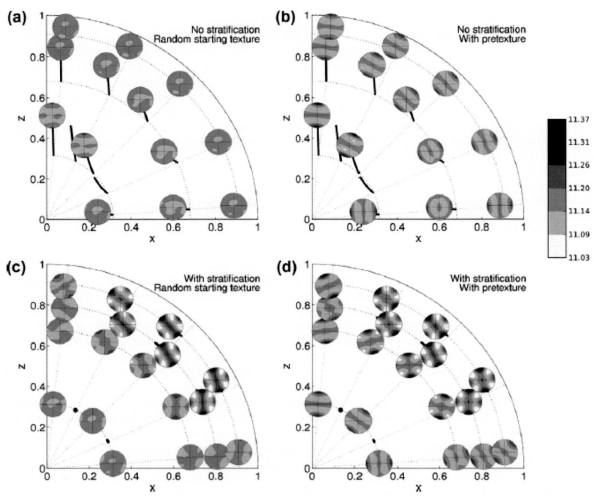

FIG. 7.33 A section of a meridional plane in the inner core showing present day P wave anisotropy for topographical relaxation without (A, B) and with (C, D) stable stratification, and without (A, C) and with (B, D) a solidification pretexture. Velocity scale in km/s. The calculated anisotropy assumes dislocation creep in hcp Fe, primary basal slip with secondary prismatic and pyramidal slip, and elastic constants similar to those in Fig. 2.27B. *From Deguen, R., Cardin, P., Merkel, S., Lebensohn, R.A., 2011. Texturing in Earth's inner core due to preferential growth in its equatorial belt. Phys. Earth Planet. Inter. 188, 173–184, where further details can be found.*

As we have also seen in Chapter 6, a seismic attenuative anisotropy has also been inferred, with the fast, spin axis direction being more attenuating. The fast direction being more attenuating is opposite to what is observed in the mantle, and to what one would expect from most materials with anisotropic elastic constants. One might again argue that this could be due to anisotropic scattering such as one observes in ultrasonic laboratory studies in columnar crystals, but the attenuative anisotropy is now seen in normal modes as well as body waves. The cause for attenuative anisotropy thus remains a mystery.

As with elastic anisotropy, there are indications attenuation, attenuative anisotropy, and even isotropic seismic velocity may have hemispherical and lateral variations. The cause(s) is unknown, perhaps it is persistent mantle-controlled outer core fluid flow affecting rates of inner core solidification and/or deformation. Clearly, we have reached the frontier.

7.7 Summary

- Definitive interpretation of the seismic inferences of inner core elastic anisotropy has been the prize that has driven much of the research on inner core dynamics, which requires an understanding of solidification, deformation, and annealing.

- We discuss directional solidification in the laboratory and in the core, including the resulting microstructure and solidification texture. We also examine other growth processes: inner core translation, the "snowing" core model, and the nucleation problem.
- Post solidification, the inner core is subject to stress/strain that could be due to solid-state convection or a variety of other geodynamic sources.
- Stress/strain results in deformation, through either dislocation or diffusion creep, which we review in detail, including the important topic of slip systems. Depending on the operative deformation mechanism a deformation texture can result.
- At the high temperatures of the core, deformed materials may anneal through the competing processes of recovery, and recrystallization and grain growth.
- The geodynamics and materials processes of solidification, deformation, and annealing control the inner core grain size. The grain size and stress, along with the stable phase, self-diffusivity, shear modulus, and available slip systems of Fe, determine the deformation mechanism and hence the solid-state viscosity. Because the viscosity in part controls the geodynamics, and material properties are still uncertain, at present the best one can say is the inner core dynamic viscosity likely lies in the range 10^{15}–10^{18} Pa s, though the low inner core shear modulus leads some to prefer an even lower value.
- An explanation for inner core elastic anisotropy remains elusive. Neither solidification nor deformation or, if the latter, no source of stress/strain, has emerged as a sole viable cause. Rather, it seems plausible that a combination of core evolution processes will be required to interpret the seismic inferences, but such modeling will also require continued progress in seismology and mineral physics.

References

Alboussiere, T., Deguen, R., Melzani, M., 2010. Melting-induced stratification above the Earth's inner core due to convective translation. Nature 466, 744–747.

Aubert, J., Amit, H., Hulot, G., Olson, P., 2008. Thermochemical flows couple the Earth's inner core growth to mantle heterogeneity. Nature 454, 758–761.

Bergman, M.I., 1998. Estimates of the Earth's inner core grain size. Geophys. Res. Lett. 25, 1593–1596.

Bergman, M.I., Agrawal, S., Carter, M., Macleod-Silberstein, M., 2003. Transverse solidification textures in hexagonal close-packed alloys. J. Cryst. Growth 255, 204–211.

Brennan, M.C., Fischer, R.A., Couper, S., Miyagi, L., Antonangeli, D., Morard, G., 2021. High-pressure deformation of iron–nickel–silicon alloys and implications for Earth's inner core. J. Geophys. Res. Solid Earth 126. https://doi.org/10.1029/2020JB021077, e2020JB021077.

Buffett, B.A., 1997. Geodynamic estimates of the viscosity of the Earth's inner core. Nature 388, 571–573.

Buffett, B.A., 2009. Onset and orientation of convection in the inner core. Geophys. J. Int. 179, 711–719.

Buffett, B.A., Bloxham, J., 2000. Deformation of Earth's inner core by electromagnetic forces. Geophys. Res. Lett. 27, 4001–4004.

Buffett, B.A., Wenk, H.-R., 2001. Texturing of the Earth's inner core by Maxwell stresses. Nature 413, 60–63.

Deguen, R., 2012. Structure and dynamics of Earth's inner core. Earth Planet. Sci. Lett. 333–334, 211–225.

Deguen, R., Alboussiere, T., Brito, D., 2007. On the existence and structure of a mush at the inner core boundary of the earth. Phys. Earth Planet. In. 164, 36–49.

Gubbins, D., Masters, G., Nimmo, F., 2008. A thermochemical boundary layer at the base of Earth's outer core and independent estimate of core heat flux. Geophys. J. Int. 174, 1007–1018.

Gubbins, D., Sreenivasan, B., Mound, J., Rost, S., 2011. Melting of the Earth's inner core. Nature 473, 361–363.

Gubbins, D., Alfe, D., Davies, C.J., 2013. Compositional instability of Earth's solid inner core. Geophys. Res. Lett. 364, 37–43.

Hellawell, A., Herbert, P.M., 1962. The development of preferred orientations during the freezing of metals and alloys. Proc. R. Soc. Lond. A 269, 560–573.

Huguet, L., van Orman, J.A., Hauck, S.A., Willard, M.A., 2018. Earth's inner core nucleation paradox. Earth Planet. Sci. Lett. 487, 9–20.

Humphreys, F.J., Hatherly, M., 1996. Recrystallization and Related Annealing Phenomena. Pergamon, New York.

Jeanloz, R., Wenk, H.-R., 1988. Convection and anisotropy of the inner core. Geophys. Res. Lett. 15, 72–75.

Karato, S.-I., 1999. Seismic anisotropy of the Earth's inner core resulting from flow induced by Maxwell stresses. Nature 402, 871–873.

Karato, S.-I., 2008. Deformation of Earth Materials. Cambridge Press, Cambridge.

Kurz, W., Fisher, D.J., 1992. Fundamentals of Solidification, third ed. Trans Tech, Switzerland.

Lincot, A., Merkel, S., Cardin, P., 2015. Is inner core seismic anisotropy a marker of plastic flow of cubic iron? Geophys. Res. Lett. 42, 1326–1333. https://doi.org/10.1002/2014GL062862.

Lincot, A., Cardin, P., Deguen, R., Merkel, S., 2016. Multiscale model of global inner-core anisotropy induced by hcp alloy plasticity. Geophys. Res. Lett. 43, 1084–1091. https://doi.org/10.1002/2015GL067019.

Loper, D.E., Roberts, P.H., 1981. A study of conditions at the inner core boundary of the earth. Phys. Earth Planet. In. 24, 302–307.

Merkel, S., Wenk, H.-R., Gillet, P., Mao, H.-K., Hemley, R.J., 2004. Deformation of polycrystalline iron up to 30 GPa and 1000 K. Phys. Earth Planet. In. 145 (1–4), 239–251. https://doi.org/10.1016/j.pepi.2004.04.001.

Merkel, S., Gruson, M., Wang, Y., Nishiyama, N., Tome, C.N., 2012. Texture and elastic strains in hcp-iron plastically deformed up to 17.5 GPa and 600 K: experiment and model. Model. Simul. Mater. Sci. Eng. 20, 024005. https://doi.org/10.1088/0965-0393/20/2/024005.

Miyagi, L., Kunz, M., Knight, J., Nasiatka, J., Voltolini, M., Wenk, H.-R., 2008. In situ phase measurement and deformation of iron at high pressure and temperature. J. Appl. Phys. 104, 103510. https://doi.org/10.1063/1.3008035.

Mullins, W.W., Sekerka, R.F., 1963. Morphological stability of a particle growing by diffusion of heat flow. J. Appl. Phys. 34, 323–329.
Poirier, J.-P., Langenhorst, F., 2002. TEM study of an analogue of the Earth's inner core ε-iron. Phys. Earth Planet. In. 129 (3–4), 347–358. https://doi.org/10.1016/S0031-9201(01)00300-4.
Poirier, J.-P., Price, G.D., 1999. Primary slip system of ε-iron and anisotropy of the Earth's inner core. Phys. Earth Planet. In. 110 (3–4), 147–156. https://doi.org/10.1016/S0031-9201(98)00131-9.
Ritterbex, S., Tsuchiya, T., 2020. Viscosity of hcp iron at Earth's inner core conditions from density functional theory. Sci. Rep. 10, 6311. https://doi.org/10.1038/s41598-020-63166-6.
Roberts, P.H., 1988. Future of geodynamo theory. Geophys. Astrophys. Fluid Dyn. 44, 3–31.
Shimizu, H., Poirier, J.-P., Le Mouel, J.-L., 2005. On crystallization at the inner core boundary. Phys. Earth Planet. In. 151, 37–51.
Singh, S.C., Taylor, M.A.J., Montagner, J.-P., 2000. On the presence of liquid in Earth's inner core. Science 287, 2471–2474.
Stixrude, L., Cohen, R.E., 1995. High-pressure elasticity of iron and anisotropy of the Earth's inner core. Science 267, 1972–1975.
Sumita, I., Yoshida, S., Kumazawa, M., Hamano, Y., 1996. A model for sedimentary compaction of a viscous medium and its application to inner core growth. Geophys. J. Int. 124, 502–524.
Takehiro, S., 2011. Fluid motions induced by horizontally heterogeneous joule heating in the Earth's inner core. Phys. Earth Planet. In. 184, 134–142.
Vocadlo, L., 2007. Mineralogy of the earth—The Earth's core: iron and iron alloys. In: Schubert, G. (Ed.), Treatise on Geophysics. Mineral Physics, vol. 2. Elsevier, Amsterdam, pp. 91–120.
Weeks, W.F., Gow, A.J., 1980. Crystal alignments in the fast ice of Arctic Alaska. J. Geophys. Res. Oceans 85, 1137–1146.
Wenk, H.-R., Takeshita, T., Jeanloz, R., Johnson, G.C., 1988. Development of texture and elastic anisotropy during deformation of hcp metals. Geophys. Res. Lett. 15, 76–79. https://doi.org/10.1029/GL015i001p00076.
Wenk, H.-R., Matthies, S., Hemley, R.J., Mao, H.-K., Shu, J., 2000. The plastic deformation of iron at pressures of the Earth's inner core. Nature 405, 1044–1047.
Yamazaki, D., Tsujino, N., Yoneda, A., Ito, E., Yoshino, T., Tange, Y., Higo, Y., 2017. Grain growth of ε-iron: implications to grain size and its evolution in the Earth's inner core. Earth Planet. Sci. Lett. 459, 238–243.
Yoshida, S., Sumita, I., Kumazawa, M., 1996. Growth model of the inner core coupled with outer core dynamics and the resulting elastic anisotropy. J. Geophys. Res. 101, 28085–28103.

Further reading

Solidification

Porter, D.A., Easterling, K.E., 1992. Phase Transformations in Metals and Alloys, second ed. Chapman and Hall, London (A thorough and readable introduction to solidification).

Deformation

Courtney, T.H., 1990. Mechanical Behavior of Materials. McGraw-Hill, New York (One of many general introductions to the subject).
Karato, S.-I., 2008. Deformation of Earth Materials. Cambridge Press, Cambridge (A very detailed monograph, though more focused on the mantle).
Poirier, J.-P., 1985. Creep of crystals. Cambridge Press, Cambridge (A great title. Also a good book, though very little on Fe under inner core conditions).
Wang, Y.N., Huang, J.C., 2003. Texture analysis in hexagonal materials. Mater. Chem. Phys. 81, 11–26 (A useful review of textures in hcp materials).
Weertman, J., Weertman, J.R., 1992. Elementary Dislocation Theory. Oxford Press, Oxford (A nicely illustrated little book).

Annealing

Humphreys, F.J., Hatherly, M., 1996. Recrystallization and Related Annealing Phenomena. Pergamon, New York (Very comprehensive).

Reviews

Deguen, R., 2012. Structure and dynamics of Earth's inner core. Earth Planet. Sci. Lett. 333–334, 211–225 (Highly recommended for its clarity).
Sumita, I., Bergman, M.I., 2015. Inner core dynamics. In: Schubert, G. (Ed.), Treatise on Geophysics, second ed. Core Dynamics, vol. 8. Elsevier, Amsterdam, pp. 297–316 (Significant overlap with this chapter, but with a more extensive reference list).

CHAPTER

8

Formation and evolution of the core

In this chapter, we present current theories on how the Earth and its core were formed, and how the core evolved between the time of formation and the present day. We begin by summarizing the enormous amounts of energy expended in Earth's formation. We briefly discuss the genesis of core-forming elements and the origin and early evolution of the solar system, setting the stage for a discussion of Earth's accretion, core formation models, and the geochemical evidence behind them. We argue that the earliest history of the core was largely over written by the catastrophic Moon-forming event. The next part deals with evolution of the core after its formation, with a focus on the timing of inner core nucleation. We then consider the long-term history of the geodynamo, including dynamo initiation.

8.1 Formation of the core

The origin of the core, like the origin of the Earth itself, has been the object of conjecture seemingly forever. It is fair to say no other subject involving the deep Earth has undergone more revisions. And as new information comes available on the core, the early Earth, and the start of the solar system, there is every prospect for still more revisions.

Why is this subject so volatile? The fundamental reason is that, although the formation of the Earth and its core were primarily astrophysical and geophysical events, much of the critical data on these events is cosmochemical and geochemical in nature, derived from the compositions of the Sun, meteorites, and the other planets, in addition to the Earth. Furthermore, these data are almost entirely after-the-fact. It is a very daunting challenge to correctly interpret cosmochemical and geochemical data in terms of physical processes that happened long ago, some of which are no longer active in the solar system. In other words, although the chemical evidence offers vital constraints, it does not provide a unique picture of what went on four-and-one-half billion years ago. And because there are many possible ways to interpret the chemical data, there are many possible scenarios by which the Earth and its core could have formed. Coupled with the fact that astronomical observations of distant planetary systems are just beginning, it seems inevitable that our ideas about the early Earth will be in a state of flux for the foreseeable future.

In response to this situation, theory-based models of planetary formation have played outsized roles in guiding our ideas about how the Earth's core formed. It is beyond the scope of this book to describe how and why these theories came to prominence, or to fully dissect their contents. Instead, we refer the reader to excellent review articles where these details can be found. Similar comments apply to the chemical and isotopic evidence, as well as to core formation mechanics. Our approach in this chapter is to summarize the theories and the evidence along with their implications, and leave the full discussion of their methods, uncertainties, and subtle caveats to more specialized reviews.

8.1.1 Energetics of Earth formation

The simplest way to characterize the physical setting under which the core formed is by considering the energetics of the Earth during and shortly after accretion. As the illustration in Fig. 8.1 implies, this was an extremely disruptive time in Earth history. We can estimate the major energy sources and losses during Earth's formation by applying the results of terrestrial planet accretion models (Chambers, 2016; Nimmo et al., 2018; Carter et al., 2020). Fig. 8.2 summarizes the most important contributors to Earth's energy budget at that time, according to these models.

The single largest source of energy is the kinetic energy of the matter that went into building the Earth. Labeled as *accretion energy* in Fig. 8.2 and sometimes referred to as *binding energy*, it is based on the kinetic energy of accreting

FIG. 8.1 Artist's conception of the accretion of the proto-Earth, showing the early Sun, the solar nebula, and impactors of all sizes.

matter landing on the growing proto-Earth, assuming that matter started from rest infinitely far away. The second-largest energy source is core formation, the release of gravitational potential energy as core-forming metals segregate from mantle silicates and descend into the proto-core. Two additional but smaller energy sources are shown in Fig. 8.2, representing the sensible heat content of the impactors and heat production from short-lived radionuclides, primarily aluminum-26 and iron-60. Here, *sensible heat content* has the same meaning as in Chapter 2, namely, the heat content of a body proportional to its temperature. There is possibly some double counting involved in these last two sources, because it is almost certain that some of the heat attributed to the short-lived radioactive sources was deposited into the impactors, contributing to their sensible heat content.

In terms of energy losses, heat radiated from the surface or from the atmosphere back into space is by far the largest item in this category. Another significant energy that appears in the loss category in Fig. 8.2 is the reduction in kinetic energy of Earth's rotation due to *tidal friction*. Tidal friction losses were probably very high in the immediate aftermath of Moon formation, when the Moon was far closer to Earth than today. A possible role for tidal deformation and tidal energy loss in homogenizing the core and initiating dynamo action is discussed later in this chapter.

Finally, the two most important forms of sequestered energy, that is, energy stored for an appreciable time after Earth's accretion ended, are the sensible heat content of the mantle and core plus their latent heat content due to melting. Here, we are assuming that the whole of the proto-Earth's interior, including its core, was entirely molten, at least during one stage of the accretion process, and perhaps multiple times. We have not included the contribution from long-lived radioactive heating in this tally of sequestered sources, as this heat has been produced continuously over the whole age of the Earth, and has its greatest effects on the long-term, postaccretion evolution of the core.

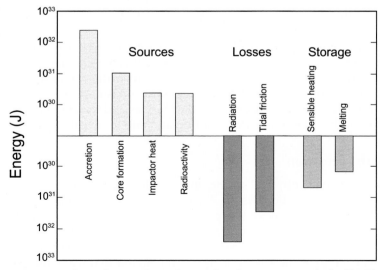

FIG. 8.2 Tabulation of major energy sources (inputs), energy losses (outputs), and energy storage during Earth's accretion, all measured in Joules.

Perhaps the most significant implication of the energy balance in Fig. 8.2 is that the inputs and outputs of power were orders of magnitude higher during and shortly after the Earth and its core were forming, compared to the present day. For example, if we assume a nominal value of 100 Myr for the duration of the energy source and loss processes in Fig. 8.2, then the proto-Earth system absorbed and surrendered nearly a million terrawatts of power over that interval of time. Most of this power did not affect the core directly, but even if a fraction of 1% did, it implies that the rate of energy flow into and out of the proto-core was measured in petawatts (10^{15} Watts), rather than terawatts (10^{12} Watts) today.

8.1.2 Genesis of the core-forming elements

As a prelude to a discussion of the events that led to the formation of Earth's core, we begin by stepping farther back in time, examining the origin and composition of the materials from which the planets grew, and identifying the important events in the earliest part of solar system history that governed the composition and structure of all the terrestrial planets, the Earth in particular. A good starting point is to consider the processes responsible for creating the elements in the periodic table (Fig. 8.3), their relative abundances in our solar system, and why some of these are abundant in the core.

Much of the solar inventory of the lightest elements, most abundantly hydrogen and helium but also some amounts of lithium and beryllium, were created in the immediate aftermath of the Big Bang (Cyburt et al., 2016). Based on the measurements of the *Hubble constant*, these primordial elements were formed at about 13.8 Ga, significantly earlier than the 4.567 Ga age assigned to our solar system based on uranium-lead systematics. The remaining naturally occurring elements (88 or 90; the exact number in this category depends on whether or not the exceedingly rare technetium and promethium are included) were synthesized later, in stellar interiors and during explosive *supernova* events. Stellar nucleosynthesis produces elements with atomic masses between carbon and iron; supernova and other types of cosmic explosions produce the elements with atomic masses ranging from iron to uranium and plutonium. Fragmentation of carbon, nitrogen, and oxygen by *cosmic ray spallation*, the breakup of nuclei by cosmic radiation, is the major natural source for boron and is also important for producing lithium and beryllium.

Our current understanding of the sequence of events in the first 1 Gyr following the Big Bang includes an initial phase of galaxy and star formation by radiation cooling, condensation, and gravitational accretion in the expanding

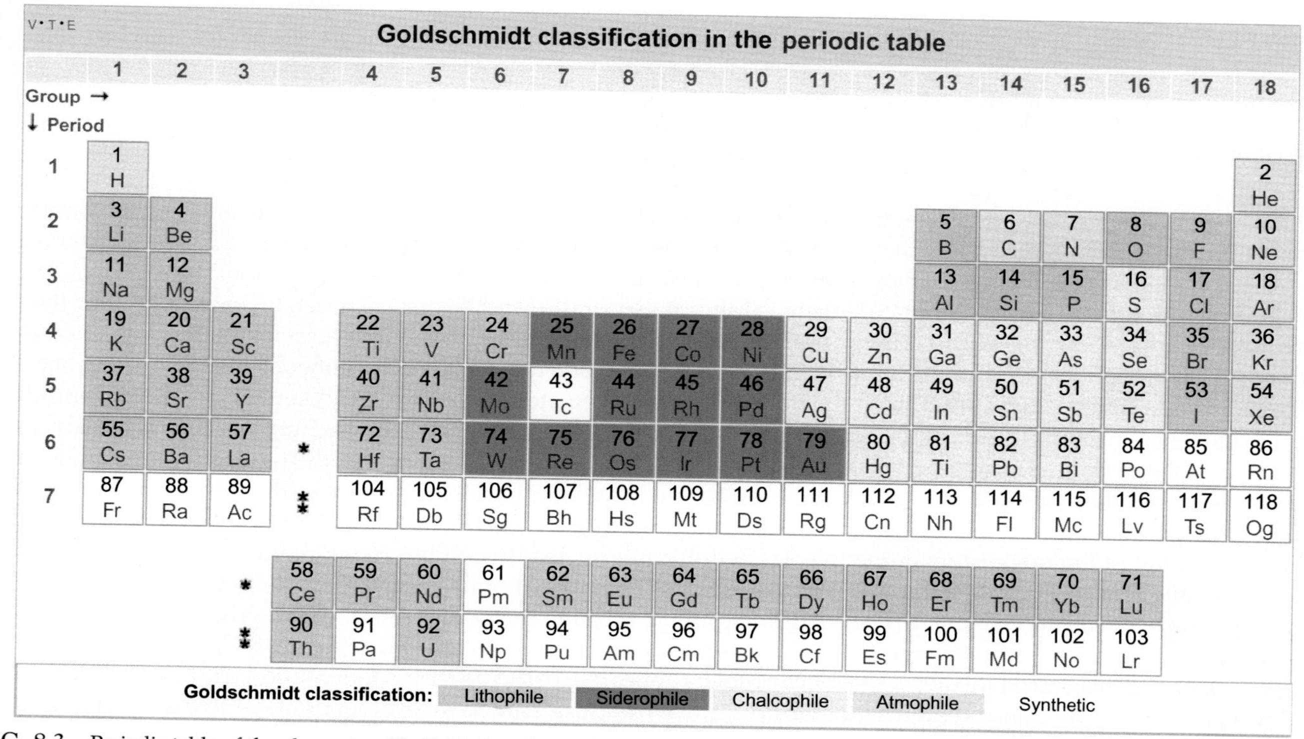

FIG. 8.3 Periodic table of the elements with Goldschmidt classifications indicated by shades.

universe (Stark, 2016). Fusion of hydrogen (H) to helium (He) dominated the interior of these early stars. The most massive and shorter lived of these first-generation stars followed an evolutionary path in which they eventually exploded as supernova, leaving behind masses of hot gas populated by all the naturally occurring elements but particularly enriched in H and He. Modifications of this sequence include the possibility of multiple generations of stellar fusion, super nova explosion, and more recently, the possibility of heavy element formation from the merging of massive neutron stars.

8.1.3 Origin, composition, and early evolution of the solar system

In the aftermath of a supernova explosion, much of the remaining matter is in the form of an extended gaseous *molecular clouds*, so named because they consist primarily of molecular hydrogen, in addition to all the other natural elements in smaller amounts. As the molecular cloud cools and contracts, in certain regions called *stellar nebula* a critical density is exceeded, whereupon gravity takes over, leading to further collapse and densification.

Through the combined action of random collisions among gas molecules and inward contraction, stellar nebula typically acquire a net angular moment about one axis, evolving from a diffuse shape into a more spatially compact, flattened structure that astronomers call a *protoplanetary disk*, as illustrated in Fig. 8.4. Through loss of angular momentum, matter converges toward the mass center of the protoplanetary disk, further increasing the density there. Once a critical density is reached and the central temperature is high enough, nuclear fusion begins and a central star is formed.

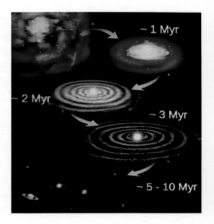

FIG. 8.4 Artist's conception of the early steps solar system formation, including collapse of a gas cloud to form the protoplanetary disk (a.k.a. the solar nebula), formation of the Sun and concentrations of dust in the nebular gas, planetesimal formation and their accretion to make the planets.

In planetary science, the protoplanetary disk from which the Sun and the rest of solar system emerged is commonly referred to as the *solar nebula*. Early in its evolution, the Sun acquired most of the solar nebula gas, and now contains 99.9% of the solar system by mass. The elemental abundances in the Sun are shown in Fig. 8.5. On a molar basis, hydrogen and helium far and away dominate, whereas lithium, beryllium, and boron are much less abundant, for the reasons given earlier. Starting with carbon, solar abundances decrease approximately exponentially with increasing atomic number, with the even atomic number elements forming one trend, the odd number elements forming another. Carbon, oxygen, silicon, and sulfur lie on the more abundant even atomic number trend, and all are major candidates for the light element composition of the core. The even atomic number trend continues to iron, which is relatively abundant in the solar system as a consequence of its anomalous binding energy. Elements with atomic numbers higher than iron are less abundant because the energy required to produce them increases rapidly with atomic number. For this reason, nickel is the only element heavier than iron that is a major constituent in the core.

The evolution of the matter remaining in solar nebula after nuclear fusion began was controlled by its interactions with the young Sun and how fast the solar nebula cooled. The first materials known to have condensed from the solar nebular gas are *calcium-aluminum inclusions* (CAIs), found in some meteorites. As these are the oldest known solar system solids, their uranium-lead age of 4.567 Ga is traditionally used as the starting time for planetary evolution.

As the solar nebula cooled further, a sequence of solids appeared, based on their condensation temperatures. Elements and compounds that condense at high temperature, termed *refractory*, tended to form solids closer to the Sun, while those that condense at lower temperature, termed *volatile*, tended to form solids further from the Sun. The resulting sequence of

FIG. 8.5 Element abundances in the Sun, normalized by silicon. Elements that are important for the core are highlighted. *Abundance data from Lodders, K., Palme, H., Gail, H.P., 2009. Abundances of the elements in the solar system. In: Trumper, J.E. (Ed.), Landolt-Bornstein, New Series, Astronomy and Astrophysics. Springer Verlag, Berlin, pp. 560–630.*

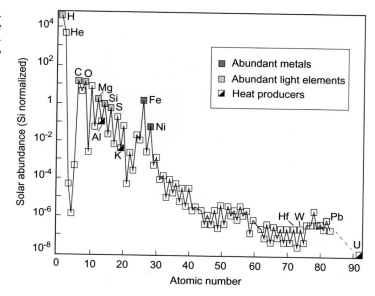

condensation, illustrated in Fig. 8.6, accounts for much of the variation in the bulk composition of the planets, including such basic differences as the larger size of iron-rich core of Mercury compared to the cores of Earth and Venus, the abundance of helium and hydrogen in the gas giants Jupiter and Saturn, and the abundance of low-temperature condensates in the ice giants Uranus and Neptune.

FIG. 8.6 The bulk composition of planets in the solar system is consistent with decreasing temperature with distance from the Sun in the protoplanetary disk, which controls the ability of elements and compounds to condense.

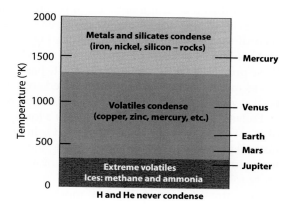

Most of the solids that condensed from the solar nebula began as minute dust particles, generally in the 1–10 μm size range. Astronomical observations of nebula surrounding young solar-mass stars (Ansdell et al., 2017) suggest that the mass ratio of gas to dust in the solar nebula was likely in the range of 10/1, and perhaps as large as 100/1. In other words, the planets of the solar system, including the Earth and it core, grew from a dusty yet gas-rich environment. This dust consisted of all of the naturally occurring elements; the abundant ones making up the core are highlighted in Fig. 8.5.

8.1.4 Evidence from meteorites

As currently understood, the process of forming terrestrial planets from nebular dust involves multiple steps and several different pathways. Key evidence for these steps comes from meteorites derived from the asteroid belt, whose parent bodies were formed in the earliest stages of solar system history. Other evidence comes from lunar samples and from Martian meteorites.

FIG. 8.7 Meteorites specimens: (A) Allende carbonaceous chondrite, with spheroidal chondrules and calcium-aluminum inclusions (CAIs). (B) Acomita pallasite (stony iron), with olivine crystals in an iron-nickel matrix. *From the Institute of Meteoritics, University of New Mexico collection, with permission.*

A large number of mineralogical types of meteorites are recognized (Brearley and Jones, 1998; Mittlefehldt et al., 1998) but for core formation, only four are of primary importance. The first type are *chondrites*, mixtures of dust, metal, and glass from the solar nebula, with various silicates dominating their mineralogy. Most chondrites include small, nearly spherical structures called *chondrules*, thought to have been formed through rapid quenching from a molten state in the presence of nebular gas. Refractory CAI inclusions are found in chondrites, and some also have abundant carbon and are called *carbonaceous chondrites*. A famous example of a carbonaceous chondrite is shown in Fig. 8.7A. A second type, the *achondrites* are silicate-bearing meteorites similar in appearance and chemistry to ultramafic igneous terrestrial rocks.

The third common type, metallic meteorites or *irons*, are rich in nickel as well as iron, and sometimes contain iron sulfides. Several metallic meteorites are shown in Fig. 7.6. Compelling evidence indicates that most iron meteorites were produced by fragmentation of parent bodies that had already differentiated into silicate and metal-rich layers. Furthermore, this differentiation occurred very early in solar system history, generally within 1 Myr of CAI formation, according to the radiometric dating (see Fig. 8.8). Mechanisms responsible for this early differentiation include impacts and heating by decay of aluminum-26, a radioactive isotope with a half-life of just 0.73 Myr. The size distribution of these previously differentiated meteorite parent bodies is not known, but estimates based on the required gravitational forces range from a few tens to hundreds of kilometers in diameter. In addition, we know that some of these parent bodies were magnetized, a fact that bears on the age of the geodynamo, an issue we address later in this chapter.

Pallasites are interesting members of a fourth meteorite type, the *stony irons*. As shown in Fig. 8.7B, pallasites consist of large crystals of olivine $(Mg, Fe)_2SiO_4$ in an iron-nickel matrix, iron-nickel being the continuous (physically connected) phase, the olivine crystals being the discontinuous phase. Pallasites may represent the core-mantle boundary region of differentiated parent bodies, or alternatively, fragments assembled from a previously differentiated parent body following a destructive impact.

In contrast to iron meteorites, most chondrules formed a bit later, generally 2–4 Myr after CAIs, as indicated in Fig. 8.8. By this time, solar nebular dust had accumulated into dense clumps, forming not just chondrules but also larger aggregates called *pebbles*, and still larger bodies called *planetesimals*. What steps were involved in getting dust to clump and for the clumps to grow?

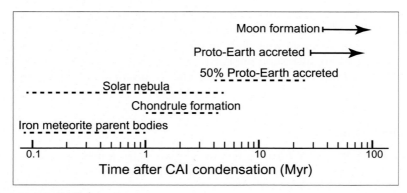

FIG. 8.8 Time line of major events in Earth formation.

The formation of millimeter and centimeter-sized pebbles relies on dust particles sticking together. This process is easier far out in the solar nebula, beyond the so-called *snow line*, where the presence of various ices allow dust grains to adhere to one another. The orbital radius of the proto-Earth mostly lay inside the snow line, where grain sticking is more problematic, although still possible because of electrostatic and surface forces. A bigger problem is how to grow from pebbles to planetesimals of 1 km or larger, objects massive enough to begin to attract material with their own gravity, before the pebbles fragment or spiral into the Sun. This problem is called the *meter-sized barrier*.

Theory and experiments indicate that solar nebula gas and dust were subject to a variety of processes that allowed pebbles to grow across the meter-sized barrier. One of these, the so-called *gravitational instability* is essentially a finite amplitude compression wave made unstable by gravitational forces acting between the compressed and decompressed parts of the wave. Through the action of this instability, pebbles become concentrated in the wave compressions, allowing them to coagulate there. Nebular solids sorted into rings by gravitational instability are illustrated schematically in Fig. 8.4. In addition, there are several types of *streaming instabilities* that can draw pebbles into the wake of an already-formed pebble cluster. Once sufficiently concentrated, pebble clusters will collapse by mutual gravitational attraction to form a new planetesimal. Although very initially weak, planetesimal gravity is strong enough to attract nearby dust and pebbles, affording some the opportunity to grow further. As they continue to grow, gravitational differentiation occurs within some of the larger planetesimals, with the denser metals sinking to the center and the lighter silicates and oxides forming a mantle. The differentiation of these early-formed objects mimics some of the later processes involved in Earth's core formation, but at much smaller scale.

As solar system objects began to increase in size, a complex sequence of ever more violent collisions and orbital disruptions occurred that ultimately resulted in the current distribution of terrestrial (Earth-like) planets close to the Sun, with the gas giants and ice giants further out, all orbiting in nearly the same plane with generally low eccentricities. A major hiccup in this arrangement may have occurred within the first 10 Myr, when Jupiter executed its so-called *Grand Tack*. According to the Grand Tack hypothesis (Walsh et al., 2011), Jupiter first accreted around 3.5 AU (1 AU = Earth's mean orbital radius) then migrated in toward the Sun, reaching perhaps 1.5 AU before reversing direction and migrating out to its present location at 5.2 AU. Meanwhile, the remnants of the solar nebula were being swept away by the action of the *solar wind*, a plasma consisting of electrons, protons, and α-particles (helium-4 nuclei), and also through photoevaporation by the elevated amounts of *XUV* (extreme ultraviolet radiation) emanating from the young Sun, leaving a near vacuum between the formed planets, as depicted in Fig. 8.4.

If this sequence of events is correct, then why are the Earth and other terrestrial planets depleted in hydrogen and helium, compared to the giant planets? One explanation is that nebular gas was driven from the inner solar system prior to terrestrial planet formation. We know from astronomical observations that the lifetime of the gas in a protoplanetary disk is short, typically just a few million years, although in a small minority gas persists for as long as 10 Myr (Williams and Cieza, 2011). In order to acquire a hydrogen-helium atmosphere of its own, a planet must grow to about a quarter of Earth's present-day mass (i.e., $\sim 0.25 M_{EAR}$) while immersed in nebular gas (Ikoma and Genda, 2006). As discussed here, such a fast initial growth rate is possible for the proto-Earth, although the evidence indicates that it took far longer for the Earth to fully form. And assuming it had captured a nebular-type atmosphere, the Earth is not massive enough to have retained such an atmosphere for very long, faced with photoevaporation by the XUV coming from the young Sun.

8.1.5 Terrestrial planet formation models

The most widely used model for terrestrial planet formation highlights the role of accreting planetesimals (Chambers, 2004). According to this model, increasingly frequent planetesimal collisions occurred as these objects began to populate the protoplanetary disk, leading to a stage of runaway growth for some. Out of a far larger initial population, a small minority of plantesimals grew by collisions to sizes of 10^2-10^3 km. In the next stage, termed *oligarchic growth*, a few hundred of the largest planetesimals continued their growth, accreting smaller planetesimals from nearby regions of the protoplanetary disk. The outcome of this stage is several dozen larger objects called *planetary embryos*. Mars is often cited as an example of a planet that halted its accretion at the embryo stage. The final accretion stage is a period in which planetary embryos merge via giant collisions, yielding just a handful of terrestrial planets. The Moon-forming impact (assuming it indeed happened) was presumably the last such giant impact in Earth's accretion history. Although the exact time scale for each step in this model is imprecise, planetesimals are hypothesized to grow to substantial size in about 1 Myr, as indicated in Fig. 8.8. The next two stages require far more time, and in the case of the Moon-forming impact, the evidence indicates it occurred at least 30 Myr after CAI formation, and quite possibly long after that.

There is an emerging competitor for the planetesimal model, called *pebble accretion*. Originally applied to the formation of the rocky cores of gas giant planets (Lambrechts and Johansen, 2012), it has recently been extended to encompass terrestrial planet accretion (Levison et al., 2015; Johansen et al., 2021). This model is motivated by astronomical observations suggesting that protoplanetary disks are filled with pebbles, and is also motivated by the need to circumvent the meter-sized barrier.

The pebble accretion model begins with an initial stage similar to the planetesimal model: dust grains in the solar nebula become concentrated by adhesion, forming pebble aggregates the size of millimeters, centimeters, or a bit larger (Nimmo et al., 2018). At this stage, according to the pebble accretion model, the protoplanetary disk is dominated by pebbles, with only a few large objects present. However, in the presence of nebular gas, pebbles are readily captured by the gravitational attraction of the few large objects, which act as seeds. The aerodynamic drag that the nebular gas exerts on pebbles prevents their escape from the gravity well of the larger object. Through this mechanism, a small number of isolated planetesimals or planetary embryos can rapidly and efficiently capture pebbles in their neighborhoods, thereby allowing them to grow faster than if they had to rely solely on random collisions with other planetesimals.

Supposing the proto-Earth grew mostly by pebble accretion, it could have accreted as much as 60% of its mass within the lifetime of the solar nebula (Johansen et al., 2021). In contrast, models of growth dominated by planetesimal accretion generally find that 20–40 Myr is needed to reach this stage (Raymond et al., 2014; Chambers, 2016). By that time the solar nebula was probably long gone, and with it, the mechanism to efficiently capture pebbles. For the core, the consequences of short versus long accretion times are twofold: First, rapid accretion implies greater rates of energy dissipation and hence higher temperatures in the proto-core. Second, rapid accretion is more likely to incorporate volatiles from the solar nebula into the proto-core. With slower accretion, volatiles such as hydrogen (needed for water) and the noble gases would mainly have come from meteorites and comets during a *late addition* phase, after core formation was essentially complete.

8.1.6 Core formation processes

Two sets of generic hypotheses that are often used as end members in discussions of core formation are called *single-stage models* and *multistage models*, respectively. Traditional single-stage models start with a heterogeneously accreted Earth, a mixture of metals and silicates, presumably close to carbonaceous chondrites in its bulk composition. According to these models, depending on relative melting temperatures and the chemical equilibration between elements and compounds, iron-rich melts separated from silicates and sank to form the core in a single event. Here, the core formation event is usually assumed to have occurred either after accretion or near its end.

In contrast, multistage models suppose a sequence of core-forming events in time, affecting both the impactors and the proto-Earth. Consequently, these models have many variants, involving different assumptions about the history of accretion, impactor sizes, compositions, temperatures, and the ability of impactors to melt the proto-Earth. A popular version of the multistage model for Earth's core formation assumes continuous accretion, typically at a rate that decreases exponentially with time, plus a final Moon-forming event at a later time. Most of these continuous accretion models also assume that core formation kept pace with the accretion process, so that the proto-core grew in step with the growth of the proto-Earth. The pros and cons of various core formation models are discussed in the AGU monograph edited by Badro and Walter (2015) and in a review article by Rubie et al. (2015) listed at the end of this chapter.

The physical separation of core-forming metals from mantle-forming silicates can happen through a variety of mechanisms, including (1) percolation at relatively low temperatures, in which the silicates are solid but the metals are molten and able to flow between silicate grains; (2) diapirism at intermediate temperatures, with large metal diapirs (e.g., iron-rich blobs) sinking through partially molten silicate; or (3) iron rain at high temperatures, with fragmented iron droplets precipitating through a molten silicate magma ocean. The first mechanism, percolation, finds its strongest application to core formation in planetesimals and at shallow depths in the proto-Earth; the second two mechanisms are strong candidates for silicate-metal segregation deep in the proto-Earth.

Among the various metal-silicate separation mechanisms, percolation of molten metals through a solid silicate-oxide matrix is perhaps the best studied. Percolation requires the solid silicate phase to be compressible, so it will compact as the liquid metal drains out, and it also requires the liquid metal to be continuous, that is, to reside within interconnected channels between silicate grains. This latter requirement is satisfied if the wetting angle between solid grain boundaries in contact with the melt is less than 60 degrees, as shown in Fig. 8.9. Equilibrium wetting angles depend on the compositions of both solid and liquid, along with temperature and pressure, all of which affect the surface energy of the solid-liquid interface. Much experimental work has gone into measuring wetting angles in metal-silicate systems, and the overall picture is that during core formation, percolation is generally efficient at low pressures

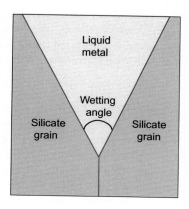

FIG. 8.9 The equilibrium wetting angle between solid crystalline grains and a liquid is determined by the surface energy at the liquid solid interface. Small wetting angles are needed in order for liquid iron to percolate between solid silicate grains.

(<3 GPa), becomes far less efficient at intermediate pressures (3–25 GPa), but increases again at higher pressures, particularly where silicate perovskite is the solid phase (Shannon and Agee, 1998; Takafuji et al., 2004; Shi et al., 2013).

The isotopic evidence summarized in the following section indicates that partial equilibration was established between mantle-forming silicates and core-forming metals during Earth's accretion, before the core was fully formed. This is an important constraint on core formation models and the metal-silicate separation mechanisms listed earlier. Establishing metal-silicate equilibrium through iron percolation seems rather easy. The same goes for iron rain in a magma ocean, as might occur during pebble accretion. But is metal-silicate equilibrium likely with planetesimal and planetary embryo accretion? In other words, do we expect that partial metal-silicate equilibrium is established following large impacts, when the impactor already has a massive preformed core? In order for chemical and isotopic equilibrium to occur, large volumes of metals derived from the impactor core must mix with molten silicates down to very small scales. This probably would occur only in a magma ocean environment, in which case the equilibrium must be established quickly. Accordingly, a more precise question is: What physical mechanism can reduce the size of a preformed impactor core by many orders of magnitude in a magma ocean, and do so in short order?

The answer to this question lies in the extreme levels of turbulence that result from such an impact. A large impact generates turbulence in a magma ocean characterized by Reynolds numbers $Re = vR/\nu$ of order 10^{10} or larger, where v, R, and ν are the characteristic sinking velocity of the impactor core, radius of that core, and kinematic viscosity of the magma, respectively. Since turbulence typically sets in at $Re \sim 10^3$, large impactor cores generate very intense turbulence in magma oceans.

The action of this intense turbulence shown in Fig. 8.10 is closely analogous to the classical phenomenon entrainment in a turbulent thermal, in which the rate of entrainment of the ambient fluid into a negatively buoyant turbulent blob is proportional to the descent speed of the blob. Both theory (Dahl and Stevenson, 2010) and experiments (Deguen et al., 2014) predict that the effective radius of a turbulent metal blob R grows with depth of descent z according to $R = R_0 + \alpha z$, where

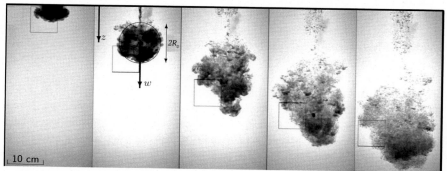

FIG. 8.10 A fluid dynamics experiment simulating metal-silicate turbulent mixing and fragmentation in a magma ocean. Time sequence (left to right) shows the turbulent development and fragmentation of a nearly spherical blob of dense NaI salt solution, representing core-forming metals, falling in less dense silicone oil, representing silicate magma. R_0 is initial radius of the blob and w is its descent velocity. Turbulent entrainment increases the effective radius of the blob. Fragmentation of the turbulent blob into droplets occurs between the third and fourth snapshots, as indicated by dye structures in the boxed regions. *Reproduced from Fig. 1 of Deguen, R., Landeau, M., Olson, P., 2014. Turbulent metal-silicate mixing, fragmentation, and equilibration in magma oceans. Earth Planet. Sci. Lett. 391, 274–287.*

α is the *entrainment coefficient*, which, as revealed by experiments and field measurements, is approximately equal to 0.25. Similarly, theory and experiments also show that the travel time $t(z)$ required for the blob to reach a depth z under turbulent entrainment conditions is independent of the viscosity of the ambient magma ocean fluid and is given by a ballistic-type formula with the form

$$t(z) \sim z^2 / (g' R_0^3)^{1/2}, \tag{8.1}$$

where g' is the sinking buoyancy of the metallic blob in the magmatic fluid.

Following the collision between the proto-Earth and a large impactor with a preformed metallic core, a silicate magma ocean is generated by shock-wave heating beneath the impact point, and a turbulent metal blob descends through the magma ocean. Increasingly fine-scale turbulence is produced at the metal-silicate interface, entraining silicate magma into the blob, eventually leading to fragmentation of the blob into small drops, as illustrated in the last two panels of Fig. 8.10.

We can apply these experimental results to the process of metal-silicate equilibration in a magma ocean in the aftermath of an impact with a previously differentiated planetesimal or planetary embryo. If a metal blob derived from the impactor core has freedom to fall a distance greater than three to four times its initial diameter through the magma ocean, turbulence causes the blob to entrain silicate melts and ultimately fragment into metal droplets. The size of these droplets depends on the interfacial tension between the metal and the silicate magma, but is likely to be in the centimeter range (Clesi et al., 2020). At this small scale, metal droplets easily come into local equilibrium with silicates as they rain through the magma ocean. According to this scenario, metal-silicate equilibration is expected to be nearly complete. Note that the durations implied for these fluid dynamical processes are quite short. According to Eq. (8.1), a metal blob derived from a $R_0 = 20$ km radius impactor core has a sinking buoyancy of $g' = 10$ m s^{-2} in silicate magma, and will descend 10^3 km through a magma ocean in about 1 day. Nevertheless, because of entrainment, and because it likely fragmented into centimeter-scale droplets within the first few hundred kilometers of its descent, it may approach full equilibrium with the silicate magma before reaching the proto-core. On the other hand, the core of a large planetary embryo would entrain relatively less silicate magma before reaching the proto-core, with only partial metal-silicate equilibration.

8.1.7 Isotopic evidence

Short-lived isotopes provide important evidence on core formation processes. In particular, the radioactive isotope hafnium-182 decays to the stable isotope tungsten-182 with a half-life of just 8.9 Myr, making this isotope system particularly well suited to investigate Earth formation and differentiation processes. In metal-silicate systems, tungsten (W) strongly partitions into metal portion, whereas hafnium (Hf) partitions into silicate portion. Accordingly, it is possible to use Hf-W systematics to place bounds on both the time when mantle-forming silicates were last in equilibrium with core-forming metals, and also on the degree of equilibration that took place.

The Hf-W chronometer basically works as follows: As metals segregate from silicates and descend into the proto-core, the silicate portion of the proto-Earth continues to accumulate W-182 through the decay of Hf-182, giving the silicate portion an excess of W-182 relative to the tungsten content in an undifferentiated planet of the same size. It is observed that rocks from Earth's crust and mantle do have excess W-182 compared to undifferentiated meteorites (Kleine and Walker, 2017). This implies that partial metal-silicate equilibrium occurred, and also that a substantial fraction of the core was formed within a few half-lives of Hf-182.

But in order to go further, we need to quantify what is meant by "a substantial fraction of the core," and also, what is meant by "a few Hf-182 half-lives." In order to do this, it is necessary to invoke a model for Earth's accretion. As discussed in the previous sections, there are several competing models for how Earth accreted, each model implying somewhat different rates and durations. In order to simplify the discussion here, we consider only continuous accretion, in which the accretion rate (i.e., the rate of mass addition) decreases exponentially with a fixed time constant τ. Furthermore, in order to account for mass addition by the Moon-forming event, we truncate the exponential growth when the proto-Earth reaches 90% of the final mass, that is, $0.9 M_{EAR}$, the remaining 10% being added through a later Moon-forming event.

Fig. 8.11 taken from a study by Kleine and Rudge (2011) shows two multistage accretion/core formation models with continuous exponential accretion, both of which satisfy the isotopic constraints but lead to different interpretations of the early history of the core and its light element composition. Fig. 8.11A and C depicts a model with fast accretion and full equilibration, as might result from pebble accretion, for example. In this model, complete metal-silicate equilibrium is established during accretion because the accreting metals either originate as or fragment into droplets small enough to equilibrate with mantle silicates, particularly where the mantle is molten (i.e., in a magma ocean). According to this model, the mantle acquires an excess of radiogenic W-182 primarily because of early, rapid core formation. And because metal-silicate equilibrium is complete, the light element content of the core mostly reflects conditions in the proto-mantle.

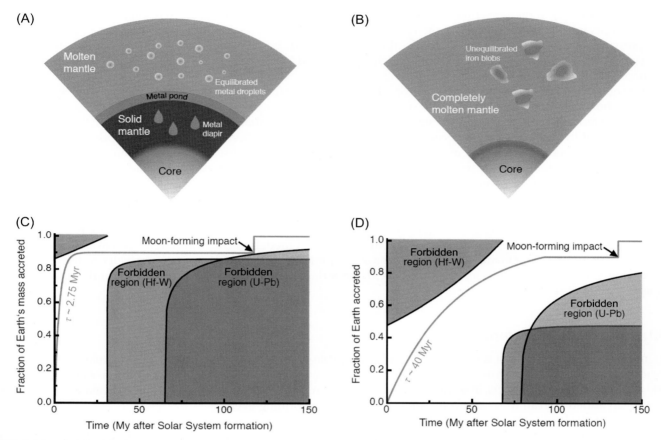

FIG. 8.11 Models of Earth accretion and core formation that satisfy constraints from the hafnium-tungsten (Hf-W) and uranium-lead (U-Pb) isotopic systems. (A) Fast accretion with full metal-silicate equilibration in a magma ocean. (B) Slower accretion with partial metal-silicate equilibration. (C) and (D) White regions are bounds on acceptable accretion histories for models (A) and (B), respectively. The *curves* show hypothetical exponential growth accretion models, identified by their exponential time constants τ. *Reproduced from Fig. 4 of Kleine, T., Rudge, J.F., 2011. Chronometry of meteorites and the formation of the Earth and Moon. Elements 7, 41–46.*

The disequilibrium model depicted in Fig. 8.11B and D satisfies the same isotopic constraints, but does so by restricting metal-silicate equilibrium. The easiest way to do this is by forming the core out of large-scale metal fragments, as would be the case with accretion by large differentiated planetesimals and planetary embryos. Such large preformed cores would experience limited interaction with the proto-mantle while on route to the proto-core. In this case, W-182 accumulates by radioactive decay in the proto-mantle, but in order to limit its excess to what is observed, the accretion process must be slow, extending over many Hf-182 half-lives. As for the composition of the core implied by this model, its light element inventory would mostly reflect the compositions of the impactors.

8.1.8 Evidence from siderophile elements

The elemental compositions of terrestrial rocks also offer clues to core formation. The search for signatures of core formation in samples from the mantle and crust is based on whether a specific element is either *siderophile* (iron-loving), *lithophile* (rock-loving), *calcophile* (sulfide ore-loving), or *atomphile* (gas-loving). These terms follow a classification system developed early in the 20th century by the pioneering geochemist V.M. Goldschmidt, and groups elements according to the compounds or solutions for which they have most affinity (Fig. 8.3). Hafnium and tungsten are lithophile and siderophile, respectively, according to this scheme.

Refractory lithophile elements such as aluminum and calcium are present in the mantle in about the same concentrations as in carbonaceous chondrites. In contrast, by this same measure, the mantle is strongly depleted in siderophile elements, implying that this group of elements was partitioned into core-forming metals. This group includes *moderately siderophile elements* such as iron, nickel, cobalt, and tungsten, as well as *highly siderophile elements* like platinum and gold.

A quantitative measure of siderophile behavior is the *metal-silicate partition coefficient*, the ratio of the mass concentration of an element in the metal to its concentration in the silicate at equilibrium. Experimentally, the moderately siderophile elements have metal-silicate partition coefficients between 1 and 10^4 at standard pressure, whereas highly siderophile elements at the same pressure typically have metal-silicate partition coefficients in the range 10^6–10^{10}. Yet in the mantle, the observed ratios are in the range 10–30 and 500–1500, respectively, far lower than predicted for surface pressure conditions (Rubie et al., 2015). So although the mantle is depleted in siderophiles relative to chondrites, it apparently has excess siderophiles compared to what is expected from core formation, had the metal-silicate partitioning occurred at surface pressure.

One solution to this apparent "excess siderophile problem" is to factor in the dependence of pressure, temperature, and oxidation state, all of which affect the partition coefficients. This approach has led to the particular metal-silicate equilibrium model of core formation shown in Fig. 8.11A. It postulates that, during Earth's accretion, the core-forming metals pond at the base of a magma ocean, with metal-silicate equilibrium occurring at the depth level of the pond base, but not deeper. A particularly nice attribute of this model is that the observed mantle siderophile element abundances can be used to infer the pressure-temperature-oxidation conditions at the base of the metal pond.

Initially, comparisons between high-pressure metal-silicate partition experiments and the observed mantle siderophile abundances indicated equilibrium occurred at pressures around 20–30 GPa and temperatures around 3000 K (Li and Agee, 1996), that is, at mid-mantle conditions. However, more recent data on siderophile partitioning suggest the equilibration may have taken place at more extreme equilibrium conditions, up to 60 GPa and 4000 K (Righter et al., 2016). Furthermore, it has since been shown that disequilibrium models of core formation, such as the core-merging model in Fig. 8.11B, can be made to fit the observed siderophile element abundances equally well (Rudge et al., 2010). Accordingly, mantle siderophile abundances provide important constraints on core formation, but they do not give a unique picture of the process.

8.1.9 Effects of Moon formation

In the core formation models shown in Fig. 8.11, the proto-Earth and its core grow to approximately 80%–90% of their present-day masses at an exponentially decreasing rate, which represents the combined effects of pebble accretion and accretion by larger impacts, in some unknown proportion. The remaining 10%–20% of Earth's mass is assumed to have been added later in time, as a consequence of the Moon-forming event.

Currently, the leading hypothesis for the origin of the Moon is that it was formed from debris left over from a giant impact between the proto-Earth and a planetary embryo or an exceptionally large planetesimal. The physical details as well as the timing of this event remain highly uncertain, although isotopic evidence requires it to have occurred at least 30 Myr after CAI formation, as indicated in Fig. 8.8. By this time, the mass of the proto-Earth was already a large fraction of Earth's present mass, and according to the giant impact hypothesis, grew to within a percent or so of its present-day mass as a consequence of this event.

Apart from a handful isotopic constraints (such as the similarity between Earth and Moon in terms of oxygen isotopes), much of the insight we have on the nature of this event comes from giant impact simulations, which seek to infer the properties of the impactor (now called *Theia*) and its target (the proto-Earth) at the time of collision, by matching properties of the resulting two-body system with those of the Earth-Moon system extrapolated backward to the time of the event. A wide range of sizes for Theia have been proposed, from Moon to Mars and larger (Canup, 2004, 2012; Ćuk and Stewart, 2012), multiple giant impacts (Citron et al., 2018) and for the proto-Earth, additional considerations such as a preexisting magma ocean have also been examined (Hosono et al., 2019). Irrespective of these differences, however, there is little doubt that an impact great enough to form the Moon must have had first-order effects on proto-Earth's core. Some of these effects are depicted schematically in Fig. 8.12.

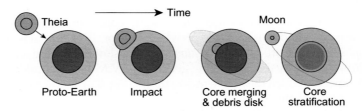

FIG. 8.12 Schematic depiction of the stages in the Moon-forming event, according to the giant impact hypothesis. Time progresses from left to right. Colors of the silicate mantles (*blue-green; light gray in print version*) and metallic cores (*red-orange; dark gray* in in print version) indicate differing compositions.

Suppose, as is very likely, that the compositions of the mantle and core of the proto-Earth differed from those of Theia. Impact simulations (Nakajima and Stevenson, 2015), mixing theory (Dahl and Stevenson, 2010), and laboratory experiments (Landeau et al., 2021) indicate that Theia's core would not have fully fragmented on collision, so that some portion would have sunk through the mantle of the proto-Earth, merging with the proto-core without equilibrating with the mantle. By not equilibrating with mantle silicates, the metal alloys from Theia's core would then differ in composition from the bulk composition of the proto-Earth's core, which according to the isotopic data, was assembled under conditions of partial metal-silicate equilibrium (Rudge et al., 2010). In particular, if Theia's core was more enriched in light elements and hence less dense than the proto-core, upon merging that material would spread out beneath the core-mantle boundary, creating a stable, compositionally stratified layer (Landeau et al., 2016) as depicted in Fig. 8.12. Furthermore, calculations show that very high temperatures are generated in Theia's core at impact by shock waves (Carter et al., 2020), and in addition, shock-wave heating also perturbs the temperature structure of the proto-Earth's core. As a result of this heating, thermal stratification also develops below the core-mantle boundary (Arkani-Hamed, 2017), further increasing the stability there.

In summary, in the aftermath of the Moon-forming giant impact, the core was certainly a bit larger, probably more enriched in light elements, significantly hotter, and quite possibly stratified, particularly in the region just below the core-mantle boundary. Although it is possible that the sum of these changes enhanced convection and dynamo action in the early core, on balance it is far more likely they inhibited convection and dynamo action there. In other words, it is probable that the Moon-forming event scrubbed the slate clean, resetting the thermochemical state of the core and forcing the geodynamo process to begin from scratch.

8.2 Core evolution

In the following sections, we examine how the core has evolved with time since its formation. As described previously in this chapter, core formation was likely complete within the first 200 million years of solar system history, and probably before then. We now consider the evolution of the core since that time, with an emphasis on its thermal history, using the gross energetics of the core from Chapter 2 as a basis. There may well have been mass added to or subtracted from the core over this time, but these entail minor changes compared to the present-day mass of the core. Accordingly, when modeling the evolution of the core since its formation, changes in total mass will be treated as small perturbations of the present-day core mass, and changes in core chemistry are dealt with only insofar as they affect its gross energetics. For example, there is renewed interest in possible long-term changes in the light element chemistry of the core that predate the nucleation of the inner core. We briefly describe how these compositional changes may have occurred, and what their effects may be on core energetics and the geodynamo process.

Before describing thermal history models, it is fair to ask: Why should we care about the evolution of the core in the deep past? This question can be answered several ways. First, the core is a massive thermal reservoir; over time it has transferred a significant amount of heat to other parts of the Earth system. Because core heat transfer controls the geodynamo process, the paleomagnetic record offers clues about how this transfer has varied over time. Second, we expect that the thermal evolution of the core partly reflects the conditions under which the core formed. Because those formative conditions remain obscure, we need to understand how the later history of the core connects to its formation. And finally, the evolution of Earth's core serves as a useful guide for understanding the history of other terrestrial planets.

8.2.1 The equilibrium model

Fig. 8.13 illustrates the main properties of the core that control its recent thermal evolution. Using the notation from Chapter 2, the solid curves represent the present-day adiabatic temperature profile T_{ad} and light element concentration C, and the dotted curves represent the same variables at the time of inner core nucleation, hereafter abbreviated as ICN. The dashed curve is the melting curve of the core denoted by T_M, the total heat loss from the core to the mantle at the core-mantle boundary (labeled CMB in Fig. 8.13) is denoted by Q_{CMB}, and the total heat production within the core from radioactive decay is denoted by Q_R.

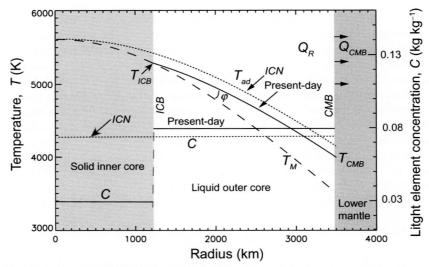

FIG. 8.13 The equilibrium model of core evolution since inner core nucleation. Symbols defined in the text. Temperatures and light element concentrations are approximate.

As Fig. 8.13 implies, the inner core boundary (labeled ICB) is assumed to be an equilibrium phase boundary between the solid inner core and the liquid outer core, so that $T_{ICB} = T_M$ at r_{ICB}. We also assume, consistent with the seismic data in Chapter 1, that the geotherm in the bulk of the outer core approximately follows an adiabatic temperature profile T_{ad}, the light element concentration is uniform except for small perturbations related to convection (in particular, we ignore compositional variations in the inner core and in the E' and F regions), and that the adjustment time of the dynamics in the outer core is short compared to the time scale for changes in the thermal structure of the lower mantle and core. Some amount of stable stratification is permitted in the outer core, but not so much as to violate the other model assumptions. Under these conditions, the outer core remains in a state of statistical thermal and compositional equilibrium with respect to the radius of the outer core r_{CMB} as well as Q_{CMB} and Q_R.

Based on these assumptions, we can use the heat balance derived in Chapter 2 to construct a model for the thermal history of the core in terms of inner core growth. Factoring out the rate of inner core growth $\dot{r}_{ICB}$, the heat balance equation (Eq. 2.33) can be rewritten as

$$\dot{r}_{ICB} = \frac{Q_{CMB} - Q_R}{\mathcal{R}_{LH} + \mathcal{R}_C + \mathcal{R}_{SC}}, \tag{8.2}$$

in which the denominator consists of the sum of the contributions to the core heat budget from latent heat release ($\mathcal{R}_{LH}$), compositional energy release ($\mathcal{R}_C$), and release of sensible heat (secular cooling, $\mathcal{R}_{SC}$), respectively. These $\mathcal{R}$-factors are defined in Eqs. (2.18), (2.21), (2.29) of Chapter 2.

A parallel approach can be used for the light element history of the core. Assuming that the inner core light element concentration remains constant in time, the average light element concentration in the outer core evolves during inner core growth according to Eq. (2.28) of Chapter 2. This can be rewritten as

$$\dot{C}_{OC} = \frac{3\rho_{IC} r_{ICB}^2 \dot{r}_{ICB}}{\rho_{OC} r_{CMB}^3}(C_{OC} - C_{IC}). \tag{8.3}$$

Here, the subscripts OC and IC are used to distinguish the outer and inner core mean densities and their light element concentrations. In deriving Eq. (8.3), we have ignored small terms proportional to the ratio of inner to outer core volumes.

8.2.2 Core evolution inputs

The key model input properties for Eq. (8.2) include $Q_{CMB} - Q_R$ plus the individual energy sources that make up the $\mathcal{R}$-factors, and for composition, the light element partitioning at the inner core boundary, represented in Eq. (8.3) by $C_{OC} - C_{IC}$. Accordingly, the thermochemical history of the core depends on a large number of dynamical and thermodynamic properties. But overall, it is most sensitive to the difference between Q_{CMB} and Q_R. Inner core growth is also sensitive to the difference between the gradients of the core adiabat T_{ad} and the melting curve T_M at the inner core boundary, that is,

$$\varphi = \frac{dT_{ad}}{dr} - \frac{dT_M}{dr} \tag{8.4}$$

evaluated at the present-day inner core boundary. This property appears in $\mathcal{R}_{SC}$.

These two sensitivities can be seen graphically in Fig. 8.13. A large value of $Q_{CMB} - Q_R$ implies a rapid decrease in the geotherm represented by T_{ad} and a correspondingly rapid increase in r_{ICB}, the radius where T_{ad} and T_M intersect. Likewise, a narrow angle φ implies that T_{ad} and T_M are nearly parallel, so that a slight decrease in T_{ad} results in a large increase in r_{ICB}. Under these conditions, the inner core radius increases relatively fast and the inner core is, therefore, relatively young. Conversely, if $Q_{CMB} - Q_R$ is small and/or the angle φ is large, r_{ICB} increases more slowly with time and the inner core is relatively old. There is also some sensitivity to the central temperature of the core T_c, and as discussed in Chapter 2, there is broad interest in the effects of thermal conductivity on core evolution.

Sensitivity to all these properties is examined in the next subsection. Unfortunately, the properties on which core history is most sensitive have generous uncertainties. For example, as documented in Chapter 2, the uncertainty in the present-day Q_{CMB} is a factor of 3 and the uncertainty is even greater for the thermal conductivity k. The situation worsens going back into the past, where the uncertainty in Q_{CMB} is further magnified. So, in order to give a full picture of these property uncertainties and how they map into uncertainties in core history, we adopt the following wide ranges for present-day property values: Q_{CMB} 6–18 TW; $Q_R = 0$–1.4 TW by radioactive K-40; $T_c = 5000$–6000 K; and $\varphi = 0.05$–0.27 K km^{-1}, along with k ranging from 20 to 140 W m^{-1} K^{-1}. We also consider the effects of an increased Q_{CMB} in the deep past.

8.2.3 Inner core growth histories

We apply the following modeling strategy for calculating the thermochemical evolution of the core and the age of the inner core. First, we adopt the values of the fixed properties in Table 8.1 and we select values of the variable properties within the ranges given earlier. Second, we integrate Eqs. (8.2), (8.3) backward in time (forward in age), starting from present-day conditions, tracking the evolution of the core to determine the changes in temperature, outer core light element concentration, and inner core radius. Analytical formulas for T_{ad}, T_M, $\mathcal{R}_{SC}$, $\mathcal{R}_{LH}$, and $\mathcal{R}_C$ between the present day and the time of ICN are given in Appendix 8.1; similar expressions can be found in various published core evolution studies (Nimmo, 2015; Labrosse, 2015; Davies, 2015). Prior to ICN, we assume heat flow at the core-mantle boundary is balanced by the sum of secular cooling of the core plus heat generated internally by radioactive decay. For now, we assume the light element concentration of the entirely molten core remains constant in time; later on we relax this assumption and consider chemical exchange with the mantle. For ages greater than the ICN, we use the heat balance described earlier as given in Appendix 8.1 to track the change in core-mantle boundary temperature over time.

TABLE 8.1 Core evolution model properties.

Fixed property	Notation	Adopted value(s)
Density, core center	ρ_c	12,500 kg m^{-3}
Density, zero pressure	ρ_0	7500 kg m^{-3}
Compositional density jump, ICB	$\Delta\rho_{ICB}$	500 kg m^{-3} [a]
Incompressibility, zero pressure	K_0	4.75×10^{11} Pa
Latent heat, melting	L	6.6×10^5 J kg^{-1}
Grüneisen parameter	γ	1.5
Heat capacity	c_P	850 J kg^{-1} K^{-1}
Thermal expansivity	α_T	1.5×10^{-5} K^{-1}
Compositional expansivity	α_C	1
Density length scale	r_ρ	7400 km

Continued

TABLE 8.1 Core evolution model properties—cont'd

Fixed property	Notation	Adopted value(s)
Temperature length scale	r_T	6040 km
ICB radius	r_{ICB}	1220 km[a]
CMB radius	r_{CMB}	3480 km
Light element concentration, outer core	C_{OC}	8 wt%[a]
Light elements concentration, inner core	C_{IC}	3 wt%
Variable property	**Notation**	**Range**
Core heat flow	Q_{CMB}	6–18 TW[a]
Core heat production	Q_R	0–1.4 TW[a]
Melting-adiabatic gradient	φ	0.05–0.27 K km^{-1}
Central temperature	T_c	5000–6000 K[a]
Thermal conductivity, CMB	k	20–140 W m^{-1} K^{-1}

[a] *Present-day values.*

To begin with, assume that the total core-mantle boundary heat flow is either constant in time or changes so little that its time average is what dictates the age of ICN. Fig. 8.14 shows the variation of core-mantle boundary temperature T_{CMB} and inner core radius r_{ICB} with age for two core-mantle boundary heat flow choices, $Q_{CMB} = 12$ and 14 TW, zero radioactive heating, $Q_R = 0$, melting gradient parameter $\varphi = 0.17$ K km^{-1}, and present-day inner core central temperature $T_c = 5800$ K. Backward integration of Eq. (8.2) yields ICN ages are 770 and 660 Ma for $Q_{CMB} = 12$ and 14 TW, respectively, and a decrease in T_{CMB} since ICN of approximately 95 K in each case. With constant Q_{CMB}, the inner core mass increases at a nearly constant rate, so that its radius grows with time t roughly like $r_{ICB} \propto t^{1/3}$. According to Eq. (8.3), the corresponding increase in the light element concentration of the outer core is approximately 0.75% between ICN and the present day.

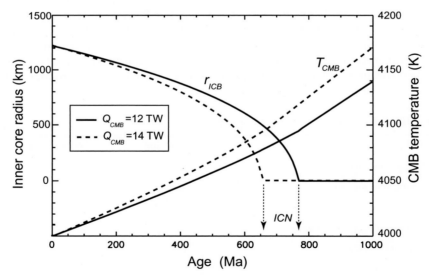

FIG. 8.14 Inner core radius and core-mantle boundary temperature versus age for various assumed total core-mantle boundary heat flows with zero radioactive heating, from thermal history calculations. ICN indicates inner core nucleation times.

8.2.4 Inner core nucleation age

As discussed in Chapter 2, the timing of inner core nucleation is a benchmark event for the evolution of the core (indeed, for the entire Earth) and as described in Chapter 3, so far it has eluded detection in the paleomagnetic record. Fig. 8.15 shows the ICN age in Ma according our core evolution model, contoured as a function of Q_{CMB} and the melting gradient parameter φ, with and without radioactive heat production, calculated using the same procedure and the other core properties from Fig. 8.14. For the calculations with radioactive heating, the decay rate of radioactive K-40 is applied backward in time. The boxes outlined by dashed lines in Fig. 8.15 delineate the (Q_{CMB}, φ) combinations that are considered more likely, according to the analysis in Chapter 2.

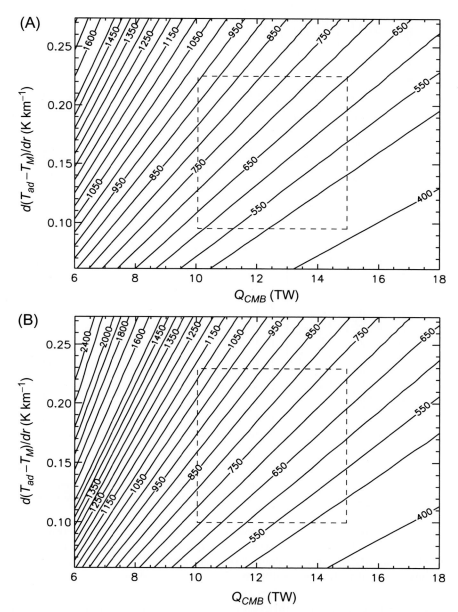

FIG. 8.15 Contours of inner core age in Ma versus total core-mantle boundary heat flow and the difference between the melting curve and adiabatic temperature gradients at the inner core boundary, from thermal history calculations with various amounts of present-day radioactive heating Q_R. (A) $Q_R = 0$; (B) $Q_R = 1$ TW by K-40. *Dashed boxes* indicate likely bounds.

With no radioactive heating, ICN ages in Fig. 8.15A range from slightly more than 1600 Ma for $Q_{CMB} = 6$ TW to less than 400 Ma for $Q_{CMB} = 18$ TW, but for the preferred combinations of Q_{CMB} and φ, the predicted ICN ages range from about 440 to 1000 Ma. Including significant radioactive heating has only minor influences on ICN ages if Q_{CMB} is large, but significantly increases ICN ages if Q_{CMB} is small. Fig. 8.15B shows calculated ICN ages for the same cases as shown in Fig. 8.15A but including 1 TW of present-day radioactive heating by potassium-40. Comparing Fig. 8.15A and B, there is little difference between predicted ICN ages with and without radioactivity for Q_{CMB} larger than about 12 TW. But for $Q_{CMB} \simeq 6$ TW, maximum ICN ages increase from about 1600 to 2400 Ma when radioactive heating is included.

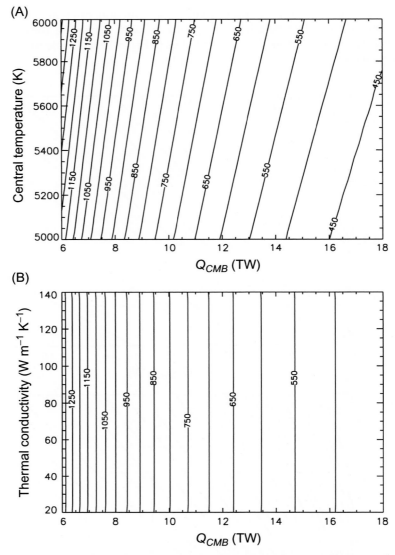

FIG. 8.16 Contours of inner core age in Ma versus total core-mantle boundary heat flow and (A) present-day central core temperature and (B) thermal conductivity, from thermal history calculations with zero radioactive heating. Other core properties the same as shown in Fig. 8.15.

What additional factors could affect the ICN age ranges shown in Fig. 8.15? Supercooling of the core would delay nucleation and reduce the ICN age. Conversely, excessive radioactive heating would increase it. Less likely (but still possible) is that the core heat flow Q_{CMB} and the energy requirements of the geodynamo in Chapters 2 and 3 have all been overestimated, or have radically changed with time. Finally, what about thermal conductivity and core temperatures? How do the uncertainties in these properties affect the estimates of inner core age? Fig. 8.16 shows ICN age contoured as functions of present-day central temperature T_c and core thermal conductivity k, with Q_{CMB} again ranging between 6 and 18 TW. Fig. 8.16A shows that there is some sensitivity to the central core temperature, but it is rather weak. For example, with $Q_{CMB} = 12$ TW, the predicted ICN age increases from about 600 to 700 Ma as the core central temperature is raised from 5000 to 6000 K. In addition, Fig. 8.16B shows the ICN age has no sensitivity to the thermal conductivity of the core. This result is perhaps surprising, in light of the importance of thermal conductivity to the thermal structure of the outer core. ICN age independence from conductivity is a consequence of the thermal equilibrium assumption, along with the notion that Q_{CMB} is controlled by the mantle, irrespective of the transport properties of the core.

To summarize: Thermochemical evolution scenarios for the core can only place broad bounds on the ICN age. According to the equilibrium model, the inner core is rather young, probably less than 1600 Ma, and possibly as young as 400 Ma. However, more extreme ages remain possible within the framework of most core evolution scenarios. In addition, the currently available observational and experimental evidence fails to clarify this picture very much. Based on the recent interpretations of the paleomagnetic record, ICN ages ranging from 1.5 to 0.56 Ga (Fig. 3.14) have been inferred (Biggin et al., 2015; Driscoll, 2016), and in the other extreme, ICN ages exceeding 2 Ga (Fig. 2.18) continue to be proposed (Hsieh et al., 2020). In short, the long-sought goal of calibrating the rate of Earth's evolution based on the age of the inner core has yet to be realized.

8.2.5 Dynamo model predictions

In light of the fact that the ICN has not been pinpointed by core evolution models or the existing paleomagnetic record, it is worthwhile adding a theoretical element to the search, using numerical dynamos to predict how the geodynamo responded before, during, and after this event. The numerical dynamo scaling laws described in Chapter 4 predict that B, the geomagnetic field intensity within the core, scales with the outer core buoyancy flux F approximately as $B \propto F^{1/3}$. According to this scaling, the onset of inner core solidification could produce a substantial increase in B, as light element partitioning adds a compositional buoyancy flux F_C in the outer core. This is particularly expected if the thermal buoyancy flux F_T, which presumably supported the geodynamo just prior to ICN, was relatively small. Suppose, for example, that at the time of ICN $F_T = 1 \times 10^{-13}$ m^2 s^{-3} (corresponding to about 1.5 TW of superadiabatic heat flow at the core-mantle boundary) and that ICN added a compositional buoyancy flux of $F_C = 1.5 \times 10^{-12}$ m^2 s^{-3}. Then, F would have increased by a factor of 16 due to inner core solidification, and according to the dynamo scaling, the geomagnetic field intensity within the core would have increased by a factor of 2.5 on average. Indeed, a study by Aubert et al. (2009) that combined a core evolution model with dynamo scaling laws predicts a slightly larger change, with the core field intensity increasing by a factor of 3–4 following the ICN.

One problem with using dynamo scaling laws for predicting the change in geomagnetic intensity associated with inner core nucleation is that most dynamo scaling laws, including those described in Chapter 4, apply to the r.m.s. magnetic field intensity inside the geodynamo region. They do not necessarily apply in the same way to the dipole part of the geomagnetic field. This means that the observable geomagnetic field, which for paleomagnetic intensity is essentially just the dipole part, may behave differently than dynamo scaling laws predict, especially as the structure of the core and the sources of outer core buoyancy change, as certainly happened before and after the ICN.

Fig. 8.17 from a numerical dynamo study of inner core nucleation by Landeau et al. (2017) illustrates this problem. Fig. 8.17A is the timeline of a core evolution model showing the decrease of inner core radius with increasing age, starting from the present day and going backward to before the time of ICN, which is approximately 670 Ma in this case. Fig. 8.17B and C shows the radial magnetic field intensity at the Earth's surface and on the core-mantle boundary at three stages of this core evolution, obtained from a numerical dynamo driven by codensity fluxes (combined heat and light element fluxes) that correspond to each stage of the inner core growth, the codensity fluxes derived from the core evolution model.

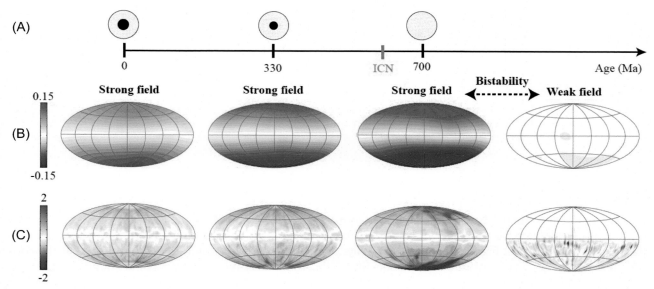

FIG. 8.17 Effects of inner core nucleation on the geodynamo. Numerical dynamo structure at 0 Ma (first column), 330 Ma (second column), and 700 Ma (third and fourth columns) for a constant heat loss equivalent to 13 TW in the core. (A) Core evolution timeline with *black filled circles* indicating inner core size. (B, C) Radial magnetic field at Earth's surface (B) and at the core-mantle boundary (C). The first three columns of maps are time averages of strong magnetic fields; the fourth column of maps is snapshots of weak magnetic fields. *Modified from Fig. 4 of Landeau, M., Aubert, J., Olson, P., 2017. The signature of inner-core nucleation on the geodynamo. Earth Planet. Sci. Lett. 465, 193–204.*

The codensity buoyancy flux driving convection in this numerical dynamo increases by more than an order of magnitude from pre-ICN times (700 Ma) to the present day (0 Ma), yet the most stable state of the pre-ICN dynamo (labeled "Strong field" in Fig. 8.17) features a surface magnetic field that is actually somewhat more intense than the present-day surface magnetic field predicted for the same dynamo. In other words, the reduced buoyancy flux at pre-ICN times does not give rise to a weaker surface magnetic field then. The explanation for this seemingly anomalous behavior is that the post-ICN dynamo is primarily driven by buoyancy produced at the inner core boundary and is therefore deep seated, whereas the pre-ICN dynamo is driven entirely by buoyancy produced at the core-mantle boundary, and is therefore shallow seated. By positioning the magnetic field generation region closer to the surface, the pre-ICN dynamo compensates for its reduced power, and produces a surface magnetic field with intensity comparable to that of the post-ICN dynamo.

There is, however, a further complication to this picture, because dynamos without an active inner core may exhibit bistability, with a weak multipolar magnetic field state competing with a state having a strong, mostly dipolar magnetic field. This bistability, which is depicted schematically in Fig. 8.17, is asymmetric for this particular dynamo, in that the weak field state spontaneously transitions to the strong field state, but not vice versa. Nevertheless, the possibility that weak magnetic field states might have persisted in the absence of the inner core leaves open the possibility of detecting the ICN using time variations in paleomagnetic intensity.

An example of this is shown in Fig. 8.18, taken from a study by Driscoll (2016) in which numerical dynamos were sequenced along a 0–2 Ga time line, with the inner core size and the outer core buoyancy production in each dynamo determined from a core evolution model. Fig. 8.18A shows the time variations of inner core size and outer core buoyancy production relative to the time of ICN, which occurs around 650 Ma according to this core evolution model. Buoyancy production decreases steadily with time until ICN, then increases suddenly with the advent of light element release from the inner core boundary. The effects of these time variations in convective power on the dipole moment are shown in Fig. 8.18B. Prior to 1.7 Ga, the persistent dynamo state is multipolar, with a relatively weak dipole moment. The dynamo then assumes a strong field, dipole-dominant configuration, lasting until about 1 Ga, when a weak dipole moment state returns, persisting until the time of ICN. Subsequently, the post-ICN dynamo remains entirely in a strong field, dipole-dominant state. The numerical dynamo histories in Figs. 8.17 and 8.18 are comparable during post-ICN times, but they differ pre-ICN because of differences in the stability of the weak field state.

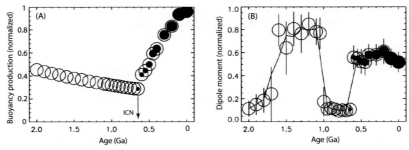

FIG. 8.18 Core and geomagnetic field evolution 0–2 Ga from a coupled thermal evolution-numerical dynamo study. (A) Normalized buoyancy production in the core and (B) dipole moment variations. Outer core size indicates by *open circles*; inner core size indicated by *filled circles* and magnetic intensity variations indicated by *error bars*. Modified from Fig. 2 of Driscoll, P., 2016. Simulating 2 Ga of geodynamo history. Geophys. Res. Lett. 43, 5680–5687.

8.2.6 Long-term core cooling

The thermal history methods used for the inner core age in the previous sections can be extended deep into the geologic past to calculate the increases in core temperatures with age, based on the assumption that prior to ICN the entire core was molten. Fig. 8.19 shows the results of such calculations, using the assumption that the core-to-mantle heat flow Q_{CMB} was constant in time. This figure shows the evolution of core-mantle boundary temperature T_{CMB} back to 4.3 Ga for choices of Q_{CMB} ranging from 6 to 16 TW, with no contribution from radioactive heating in every case. ICN times are marked with +-signs. With $Q_{CMB} = 6$ TW, T_{CMB} increases by 350 K back to 4.3 Ga, whereas it increases by slightly more than 1000 K with $Q_{CMB} = 16$ TW. In every case shown in Fig. 8.19, most of the increase in

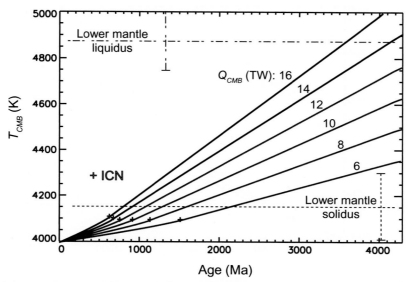

FIG. 8.19 Core-mantle boundary temperature versus age (0–4300 Ma) from thermal evolution models with various assumed total core-mantle boundary heat flows Q_{CMB}, assuming zero radioactive heating. Estimated ranges of lower mantle solidus and liquidus shown by *dashed and dash-dot lines*, respectively; plus signs (+) mark inner core nucleation times. Solidus and liquidus ranges according to Andrault et al. (2011) are for the mantle just above the core-mantle boundary.

T_{CMB} with age occurs prior to ICN, when the core heat loss is entirely sensible heat, with no contributions from latent or radioactive heats. In addition, for every case in Fig. 8.19, T_{CMB} exceeds the lower mantle solidus at some point in the deep past, and for $Q_{CMB} > 13$ TW, T_{CMB} exceeds the lower mantle liquidus before 4 Ga.

Fig. 8.20 shows the effects of radioactive heating from K-40 on the long-term evolution of T_{CMB}. With a constant $Q_{CMB} = 13$ TW, T_{CMB} increases by nearly 850 K back to 4.3 Ga. That increase occurs at nearly a steady rate prior to ICN. In contrast, radioactive heating by K-40 decay increases steadily with increasing age, tending to suppress the rise in T_{CMB} with age. For the case with 100 ppm K, corresponding to 0.7 TW present-day radioactive heat production, the increase in T_{CMB} back to 4.3 Ga is reduced to 650 K, and for the case with 200 ppm K, corresponding to 1.4 TW present-day radioactive heat production, the increase in T_{CMB} back to 4.3 Ga is just 450 K. In that case, T_{CMB} reaches its maximum around 4 Ga and then decreases slightly at older ages. This temperature maximum corresponds to the time when the heat production by K-40, Q_R, equals Q_{CMB}; at greater ages Q_R exceeds Q_{CMB} and the core is heating up with time. Like the thermal evolution cases in Fig. 8.19, in all of the cases with radioactive heating in Fig. 8.20, T_{CMB} exceeds the lower mantle solidus in the deep past. But unlike the highest heat flow cases in Fig. 8.19, the core-mantle boundary temperatures in the radioactive cases in Fig. 8.20 do not reach the lower mantle liquidus. This is a significant difference, because it implies two distinct conditions for the early core-mantle boundary region: entirely molten without core radioactivity, but only partially molten with abundant core radioactivity.

A major restriction on the results in Figs. 8.19 and 8.20 is the assumption that the core-mantle heat flow Q_{CMB} is constant in time. In reality, Q_{CMB} must adjust to the changing conditions in the mantle as well as in the core. A large number of studies have considered the combined thermal evolution of the core-mantle system. The overarching conclusion from these studies is that core temperatures were probably higher in the deep past than shown in Figs. 8.19 and 8.20, because the heat loss from the core was higher then.

A shorthand approach to characterize the joint thermal evolution of the core and mantle is to consider heat transfer across the basal thermal boundary layer in the lowermost mantle, in the D'' region just above the core-mantle boundary. Laminar thermal boundary layer theory applied to the lowermost mantle predicts that the heat flow at the core-mantle boundary varies with time t according to

$$\frac{Q_{CMB}(t)}{Q_{CMB}(0)} = \left(\frac{T_{CMB}(t) - T_{LM}(t)}{T_{CMB}(0) - T_{LM}(0)}\right)^{1+\beta}, \tag{8.5}$$

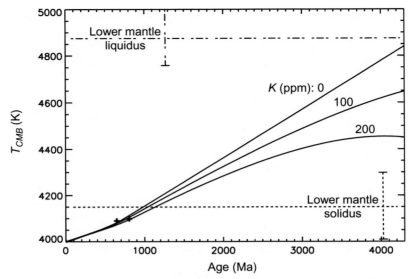

FIG. 8.20 Core-mantle boundary temperature versus age (0–4300 Ma) for various concentrations of potassium (labeled as K in ppm) that include the radioactive heat producer K-40, from thermal history models assuming 13 TW of core-mantle heat flow. Estimated ranges of lower mantle solidus and liquidus shown by *dotted and dash-dot lines*, respectively; plus signs (+) mark inner core nucleation times.

where T_{LM} is the mantle temperature above its basal thermal boundary layer, 0 denotes present day, and β is the exponent in the heat transfer scaling law for the thermal boundary layer. For constant viscosity lower mantle flow, $\beta \simeq 1/3$, but because of the strong temperature dependence of mantle viscosity, it is more likely that $\beta < 1/3$, in which case (8.5) predicts that Q_{CMB} is essentially linearly proportional to T_{CMB}, both increasing with age. The actual sensitivity of these increases depends on $T_{LM}(t)$, the time history of lower mantle temperature. If T_{LM} keeps pace with T_{CMB} then Q_{CMB} hardly changes, according to Eq. (8.5). More likely, T_{LM} changes more slowly than T_{CMB}, because of the greater thermal mass of the mantle compared to the core and also because of latent heat effects from mantle melting and solidification.

For purposes of illustration, we can model (8.5) for core thermal evolution by rewriting Eq. (8.5) in the form

$$Q_{CMB}(t) = Q_{CMB}(0) + \dot{Q}_{CMB} t, \tag{8.6}$$

where $\dot{Q}_{CMB}$ denotes the rate of change of core-mantle boundary heat flow with time, or alternatively, the rate of core-mantle boundary heat flow change with age.

Fig. 8.21 shows the result of applying Eq. (8.6) to the same thermal evolution model as shown in Fig. 8.20, using $Q_{CMB}(0) = 13$ TW for the present-day core-mantle boundary heat flow along with the radioactive heating from 100 ppm potassium. Core-mantle boundary temperature histories are shown for three cases, corresponding to core-mantle boundary heat flow variations of $\dot{Q}_{CMB} = 0, -1.6$, and -3 TW Gyr^{-1}, respectively. With constant heat flow, the $\dot{Q}_{CMB} = 0$ case, the core-mantle boundary temperature remains below the lower mantle liquidus, but for the other two cases, the higher rate of heat flow in the past raises the core-mantle boundary temperature into or above the estimated range of the lower mantle liquidus, overcoming the effects of radioactive heating. In particular, the highest-temperature case in Fig. 8.21 shows that if the core-mantle boundary heat flow at 4.3 Ga was approximately twice what it is today, as many whole-Earth thermal evolution models predict (Labrosse, 2015; Driscoll and Bercovici, 2014), it is almost certain that the base of the mantle was liquid at that time. More generally, partial melting or full melting of the mantle immediately above the core-mantle boundary is a fundamental prediction of many thermal history models of the early Earth.

8.2.7 Convection prior to inner core nucleation

The time at which Earth's inner core nucleated marked a major transition in Earth history, not only affecting the structure of the core but also inducing a sharp change in core energetics and the powering of the geodynamo. Just before the

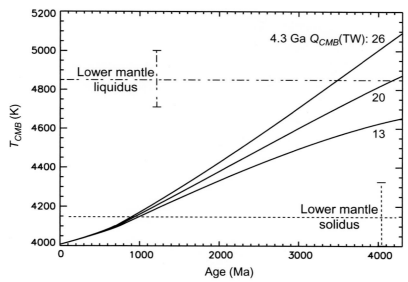

FIG. 8.21 Core-mantle boundary temperature versus age (0–4300 Ma) for various linear rates of decrease of the core-mantle boundary heat flow from thermal history models. Each case assumes radioactive heating from 100 ppm potassium and a present-day core-mantle heat flow of 13 TW. Initial core-mantle heat flows at 4300 Ma (4.3 Ga) are labeled on each curve. Estimated ranges of lower mantle solidus and liquidus shown by *dotted and dash-dot lines*, respectively.

nucleation of a solid inner core, convection in the core would have been driven by heat loss to the mantle, possibly augmented by radioactive heating within the core and loss of light elements to the mantle. Although dynamo action at this time may have been assisted by tidal instabilities (Le Bars et al., 2011) and by instabilities related to precession (Tilgner, 2005), convection likely provided much of the power for the geodynamo, at least in the immediate pre-ICN times.

As shown earlier in this chapter, it is almost certain that core heat loss exceeded heat production by radioactive decay for much of this time, and that secular cooling of the core resulted. Core heat loss to the mantle generates a destabilizing thermal buoyancy flux at the core-mantle boundary, and provided this heat loss exceeds the adiabatic heat loss, thermal convection would have occurred in the entirely liquid core.

Secular cooling of a multicomponent liquid can also produce several forms of compositional convection, if the cooling leads to saturation and precipitation of one or more of the liquid components. As described in Chapter 2, the Earth's outer core contains abundant light elements, so that in its molten state it qualifies as a multicomponent liquid. Furthermore, the solubility of many (if not all) of its light element constituents increases with increasing temperature. Accordingly, it is reasonable to assume that some amount of light elements in the molten core became insoluble as the Earth cooled over geologic time. And provided this happened, it would almost certainly have taken place near the core-mantle boundary, where core temperatures at any given time are lowest. In that environment, light elements with limited solubility in iron (such as magnesium, silicon, and possibly others) would precipitate from the outer core liquid and rise to the core-mantle boundary, leaving the liquid from which they were extracted more enriched in heavy metals and, therefore, denser than the underlying core. The compositionally driven instability that results from this action is termed as *exsolution convection* and is illustrated schematically in Fig. 8.22, from Olson et al. (2017). Analyses of the thermodynamics of light element exsolution using the same formalism as in Chapter 2 reveal that it could have been a dominant form of convection in the core over much of its early history (O'Rourke and Stevenson, 2016; Badro et al., 2016; Mittal et al., 2020).

In addition to light element exsolution, another form of compositional convection might have occurred prior to ICN, in this case driven by iron precipitation. As the molten core cooled on approach to solidification, depending on the relative positions of the adiabat and the melting curve, pure iron crystals may have solidified in a restricted depth interval. Because of their high density, these crystals precipitate downward through the core as *iron snow*, as it is called (see Chapter 7). Falling into the deeper, hotter portions of the core, the iron-snow then melts, increasing the density and destabilizing the fluid there. Iron snow has been proposed for convection in the cores of smaller terrestrial planets and moons (Dumberry and Rivoldini, 2015; Christensen, 2015).

Although the compositions of both the precipitate and the residual core fluid differ between these two forms of convection, the underlying fluid mechanics are fundamentally the same. First, cooling of the planet leads to saturation

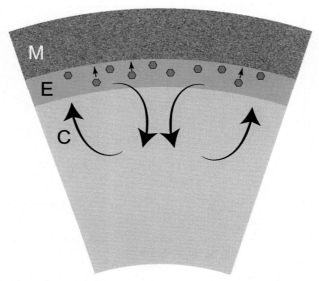

FIG. 8.22 Schematics of the magnesium/silicon exsolution convection mechanism. M, E, C denote the mantle, the exsolution-dominated region in the core, and the convection-dominated region in the core, respectively. *Small arrows* indicate precipitation directions; *large arrows* indicate convection. *Dark gray shading* denotes higher metal (iron-nickel) concentration. *Modified from Fig. 1 of Olson, P., Landeau, M., Hirsh, B., 2017. Laboratory experiments on rain-driven convection: implications for planetary dynamos. Earth Planet. Sci. Lett. 457, 403–411.*

of one or more components of the conducting fluid. Nucleation of that component produces either liquid drops or solid grains, which then precipitate. Removal of the light precipitate destabilizes the core in the exsolution mechanism, whereas dissolution of the heavy precipitate destabilizes the core in the iron snow mechanism. Compositional convection results by both mechanisms.

One major difference between these two mechanisms is the depth where the convective buoyancy is produced. Convection driven by light element exsolution would likely originate very close to the core-mantle boundary. In contrast, iron snow precipitation close to the inner core boundary is sometimes proposed as a mechanism for producing low seismic velocities in the F region (but see the discussions in Chapters 5 and 7). In this mechanism, the snow layer is stably stratified; destabilization and convection occur either below the snowing layer (if the snow melts) or above (if the snowfall increases the light element concentration). The former situation is unlikely in the F region, since the iron snow hypothesis calls on deposition of the iron snow at the inner core boundary as the mechanism for inner core growth. In the latter situation, light element enrichment above iron snow, there is little difference between light element fractionation at the inner core boundary versus the same light element fractionation in a thin snow layer just above the inner core boundary, at least insofar as outer core convection is concerned.

8.2.8 Buoyancy flux from light element exsolution

The rate of gravitational potential energy release by the light element exsolution process is proportional to the cooling rate of the core. Given the cooling rates shown in Fig. 8.21, light element exsolution is expected to release substantial amounts of gravitational potential energy per unit time, provided the core was either saturated or close to saturation in light elements as it formed. Provided these conditions were met, light element exsolution offers a plausible power source for dynamo action in the early core (Badro et al., 2019).

We can estimate the amount of light element exsolution needed using the expressions for buoyancy flux and the dynamo scaling laws derived in Chapter 4. By analogy with thermal buoyancy generated by heat loss at the core-mantle boundary, compositional buoyancy is generated by exsolution near the core-mantle boundary. Using the notation from Chapter 4, the volume averaged compositional buoyancy flux generated in this way in the fully liquid core is given by

$$F_C = -\frac{\alpha_C g_{CMB} r_{CMB}}{5} \dot{C}_0, \tag{8.7}$$

where α_C is the compositional expansivity, g_{CMB} is the gravity at radius r_{CMB}, and $\dot{C}_0$ is the rate of light element concentration change in the core from exsolution.

Let us assume that F_C is constant over an extended interval of core history, this time interval denoted by Δt. Replacing $\dot{C}_0$ with $\Delta C_0/\Delta t$ and inverting Eq. (8.7) yields, for the change in light element concentration,

$$\Delta C_0 \simeq -\frac{5 F_C \Delta t}{\alpha_C g_{CMB} r_{CMB}}. \tag{8.8}$$

If we further assume that the ancient geodynamo was supported by a compositional buoyancy flux from exsolution that was comparable to the present-day thermochemical buoyancy flux, we have from Chapter 4 that $F_C \simeq 2 \times 10^{-12}\,\mathrm{m^2\,s^{-3}}$. With the other properties from Tables S1–S4 from Core Properties and Parameters, the change in light element concentration in the core over $\Delta t = 3.15$ Gyr (10^{17} s) is, according to Eq. (8.8), $\Delta C_0 = -2\%$. Many models of the light element abundances call for more than 2% of SiO_2 in the present-day core, as well as during core formation. An unresolved question is whether the core was ever saturated in SiO_2. If the answer to this question is yes, then exsolution of SiO_2 as the core cools seems, on this basis, to be a plausible dynamo power source. The situation is a bit different for MgO, because it does not appear to be soluble at the 2% level in the core today. However, its solubility may have been much higher in the past, when the core was hotter. For example, Badro et al. (2019) infer an early core MgO abundance of 0.75%, decreasing to 0.5% today. This would generate an average compositional buoyancy flux of about $F_C \simeq 2 \times 10^{-13}\,\mathrm{m^2\,s^{-3}}$ over 3.5 Gyr. This is still marginally adequate to explain the ancient paleomagnetic dipole intensity, in light of the results from earlier in this chapter showing efficient dipole generation in numerical dynamos in which the dynamo-generating region is close to the outer boundary.

One final caveat in regard to the exsolution mechanism: Although this mechanism implies negligibly small magnesium and silicon additions to the mantle, which would not substantially affect mantle geochemical tracers such as Hf-W isotopic ratios (Rudge et al., 2010) or mantle siderophiles (Siebert et al., 2011), it is advantageous if the mantle circulation remove them from the core-mantle boundary region (Buffett et al., 2000). For if not, they might dissolve back into the core over time, shutting down the exsolution process.

8.3 The geodynamo in the deep past

8.3.1 How old is the geodynamo?

The ancient history of the geomagnetic field is not very well characterized. For example, it is unclear how deep into the past the geodynamo has operated, and whether its operation has been continuous or discontinuous in time. Evidence for a geomagnetic field of similar strength as the present-day core field extends back to about 3.4 Ga (Biggin et al., 2011; Tarduno et al., 2010) and possibly as far back as 4.2 Ga (Tarduno et al., 2015), although this latter interpretation has been vigorously challenged (Weiss et al., 2015; Borlina et al., 2020). Even so, there is reason to suspect the geodynamo may be nearly as old as the core itself. Remnant magnetization found in meteorites provides evidence that many, although not all, meteorite parent bodies hosted a magnetic field within a few million of the start of the solar system (Weiss et al., 2010).

While there are multiple unresolved questions about the early history of the geodynamo, the issue that has received the most attention is whether inner core nucleation (ICN) induced changes in magnetic field properties, leaving footprints in the paleomagnetic record. Although locating the ICN in time would provide a benchmark for the evolution of the core and for the whole Earth, another issue, perhaps more fundamental, is how to explain the enigmatic lack of a clear trend in the paleomagnetic intensity record during the last three billion years (Smirnov et al., 2016), a period that comprises both the ICN, and as discussed previously, a major change in the power sources available to the geodynamo.

A few investigations have claimed to detect a decrease in field intensity with age, interpreted as a footprint of ICN, but the location in time of this event varies widely from one study to the next. As examples, Biggin et al. (2015) identified an apparent intensity decrease between 1 and 1.5 Ga, with a field minimum around 1.5 Ga, Macouin et al. (2004) argued for a decrease within 0.3–1 Ga, and Valet et al. (2014) point to an intensity minimum within 1.8–2.3 Ga. Other paleomagnetic investigations argue for the absence of clear long-term trend in the paleomagnetic intensity record within the Precambrian period (Tauxe, 2006; Smirnov et al., 2016). Here, we note that the average of the Precambrian dipole moment and its variability, 49 ± 31 ZAm^2 according to the PINT database, are indistinguishable, within uncertainties, with 0–300 Ma dipole moment of 46 ± 32 ZAm^2 determined by Selkin and Tauxe (2000), as well as a 49 ZAm^2 average dipole moment determined by Wang et al. (2015) for 0–5 Ma. On balance, it is clear that the interpretation of the paleomagnetic intensity record in terms of core evolution in the deep past is still unsettled. Another possible observational metric for the geodynamo in the deep past is the morphology of the surface geomagnetic field. Although most paleomagnetic investigations suggest that the surface geomagnetic field has been dominated by an axial dipole for the last two billion years (Evans, 2013), anomalies in paleomagnetic inclinations have been attributed to a stronger octupole field during the Precambrian (Veikkolainen and Pesonen, 2014).

8.3.2 Numerical models of the ancient geodynamo

Heimpel and Evans (2013) have examined possible geodynamo states in the deep past using numerical dynamos driven by various amounts of buoyancy produced at the core-mantle boundary, both with and without active inner cores. They consider three likely evolutionary states of the core: a post-ICN scenario 1 with thermochemical convection and most of the buoyancy production at the inner core boundary; a scenario 2 with anomalously high buoyancy production at the core-mantle boundary, as would result from a transient surge in mantle circulation, and a pre-ICN scenario 3 without inner core activity and all of the buoyancy production at the core-mantle boundary.

Heimpel and Evans (2013) found that, although latitudinal variations of heat flux over the core-mantle boundary affects the frequency of geomagnetic polarity reversals, variations in the total heat flow at the core-mantle boundary have the greatest influence on the geomagnetic field structure. As shown in Fig. 8.23, the magnetic field structures in their post-ICN scenario 1 dynamos that are driven mainly by release of buoyancy at the inner core boundary are persistently dipolar with occasional polarity reversals. In the dynamos in scenario 2, the elevated core-mantle boundary heat flow precipitates hyperreversal activity and produces strong nondipole field components. Lastly, the lack of inner core buoyancy production in the scenario 3 dynamos produces a persistent octupole magnetic field coexisting with the

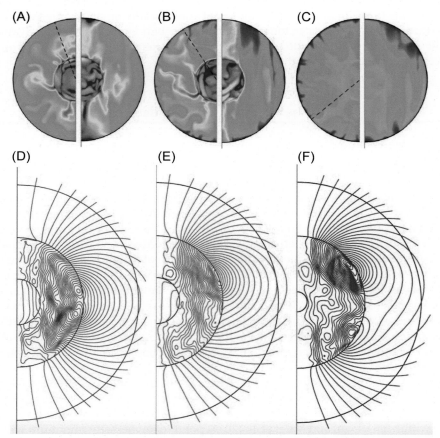

FIG. 8.23 Snapshots of numerical dynamo thermochemical buoyancy and magnetic fields for three core evolution scenarios. First column (scenario 1) corresponds to the post-ICN core; the second column (scenario 2) corresponds to the post-ICN core with elevated core-mantle boundary heat flow; and the third column (scenario 3) corresponds to the pre-ICN core. *Top row images* (A)–(C) show equatorial slices (*left image*) and meridional slices (*right image*) of outer core thermochemical buoyancy, with *red and blue* corresponding to positive and negative buoyancy, respectively. The *dashed lines* indicate the orientations of the meridional slices. *Bottom row images* (D)–(F) show axisymmetric magnetic field lines, with *red* and *blue* indicating normal and reversed poloidal magnetic fields, respectively. *Modified from Fig. 3 of Heimpel, M.H., Evans, M.E., 2013. Testing the geomagnetic dipole and reversing dynamo models over Earth's cooling history. Phys. Earth Planet. Inter. 224, 124–131.*

dipole field. The differences in the core field morphology between scenarios 1 and 3 in Fig. 8.23, particularly the reduced magnetic inclinations connected to the absence of inner core buoyancy production in scenario 3, offer an observational basis for testing models of core evolution that predict rapid cooling and light element exsolution in the pre-ICN core.

8.4 Seeding the early geodynamo

8.4.1 Seed magnetic fields

There is no shortage of possible magnetic fields external to the Earth from which the geodynamo could have been seeded. Magnetic fields are found on all cosmic scales, not just those originating from within planets and stars, but also distributed across galaxies and galaxy clusters. These latter magnetic fields typically have intensities measured in fractions of a nanoTesla (10^{-9} T), but are nevertheless coherent over distances measured in kiloparcecs, that is, distances of order 10^{17} km (Clarke et al., 2001). Furthermore, there is evidence for even weaker but spatially coherent magnetic fields permeating the intergalactic medium, with intensities of order 10^{-20} T (Neronov and Vovk, 2010). Speculation as to the origin of these intergalactic fields centers around the possibility they are primordial, relics of the early expansion of the universe that were subsequently amplified by dynamo action within galaxies and stars (Subramanian, 2016).

8.4.1.1 Astrophysical batteries

In addition to primordial magnetic fields, seed magnetic fields can be generated at any time in cosmic history through the action of astrophysical batteries. Most astrophysical batteries operate on the difference in dynamical response of light electrons versus heavy ions in thermal plasmas. Subject to the same pressure gradient, thermal plasma electrons are accelerated relative to the heavier ions; wherever this acceleration produces charge separation, an electric field is generated. If, because of its geometry, this electric field has a nonzero curl, then by Faraday's law, a magnetic field will grow. This mechanism is called the *Biermann battery*, first proposed in 1950 by L. Biermann. The presence of the Biermann battery effect in a thermal plasma introduces a source term into the magnetic induction equation (see Appendix 2 in Chapter 3), which then becomes

$$\frac{\partial \mathbf{B}}{\partial t} - \nabla \times (\mathbf{v} \times \mathbf{B}) - \kappa_B \nabla^2 \mathbf{B} = \frac{\nabla P_e \times \nabla n_e}{q_e n_e^2}, \tag{8.9}$$

where q_e and n_e are the charge and number density of the electron, and P_e is the electron pressure. According to Eq. (8.9) wherever the plasma is *baroclinic*, that is, wherever plasma electron density and pressure gradients are misaligned, this source term is nonzero and magnetic fields can be generated spontaneously.

8.4.1.2 Solar nebula seeding

Paleomagnetic studies of meteorites have shown that the solar nebula was likely magnetized, and that some early planetary bodies generated dynamo magnetic fields in their liquid or partly liquid metallic cores. Evidence for strong magnetic fields in the solar nebula includes magnetized chondrules (Fu et al., 2014) along with observations of magnetic fields in other protoplanetary disks (Donati et al., 2005; Stephens et al., 2014). In addition, there are long-standing theoretical arguments favoring dynamo action and angular momentum transfer by large-scale magnetic fields in the solar nebula (Hayashi, 1981; Stepinski, 1992). Because the solar nebula eroded away within the first few million years of solar system history, it is expected that seeding by the nebular magnetic field was limited to those planetesimals and protoplanetary bodies that had experienced early core formation (Weiss et al., 2010). Nevertheless, it is entirely possible that the proto-Earth had some sort of dynamo at this time.

8.4.2 Geodynamo seeding times

The time required to seed the geodynamo starting from a weak external field can be estimated using the α^2-dynamo model in Chapter 3. As shown in Appendix 8.2 starting from an initial seed magnetic field with intensity B_{seed}, the time required for this kinematic dynamo to amplify its magnetic field to intensity B is given by the product of the exponential growth time $\tau_{grow} = (r_{CMB}^2/20\kappa_B)(Rm_{crit}/Rm)$ and the natural log of the magnetic field amplification factor B/B_{seed}:

$$\tau_{seed} = \tau_{grow} \ln\left(\frac{B}{B_{seed}}\right) \simeq \frac{r_{CMB}^2}{20\kappa_B}\left(\frac{Rm_{crit}}{Rm}\right)\ln\left(\frac{B}{B_{seed}}\right). \tag{8.10}$$

Here, Rm/Rm_{crit} represents the supercriticality of the dynamo expressed in terms of magnetic Reynolds number.

We can use Eq. (8.10) to calculate geodynamo seeding times based on the estimated intensities of the various magnetic fields that have been proposed as geomagnetic seeds. For the early core, we will assume $B = 5$ mT for the r.m.s. geomagnetic intensity. Using $\kappa_B = 1$ m^2 s^{-1} for the magnetic diffusivity gives $r_{CMB}^2/20\kappa_B \simeq 20$ kyr. In the present-day outer core, $Rm/Rm_{crit} \simeq 100$ (see Chapter 3). For purposes of estimation, we will assume one low and one high value for this ratio at the time of seeding, that is, 1 and 100, respectively.

Seeding the geodynamo from the intergalactic magnetic field with $B_{seed} \simeq 10^{-20}$ T implies $\ln(B/B_{seed}) \simeq 41$; substituting into Eq. (8.10) yields $\tau_{seed} = 820$ kyr for $Rm/Rm_{crit} = 1$ and only 8.2 kyr for $Rm/Rm_{crit} = 100$. All the other proposed seed magnetic fields (induction by nebular magnetic fields, electromagnetic implantation by core merging, etc.) imply even shorter seeding times.

To be sure, the kinematic growth assumption used in these estimates only applies when the magnetic intensity is so low that the Lorentz force is negligible in the dynamics. Once the magnetic intensity becomes large enough to produce an appreciable Lorentz force, the magnetic equilibration process begins and the exponential growth phase of the magnetic field is over. That would lengthen the seeding time somewhat, but even if it lengthens the seeding time by a factor of 10 (to 8 Myr, say), that is still very short compared to the age of the geodynamo and is also far shorter than the time resolution we now have for the ancient paleomagnetic field.

In summary, seeding times for the geodynamo and the various candidates for seed magnetic fields are astrophysically interesting questions, but seeding times of the core were likely too short to impact our interpretation of the paleomagnetic record. Nevertheless, short seeding times and the abundance of candidate seed magnetic fields are geophysically significant in one respect: they are consistent with the idea that the geodynamo may have been intermittent. Because seeding times are short and seed magnetic fields are always available, once the conditions in the core became favorable, the geodynamo would have started quickly. Likewise, if the geodynamo had ceased to operate at some time in the past, it would have quickly restarted once dynamo-favorable conditions in the core returned. Therefore, better questions to ask are: when and how did dynamo-favorable conditions first appear, and did dynamo-unfavorable conditions ever develop?

8.4.3 Geodynamo initiation

We argued previously in this chapter that a probable outcome of the Moon-forming event was stratification of the core. This stratification would tend to inhibit convective dynamo action, thereby either extinguishing a preexisting dynamo or delaying its start. What mechanisms were around in the newly completed core to overcome this inhibition?

8.4.3.1 Electromagnetic implantation

As previously noted, numerical and laboratory experiments (Nakajima and Stevenson, 2015; Landeau et al., 2016) show that a Mars-size core would be large enough to merge with the proto-Earth core without entirely fragmenting, the merging process being accomplished within a matter of weeks if not days. This opens the possibility that the proto-Earth core could have acquired a set of electric currents from the impactor Theia's core. Helical turbulence in the core, assuming the core was molten, could potentially amplify the implanted electric currents and their associated magnetic fields. Furthermore, it has been argued that the magma ocean through which the impactor core descended had a very high electrical conductivity of its own (Stixrude et al., 2020), allowing the implanted electric currents to spread into the molten mantle. This scenario for re-starting the geodynamo is not without its problems, however. The most serious problem is the one noted earlier: thermal and compositional stratification of the core and magma ocean produced by the impact. Both of these effects would tend to inhibit dynamo action until most of the stratification was removed.

8.4.3.2 Tidal stirring

Another possible mechanism for homogenizing the core and initiating dynamo action is the stirring and mixing that is expected to have occurred through the action of tidal forces just after the Moon formed, when it was far closer to the Earth. A plausible energy source for stirring the core at this time are several types of inertial instabilities, generated by the very strong tidal forces exerted on the core by the nearby Moon, particularly if the Moon formed at its Roche limit, approximately 1.85×10^4 km distant from the Earth.

Because the tide-raising force depends inversely on distance between Earth and Moon raised to the third power, this force was far more effective when the Moon was closer. Today, most of the rotational energy dissipation by tidal

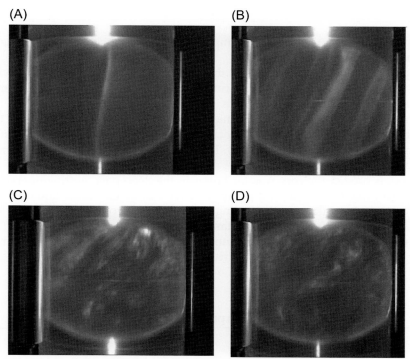

FIG. 8.24 Elliptical instability. Flows produced by elliptical instability in a rotating oblate spheroid of Rheoscopic fluid at Ekman number $Ek = 9.5 \times 10^{-6}$ and spheroidal eccentricity $e = 0.09$, for different values of frequency ratio of the elliptical deformation relative to rotation ϵ. The instability is driven by outer boundary deformation provided by two rollers located 180 degrees apart. Note that the pattern of the instability is slightly inclined to the rotation axis of the fluid. (A) $\epsilon = 0.03$, (B) $\epsilon = 0.04$, (C) $\epsilon = 0.05$, (D) $\epsilon = 0.06$. *Reproduced from Fig. 5 of Le Bars, M., Lacaze, L., Le Dizes, S., Le Gal, P., Rieutord, M., 2010. Tidal instability in stellar and planetary binary systems. Phys. Earth Planet. Inter. 178, 48–55.*

friction takes place in shallow seas; in the deep past, however, without shallow seas, tidal dissipation was likely more concentrated in the mantle and in the core. In particular, the deformation of the core-mantle boundary is predicted to have been greater then, which is precisely the condition under which inertial instabilities are excited.

As mentioned in Chapter 4, deviations from uniform rotation can induce fluid motions in the outer core through inertial instability. This category includes inertial instabilities that occur when the streamlines of the rotating fluid have an elliptical shape, called *elliptical instabilities*. Elliptical instabilities are related to a parametric resonance of a pair of inertial waves generated in a rotating fluid by the elliptical motion (Kerswell, 2002) and are also related to the so-called *tilt-over mode*, rotation of the core about an axis slightly inclined to the mantle rotation axis. The deformation of the core-mantle boundary by tidal forces produces consists of a spherical harmonic degree 2 pattern, which leads to elliptical streamlines in the outer core. These can be highly unstable, leading to intense bursts of quasigeostrophic turbulence capable of homogenizing the early core.

The experimental techniques for exciting elliptical instabilities were pioneered by W.V.R. Malkus in 1969 and have been developed further by M. Le Bars (Le Bars et al., 2010). As illustrated in Fig. 8.24, a flexible rotating fluid-filled sphere or spheroid is deformed by a pair of rollers that are either fixed in the laboratory frame or rotate differentially, producing an elliptical wave in the fluid. The dimensionless control parameters are the Ekman number Ek of the rotation, the eccentricity of the elliptical deformation e, and the frequency ratio of the deformation relative to rotation, ϵ.

Fig. 8.24 shows that the flows driven by the elliptical instability are quasigeostrophic but typically more turbulent and less strongly columnar than rotating convection at the same Ekman number. Nevertheless, the interaction between convective and elliptical instabilities within the same fluid is important, since one type of motion may have led to the other. Lavorel and Le Bars (2010) studied coexisting convective and elliptical instabilities experimentally and found that elliptical instabilities can be excited in the presence of stable as well as unstable thermal gradients. They also found that the superposition of elliptical instabilities increases the heat transported by the convection and that the kinematic helicity from elliptical instabilities is similar to the helicity from rotating convection. This suggests there may have been a smooth transition from rotational power to convective power for the early geodynamo, possibly with little observable effect on the early geomagnetic field.

8.5 Summary

- Earth accreted from solar nebula material, through impacts among successively larger planetesimals and planetary embryos, through pebble accretion, or both.
- Core formation was probably synchronous with Earth accretion, with some impactors having preformed cores of their own. Isotopic evidence indicates partial equilibrium was established between core-forming metals and mantle silicates prior to the metals segregating into the core.
- Although the paleomagnetic record confirms a geomagnetic field back to 3.4 Ga, the onset time of the geodynamo has not been established. However, meteorite evidence indicates possible dynamo action in protoplanetary bodies in the first few million years of solar system history. Disruptive events such as Moon formation likely inhibited convective dynamo action in the proto-Earth.
- Thermal evolution models predict the core was far hotter in the deep past, with temperatures at the core-mantle boundary above the lower mantle solidus and possibly above the lower mantle liquidus.
- The age of the inner core remains unknown. According to the thermal evolution model predictions, the time of inner core nucleation may be in the range 1600–400 Ma, far younger than the core itself.
- Proposed energy sources for the geodynamo before inner core nucleation include convection driven by exsolution of light elements as the core cooled, along with fluid instabilities excited by tidal forces.

Appendix

Appendix 8.1 Core evolution model

Using only the major terms in the heat balance for the core derived in Chapter 2, the growth rate of the inner core can be written as

$$\dot{r}_{ICB} = \frac{Q_{CMB} - Q_R}{\mathcal{R}_{LH} + \mathcal{R}_C + \mathcal{R}_{SC}}, \tag{8.11}$$

where $\dot{r}_{ICB}$ is the rate of change of the inner core radius and Q_{CMB} and Q_R are total heat flow at the core-mantle boundary and radioactive heat production in the core, respectively. The denominator in Eq. (8.11) is the sum of contributions to the energy budget from latent heat release (LH), gravitational energy release (C), and release of sensible heat (secular cooling, SC). Analytical representations of these factors are given below.

Assuming that the inner core light element concentration remains constant during inner core growth, the average light element concentration in the outer core evolves during inner core growth according to

$$\dot{C}_{OC} = \frac{3\rho_{IC} r_{ICB}^2 \dot{r}_{ICB}}{\rho_{OC} r_{CMB}^3}(C_{OC} - C_{IC}), \tag{8.12}$$

where the subscripts OC and IC represent volume averages for the outer and inner core, respectively. In deriving Eq. (8.12), we have ignored small terms proportional to the ratio of inner to outer core volumes.

By integrating Eqs. (8.11), (8.12) backward in time starting from present-day conditions and using the expressions for the $\mathcal{R}$-factors given below, the physical properties in Tables S1 and 8.1 from Core Properties and Parameters, along with assumed values for Q_{CMB} and Q_R, we obtain the history of inner core growth and the simultaneous evolution of outer core light element concentration.

The next step is to derive analytical representations of the individual $\mathcal{R}$-factors that are consistent with the radial variations of density, adiabatic temperature, and melting temperature through the core (Anzellini et al., 2013). Using parameterizations of these variables developed by Labrosse (2003) yields

$$\mathcal{R}_{LH} = 4\pi r_{ICB}^2 \rho_{ICB} L, \tag{8.13}$$

$$\mathcal{R}_C = \frac{8\pi^2}{3}\mathcal{G}\Delta\rho\,\rho_c r_{ICB}^2 r_{ICB}^2 \left(\frac{3}{5} - \frac{r_{ICB}^2}{r_{ICB}^2}\right), \tag{8.14}$$

$$\mathcal{R}_{SC} = 4\pi H^3 \rho_c c_P T_M(0)\left(1 - \frac{2}{3\gamma}\right)\frac{r_{ICB}}{r_T^2}\exp\left[\left(\frac{2}{3\gamma} - 1\right)\frac{r_{ICB}^2}{r_T^2}\right]I(\mathcal{H}, r_{ICB}), \tag{8.15}$$

in which the radial profiles of density ρ, gravity g, melting temperature T_M, and temperature T (assumed to be adiabatic) in the outer core are given by

$$\rho = \rho_c \exp\left(-\frac{r^2}{r_\rho^2}\right),\tag{8.16}$$

$$g = \frac{4\pi}{3}\mathcal{G}\rho_c r\left(1 - \frac{3r^2}{5r_\rho^2}\right),\tag{8.17}$$

$$T_M = T_M(0)\exp\left[-2\left(1 - \frac{1}{3\gamma}\right)\frac{r^2}{r_T^2}\right],\tag{8.18}$$

$$T = T_M(r_{ICB})\exp\left(\frac{r_{ICB}^2 - r^2}{r_T^2}\right),\tag{8.19}$$

in which

$$r_\rho = \sqrt{\frac{3K_0}{2\pi\mathcal{G}\rho_0\rho_c}\left(\ln\frac{\rho_c}{\rho_0} + 1\right)}, \quad r_T = \sqrt{\frac{3c_P}{2\pi\alpha_{Tc}\rho_c\mathcal{G}}}.\tag{8.20}$$

Here, $T_M(0)$ is the melting temperature at the center of the core, r_{ICB} is the radius of the inner core, γ is the Grüneisen coefficient assumed constant, ρ_0 and ρ_c are the density of liquid core material at zero pressure and at the center of the core, respectively, K_0 is the incompressibility (bulk modulus) at zero pressure, $\mathcal{G}$ is the gravitational constant, c_P is the heat capacity assumed constant, and α_{Tc} is the coefficient of thermal expansion of liquid core material at the center of the core, L is the latent heat of melting, $\Delta\rho$ is the density difference between inner and outer core due to differences in their light element contents, $\mathcal{H} = \left(1/r_\rho^2 + 1/r_T^2\right)^{-1/2}$, and

$$I(\mathcal{H}, r_{CMB}) = \frac{\sqrt{\pi}}{2}\mathrm{erf}\left(\frac{r_{CMB}}{\mathcal{H}}\right) - \frac{r_{CMB}}{\mathcal{H}}\exp\left(-\frac{r_{CMB}^2}{\mathcal{H}^2}\right),\tag{8.21}$$

where erf is the error function.

Prior to inner core nucleation, the temperature profile through the entirely liquid core is given by

$$T = T_c \exp\left(-\frac{r^2}{r_T^2}\right),\tag{8.22}$$

where T_c is the central temperature at $r = 0$. Combining a global heat balance of the form

$$c_P \int \rho \dot{T} dV = -(Q_{CMB} - Q_R)\tag{8.23}$$

along with Eq. (8.22), the previous expression for ρ yields the following equation for the change in central temperature with time:

$$\dot{T}_c = -\frac{4\pi\rho_c c_P}{Q_{CMB} - Q_R}\int \exp\left(-\frac{r^2}{r_\rho^2} - \frac{r^2}{r_T^2}\right)r^2 dr.\tag{8.24}$$

Appendix 8.2 Dynamo seeding times

The time required for seeding the geodynamo starting from a weak seed field can be estimated using the α^2-dynamo model in Chapter 3. In that dynamo, the toroidal magnetic field potential $\mathcal{T}$ obeys, in its diffusion-free limit

$$\left(\frac{\partial^2}{\partial t^2} + \alpha^2 \nabla^2\right)\mathcal{T} = 0.\tag{8.25}$$

Here we are interested in the time scale of the kinematic growth phase of the dynamo, when the α-parameter is larger than critical but the magnetic field is too weak to provide an appreciable Lorentz force. Under these conditions, we can write $\alpha = \alpha_{crit} Rm/Rm_{crit}$, where Rm is the magnetic Reynolds number of the core flow, and from Chapter 3, $\alpha_{crit} \simeq 4.49 \kappa_B/r_{CMB}$ is the critical value of the α-parameter for a dipolar magnetic field.

Dynamo solutions to Eq. (8.25) then have the form

$$\mathcal{T} = \mathcal{T}_{seed} \exp(t/\tau_{grow}), \tag{8.26}$$

where $\mathcal{T}_{seed}$ is the seed magnetic field potential and τ_{grow} is the exponential growth time scale during the seeding process. Substituting Eq. (8.26) into Eq. (8.25) and application of the boundary condition $\mathcal{T}=0$ at r_{CMB} yields

$$\tau_{grow} \simeq \frac{r_{CMB}^2}{20\kappa_B}\left(\frac{Rm_{crit}}{Rm}\right) \tag{8.27}$$

for the exponential growth time scale during seeding. Starting from an initial seed magnetic field with intensity B_{seed} proportional to $\mathcal{T}_{seed}$, the time required to seed the dynamo, that is, amplify its magnetic field to intensity B proportional to $\mathcal{T}$ is given by

$$\tau_{seed} = \tau_{grow} \ln\left(\frac{B}{B_{seed}}\right) \simeq \frac{r_{CMB}^2}{20\kappa_B}\left(\frac{Rm_{crit}}{Rm}\right)\ln\left(\frac{B}{B_{seed}}\right). \tag{8.28}$$

References

Andrault, D., Bolfan-Casanova, N., Lo Nigro, G., Bouhifd, M.A., 2011. Solidus and liquidus profiles of chondritic mantle: implication for melting of the earth across its history. Earth Planet Sci. Lett. 304, 251–259.
Ansdell, M., Williams, J.P., Manara, C.F., 2017. An alma survey of protoplanetary disks in the σ Orionis cluster. Astrophys. J. 153, 240.
Anzellini, S., Dewaele, A., Mezouar, M., 2013. Melting of iron at Earth's inner core boundary based on fast X-ray diffraction. Science 340, 464–467.
Arkani-Hamed, J., 2017. Formation of a solid inner during earth accretion. J. Geophys. Res. 122, 3248–3285.
Aubert, J., Labrosse, S., Poitou, C., 2009. Modelling the palaeo-evolution of the geodynamo. Geophys. J. Int. 179, 1414–1428.
Badro, J., Walter, M.J., 2015. The Early Earth, Accretion and Differentiation. AGU Geophysical Monograph Series.
Badro, J., Siebert, J., Nimmo, F., 2016. An early geodynamo driven by exsolution of mantle components from Earth's core. Nature 536, 326–328.
Badro, J., Aubert, J., Hirose, K., Nomura, R., 2019. Magnesium partitioning between Earth's mantle and core and its potential to drive an early exsolution geodynamo. Geophys. Res. Lett. 45, 13240–13248.
Biggin, A.J., de Wit, M.J., Langereis, C.G., Zegers, T.E., 2011. Palaeomagnetism of Archaean rocks of the Onverwacht group, Barberton Greenstone Belt (Southern Africa): evidence for a stable and potentially reversing geomagnetic field at ca. 3.5 Ga. Earth Planet. Sci. Lett. 302, 314–328.
Biggin, A.J., Piispa, E.J., Pesonen, L.J., Holme, R., Paterson, G.A., Veikkolainen, T., Tauxe, L., 2015. Palaeomagnetic field intensity variations suggest mesoproterozoic inner-core nucleation. Nature 526, 245–248.
Borlina, C.S., Weiss, B.P., Lima, E.A., Tang, F., et al., 2020. Re-evaluating the evidence for a Hadean-Eoarchean dynamo. Sci. Adv. 6, eaav9634.
Brearley, A.J., Jones, R.H., 1998. Chondritic meteorites (Chapter 3). In: Papike, J. (Ed.), Reviews in Mineralogy 36: Planetary Materials. Mineralogical Society of America, Washington, DC, pp. 1–398.
Buffett, B.A., Garnero, E.J., Jeanloz, R., 2000. Sediments at the top of the core. Science 290, 1338–1342.
Canup, R.M., 2004. Simulations of a late lunar-forming impact. Icarus 168, 433–456.
Canup, R.M., 2012. Forming a moon with an Earth-like composition via a giant impact. Science 227, 1052–1055.
Carter, P.J., Lock, S.J., Stewart, S.T., 2020. The energy budgets of giant impacts. J. Geophys. Res. Planets 125. e2019JE006042.
Chambers, J.E., 2004. Planetary accretion in the inner solar system. Earth Planet. Sci. Lett. 223, 241–252.
Chambers, J.E., 2016. Pebble accretion and the diversity of planetary systems. Astrophys. J. 825, 63.
Christensen, U.R., 2015. Iron snow dynamo models for ganymede. Icarus 247, 248–259.
Citron, R.I., Perets, H.B., Aharonson, O., 2018. The role of multiple giant impacts in the formation of the Earth-Moon system. Astrophys. J. 862, 5.
Clarke, T.E., Kronberg, P.P., Bohringer, H., 2001. A new radio-X-ray probe of galaxy cluster magnetic fields. Astrophys. J. Lett. 547, L111.
Clesi, V., Monteux, J., Qaddah, B., Le Bars, M., 2020. Dynamics of core-mantle separation: influence of viscosity contrast and metal/silicate partition coefficients on the chemical equilibrium. Phys. Earth Planet. Inter. 306, 106547.
Ćuk, M., Stewart, S., 2012. Making the Moon from a fast-spinning Earth: a giant impact followed by resonant despinning. Science 338, 1047–1052.
Cyburt, R., Fields, B.D., Olive, K.A., Yeh, T.H., 2016. Big bang nucleosynthesis: present status. Rev. Mod. Phys. 88, 015004.
Dahl, T.W., Stevenson, D.J., 2010. Turbulent mixing of metal and silicate during planet accretion and interpretation of the Hf-W chronometer. Earth Planet. Sci. Lett. 295, 177–186.
Davies, C.J., 2015. Cooling history of Earth's core with high thermal conductivity. Phys. Earth Planet. Inter. 247, 65–79.
Deguen, R., Landeau, M., Olson, P., 2014. Turbulent metal-silicate mixing, fragmentation, and equilibration in magma oceans. Earth Planet. Sci. Lett. 391, 274–287.
Donati, J.F., Paletou, F., Bouvier, J., Ferreira, J., 2005. Direct detection of a magnetic field in the innermost regions of an accretion disk. Nature 438, 466–469.
Driscoll, P., 2016. Simulating 2 Ga of geodynamo history. Geophys. Res. Lett. 43, 5680–5687.
Driscoll, P., Bercovici, D., 2014. On the thermal and magnetic histories of Earth and Venus: influences of melting, radioactivity, and conductivity. Phys. Earth Planet. Inter. 236, 36–51.

References

Dumberry, M., Rivoldini, A., 2015. Mercury's inner core size and core-crystallization regime. Icarus 248, 254–268.

Evans, D., 2013. Reconstructing pre-Pangean supercontinents. Geol. Soc. Am. Bull. 125, 1735–1751.

Fu, R.R., Weiss, B.P., Lima, E.A., 2014. Solar nebula magnetic fields recorded in the Semarkona meteorite. Sci. Express Rep. https://doi.org/10.1126/science.1258022.

Hayashi, C., 1981. Structure of the solar nebula, growth and decay of magnetic fields and effects of magnetic and turbulent viscosities on the nebula. Progr. Theor. Phys. 70, 35–53.

Heimpel, M.H., Evans, M.E., 2013. Testing the geomagnetic dipole and reversing dynamo models over Earth's cooling history. Phys. Earth Planet. Inter. 224, 124–131.

Hosono, N., Karato, S., Makino, J., 2019. Terrestrial magma ocean origin of the moon. Nat. Geosci. 12, 418–423.

Hsieh, W.P., Goncharov, A.F., Labrosse, S., 2020. Low thermal conductivity of iron-silicon alloys at Earth's core conditions with implications for the geodynamo. Nat. Commun. 11, 3332.

Ikoma, M., Genda, H., 2006. Constraints on the mass of a habitable planet with water of nebular origin. Astrophys. J. 648, 696–706.

Johansen, A., Ronnet, T., Bizzarro, M., Schiller, M., et al., 2021. Pebble accretion model for the formation of the terrestrial planets in the solar system. Sci. Adv. 7, eabc0444.

Kerswell, R.R., 2002. Elliptical instability. Ann. Rev. Fluid Mech. 34, 83–113.

Kleine, T., Rudge, J.F., 2011. Chronometry of meteorites and the formation of the Earth and Moon. Elements 7, 41–46.

Kleine, T., Walker, R.J., 2017. Tungsten isotopes in planets. Ann. Rev. Earth Planet. Sci. 45, 389–417.

Labrosse, S., 2003. Thermal and magnetic evolution of the Earth's core. Phys. Earth Planet. Inter. 140, 127–143.

Labrosse, S., 2015. Thermal evolution of the core with a high thermal conductivity. Phys. Earth Planet. Inter. 247, 36–55.

Lambrechts, M., Johansen, A., 2012. Rapid growth of gas-giant cores by pebble accretion. Astron. Astrophys. 544. 13 pp.

Landeau, M., Olson, P., Deguen, R., Hirsh, B., 2016. Core merging and stratification after giant impacts. Nat. Geosci. 9, 786–789.

Landeau, M., Aubert, J., Olson, P., 2017. The signature of inner-core nucleation on the geodynamo. Earth Planet. Sci. Lett. 465, 193–204.

Landeau, M., Deguen, R., Phillips, D., Neufeld, J.A., Lherm, V., Dalziel, S.B., 2021. Metal-silicate mixing by large Earth-forming impacts. Earth Planet. Sci. Lett. 564, 116888.

Lavorel, G., Le Bars, M., 2010. Experimental study of the interaction between convective and elliptical instabilities. Phys. Fluids 22, 11401.

Le Bars, M., Lacaze, L., Le Dizes, S., Le Gal, P., Rieutord, M., 2010. Tidal instability in stellar and planetary binary systems. Phys. Earth Planet. Inter. 178, 48–55.

Le Bars, M., Wieczorek, M., Karatekin, O., Cébron, D., Laneuville, M., 2011. An impact-driven dynamo for the early Moon. Nature 479, 215–218.

Levison, H.F., Kretke, K.A., Walsh, K.J., 2015. Growing the terrestrial planets from the gradual accumulation of sub-meter-sized objects. Proc. Natl. Acad. Sci. U.S.A. 112. https://doi.org/10.1073/pnas.1513364112.

Li, J., Agee, C.B., 1996. Geochemistry of mantle-core differentiation at high pressure. Nature 381, 686–689.

Macouin, M., Valet, J., Besse, J., 2004. Long-term evolution of the geomagnetic dipole moment. Phys. Earth Planet. Inter. 147, 239–246.

Mittal, T., Knezek, N., Arveson, S.M., 2020. Precipitation of multiple light elements to power Earth's early dynamo. Earth Planet. Sci. Lett. 532, 116030.

Mittlefehldt, D.W., McCoy, T.J., Goodrich, C.A., Kracher, A., 1998. Non-chondritic meteorites from asteroidal bodies. In: Papike, J. (Ed.), Reviews in Mineralogy 36: Planetary Materials. Mineralogical Society of America, Chantilly, VA, pp. 44–195 (Chapter 4).

Nakajima, M., Stevenson, D.J., 2015. Melting and mixing states of the Earth's mantle after the moon-forming impact. Earth Planet. Sci. Lett. 427, 286–295.

Neronov, A., Vovk, I., 2010. Evidence for strong extragalactic magnetic fields from Fermi observations of TeV Blazars. Science 328, 73–75.

Nimmo, F., 2015. Thermal and compositional evolution of the core. In: Gerald, S. (Ed.), Treatise on Geophysics. vol. 9. Elsevier, Oxford, pp. 201–219.

Nimmo, F., Kretke, K., Ida, S., et al., 2018. Transforming dust to planets. Space Sci. Rev. 214. https://doi.org/10.1007/s11214-11018-10533-11212.

Olson, P., Landeau, M., Hirsh, B., 2017. Laboratory experiments on rain-driven convection: implications for planetary dynamos. Earth Planet. Sci. Lett. 457, 403–411.

O'Rourke, J., Stevenson, D., 2016. Powering Earth's dynamo with magnesium precipitation from the core. Nature 529, 387–389.

Raymond, S.N., Kokubo, E., Morbidelli, A., Morishima, R., Walsh, K.J., 2014. Terrestrial planet formation at home and abroad. In: Beuther, H., Klessen, R.S., Dullemond, C.P., Henning, T. (Eds.), Protostars and Planets VI. University of Arizona Press, Tucson, AZ.

Righter, K., Sanielson, L.R., Pando, K.M., Shofner, G.A., et al., 2016. Valence and metal/silicate partitioning of Mo: implications for conditions of Earth accretion and core formation. Earth Planet. Sci. Lett. 437, 89–100.

Rubie, D.C., Nimmo, F., Melosh, H.J., 2015. Formation of Earth's core. In: Schubert, G. (Ed.), Treatise on Geophysics. vol. 9. Elsevier, Oxford.

Rudge, J.F., Kleine, T., Bourdon, B., 2010. Broad bounds on Earth's accretion and core formation constrained by geochemical models. Nat. Geosci. 3, 439–443.

Selkin, P., Tauxe, L., 2000. Long-term variations in palaeointensity. Philos. Trans. R. Soc. Lond. A358, 1065–1088.

Shannon, M.C., Agee, C.B., 1998. Percolation of core melts at lower mantle conditions. Science 280, 1059–1061.

Shi, C., Zhang, L., Yang, W., 2013. Formation of an interconnected network of iron melt at Earth's lower mantle conditions. Nat. Geosci. 6, 971–975.

Siebert, J., Corgne, A., Ryerson, F.J., 2011. Systematics of metal-silicate partitioning for many siderophile elements applied to Earth's core formation. Geochim. Cosmochim. Acta 75, 1451–1489.

Smirnov, A., Tarduno, J.A., Kulakov, E., McEnroe, S., Bono, R., 2016. Paleointensity, core thermal conductivity and the unknown age of the inner core. Geophys. J. Int. 205 (2), 1190–1195.

Stark, D.P., 2016. Galaxies in the first billion years after the Big Bang. Ann. Rev. Astron. Astrophys. 54, 761–803.

Stephens, I., Looney, W., Kwon, L.W., Fernandez-Lopez, W.M., et al., 2014. Spatially resolved magnetic field structure in the disk of a T Tauri star. Nature 514, 597–599.

Stepinski, T.F., 1992. Generation of dynamo fields in the primordial solar nebula. Icarus 97, 130–141.

Stixrude, L., Scipioni, R., Desjarlais, M.P., 2020. A silicate dynamo in the early Earth. Nat. Commun. 11, 935.

Subramanian, K., 2016. The origin, evolution and signatures of primordial magnetic fields. Rep. Progr. Phys. 79, 076901.

Takafuji, N., Hirose, K., Ono, S., 2004. Segregation of core melts by permeable flow in the lower mantle. Earth Planet. Sci. Lett. 224, 249–257.

Tarduno, J.A., Cottrell, R.D., Watkeys, M.K., Hofmann, A., 2010. Geodynamo, solar wind, and magnetopause 3.4 to 3.45 billion years ago. Science 327, 1238–1240.

Tarduno, J., Cottrell, R., Davis, W., Nimmo, F., Bono, R., 2015. A Hadean to paleo-Archean geodynamo recorded by single zircon crystals. Science 349, 521–524.

Tauxe, L., 2006. Long-term trends in paleointensity: the contribution of DSDP/ODP submarine basaltic glass collections. Phys. Earth Planet. Inter. 156 (3), 223–241.

Tilgner, A., 2005. Precession driven dynamos. Phys. Fluids 17 (3), 034104.

Valet, J.P., Besse, J., Kumar, A., 2014. The intensity of the geomagnetic field from 2.4 Ga old Indian dykes. Geochem. Geophys. Geosyst. 15, 2426–2437.

Veikkolainen, T., Pesonen, L., 2014. Palaeosecular variation, field reversals and the stability of the geodynamo in the Precambrian. Geophys. J. Int. 199 (3), 1515–1526.

Walsh, K.J., Morbidelli, A., Raymond, S.N., O'Brien, D.P., Mandell, A.M., 2011. Sculpting of the inner solar system by gas-driven orbital migration of Jupiter. Nature 475, 206–209.

Wang, H., Kent, D.V., Rochette, P., 2015. Weaker axially dipolar time-averaged paleo-magnetic field based on multidomain-corrected paleointensities from Galapagos lavas. Proc. Natl. Acad. Sci. U.S.A. 112, 15036–15041.

Weiss, B.P., Gattacceca, J., Stanley, S.S., et al., 2010. Paleomagnetic records of meteorites and early planetesimal differentiation. Space Sci. Rev. 152, 341–390.

Weiss, B.P., Maloof, A.C., Tailby, J., 2015. Pervasive remagnetization of detrital zircon host rocks in the Jack Hills, Western Australia and implications for records of the early geodynamo. Earth Planet. Sci. Lett. 430, 115–128.

Williams, J.P., Cieza, L.A., 2011. Protoplanetary disks and their evolution. Ann. Rev. Astron. Astrophys. 49, 67–117.

Cosmological topics

Andrews, S.M., 2015. Observations of solids in protoplanetary disks. Publ. Astron. Soc. Pacific 127, 961–993.

Cyburt, R., Fields, B.D., Olive, K.A., Yeh, T.H., 2016. Big bang nucleosynthesis: present status. Rev. Mod. Phys. 88, 015004.

Lodders, K., Palme, H., Gail, H.P., 2009. Abundances of the elements in the solar system. In: Trumper, J.E. (Ed.), Landolt-Bornstein. New Series, Astronomy and Astrophysics, Springer Verlag, Berlin, pp. 560–630.

Morbidelli, A., Lambrechts, M., Jacobson, S., Bitsch, B., 2015. The great dichotomy of the Solar System: small terrestrial embryos and massive giant planet cores. Icarus 258, 418–429.

Subramanian, K., 2016. The origin, evolution and signatures of primordial magnetic fields. Rep. Progr. Phys. 79, 076901.

Weiss, B.P., Gattacceca, J., Stanley, S., et al., 2010. Paleomagnetic records of meteorites and early planetesimal differentiation. Space Sci. Rev. 152, 341–390.

Earth accretion

Bottke, W.F., Walker, R.J., Day, J.M., Nesvorny, D., Elkins-Tanton, L., 2010. Stochastic late accretion to Earth, the Moon, and Mars. Science 330 (6010), 1527–1530.

Jacobson, S.A., Morbidelli, A., 2014. Lunar and terrestrial planet formation in the Grand Tack scenario. Phil. Trans. R. Soc. A 372, 20130174.

Johansen, A., Lambrechts, M., 2017. Forming planets via pebble accretion. Ann. Rev. Earth Planet. Sci. 45, 359–387.

Rubie, D.C., Jacobson, S.A., Morbidelli, A., O'Brien, D.P., Young, E.D., et al., 2015. Accretion and differentiation of the terrestrial planets with implications for the compositions of early-formed Solar System bodies and accretion of water. Icarus 248, 89–108.

Suer, T.-A., Siebert, J., Remusat, P., Day, J.M.D., et al., 2021. Reconciling metal-silicate partitioning and late accretion in the Earth. Nat. Commun. 12, 2913.

Core formation and core evolution

Fischer, R.A., Campbell, A.J., Ciesla, F.J., 2017. Sensitivities of Earth's core and mantle composition to accretion and differentiation processes. Earth Planet Sci. Lett. 458, 252–262.

Hirose, K., Labrosse, S., Hernlund, J.W., 2013. Composition and state of the core. Ann. Rev. Earth Planet. Sci. 41, 657–691.

Kendall, J.D., Melosh, H.J., 2016. Differenitiated planetesimal impacts into a terrestrial magma ocean: fate of the iron core. Earth Planet. Sci. Lett. 448, 24–33.

Lay, T., Hernlund, J., Buffett, B.A., 2008. Core-mantle boundary heat flow. Nat. Geosci. 1, 25–32.

Rubie, D., Jacobson, S.A., 2016. Mechanisms and geochemical models of core formation. In: Deep Earth: Physics and Chemistry of the Lower Mantle and Core. AGU Monograph 217, 181–190. John Wiley and Sons, Inc., Hoboken NJ.

Wood, B.J., Walter, M.J., Wade, J., 2006. Accretion of the Earth and segregation of its core. Science 441, 825–833.

Magma oceans and volatiles

Elkins-Tanton, L.T., 2012. Magma oceans in the inner solar system. Ann. Rev. Earth Planet. Sci. 40, 113–139.

Sharp, Z.D., 2017. Nebular ingassing as a source of volatiles to the terrestrial planets. Chem. Geol. 448, 137–150.

CHAPTER 9

Future research goals

9.1 Introduction

The discoveries and the insights documented in this book are outcomes of several decades of geophysical and geochemical research directed at the core. Enormous progress has been made over this time period, including a number of breakthroughs that qualify as Nobel Prize worthy. Yet, it is also fair to say that many of the longest-standing questions about the core remain unanswered, or answered only in the most vague sense. In short, there is much work to be done, first to thoroughly understand Earth's core, and second to extend our knowledge to the cores in other planets across the solar system and the cosmos. We authors make no claim to know the directions that research on the core will take in the future. Nevertheless, it might prove helpful for us to identify several areas that we think are particularly ripe for progress, by virtue of advances in the enabling technologies that are now on the horizon, and also because of the growing interest in finding answers to these questions.

9.2 Seismology

Technical advances that are both incremental and transformative will drive future seismic investigations of Earth's core. The most important incremental advances will be the density of seismic stations, data distribution, and ease of data accessibility.

9.2.1 Station coverage

The infrastructure for data distribution and accessibility is already in place with data centers such as IRIS-DMC and network standards implemented by FDSN. The value of dense wavefield sampling is well proven at the small spatial scales needed for resource exploration, making it possible to exploit continuing advances in data processing and inversion. Ocean coverage is critical for increasing the resolution of core-mantle boundary seismic anomalies from spherical harmonic degree 2 to the much finer scale needed to image mantle plume and slab fragments. Imaging of these small-scale structures along the core-mantle boundary (CMB) will be important for constraining conductive heat transport in the thermal boundary layer above the CMB and for determining the relative contribution of temperature versus composition to seismic velocities. Increasing the collection of body waves sampling the inner core along paths parallel to Earth's rotational axis will be important for improving the resolution of the inner core anisotropy, its depth and azimuthal dependence, and exploring any correlations with spatial variations of the magnetic field intensity on the CMB and rate of geomagnetic secular variation. Since both seismicity and station coverage decrease near the poles, progress in accumulating such paths will be slow. Efforts to make permanent many of the stations in the Earthscope transportable array in Alaska and the Canadian arctic (Busby and Aderhold, 2020) will be the best short-term opportunity.

Densification of stations continues on both land and in the ocean through both temporary and long-term deployments. A rapid increase in the density of both land and ocean sensors suffers not only from limitations in funding, but also in the engineering challenges for deep ocean deployments and land deployments at polar latitudes. Nonetheless, regional short-term networks in both ocean and polar environments will enable incremental progress and may occasionally be revolutionary when short-term networks are designed to target small-scale structures having sharp lateral boundaries near the CMB and ICB.

Another class of body waves paths important to core studies are antipodal seismograms observed at distances between 170 and 180 degrees. These can provide constraints on inner core boundary structure from focused PKP-Cdiff and PKIIKP + PKIIIKP + ⋯ (Cormier, 2015; Tsuboi and Butler, 2020) and on any structure consistent with the existence of an inner core (Cormier and Stroujkova, 2005). Array designs can be targeted at antipodal distances in North Africa and China to record events from the Tonga-Kermedec and South American subduction zones, respectively (Butler and Tsuboi, 2010).

9.2.2 Seismic instrumentation

Increased coverage of the oceans and the polar regions can be enhanced by seismometer designs that are cheap, rugged, and easy to deploy in a hostile environment. Hostile environments include low temperature and high pressure in deep ocean basins and low temperature at the poles. Advances in insulating sensors from the effects of high temperature and corrosive fluids in boreholes can also be applied to insulating sensors from the effects of low temperatures. Equally important as sensor design are support and logistics to maintain even temporary networks in the polar regions. A response to these needs is the GEOIC initiative (Geophysical Earth Observatory for Ice-Covered Environments). GEOIC consists of a collaboration of a group of institutions and IRIS to design, collect, and manage a pool of seismic instruments built to withstand icy (down to $-40°C$) or wet environments and develop methods to reduce installation time and logistics in these environments (Sweet and et al., 2020).

Added to the effects of temperature, pressure, and moisture in recording environments is the need to either assign or accurately determine the orientation of sensors sensitive to three orthogonal components of motions. These requirements are quite similar to those of borehole and planetary seismology. Revolutionary advances are molecular electronic transducers (METs) for measuring ground acceleration and microelectromechanical systems (MEMS) recording. A MET transducer senses ground motion from the movement of a liquid electrolyte relative to fixed electrodes (Huang et al., 2013). Movement of orthogonally oriented fluid-filled tubes can be converted to conventional components of motion without the need of leveling at the ocean bottom or in a harsh environment.

9.2.3 Ambient noise, wavefield correlation, and coherence

Extending methods of modeling and inversion applied to small-scale exploration seismology or acoustics to deeper, large-scale structure continues. Recent examples include advances in the theory and imaging of scattered wavefields exploited in ambient noise correlations and correlation wavefields. These have increased seismic sampling of the core, overcoming limitations in station coverage, and have enabled the detection and measurement of weaker phases sampling the core, such as PKJKP (Section 6.5).

Observation of wavefield coherence across dense seismic arrays can be inverted for depth-dependent models of elastic structure (Zheng and Wu, 2008), revealing small-scale heterogeneity concentrated in narrow depth regions corresponding to lattice phase transformations in the upper mantle (Cormier et al., 2020). It may be possible to extend these stochastic inversion techniques to locate the postperovskite phase transitions in the lowermost mantle, which are important to constrain the temperature profile near the core-mantle boundary and the heat flow out of the core.

9.2.4 Source-time functions

A common processing step in exploration seismology is the removal of the pulse broadening and distorting effects of the source on reflected body waves, which enhance the separation of closely spaced arrivals in time. Surprisingly, neither deconvolution nor convolution of known source-time functions (e.g., Fig. 9.1) is commonly attempted in global seismology. Motivated by earthquake hazards and nuclear test verification, models of seismic sources have become increasingly better at predicting the wavefield at higher frequencies. Currently, earthquake focal mechanisms in the form of moment tensor solutions are broadly disseminated. An easy enhancement would be to include broad dissemination of far-field source-time functions, including standards for such dissemination.

Deep Earth seismology often only relies on waveform data from either deep earthquake sources, uncontaminated by surface reflections and near-source structure, or on smaller earthquake sizes to avoid the complexity of larger, more time-extended, earthquake sources. In the free-oscillation band, a simple step function in displacement has often been assumed as a source representation, which has generally been deemed sufficiently accurate at low frequency. An incorporation of some source-size finiteness in time and space, however, would ensure improved accuracy in resolving split and cross-coupled modes. In the body wave band, source-time function modeling could improve the measurement of differential travel times of closely arriving waves, for example, the times of PKIKP and PKiKP between 120 and 130 degrees.

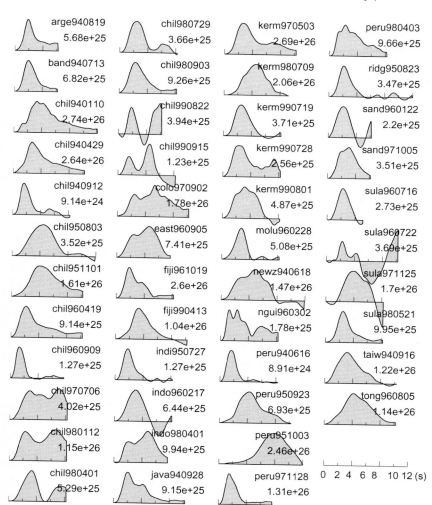

FIG. 9.1 Far-field source-time functions obtained from inverting P waves in the 30–90 degrees range from 46 intermediate and deep-focus earthquakes (Li and Cormier, 2002). All inversions also included a refined moment tensor description of the P-wave radiation pattern using the values of the Harvard CMT catalog (Ekström et al., 2012) as starting solutions (Li and Cormier, 2002).

9.2.5 Numerical synthetic seismograms

Advances in computational hardware and community software for synthesizing seismograms in three-dimensional (3D) models of Earth are currently on the cusp of being able to handle the maximum frequency and distance range of interest to core studies (up to 20,000 km great circle range and 4 Hz). Current capability with the SPECFEM3D-GLOBE (Komatitsch and Tromp, 2002a, b) community code is approaching an order of 0.1 Hz. AxiSEM (Nissen-Meyer et al., 2014), a community code originally initiated for application to 2.5D models, which assume structures that are azimuthally symmetric, can now routinely synthesize a 1 Hz wavefield with cluster hardware available in many universities. With extensions to handle local 3D perturbations and general anisotropy, AxiSEM3D (Leng et al., 2019) can be applied to investigate complex structures in the D″ region. In the not too distant future progress in imaging complex, 3D structure near the inner core boundary, the CMB, and in the inner core will be primarily limited by data availability and not by computational resources or algorithms.

9.3 Mineral physics

The three main approaches of mineral physicists who study the core have been experiments at high pressure, ab initio calculations, and use of analog materials. Modeling also plays a role. Each of the three approaches has something complementary to offer, but are not without their own challenges. We find that progress is best achieved when the approaches are used together, both to verify previous results and to spur new discoveries.

In an ideal world, it would make sense to first determine the bulk composition of the core, then the solid-phase structure of that Fe alloy under the pressure-temperature conditions of the core. With that knowledge, one could then concentrate on transport properties such as the thermal and electrical conductivities, the elastic properties, and the inner core deformation mechanism and viscosity. The latter is the most challenging because it is not solely a material property, but depends on the dynamics. Although this might be the preferred sequence of investigation, the nature of scientific research is understandably that progress on all of these will occur in parallel.

9.3.1 High-pressure experiments

A primary experimental tool for studying the core has been the diamond anvil cell (DAC), often with laser heating (LH-DAC). Because of the static nature of the diamond anvil setup, one can obtain X-ray diffraction (XRD) spectra and, more recently, inelastic X-ray scattering (IXS) data, for determining the structure and elasticity of Fe alloys. The DAC has also been useful for measuring the melting temperature, and properties such as the electrical and thermal conductivities. Nevertheless, obtaining the pressures of the inner core with stable high temperatures is nontrivial, and some extrapolation has been necessary, though inner core conditions have been reached (Tateno et al., 2010). Large temperature gradients have also plagued the interpretation of data, in particular, on the stable crystalline phase and the thermal conductivity. Using a DAC and nonhydrostatic stresses, researchers have also attempted to obtain the anisotropic elastic properties and deformation mechanism of Fe alloys under inner core conditions, but imprecise knowledge of the lattice-preferred orientation that might be present have made the results somewhat untrustworthy. A limitation of DAC studies is the small size of samples, typically on the order of microns, which is unfortunate because the deformation mechanism and hence viscosity can depend on large-scale microstructure. Despite these caveats, it is reasonable to expect steady progress on a range of key properties of Fe alloys under core conditions, and also from extrapolation of DAC results from less technically challenging pressures and temperatures (Anzellini and Boccato, 2020), particularly if and when a clear consensus on the composition and stable phase(s) of the inner core emerges.

Shock compression experiments can more readily achieve the high pressures and temperatures of the inner core, and have been used to determine the melting temperature of Fe alloys (Duffy and Smith, 2019). In theory, dynamical experimental devices such as the Z-machine in Fig. 9.2 enable one to obtain any pressure and temperature between the isotherm and the Hugoniot (the shock wave equation of state), but they are expensive to operate and not easy to control. An issue with these dynamic experiments has been the difficulty of obtaining data fast enough. New systems, however, are now becoming available with 50 ps (picosecond) electronics that will allow collection of XRD spectra as good as that of static experiments, and perhaps promise a new tool to explore the structure of Fe alloys. As with other high-pressure experimental techniques, synchrotrons can be used to produce so-called brilliant sources of X-rays. Somewhat larger samples, several millimeters in cross section, can be used for deformation studies, though the strain rates are very much higher than those relevant to the inner core.

Other experimental tools available for high-pressure studies are the multianvil apparatus (MAA, with some varieties referred to as the DIA, apparently from a 1965 Kobe Steel engineering report) and the piston-cylinder apparatus,

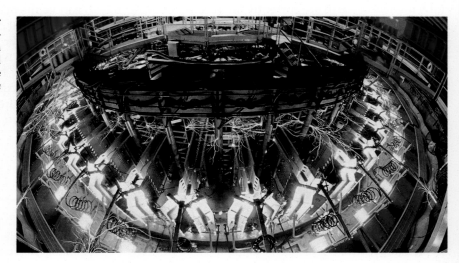

FIG. 9.2 The Z pulsed power facility, or Z machine, operated by the Sandia National Laboratories uses fast capacitor discharge into a plasma to generate extreme temperatures and pressures exceeding 500 GPa, far beyond those at the center of Earth's core. *Credit: Sandia Laboratories.*

though these cannot yet reach even lowermost mantle pressures. The primary advantage of the MAA is the larger size of samples, making it advantageous for deformation studies (then referred to, in an effort to increase the number of acronyms, as the deformation DIA or D-DIA), including determinations of active slip systems and measurements of diffusivities. Because high-pressure/temperature forms of Fe are not stable at ambient conditions, making transmission electron microscopy studies of dislocations unfeasible, slip systems are typically inferred from modeling the resulting deformation textures. The primary disadvantage of the MAA is its limited pressure range.

For instance, in the range of pressures achievable in an MAA, typically less than 30 GPa, one is barely able to reach the hcp Fe stability field, with a phase transition to fcc occurring at higher temperatures at that pressure (Fig. 2.25). Using 14 mm sintered diamond rather than tungsten carbide anvils, however, 90 GPa has now been achieved in an MAA (Zhai and Ito, 2011). The cost of sintered diamond anvils is not insignificant, but recent refinements to the MAA with tungsten carbide anvils have allowed mineral physicists to reach 65 GPa at room temperature, and 48 GPa at 2000 K, albeit with 1.5 mm anvils (Ishii et al., 2019). Although not suited for studying melting temperatures or stable phases of Fe alloys under core conditions, with advances such as these the MAA becomes increasingly attractive as a tool for exploring the elasticity and deformation of hcp Fe alloys.

9.3.2 Ab initio calculations

Ab initio methods (sometimes referred to as first principles or quantum mechanical methods) involve solving Schrodinger's equation to find the properties of a material, typically using density functional theory (Alfe, 2010). Density functional theory reduces an intractable n-electron many-body problem to the easier problem of solving for the electron densities of n noninteracting electrons, through the choice of effective potentials that account for exchange and correlation interactions. The ground-state properties depend on these electron densities. One can then use molecular dynamics (MD) to incorporate the effects of high homologous temperature.

Advantages of these methods are that they require no material property inputs, and are relatively cheap computationally, so that they are ideal for exploring parameter space. Disadvantages are that one must choose potentials appropriately, and there are often other required approximations that are not transparent to the nonspecialist. Moreover, the electronic structure of Fe is notoriously difficult to work with computationally. For this reason, it is important to verify ab initio methods by benchmarking against the transport and elastic properties of transition metals at ambient conditions, for which ample experimental data is available.

Ab initio calculations have delivered some results that reasonably well confirm or have been confirmed by experimental methods, such as the solubility of light elements in liquid and solid Fe under core conditions, and the melting temperature of Fe alloys. They have also delivered some provocative and controversial results that have not yet been confirmed by, and even disagree with, experiments or each other: the stability of bcc Fe, the extremely low shear modulus and viscosity, and high elastic anisotropy of bcc Fe; the high thermal conductivity of Fe; and hcp elastic anisotropy whose sense and magnitude have varied widely by study. There have also been some studies on self-diffusivities and stacking fault energies, more of which will be helpful for determining the inner core deformation mechanism.

Because experiments at the extremely high pressures and temperatures of the core are difficult and often yield inconsistent results, we again stress the need to better benchmark ab initio calculations with each other and with experiments in parameter spaces that overlap, perhaps at lower pressure and temperature, but still in the hcp stability field. This will test the validity of the assumptions and approximations that enter into ab initio calculations for hcp (and bcc) Fe, and give greater confidence in the results in regimes that are experimentally less accessible. As computational power increases, the ability to do MD calculations with a greater number of atoms will also increase confidence in their application to the core.

9.3.3 Analogs and processes

Analogs are most useful as a tool to study processes whose length scales exceed that which is possible to attain computationally or experimentally in a high-pressure device. While some processes, such as nucleation, may be amenable to atomic-scale simulations (Davies et al., 2019), others, such as solidification and deformation (Chapter 7), which may involve large crystals, polycrystals, and texture and microstructure, are not. The difficulty, of course, is choosing an appropriate analog for the process of interest. Factors to consider typically include the crystal structure, lattice constants, stacking fault energies, phase transitions, alloy phase diagrams, and perhaps practical considerations. Other processes, such as core formation, examined earlier in this chapter, may involve suitable choice of density and viscosity ratios. Although the use of experimental analogs (and computational simulations) to model large-scale processes may require many, many orders of magnitude extrapolation, they will continue to play a role in studying a planet's deepest interior.

9.4 Core dynamics

As with seismology and mineral physics, it is useful to divide future research topics in core dynamics into those that require only incremental steps forward and those that will require substantial leaps in our abilities, yet would be more transformative in terms of knowledge gained.

9.4.1 Liquid metal numerical dynamos

In the first category, numerical dynamos continue to progress at a steady rate, paced by the combination of more efficient numerical techniques including massive parallelization and advances in computational hardware. Fig. 9.3 depicts this rate of progress, showing the trade-offs between numerical dynamo resolution and simulated time at two epochs, 2005 and 2020, respectively. Numerical resolution is given in terms of the maximum spherical harmonic degree used in the calculation, n_{max}. The type of applications to geomagnetic phenomena that correspond to each increment of simulated time is also indicated.

In less than two decades, the peak resolution attained by numerical dynamos has increased by more than a factor of 4, and now stands near $n_{max} \simeq 2^{11} = 2048$. Even so, substantial increases will be needed in order for these models to qualify as *direct numerical simulation*, or DNS, of outer core dynamics. Roughly speaking, at least another $2^5 = 32$-fold increase in numerical resolution will be required, in order to resolve the smaller length scales that have been predicted for quasigeostrophic convection in the outer core. Furthermore, there will always be a trade-off between numerical resolution and simulated time. Fig. 9.3 reveals that this trade-off is very severe. Low-resolution numerical dynamos already can simulate large chunks of core history, whereas the best-resolved dynamos can simulate just a few thousand years. So, the prospect of a DNS for modeling core evolution is a long way off indeed.

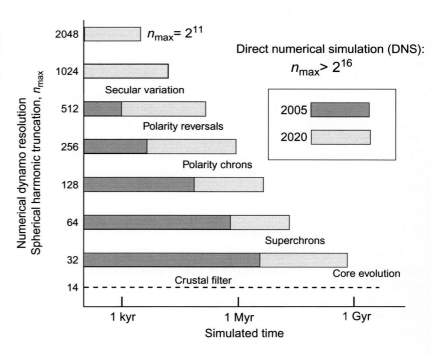

FIG. 9.3 Comparison of numerical dynamo capabilities at epochs 2005 and 2020, in terms of spherical harmonic truncation. *Bar labels* indicate the characteristic geodynamo processes corresponding to the simulation times. Resolution needed for a hypothetical geodynamo direct numerical simulation (DNS) is also shown.

In light of this situation, it is worthwhile reexamining the fundamental goals of numerical dynamos, focusing on where advances can be made with greatest impact and least expenditure. The traditional metric used for dynamo model progress has been the lowest attainable value of the Ekman number Ek, defined in Chapter 4 as the ratio of viscous to Coriolis effects. In the outer core, the Ekman number is in the range $Ek = 10^{-13}$–10^{-15}, whereas state-of-the-art, fully 3D numerical dynamos are now limited to $Ek > 10^{-8}$.

However, reaching the absurdly low Ekman number of the outer core is not really a practical objective, or even a wise one. As discussed in Chapter 4, the highest-resolution numerical dynamos already show effects of wide dynamical scale separation, ranging from global-scale flows and magnetic fields down to localized flows and fields with length scales of tens of kilometers. The force balances in these flows also vary enormously with length scale, changing from geostrophic to buoyancy plus magnetic and finally to viscous at the smallest length scales. In addition, there is the crustal filter to consider, which as Fig. 9.3 shows, cut-off magnetic signals from the core at spherical harmonic degrees not far above $n = 14$, making the small-scale properties of these dynamos effectively invisible to geomagnetic observations, and therefore probably not testable.

A more pressing challenge for numerical dynamos, one that has mostly flown under the radar so far, is to better simulate the electrical conduction properties of the core. This means approaching more realistic values of the magnetic Prandtl number Pm, the ratio of kinematic viscosity to magnetic diffusivity. As discussed in Chapter 4, Pm is of order 10^{-5} in the outer core, whereas most numerical dynamos assume $Pm \simeq 1$, with only a handful of explorations in the range $Pm = 0.1$–0.01. Keep in mind that such large magnetic Prandtl numbers, approaching superconducting behavior, are inconsistent with liquid iron alloy properties. This mismatch leaves open the possibility, however remote, that the numerical dynamos we are now using are fundamentally in error, insofar as their ability to reproduce the Earth's core dynamics over all of the important length and time scales. In addition, the failure to properly model electrical conductivity represents a substantial limitation on our ability to incorporate new results from mineral physics into models of core dynamics and the geodynamo.

Is there independent evidence that reducing the magnetic Prandtl number will make a big difference? Indeed there is. The difficulties experienced by laboratory experiments in reaching self-sustaining dynamo action can be traced, in large part, to the fact that they are required to use very low magnetic Prandtl number fluids, such as the giant University of Maryland liquid sodium sphere shown in Fig. 9.4. In contrast, numerical dynamos are free to assume higher magnetic Prandtl numbers, no matter how unrealistic these may be. And when the magnetic Prandtl number is reduced too far in a numerical dynamo, that dynamo usually fails. Accordingly, there is much room for progress here, including plenty of room for new discoveries.

FIG. 9.4 The 3-m diameter rotating spherical shell containing liquid sodium at the University of Maryland, a laboratory apparatus used for experiments on core dynamics and geodynamo processes. *Credit: D. Lathrop.*

9.4.2 Laboratory experiments on thermochemical convection

Prior to the development of fully 3D numerical models, laboratory fluid dynamics experiments were the primary means for investigating the dynamics of finite amplitude convection in the outer core, as documented in the review by Cardin and Olson (2015). Since that time, 3D numerical models of convection and dynamo action have become the tools of first choice. But because of the limitations of these numerical models, a number of unresolved issues that are important in the core have not received the attention they are due. Chief among these are the true nature of thermochemical convection and the role of convective turbulence in structuring the density and seismic velocity profiles in the outer core.

In dealing with these issues, numerical models rely on the validity of the codensity concept described in Chapter 4, which assumes that the thermal and compositional parts of thermochemical convection are interchangeable. This is clearly an oversimplification, as it ignores their vast differences in diffusivity as well as the differences in the ways that thermal and compositional buoyancies are generated at the inner core boundary and at the core-mantle boundary. Ignoring these differences has repercussions for questions such as the nature and origin of stratification in the E' and F regions, the scale of convection in the outer core, and the relative contributions of thermal versus compositional buoyancy to the generation of the geomagnetic field.

While numerical models will continue to struggle to with the differences between compositional versus thermal properties, laboratory experiments can readily include them. Accordingly, there is an untapped opportunity to combine the recent advances in material fabrication with the recent advances in experimental imaging techniques, to construct a laboratory-scale model of thermochemical convection that captures what is now missing from 3D numerical models.

The basic setup for outer core convection is relatively simple. It consists of a rotating spherical shell in which a thermal buoyancy flux is produced at the outer boundary (either homogeneously or, more likely, heterogeneously) and a large compositional buoyancy flux with a bit of thermal buoyancy is produced at the inner boundary. All of the major pieces of an experimental device for this system have already been tested and employed. Improvements lie in areas of controlled buoyancy production, imaging, and data acquisition.

For example, advances in printed circuitry now allow for virtually any pattern of heat flux to be applied to the outer boundary. In addition, advances in porcelain filter technology allow for a controlled compositional buoyancy flux to be applied to the inner boundary. Similar advances have been made in ultrasonic imaging of the resulting convection (Gillet et al., 2007) and of the density profile resulting from temperature and composition (Zimmerman et al., 2014), as well as magnetic mapping of in electrically conducting fluids (Nataf, 2013). All of these imaging techniques bear directly on the geophysical methods for probing the core described in Chapters 1, 3, and 5. In short, the components of a self-consistent thermochemical core experiment already exist, waiting to be put together.

9.4.3 Core-mantle boundary heat flow and the age of the inner core

The second category of future research topics in core dynamics include those that will require substantial advances in our capabilities. Accordingly, these topics are more aspirational, but they also have the potential to be more transformative. Two high priority but closely related topics in this category are the heat flow at the core-mantle boundary and the age of the inner core. Heat flow at the core-mantle boundary, a central theme in this book, remains in a state of broad uncertainty, in spite of two decades work to place better constraints on it. The same can be said for the age of the inner core, that is, the time of inner core nucleation (ICN). Both of these represent benchmarks for calibrating the evolution of the core, and indeed, the evolution of the Earth as a whole. And in spite of the daunting challenges they present, there is a geophysical pathway forward in each case.

Consider, for example, the search for ICN in the paleomagnetic record. This search is hamstrung on two counts. First, it is not obvious what the primary magnetic signal of the ICN actually is. This shortcoming can be overcome by a more comprehensive effort to model geodynamo behavior around the time of the ICN. The developments described earlier with regard to iron alloy properties, coupled with the enhanced capability to model dynamo action in a fully liquid core under approximately realistic conditions, should be able to resolve this part of the puzzle.

The second part of the puzzle will require more sustained effort and better organization. Even a cursory inspection of the available paleomagnetic data during the period of time in question shows that this data set is far from adequate to detect the ICN, unless we happen to be very lucky. Paleomagnetic intensity measurements of the ancient geomagnetic field exist in clusters with large intervening gaps. So in order to spot the ICN, it would need to have fortuitously occurred at the time of one of the few data clusters. Before we can construct a more comprehensive history of the geomagnetic field in the deep past, capable of pinpointing the ICN, a very large increase in the temporal coverage of paleomagnetic data is needed, probably on the order of $10\times$ compared to what is available today. This will only happen by ramping up the exploration and sampling of ancient crustal terranes, and similarly ramping up the speed by which reliable paleomagnetic intensities can be acquired in the lab. Such increases are very unlikely to happen without the impetus provided by a large international organization.

Are the benefits of an activity on this scale likely to be worth the time and money invested? If only the immediate benefits to the geomagnetic and geodynamic communities are counted, the answer might be no. But these are not the only communities who will benefit. The ability to model the ancient geomagnetic field coupled with a detailed record of its behavior will provide a powerful toolset for understanding the magnetic history of other planets, including extrasolar planets, assuming that magnetic fields will eventually be detected on some.

Of more immediate interest to Earth history, a detailed record of the ancient geomagnetic field provides the means for deciding, once and for all, what role it plays in the evolution of life. Every few years, anecdotal evidence and vague theoretical arguments are published, showing correlations between geodynamo events such as polarity reversals or polarity excursions and various faunal extinctions and climate fluctuations. Intriguing as these are, such isolated correlations will never be wholly convincing, in large part because they are found only in very limited portions of the record. In order to confidently evaluate the connections between geomagnetic field history and the history of life, the geomagnetic field history must be known in far greater detail and with far better precision, compared to what it is now.

A similar argument can be made for a coordinated attack on the problem of heat and mass transfer between the core and mantle, in particular, the present-day heat flow at the core-mantle boundary. Because there are no ways to measure core-mantle heat flow directly, improved constraints on this property require an interdisciplinary approach, with most of its focus on the mantle side of the boundary.

The starting point is to map the seismic structure of the core-mantle boundary region with better coverage and in more detail, taking advantage of the advances in instrumentation and data processing described in Section 9.2. In parallel, it is necessary to resolve the thermal conductivity in this region, particularly the heterogeneity in conductivity related to the transition from perovskite to postperovskite structures in the lowermost mantle. The next step is to better constrain the temperature structure above the core-mantle boundary. For this step, it is necessary to have equations of state for the mantle phases, and equally important, a reliable modeling tool to put the seismic and mineral physics data into a realistic geodynamic framework. For this purpose, improved mantle global circulation models would be the instruments of choice.

The payoffs from a global map of heat transfer in the core-mantle boundary region will be far reaching. For the core, it will provide the upper thermal boundary condition. For the mantle, it will provide a picture of where and how heat from the core is redistributed by lower mantle dynamics. For other terrestrial planets, it will offer a basis for scaling the heat transfer from cores to mantles of diverse sizes. And for the histories of the Earth and other terrestrial planets, it gives a much-needed calibration point on the rates of cooling of their deepest interiors.

References

Alfe, D., 2010. Iron at Earth's core conditions from first principles calculations. Rev. Mineral. Geochem. 71, 337–354.

Anzellini, S., Boccato, S., 2020. A practical review of the laser-heated diamond anvil cell for university laboratories and synchrotron applications. Crystals 10, 459. https://doi.org/10.3390/cryst10060459.

Busby, R.W., Aderhold, K., 2020. The Alaska transportable array: as built. Seismol. Res. Lett. 91 (6), 3017–3027.

Butler, R., Tsuboi, S., 2010. Antipodal seismic observations of temporal and global variation at Earth's inner-outer core boundary. Geophys. Res. Lett. 37. https://doi.org/10.1029/2019GL042908.

Cardin, P., Olson, P., 2015. Experiments on core dynamics (Chapter 8.13). In: Schubert, G. (Ed.), The Treatise on Geophysics. vol. 8. Elsevier, Oxford.

Cormier, V.F., 2015. Detection of inner core solidification from observations of antipodal PKIIKP. Geophys. Res. Lett. 42, 7459–7466. https://doi.org/10.1002/2015GL065367.

Cormier, V.F., Stroujkova, A., 2005. Waveform search for the innermost inner core. Earth Planet. Sci. Lett. 236, 96–105.

Cormier, V.F., Tian, Y., Zheng, Y., 2020. Heterogeneity spectrum of Earth's upper mantle obtained from the coherence of teleseismic P waves. Commun. Comput. Phys. 28, 74–97. https://doi.org/10.4208/cicp.OA-2018-0079.

Davies, C., Pozzo, M., Alfè, D., 2019. Assessing the inner core nucleation paradox with atomic-scale simulations. Earth Planet. Sci. Lett. 507, 1–9.

Duffy, T.S., Smith, R.F., 2019. Ultra-high pressure dynamic compression of geological materials. Front. Earth Sci. 7. https://doi.org/10.3389/feart.2019.00023.

Ekström, G., Nettles, M., Dziewonski, A.M., 2012. The global CMT project 2004–2010: centroid-moment tensors for 13,017 earthquakes. Phys. Earth Planet. Inter. 200–201, 1–9. https://doi.org/10.1016/j.pepi.2012.04.002.

Gillet, N., Brito, D., Jault, D., Nataf, H.C., 2007. Experimental and numerical study of magnetoconvection in a rapidly rotating spherical shell. J. Fluid Mech. 580, 123–143.

Huang, H., Agafonov, V., Yu, H., 2013. Molecular electric transducers as motion sensors: a review. Sensors (Basel) 13 (4), 4581–4597. https://doi.org/10.3390/s130404581.

Ishii, T., Liu, Z., Katsura, T., 2019. A breakthrough in pressure generation by a Kawai-type multi-anvil apparatus with tungsten carbide anvils. Engineering 5, 434–440.

Komatitsch, D., Tromp, J., 2002a. Spectral-element simulations of global seismic wave propagation-I. Validation. Geophys. J. Int. 149 (2), 390–412.

Komatitsch, D., Tromp, J., 2002b. Spectral-element simulations of global seismic wave propagation-II. Three-dimensional models, oceans, rotation and self-gravitation. Geophys. J. Int. 50 (1), 303–318.

Leng, K., Nissen-Meyer, T., van Driel, M., et al., 2019. AxiSEM3D: broad-band seismic wavefields in 3-D global earth models with undulating discontinuities. Geophys. J. Int. 217 (3), 2115–2146. https://doi.org/10.1093/gji/ggz092.

Li, X., Cormier, V.F., 2002. Frequency-dependent seismic attenuation in the inner core 1. A viscoelastic interpretation. J. Geophys. Res. 107 (B12). https://doi.org/10.1029/2002JB0011795.

Nataf, H.C., 2013. Magnetic induction maps in a magnetized spherical Couette flow experiment. C. R. Phys. 14, 248–267.

Nissen-Meyer, T., van Driel, M., Sthler, S.C., Hosseini, K., et al., 2014. AxiSEM: broadband 3-D seismic wavefields in axisymmetric media. Solid Earth 5 (1), 425–445.

Sweet, J., et al., 2020. A shared resource for studying extreme polar environments. EOS 101. https://doi.org/10.1029/2020EO141553.

Tateno, S., Hirose, K., Ohishi, Y., Tatsumi, Y., 2010. The structure of iron in Earth's inner core. Science 330, 359–361.

Tsuboi, S., Butler, R., 2020. Inner core differential rotation inferred from antipodal seismic observations. Phys. Earth Planet. Inter. 301. https://doi.org/10.1016/j.pepi.2020.106451.

Zhai, S., Ito, E., 2011. Recent advances of high-pressure generation in a multianvil apparatus using sintered diamond anvils. Geosci. Front. 2, 101–106.

Zheng, Y., Wu, R.S., 2008. Theory of transmission fluctuations in random media with a depth-dependent background velocity structure. In: Advances in Geophysics, pp. 21–41.

Zimmerman, D., Triana, S., Nataf, H.C., Lathrop, D., 2014. A turbulent, high magnetic Reynolds number experimental model of Earth's core. J. Geophys. Res. Solid Earth 119. https://doi.org/10.1002/2013JB010733.

Notation tables

TABLE N1 Roman symbols.

Math style	Definition
A, a	Surface area, lattice constant, and scatterer length scale
$\mathbf{B}, B$	Magnetic field, magnetic intensity
b	Burger's vector and bond spring constant
C	Light element concentration
c	Specific heat and phase velocity
D	Self-diffusivity and density of states
d	Outer core thickness and grain size
$E, \mathbf{E}, e$	Energy, electric field, efficiency, and electron charge
$F, \mathbf{F}, f$	Buoyancy flux, force, and Coriolis parameter
G, g, g'	Gibbs energy, gravity, and buoyancy
$H, \overline{H}, H_R$	Enthalpy, helicity, and reaction heat
h	Radioactive heat production and height
I	Moment of inertia
$\mathbf{J}$	Electric current density
K	Bulk modulus
k	Thermal conductivity, wavenumber, and partition coefficient
k_F, v_F	Fermi wavenumber, Fermi velocity
L, l	Latent heat, particle-in-a-box size, and vortex diameter
M, m	Mass, dipole moment, and electron mass
N, n	Electron number, electron density, and phonon occupancy
P	Pressure
Q	Heat production/heat loss, quality factor, and activation energy
q	Local heat flux and electric charge
R, r	Mean surface radius, internal radius
S, s	Entropy, cylindrical coordinate
T, t	Temperature, time and periodicity
$U, \mathbf{u}$	Displacements
$V, \mathbf{v}$	Volume and solidification rate, fluid velocity
V_P, V_S	Seismic P-wave, S-wave velocities
v_d	Drift velocity
W	Work rate
Z	Potential vorticity
x, y, z	Cartesian coordinates

Continued

TABLE N1 Roman symbols—cont'd

Calligraphic style	Definition
$\mathcal{B}$	Buoyancy production
$\mathcal{E}$	Mean electromotive force
$\mathcal{P}, \mathcal{T}$	Poloidal, toroidal magnetic potentials
$\mathcal{S}$	Inner core stability

TABLE N2 Greek symbols.

Notation	Definition
α_T, α_C	Thermal, compositional expansions
Γ, γ	Torque, Grüneisen parameter and surface energy
Δ, δ	Difference and great circle angle, thickness
ϵ_{ij}, ϵ	Strain tensor, strain, codensity source, and ellipticity
ϵ_F	Fermi energy
η	Dynamic viscosity
θ	Colatitude and angle
κ_T, κ_C	Thermal, compositional diffusivities
κ_χ, κ_B	Codensity, magnetic diffusivities
Λ, λ	Elsasser number and mean free path, Lamé parameter
μ	Shear modulus and chemical potential
μ_0	Magnetic permeability
ν	Kinematic viscosity
ρ	Density, dislocation density, and electrical resistivity
σ, σ_{ij}	Electrical conductivity and stress, stress tensor
τ	Time scale and superadiabatic temperature gradient
χ	Codensity
Φ	Magnetic flux and dissipation
ϕ	East longitude and seismic parameter
Ψ, ψ	Geomagnetic potential, gravitational potential, and streamfunction
Ω, Ω_p	Rotation, precession angular velocities
ω, ω_D	Frequency, Debye frequency
r, θ, ϕ	Spherical coordinates
s, ϕ, ζ	Cylindrical coordinates

TABLE N3 Subscripts, superscripts, and prefixes.

Subscript	Definition
OC, IC	Outer core, inner core
CMB	Core-mantle boundary
ICB	Inner core boundary
E, EAR	Eutectic, earth surface
H, L	Horizontal and high, low, liquid, and liquidus
M	Melt and melting temperature
CR	Crust and chemical reaction
F	Flux
UM, LM, D″	Upper mantle, lower mantle, D″ region
T, P, V	Temperature, pressure, volume
S	Entropy, solid, and solidus
C, χ	Composition, codensity
B, J, ν	Magnetic, Joule (Ohmic), viscous
LH, R	Latent, radioactive heats
SC	Secular cooling
SV	Secular variation (geomagnetic)
c	Core center
crit	Critical value
e	Electric
ad, k	Adiabatic, conduction
n	Spherical harmonic degree
d	Dipolar and drift
0	Reference value
rms	Root-mean-square
1,2,3,…	Model parts
ij, hkl	Tensor components, miller indices
kin, mag	Kinetic, magnetic (energies)
mantle, core	Whole mantle, whole core
x, y, z	Cartesian vector components
r, θ, ϕ	Spherical vector components

Superscript	Definition
′	Perturbation
°	Degrees
n, p, s	Stress, grain size, grain growth (exponents)

Prefix	Definition
p, n, μ, m	$10^{-12}, 10^{-9}, 10^{-6}, 10^{-3}$
k, M, G, T, P, Z	$10^3, 10^6, 10^9, 10^{12}, 10^{15}, 10^{21}$

TABLE N4 Abbreviations

Abbreviation	Definition
ADM	Axial dipole moment
AU	Astronomical unit
CAI	Calcium-aluminum inclusion
CMB	Core-mantle boundary
D″	Basal mantle region
DIA	Multi-anvil type
DAC	Diamond anvil cell
DNS	Direct numerical simulation
E′, F	Thin upper, lower outer core regions
EBSD	Electron backscatter diffraction
GCM	Global circulation model
ICB	Inner core boundary
ICN	Inner core nucleation
IXS	Inelastic X-ray scattering
LOD	Length-of-day
LH	Laser heated
LPO	Lattice preferred orientation
LLSVP	Large low shear wave velocity province
MAA	Multi-anvil apparatus
MAC	Magnetic-Archimedean-Coriolis
MD	Molecular dynamics
MORB	Mid-ocean ridge basalt
PREM	Preliminary reference earth model
SPO	Shape preferred orientation
ULVZ	Ultra-low velocity zone
VADM	Virtual axial dipole moment
XRD	X-ray diffraction
XUV	Extreme ultraviolet
bcc	Body-centered cubic
dhcp	Double hexagonal close-packed
fcc	Face-centered cubic
hcp	Hexagonal close-packed
ppm	Parts per million
rms	Root-mean-square
sc, sec	Simple cubic, second
yr, a	Years, age in years
ϵ-Fe	High-pressure hcp Fe

TABLE N5 Math and physical units.

Math style	Definition
$v, B, u, \ldots$	Scalars
$\mathbf{v}, \mathbf{B}, \vec{u}, \ldots$	Vectors
$\sigma_{ij}, \epsilon_{ij}, \ldots$	Tensors
$\bar{\rho}, \bar{T}, \ldots$	Averages
$|a|, |b|, \ldots$	Absolute values
$da/dt, \dot{b}$	Full-time derivatives
$du/dx, dw/dr, \ldots$	Ordinary derivatives
$\partial a/\partial t, \partial b/\partial x, \ldots$	Partial derivatives
$\nabla A, \nabla \times \mathbf{B}, \nabla \cdot \mathbf{E}, \ldots$	Vector operators
Physical units (SI)	**Definition**
kg, m, s	Kilogram, meter, second
K, W, J, mol	Kelvin, Watt, Joule, mole
N, A, T, S	Newton, Ampere, Tesla, Siemen
rad, deg	Radian, degree

TABLE N6 Constants and coefficients.

Symbol	Definition	Value (SI units)
π	$4\tan^{-1}(1)$	$3.14159, \ldots$
G	Gravitational constant	6.673×10^{-11} kg^{-1} m^3 s^{-2}
AU	Astronomical unit	1.496×10^8 km
R_g	Gas constant	8.315 J mol^{-1} K^{-1}
μ_0	Vacuum permeability	$4\pi \times 10^{-7}$ N A^{-2}
k_B	Boltzman constant	1.381×10^{-23} J kg^{-1}
$\hbar$	Planck constant (reduced)	1.055×10^{-34} J s
L_0	Lorenz number	2.44×10^{-8} W K^2 s^{-1}
Θ_D	Debye temperature	Variable, K
g_n^m, h_n^m	Gauss coefficients	Variable, T
R_n	Geomagnetic spectrum coefficients	Variable, T^2

Core properties and parameters

Tables S1–S4 list important physical properties and dimensionless parameters for Earth's core. Some are well constrained; for these we give a single value with uncertainty typically in the last digit. Others are less well constrained; for these we give a central value with uncertainty. Still others are very poorly constrained; for these we either give a range or use [~] notation for order of magnitude. Nondimensional properties are identified by n.d.

TABLE S1   Core physical properties.

Property	Notation	Units	Value
Earth radius (mean)	R, r_{EAR}	m	6.371×10^6
Outer core radius (mean)	r_{CMB}	m	3.480×10^6
Inner core radius (mean)	r_{ICB}	m	1.22×10^6
Outer core thickness	d	m	2.26×10^6
Core-mantle boundary (CMB) area	A_{CMB}	m^2	1.52×10^{14}
Inner core boundary (ICB) area	A_{ICB}	m^2	1.87×10^{13}
Core-mantle boundary ellipticity	ϵ_{CMB}	n.d.	2.5×10^{-3}
Outer core volume	V_{OC}	m^3	1.69×10^{20}
Inner core volume	V_{IC}	m^3	7.6×10^{18}
Earth mass	M_{EAR}	kg	5.972×10^{24}
Outer core mass	M_{OC}	kg	1.84×10^{24}
Inner core mass	M_{IC}	kg	9.8×10^{22}
Outer core density (mean)	ρ_{OC}	kg m^{-3}	1.09×10^4
Inner core density (mean)	ρ_{IC}	kg m^{-3}	1.29×10^4
Core moment of inertia	I_{core}	kg m^2	9.2×10^{36}
Density below CMB	ρ_{CMB}	kg m^{-3}	$0.98 \pm 0.03 \times 10^4$
Density above ICB	ρ_{ICB}	kg m^{-3}	$1.22 \pm 0.03 \times 10^4$
Core-mantle boundary gravity	g_{CMB}	m s^{-2}	10.68
Inner core boundary gravity	g_{ICB}	m s^{-2}	4.40
Core-mantle boundary pressure	P_{CMB}	GPa	136
Inner core boundary pressure	P_{ICB}	GPa	323
Central core pressure	P_c	GPa	364
P-wave velocity below CMB	V_P	km s^{-1}	8.05 ± 0.05
P-wave velocity above ICB	V_P	km s^{-1}	10.31 ± 0.05
P-wave velocity jump across ICB	ΔV_P	km s^{-1}	0.7 ± 0.2
S-wave velocity above, below ICB	V_S	km s^{-1}	0–3.5 ± 0.2

Continued

TABLE S1 Core physical properties—cont'd

Property	Notation	Units	Value
Poisson ratio, inner core	ν_P	n.d.	0.44
Angular velocity of rotation	Ω	rad s^{-1}	7.2921×10^{-5}
Core free nutation period	t_N	s	3.71×10^7
Geomagnetic dipole moment	m_d	A m^2	7.7×10^{22}
Geomagnetic dipole tilt	θ_d	deg	9.4
Magnetic intensity, CMB (rms)	B_{CMB}	mT	0.42 ± 0.2
Magnetic intensity, outer core (rms)	B_{rms}	mT	5 ± 2
Magnetic dipole intensity, CMB (rms)	B_{CMB}^{dip}	mT	0.26

TABLE S2 Core thermodynamic and transport properties.

Property	Notation	Units	Range
Core-mantle boundary temperature	T_{CMB}	K	4000 ± 300
Inner core boundary temperature	T_{ICB}	K	5400 ± 500
Outer core temperature (mean)	$\overline{T}$	K	4600 ± 400
Adiabatic temperature gradient, CMB	$-\left(\frac{dT}{dr}\right)_{ad}$	K km^{-1}	0.8 ± 0.2
Adiabatic temperature gradient, ICB	$-\left(\frac{dT}{dr}\right)_{ad}$	K km^{-1}	0.4 ± 0.2
Adiabatic vs. melting gradients, ICB	$\left(\frac{dT}{dr}\right)_{ad} - \left(\frac{dT}{dr}\right)_{M}$	K km^{-1}	0.15 ± 0.1
Density jump, ICB	$\Delta\rho_{ICB}$	kg m^{-3}	500–900
Density jump, composition, ICB	$\Delta\rho_{ICB} - \Delta\rho_{phase}$	kg m^{-3}	400–600
Light element concentration, outer core	C_{OC}	kg kg^{-1}	0.08 ± 0.03
Light element concentration, inner core	C_{IC}	kg kg^{-1}	0.03 ± 0.02
Thermal expansivity	α_T	K^{-1}	$1.3 \pm 0.4 \times 10^{-5}$
Light element expansivity	α_C	n.d.	0.8 ± 0.2
Specific heat	c_P	J kg^{-1} K^{-1}	850 ± 30
Latent heat, crystallization	L	J kg^{-1}	$1 \pm 0.5 \times 10^6$
Grüneisen parameter	γ	n.d.	1.2–1.5
Electrical conductivity	σ	S m^{-1}	4–20×10^5
Thermal conductivity	k	W m^{-1} K^{-1}	20–150
Thermal diffusivity	κ_T	m^2 s^{-1}	10^{-5}–10^{-6}
Magnetic diffusivity	κ_B	m^2 s^{-1}	0.4–2
Compositional diffusivity, outer core	κ_C	m^2 s^{-1}	10^{-7}–10^{-11}
Codensity diffusivity	κ_χ	m^2 s^{-1}	10^{-5}–10^{-7}
Self-diffusivity, inner core	D	m^2 s^{-1}	10^{-11}–10^{-15}
Kinematic viscosity, outer core	ν	m^2 s^{-1}	10^{-2}–10^{-6}
Dynamic viscosity, inner core	η_{IC}	Pa s	10^{10}–10^{20}

TABLE S3 Core dynamical properties.

Property	Notation	Units	Range
Total heat flux, CMB	Q_{CMB}	TW	6–18
Adiabatic heat flux, CMB	Q_{ad}	TW	3–18
Radioactive heat production	Q_R	TW	0–2
Ohmic dissipation, core	Q_J	TW	0.2–4
Buoyancy flux, outer core	F	m^2 s^{-3}	$0.5–3 \times 10^{-12}$
Fluid velocity, outer core	v	m s^{-1}	$0.3–3 \times 10^{-3}$
Alfvén velocity, outer core; CMB	c_{alfv}	m s^{-1}	$4; 0.4 \times 10^{-2}$
Outer core age	t_{OC}	Gyr	>4.4
Inner core age	t_{IC}	Gyr	0.4–1.6
Inner core growth rate	$\dot{r}_{IC}$	m year^{-1}	$0.4–1.6 \times 10^{-3}$
Outer core light element enrichment	$\dot{C}_{OC}$	s^{-1}	$0.5–2 \times 10^{-19}$
Secular variation time scale	τ_{SV}	year	400–430
Viscous diffusion time scale	τ_{visc}	Gyr	~10
Thermal diffusion time scale	τ_{temp}	Gyr	~100
Magnetic diffusion time scale	τ_{mag}	year	$1–3 \times 10^5$
Dipole diffusion time scale	τ_{dip}	year	$2–6 \times 10^4$
Circulation time scale, outer core	τ_{circ}	year	100–300

TABLE S4 Nondimensional parameters.

Outer core parameter	Notation	Definition	Value/range
Depth ratio	d^*	$r_{CMB} - r_{ICB}/r_{CMB}$	0.649
Magnetic Reynolds number	Rm	vd/κ_B	1000–3000
Prandtl number	Pr	ν/κ_T	0.1–1
Magnetic Prandtl number	Pm	ν/κ_B	$10^{-4}–10^{-6}$
Ekman number	Ek	$\nu/\Omega d^2$	$10^{-13}–10^{-15}$
Rayleigh number, ΔT	Ra_T	$\alpha_T g \Delta T d^3/\kappa_T \nu$	~10^{27}
Rayleigh number, q	Ra_q	$\alpha_T g q d^4/k\kappa_T \nu$	$10^{28}–10^{29}$
Rayleigh number, F	Ra_F	$Fd^4/\nu^2 \kappa_\chi$	~10^{29}
Dissipation number	Di	$\alpha_T g d/c_p$	0.3
Elsasser number, rms	Λ	$\sigma B_{rms}^2/\rho \Omega$	10–20
Elsasser number, CMB	Λ_{CMB}	$\sigma B_{cmb}^2/\rho \Omega$	~1
Reynolds number	Re	vd/ν	$10^7–10^8$
Rossby number	Ro	$v/\Omega d$	$2–5 \times 10^{-6}$
Local Rossby number	Ro_l	$vl/\Omega d^2$	<0.1
Inner core parameter	**Notation**	**Definition**	**Value/range**
Rayleigh number	Ra_T	$\alpha_T g \Delta T d^3/\kappa_T \nu$	$1–10^4$
Hartmann number	Ha	$\sigma B^2 r_{ICB}^2/\eta_{IC}$	~10^{-3}

Glossary

a-axis Direction from the central symmetry axis to any vertex in a hexagonal crystal.
ab initio Literally *From the beginning* or *first principles*. A calculation of the physical properties of a medium, including elastic moduli, mass density, viscosity, thermal conductivity, and electrical conductivity using quantum mechanical models of the lattice structure and potentials for the forces of interatomic bonding and electron distributions.
accretion Planet formation through impact and addition of many separate masses.
accretion energy Energy in planet formation derived from the kinetic energy of impacting masses. Also called *binding energy*.
achondrites Silicate meteorites with compositions similar to crustal ultramafic igneous rocks but not containing chondrules.
adiabatic A thermodynamic property or process in which heat or entropy is conserved.
Alfvén waves Wave motion in an electrically conducting fluid in which magnetic tension provides the restoring force.
anelastic A rheology in which there is a time delay between the applied stress and the resultant strain.
anistropic dispersion Wave motion in which the phase velocity depends on the propagation direction and frequency. Commonly applied to inertial and internal gravity waves.
annealing High-temperature processes of recovery and recrystallization that removes defects in a solid.
annihilation A process in which dislocations are removed from a crystal.
archeomagnetism Study of remnant magnetism exhibited by archeological materials.
atomphile A gas-loving element, with an affinity for the atmosphere.
azimuthal flow Fluid motion in the east-west directions.
baroclinic flow Fluid motion in which the density and pressure gradients are inclined. Applied to variable density fluids.
barotropic flow Fluid motion in which the density and pressure gradients are aligned. Usually applied to uniform density fluids.
basal plane Base of a hexagonal prism, applied to hexagonal crystals.
basal slip Slip on the basal plane in the hexagonal close-packed crystal.
Birch's law An empirical relation in which compressional wave velocity is linearly proportional to density.
boundary layer A layer whose thinness admits that a higher spatial derivative diffusive term becomes important, for example, viscous forces or thermal conduction. Also refers to a layer separating regions that do not mix, with transport by diffusion.
Bravais lattice A distinct lattice geometry. Synonym for *lattice type*.
Buckingham π-theorem Theorem specifying the number of independent dimensionless parameters in a physical system equals the number of physical properties minus the number of fundamental physical units involved.
bulk modulus Proportionality coefficient between hydrostatic or lithostatic pressure (nondeviatoric stress) and volumetric strain. Also called *incompressibility*.
buoyancy frequency Natural oscillation frequency in a stably stratified fluid, with negative buoyancy the restoring force.
Burgers vector Vector representing the distortion of a loop of atoms surrounding a dislocation line.
Busse rolls Planform at the onset of convection in a rapidly rotating fluid sphere. Applicable to the onset of outer core convection outside the inner core tangent cylinder.
CAI Refractory calcium-aluminum inclusions found in meteorites, thought to be the first solids condensed in the solar nebula.
carbonaceous chondrites Chondritic meteorites containing carbon.
caustic Three-dimensional surface in which wave energy is strongly focused. In plots of body-wave travel times as a function of great circle distance, caustics are expressed by the distances at which the travel time curves reverse direction for a monotonically increasing or decreasing sequence of take-off angles. At these points, ray-theoretical amplitudes become infinite, but actual amplitudes remain regular, frequency dependent, and larger at higher frequency.
c-axis Direction perpendicular to a basal plane in a hexagonal crystal.
centrifugal acceleration Acceleration of a body in rotation, equal to the square of the angular velocity of rotation times the distance from the rotation axis. Directed toward the rotation axis.
centrifugal force Product of the mass times the negative of the centrifugal acceleration, directed away from the rotation axis. Derived from the gradient of the *centrifugal potential*.
chondrites Meteorites containing chondrules.
chondritic The composition of an undifferentiated meteorite, reflecting the primitive composition of terrestrial bodies.
chondrules Small spheroidal inclusions in stony meteorites, thought to originate by condensation in the solar nebula.
CIA force balance Dynamical balance in a fluid between the buoyancy force and the Coriolis and inertial accelerations.
Clapeyron slope In a pressure versus temperature diagram, the slope of the curve separating two different phases at equilibrium.
climb Motion of an edge dislocation in response to tensile or compressive stress.
close-packed structure Structure with a high packing fraction. Face-centered cubic and hexagonal close packed are examples.
CMB Core-mantle boundary.
coarsening High-temperature process in which small-scale solidification features anneal out, preserving the original crystal. An example of *Ostwald ripening*.

codensity Hybrid variable representing the combined effects of temperature and composition on density.
codensity sink Sink/source term in the codensity transport-diffusion equation.
columnar dendrites Elongated microstructure resulting from directional solidification in alloys.
compositional energy Energy associated with the release of less dense component during solidification. Sometimes referred to as gravitational energy release.
constitutional supercooling Supercooling in a boundary layer with liquidus temperature increasing with distance from the solid. May result in morphological instability and dendritic crystals.
control parameters Dimensionless parameters that define forces and material properties in a physical system.
core field Portion of the geomagnetic field originating in the core.
Coriolis acceleration Acceleration of a body in a rotating reference frame, equal to the vector product of twice the angular velocity of rotation times the velocity in the rotating reference frame.
Coriolis force The product of mass times the negative of the Coriolis acceleration.
Coriolis parameter Local measure of Earth's rotation, equal to twice the local vertical component of Earth's angular velocity vector.
correlegram The cross-correlation of two seismograms. Peaks in a correlegram occur when the waveforms of cross-correlated body waves have the similar shapes. The strongest peaks will occur from pairs of body waves radiated from their source as the same wave type (P or S) and sampling the crust and mantle beneath the receiver at the same angle, measured by the apparent velocity along the Earth's surface (horizontal slowness).
creep Flow of a solid in response to a sustained stress, sometimes called *viscoelastic creep*. A synonym for *plastic flow*.
cross-slip Motion of a screw dislocation around a line or point defect by a change in slip plane, facilitating glide.
crustal field Portion of the geomagnetic field originating in the crust.
crustal filter Screening effect of the crustal field on the core field.
crystal structure Structure formed by attachment of atoms to lattice points. Examples are face-centered cubic and hexagonal close packed.
crystal system Lattice type defined by lengths and angles between the lattice translation vectors. Examples are cubic or hexagonal.
cyclone, cyclonic Geostrophic circulation around low pressure, clockwise, and counterclockwise in the southern and northern hemispheres, respectively. An *anticyclone* is the geostrophic circulation around high pressure, clockwise, and counterclockwise in the northern and southern hemispheres, respectively.
cylindrical anisotropy In application to Earth's core, a form of transverse isotropy in which the axis of symmetry is coincident with Earth's axis of rotation.
D″ region Heterogeneous region in the lower mantle extending from the core-mantle boundary upward for several hundred kilometers.
Debye model Approximation used to calculate the phonon contribution to the specific heat of a solid in which the sound velocity is assumed independent of frequency. Above the Debye frequency, corresponding to the Debye temperature, phonons are inactive.
deformation mechanism Microphysical process by which a solid flows when subjected to a sustained stress.
deformation mechanism map Diagram showing deformation mechanism regimes plotted as a function of homologous temperature and shear stress, for a particular grain size.
dendrites Tree-like crystal microstructure resulting from morphological instability during solidification.
diapir A large, isolated buoyant mass, sometimes used to describe the head of a mantle plume.
diffraction Frequency-dependent scattering of wave energy by a discontinuity or heterogeneity.
diffusion creep Flow of a solid by the movement of atoms and vacancies.
dimensionless parameters Parameters with no physical dimension, consisting of ratios and products of the properties of a system.
dipolarity Ratio of the r.m.s. dipole intensity to the r.m.s. total magnetic field intensity.
dipole moment A measure of the intensity of a dipolar magnetic field.
direct numerical simulation (DNS) Numerical simulation of a physical system using realistic property values, in which no extrapolation is necessary.
dislocation Crystalline line defect that facilitates deformation by requiring fewer broken bonds during strain. Edge and screw types.
dislocation creep Flow of a solid by the movement of dislocations.
double diffusive layer A layer with two diffusive contributions to the density, for example, from gradients in temperature and composition.
doubly diffusive convection Form of convection in which two or more variables with different diffusivities contribute to the fluid buoyancy.
α-dynamo effect Production of poloidal magnetic field from toroidal magnetic field by 3D fluid motions lacking reflection symmetry, such as helical flow. Toroidal magnetic field can also be produced from poloidal magnetic field this way.
ω-dynamo effect Production of toroidal magnetic field from poloidal magnetic field by a shear flow.
dynamo, exsolution Self-sustaining dynamo in which the fluid motion is driven by exsolution of chemical species.
dynamo, kinematic Self-sustaining dynamo in which the motion of the electrical conductor is prescribed.
dynamo, magnetohydrodynamic Self-sustaining dynamo in which the motion of the electrically conducting fluid is determined dynamically.
dynamo, self-sustaining System that sustains its own magnetic field through motions of its electrically conducting parts.
E′ region A thin region (or layer) in the outer core below the core-mantle boundary with anomalous seismic velocities.
earthquake doublet A pair of earthquakes assumed to have occurred at the same location, with the same fault mechanism and source-time function.
eccentric dipole Dipole magnetic field with a center of symmetry displaced from Earth's center.
eigenfrequency Frequency of a normal mode of oscillation of the Earth.
Ekman boundary layer Boundary layer in rotating fluid defined by a balance between the Coriolis acceleration and viscous forces.
Ekman number Dimensionless parameter, the ratio of the viscous force to the Coriolis acceleration in a fluid.
Ekman pump Upwelling or downwelling motion in a fluid induced by an Ekman boundary layer.
Ekman spiral Spiraling pattern of the velocity vector with depth through an Ekman boundary layer.
elastic A rheology in which strain is proportional to stress.
elastic anisotropy Linear elastic rheology where individual crystals exhibit a wavespeed that depends on direction.
elliptical instability Dynamical instability resulting from elliptical streamlines in a rotating fluid.
Elsasser number Dimensionless parameter characterizing the ratio of the Lorentz force to the Coriolis acceleration in a fluid.

Glossary

entrainment coefficient Proportionality factor relating the radius of a buoyant element to its distance of rise or fall in a turbulent fluid.
equation of state Equation describing how the density or volume of a material changes as a function of pressure and temperature.
equatorial undercurrent East-west directed flow at depth beneath the outer core equator.
equiaxed Microstructure whose grains exhibit no shape-preferred orientation.
equipotential surface A surface on which the sum of the gravitational potential and centrifugal potential is constant. The *geoid* is Earth's reference equipotential potential surface.
eutectic Alloy in which two or more solid phases coexist with the liquid; in equilibrium at a given temperature, each solid has a specific composition. It occurs when the enthalpy of mixing is positive.
excess ellipticity The difference between the ellipsoidal shape of the core-mantle boundary and the ellipsoid predicted by hydrodynamic equilibrium.
exsolution Supersaturation of one or more components, leading to solidification and/or precipitation.
exsolution convection Convection driven by the precipitation and segregation of light elements in the core.
FDSN Federation of Digital Seismographic Networks organization.
Fermi energy level, Fermi gas A Fermi gas is composed of noninteracting particles that obey the Pauli exclusion principle, so that even in the lowest-energy configuration all the particles cannot be in the ground state. The Fermi energy level is the energy of the highest occupied state in the lowest-energy configuration.
***f*-plane** Local approximation that assigns a constant value to the Coriolis parameter.
Frank-Read source Mechanism by which stress introduces dislocations, whose movement then facilitates strain.
Frechet derivative Derivative of an object function, usually constructed from the difference between an observed and predicted quantity, with respect to an unknown parameter; used in inverse theory.
free inner core nutation (FICN) Wobble of the inner core with respect to Earth's rotation axis.
F region A thin region (or layer) in the outer core above the inner core boundary with anomalously low seismic P-wave velocities.
Gauss coefficients Coefficients in the spherical harmonic expansion of the geomagnetic field at Earth's surface.
geocentric axial dipole hypothesis Assumption that the geomagnetic field closely approximates an axial, Earth-centered dipole as its time-average configuration.
geodynamo Set of processes that sustain the geomagnetic field inside Earth's core.
geoid Reference surface of equipotential on the Earth, corresponding to mean sea level.
geomagnetic intensity Strength of the geomagnetic field, measured in Tesla.
geomagnetic jerks Sudden, nearly discontinuous changes in the second derivative of the geomagnetic field with respect to time.
geomagnetic polarity excursions Large, transient variations in the strength and location of the geomagnetic dipole moment and other geomagnetic field components, followed by recovery of the pretransient field configuration.
geomagnetic polarity reversals Change in the geomagnetic field between two persistent statistically equal states (chrons) that reorients the dipole moment vector between the northern and southern hemispheres.
geomagnetic polarity timescale Geologic timescale determined from geomagnetic polarity chrons.
geomagnetic poles Locations where the axis of the geocentric dipole intersects Earth's surface.
geomagnetic potential Scalar potential whose negative gradient equals the geomagnetic field.
geomagnetic secular variation Temporal changes in the geomagnetic field on annual to multicentury time scales.
geostrophic contours Contours on the core-mantle boundary along which outer core fluid columns have uniform height.
geostrophic turbulence Turbulence strongly affected by the Coriolis acceleration.
geostrophy Literally *Earth turning*. Fluid motions resulting from a balance between the Coriolis acceleration and the pressure-gradient force.
geotherm The temperature variation in the Earth. In those parts of the Earth that are converting the geotherm lies close to the adiabat.
glide Motion of dislocations in response to shear stress.
grain Region with uniform crystal structure and orientation. Synonym is *single crystal*.
grain boundary Surface separating crystals with differing orientations.
grain boundary migration Processes by which grain growth occurs, often through the motion of dislocations or grain boundary sliding.
grain boundary sliding Shear deformation along grain boundaries during diffusion creep. A synonym is *superplasticity*.
grain growth Process in which new, lower energy crystals grow at the expense of older, higher energy ones.
grand tack Transient contraction and expansion of Jupiter's orbit proposed for the early history of the solar system.
gravitational differentiation Separation of components with different densities in a self-gravitating body.
gravitational instability An accretion mechanism in which the density perturbations in compression waves grow under the action of gravitational forces.
gravitational locking Torques derived from gravitational forces that keep the inner core rotating at the same rate as the mantle, preventing inner core super-rotation.
Grüneisen parameter Parameter measuring the change in crystalline vibrational frequencies as a function of volume.
Harper-Dorn creep Deformation mechanism that occurs at low stresses and dislocation densities, and results in a Newtonian viscosity.
helicity Measure of correlation between a vector field and its curl. Kinematic helicity is the correlation between velocity and vorticity in fluid motion.
helicity equator Symmetry plane for fluid helicity in the outer core.
heterogeneity spectrum The power spectrum of elastic velocity or density fluctuations.
homologous temperature The ratio of temperature to melting temperature in a material.
Hubble constant Parameter describing the rate of universe expansion, equal to the speed of recession divided by distance.
ICB Inner core boundary.
ICN Inner core nucleation.
ideal c/a ratio Ratio of the distance between bases to the distance between the base center and a vertex in a hexagonal crystal with maximum packing fraction. Approximately equal to 1.633.
inertial wave Wave motion in a rotating fluid with a balance between inertia and Coriolis accelerations.

inner core tangent cylinder Imaginary cylinder extending across the outer core, circumscribing the inner core equator and concentric with Earth's rotation axis.
internal gravity waves Wave motion in a stably stratified fluid with a balance between inertia and the buoyancy force.
IRIS Incorporated Research Institutions for Seismology organization.
iron rain Precipitation of molten iron drops in a magma ocean.
irons Meteorites consisting of iron-nickel alloys.
iron snow Precipitation of solid iron crystals in the liquid core.
ISC International Seismological Centre organization.
Large low seismic velocity province (LLSVP) Name given to the two large regions in the lower mantle with anomalously low seismic shear wave velocity, up to 10% or more negative perturbation with respect to a reference Earth model. Sometimes called *LLVP*, large low velocity province.
late addition Volatile-rich material from beyond the snow line added to the Earth after the main phases of accretion.
latent heat Heat required to convert solid into liquid or liquid to vapor without temperature change.
lattice A periodic array of points in space.
lattice preferred orientation (LPO) A polycrystal with a preferred crystallographic alignment, typically caused by deformation or solidification. Synonym for *textured solid*.
lattice rotation Rotations in a polycrystal undergoing shear strain by dislocation glide.
lever rule Graphical construction displaying the equilibrium fractions of solid and liquid in an alloy.
liquidus The temperature above which an alloy is completely liquid, which depends on the composition.
lithophile Rock-loving element, with an affinity for silicates and oxides.
LOD Length-of-day; usually in reference to irregularities in Earth's rotation that produce changes in the length of the day.
Lorentz force Force exerted by a magnetic field on moving charges. In a conducting medium, the Lorentz force is proportional to the vector product of the electric current density and the magnetic field.
Lorenz number In the Wiedemann-Franz law, the constant of proportionality between the thermal conductivity and the electrical conductivity multiplied by the temperature.
MAC force balance Dynamical balance in an electrically conducting fluid between the buoyancy force, the Lorentz force, and the Coriolis acceleration.
MAC waves Wave motion in the outer core with a MAC force balance.
magma ocean A large volume of partially or wholly molten silicates and oxides in a planet, typically a consequence of giant impacts.
magnetic poles Surface locations where the magnetic dip angle is ±90 degrees.
magnetic Prandtl number Dimensionless parameter, the ratio of kinematic viscosity to magnetic diffusivity in an electrically conducting fluid.
magnetic pressure Pressure exerted by concentrated magnetic fields, sometimes called the Maxwell pressure.
magnetic Reynolds number Dimensionless parameter, the ratio of magnetic advection to magnetic diffusion in an electrically conducting fluid.
magnetic tension Tangential forces exerted by magnetic fields, derived from Maxwell shear stresses.
magnetic wind Flow in the outer core driven by the Lorentz force.
magnetohydrodynamics Dynamics of electrically conducting fluids including electric currents and magnetic fields.
magnetostrophy Fluid motions resulting from a balance between the Coriolis acceleration and the magnetic pressure force.
mantle GCM Numerical model of the mantle global circulation, often with plate-motion constraints applied.
mantle plume A localized, convective mantle upwelling, some originating above the core-mantle boundary.
Matthiessen's rule An empirical law that electron scattering due to other electrons, lattice vibrations (phonons), and impurities and defects add in parallel. The electrical and thermal resistivity in metals due to these effects also add in parallel.
Mauersberger-Lowes spectrum Plot of mean-squared geomagnetic intensity versus spherical harmonic degree.
melting and solidification poles Symmetry points for heterogeneous melting and solidification at the inner core boundary.
meridional Property variation in radius and latitude, at constant longitude.
MET Molecular electronic transducers, state-of-the-art motion sensors.
metal-silicate partition coefficient Ratio of the mass concentration of an element in metal to its concentration in silicate at equilibrium.
meter-size barrier A limitation on pebble growth because of fragmentation or drift into the Sun; a problem for planet formation.
microstructure Morphology of crystals, including grain size, shape, and dendrites.
Miller-Bravais indices Four indices are used in the hexagonal system.
Miller indices Notation for indicating crystallographic planes and directions.
molecular cloud Diffuse region composed mainly of hydrogen molecules, from which stars form.
morphological instability The tendency of a liquid-solid interface to become corrugated during solidification. May lead to dendritic crystals.
multipolar transition Transition between dipole-dominant to multipolar magnetic field configurations in numerical dynamos.
mushy zone Solid + liquid region lying between pure solid and a liquid in which temperature and composition are linked by the liquidus.
normal mode Oscillation of the Earth at a specific frequency, characterized by a finite number of nodes on a spherical surface.
normal polarity Geomagnetic field configuration in which magnetic north orients toward geographic north. The polarity of the present-day geomagnetic field.
oligarchic growth A stage in planetary formation characterized by fast growth of the largest embryos, leading to a bimodal mass distribution.
packing fraction Fraction of space occupied by nonoverlapping spheres in a lattice structure.
paleomagnetism Study of remnant magnetism induced by the geomagnetic field and preserved in rocks and minerals.
pallasites Meteorites consisting of olivine crystals in an iron-nickel matrix.
partition coefficient The ratio of the solid solute composition to the liquid solute composition of an alloy, at a given temperature.
pebble accretion Planetary formation by accretion of pebbles.
pebbles Small aggregates of dust in a protoplanetary disk.
penetrative convection Form of convection in which the convective motions penetrate into a stably stratified region of the same fluid.

Glossary

Perovskite Magnesium silicate perovskite (Mg, Fe)SiO$_3$. In the lower mantle, also known as Bridgemanite, having the lattice structure of CaTiO$_3$. *Post-Perovskite* is a high-pressure transformation of Perovskite, tentatively identified in the lowermost mantle.

phase The state of a material, e.g., gaseous, liquid, or solid, and if solid, the particular crystal structure.

phase loop On a phase diagram, the compositional difference between the liquidus and the solidus, at a given temperature.

phonons Quantized lattice vibrations. Their long-wavelength limit are sound waves.

planetary embryos Protoplanets 1000–3500 km in radius built from planetesimals and pebbles, typically differentiated.

planetary vorticity Background vorticity due to planetary rotation, equal to twice the rotation angular velocity.

platelets Form of dendrites in hexagonal close-placked crystals.

planetesimals Small bodies of order 1–100 km in size formed in the early solar system, made of nebular dust, chondrules, pebbles, and fragments of previous bodies.

plastic deformation The deformation that occurs when a solid is subject to a sustained stress. The solid does not return to its original state when the stress is removed. It is sometimes referred to as creep.

Poisson's ratio A measure of elastic strain in the direction perpendicular to the applied stress.

polarity chrons Interval of time between geomagnetic polarity reversals. Classified as chrons, subchrons, or events, based on their duration.

polar vortex Azimuthal (east-west) recirculating flow at the top of the outer core, located inside and adjacent to the inner core tangent cylinder.

pole figure Graphical display of texture in a polycrystal, in which crystallographic axes are plotted on a unit sphere.

poloidal magnetic field Magnetic field with both radial and tangential components, such as a dipole field.

polycrystal Material composed of many crystals.

postcritical reflection At postcritical angles of reflection on a discontinuity such that the elastic wave velocity is higher in the medium on the side opposite to the incident wave. Postcritical reflections of seismic waves occur at more grazing angles of incidence on a discontinuity. The reflection of a postcritically reflected wave is total, and no ray-theoretical, transmitted, wave is excited below the boundary.

potential vorticity Hybrid variable used to characterize quasigeostrophic flow. Often the ratio of planetary plus local vorticity to fluid column height.

Prandtl number Dimensionless parameter, the ratio of kinematic viscosity to thermal diffusivity in a fluid.

precritical reflection At precritical angles of incidence on a discontinuity both a reflected and transmitted elastic wave are excited, with both the reflected and transmitted wave having less energy than the incident wave. Precritical reflections of seismic waves occur at relatively steep angles of incidence on discontinuities in Earth. They are relatively small in amplitude compared to the incident wave and the wave transmitted through the discontinuity.

primary slip Slip system that occurs most easily during dislocation creep.

prismatic plane Side of a hexagonal prism, applied to hexagonal crystals.

prismatic slip Slip on a prismatic plane in a hexagonal close-packed crystal.

proto-Earth The Earth during formation, not fully formed.

protoplanetary disk Disk-shaped mass of gas and dust from which planets are produced.

pyramidal slip Slip on a pyramidal plane in the hexagonal close-packed crystal structure.

quality factor Reciprocal of the fraction of elastic energy lost to heat during a loading cycle. Often abbreviated as Q.

quasigeostrophic Rotationally dominated flow with slight departures from pure geostrophy, often in the form of fluid column stretching.

quasi-S wave One of two possible S waves in an elastically anisotropic medium.

radiative transport In application to seismic waves, elastic energy propagation along a trajectory determined by heterogeneity.

ray theory Frequency-independent solution to the elastic equation of motion characterized by curves representing streamlines of the wave energy flux, perpendicular to wave fronts.

Rayleigh number Dimensionless parameter, the ratio of buoyancy to viscous and thermal diffusion effects in a fluid. Characterizes the onset and strength of convection.

recovery High-temperature process by which defects are removed while preserving the original crystals.

recrystallization High-temperature processes in which grains with high free energy are replaced by grains with a lower free energy. Often results in a lower density of crystallographic defects.

refractory Term referring to solids that melt and condense at very high temperature.

refrigerator action Term for compositional convection that transports heat downward.

relaxation spectrum Spectrum of a frequency dependent, complex, elastic modulus of an a viscoelastic medium, characterized by a distribution of relaxation times in which elastic energy is absorbed by damped oscillations of lattice defects. Commonly displayed by the magnitude of the frequency dependence of the attenuation parameter, $1/Q$.

response parameters Dimensionless parameters that define the response of a physical system to forces.

reverse polarity Geomagnetic field configuration in which magnetic south orients toward geographic north. Opposite the polarity of the present-day geomagnetic field.

Reynolds number Dimensionless parameter, the ratio of inertial to viscous effects in a fluid. Characterizes the onset and strength of turbulence.

Reynolds stress Stresses generated by correlations between small-scale velocity components in a fluid.

rheology Mathematical relation between stress and strain. Examples are elastic, viscoelastic, plastic, viscous, and brittle.

rms Root-mean square, calculated by taking the square root of the mean-squared amplitude of a variable.

Rossby number Dimensionless parameter, the ratio of inertia and Coriolis accelerations in a fluid.

Rossby wave Wave type in a rotating fluid that involves variations in column height.

secular cooling The difference between heat production and heat loss.

sensible heat Thermal energy whose transfer in or out of the Earth results in a temperature change.

shape-preferred orientation (SPO) Polycrystal with microstructure elongated in one or two directions.

shear modulus Proportionality coefficient between shear stress and shear strain.

shock wave A propagating high-pressure disturbance that can generate very high densities and temperatures; an equation of state based on shock waves is known as a Hugoniot.

SH polarization Polarization of an S wave in which particle motion is perpendicular to its ray, tangent to its wave front, and perpendicular to the plane defined by the source, receiver, and center of Earth.
siderophile Iron-loving element, with little affinity for oxygen, hence likely to be in the core. Subdivided into moderately and highly siderophile.
slip system A combination of crystallographic plane and direction along which dislocation movement occurs.
slurry A mixture of solid particles in a liquid.
snow line The distance from the Sun beyond which ices form in the solar nebula.
solar nebula Generic term for the gas and dust surrounding the young Sun.
solar wind Particle stream released from the Sun consisting of electrons, protons, and α particles.
solid solution An alloy whose solid phase can exist for a range of compositions within a single-crystal structure. It occurs when the enthalpy of mixing is zero.
solidus The temperature below which a material is completely solid, which depends on the composition.
spallation Nuclear breakup by bombardment.
spherical harmonic A special function of the two angles in a spherical coordinate system that satisfies the angularly separated portion of differential equations containing a Laplacian. The variation of properties on the surface of a planet or on a surface at a radius at a specific depth may be expressed by a superposition of spherical harmonic functions, similar to the way in which a function of one variable may be decomposed into a Fourier series.
spheroidal mode Normal mode of oscillation in the Earth with particle motion perpendicular to its surface.
spin-over mode Rotation of the core about an axis inclined to the rotation axis of the mantle.
splitting function Function designed to demonstrate the sensitivity of a particular normal mode of oscillation of Earth to perturbations of its elastic structure as function of latitude and longitude on the surface of Earth. The splitting function incorporates effects that remove the degeneracy from the azimuthal order number of a mode singlet, making each azimuthal order number of a mode singlet have a slightly different frequency, effectively broadening the frequency band of over which the mode is observed.
stably stratified layer A layer whose vertical temperature and composition distribution are such that it will not convectively overturn.
stacking fault A local disruption to the stacking sequence.
stacking sequence Order in which planes of atoms are layered in a crystal; hcp is ABABAB..., fcc is ABCABC...
stellar incubator Generic term for the gas clouds from which stars are produced.
stellar nebula Portion of a molecular cloud undergoing gravitational collapse, leading to star formation at the center.
Stishovite High-pressure form of quartz, SiO_2.
stony-irons Meteorites consisting of silicate crystals in an iron-nickel matrix.
streaming instability A dust coagulation mechanism in which larger particles moving faster in a gas collect smaller, slower-moving particles.
subgrain Region of a single crystal slightly misorientated, often by a low-angle grain boundary composed of dislocations.
superchrons Anomalously long-lasting geomagnetic polarity chrons.
supercooling Liquid existing beneath its equilibrium solidification temperature. Synonym for *undercooling*.
supernova Large explosion at the end of the evolutionary cycle of a star, a source of high mass elements.
super-rotation Faster rotation of the inner core with respect to the mantle.
SV polarization Polarization of an S wave in which particle motion is perpendicular to its ray, tangent to its wave front, and lies in the plane defined by the source, receiver, and the center of Earth.
tangential geostrophy Assumption that the tangential components of outer core flow are in geostrophic balance.
Taylor constraint Constraint on dynamo action in a rapidly rotating electrically conducting fluid sphere or spherical shell.
Taylor-Proudman constraint Constraint on steady laminar flow in a rapidly rotating fluid, sometimes called the geostrophic constraint.
Theia Name given to the impactor in the giant impact theory of Moon formation. From the Greek for "goddess."
thermal wind Flow driven by lateral gradients in density and affected by the Coriolis acceleration.
thermochemical pile A pile of chemically and thermally distinct material in the lower mantle resting on the core-mantle boundary.
tidal friction Friction between parts of the Earth system resulting from the deformation caused by tidal forces. Produces deceleration of Earth's rotation, sometimes called *tidal braking*.
tie line On an alloy phase diagram, the horizontal line (isotherm) that gives the composition of two phases in equilibrium.
topographical relaxation Viscous flow resulting from constant density surfaces that are inclined to equipotential surfaces.
toroidal magnetic field Magnetic field with only tangential components, associated with poloidal electric currents.
toroidal mode Normal mode of the Earth in which particle motion is tangent to its surface.
torsional oscillations Oscillations in the outer core on geostrophic cylinders, with magnetic tension as the restoring force.
transverse isotropy An elastic medium specified by five elastic constants. Possible body waves are a P wave and two quasi-S waves having different phase and group velocities, whose polarizations are all orthogonal. The trajectories of the body-wave rays are independent of azimuth around an axis of symmetry. In the case of radially varying elastic constants and a radial axis of symmetry the polarization of the quasi-S waves are equivalent SH and SV components of motion in an isotropic medium. If the axis of symmetry is instead inclined with respect to the radial direction, the two quasi-S waves will be visible on both the SH and SV components of motion.
transverse solidification texture The tendency of platelets in different crystals to align as a result of flow in the melt.
travel time triplication Three intersecting curves on a travel time plot formed by three separate waves arriving at the same distance.
twinning Plastic deformation in which atomic planes move a definite distance and direction, with the extent of movement dependent on the distance from the twinning plane.
ultra-low-velocity zone (ULVZ) Name given to small, isolated regions just above the core-mantle boundary where the seismic velocity is strongly reduced compared to adjacent regions of the lower mantle.
unit conventional cell Volume spanned by the shortest lattice translation vectors.
Veronis cells Planform at the onset of convection in a rapidly rotating fluid layer. Applicable to the onset of outer core convection inside the inner core tangent cylinder.

virtual axial dipole moment (VADM) Dipole moment inferred from magnetic intensity and direction determined at a point on Earth's surface, assuming the geomagnetic field consists of a geocentric axial dipole.
viscoelasticity Compound rheology combining viscosity and elasticity, in which stress lags applied strain in time.
viscosity Proportionality between stress and strain rate. Kinematic viscosity is dynamic viscosity divided by density. In *Newtonian viscosity*, strain rate is linearly proportional to stress.
volatile Term referring to solids and liquids that transform to gas at relatively low temperature.
vorticity The curl of a velocity field, equal to twice the angular velocity at a point.
Wiedemann-Franz law An empirical law relating the electrical and thermal conductivities of metals, with a theoretical basis in the free electron theory of metals.
XUV Extreme ultraviolet radiation, describing the short-wavelength emissions from the young Sun, sometimes called EUV.
zonal Property variation with latitude, independent of longitude.

Index

Note: Page numbers followed by *f* indicate figures and *t* indicate tables.

A

Ab initio calculation, 285
Adams-Williamson equation, 23
Age of inner core, 54, 54*t*, 55*f*
AK135-F core model, 3–4
Alfvén's theorem, 94
Alfvén waves, 135–136, 136*f*
α-effect, 99, 99*f*
Analogs and processes, 285–286
Anisotropic dispersion, 138
Anisotropy, 14–15, 15*f*, 28–29, 29*f*
Annealing
  annihilation, 235, 236*f*
  coarsening, 234–235, 237, 238*f*
  grain boundary migration, 235–237, 237*f*
  paths, 234–235, 235*f*
  recovery and recrystallization, 235–237
  sub-grain formation, 235, 236*f*
Annihilation, 235, 236*f*
Archeomagnetic intensity, 86–87
Astrophysical batteries, 273
Attenuation and isotropic heterogeneity, 243–244

B

Biermann battery, 273
Birch's law, 26–27
Body waves
  angle of incidence, 9
  bulk modulus/incompressibility, 5
  caustic, 5
  constraint on density, 5
  core-grazing wave, 8
  core-mantle boundary (CMB), 6
  core structure, 5, 9
  differential travel times, 8–9, 8*f*
  discontinuous velocity, 5
  inner core boundary, 7
  precritical to postcritical reflection, 6
  P waves polarization, 5–6
  ray paths, 8
  sensitivity, 7, 7*f*
  S waves polarization, 6–7, 7*f*
  triplication, 5
  underside reflections, 7
Boundary layers, 38–39
Boundary regions
  CMB topography, 186–187, 186*f*
  F region
    formation theory, 190
    seismic velocity anomalies, 189, 189*f*
    ICB topography, 190–191, 190–191*f*
  lowermost mantle region (D″)
    (*see* Lowermost mantle region (D″))
  uppermost outer core (E′)
    constraints from chemistry and MAC waves, 188
    formation theory, 188
    seismic velocity anomalies, 187–188, 187–188*f*
Buckingham Pi theorem, 118
Bullen parameter, 23–24
Bullen's classification, 4
Buoyancy flux, 123–124, 270–271
Buoyancy frequency, 139

C

Characteristic time scales, 117, 118*f*, 119
CMB heat flux, 51–52
Coarsening, 234–235, 237, 238*f*
Codensity, 122–123, 151, 151*f*
Composition
  Fe-Ni, 33, 34*f*
  light element, 33–35, 34*f*, 36*f*
  meteorites, 33, 34*f*
  phase diagram, 33, 34*f*
  solid-liquid phase diagrams, 37–38, 37–38*f*, 64–65, 65–67*f*
Compositional convection, 269–270
Compositional diffusivity, 40
Compositional energy, 46, 49–50, 53
Control parameters, 118–119
Convection, 268–270, 270*f*
  compositional profile, 225–226, 225*f*
  convective dynamos, 174
  cooling rate, 224
  discussion, 143
  elastic anisotropy, 226, 226*f*
  fully developed convection
    CIA force balance, 149–150
    equatorial plane planform, 148–149, 148*f*
    laboratory experiment, 146
    numerical experiment, 146
    quasigeostrophic turbulence, 146
    rotating right cylinder, 146, 147*f*
    rotating water-filled hemispherical shell, 148–149, 148*f*
    rotating water-filled spherical shell, 147–148, 147*f*
    streamfunction, 149, 149*f*
    3D numerical models, 149, 150*f*
  mantle heterogeneity effects, 150–151, 151*f*
  onset
    Busse rolls, 144, 145*f*, 146
    control parameters, 143–144
    Rayleigh number, 144
    rotating convection, 143
    thin annulus geometry, 143–144, 144*f*
    Veronis cell, 145–146, 145*f*
  Rayleigh number, 224
  regimes, 224–225, 225*f*
  thermal instability, 224
  translation, 224–225, 225*f*
Core dynamics
  core-mantle boundary heat flow and age of inner core, 288–289
  laboratory experiments on thermochemical convection, 288
  liquid metal numerical dynamos, 286–287, 286–287*f*
Core flow images, 106–107, 106*f*
Core-mantle boundary (CMB) topography, 186–187, 186*f*
Core-to-mantle heat flow, 52
Coriolis acceleration, 116, 169–170, 169*f*
Cosmic ray spallation, 249
Cretaceous normal superchron (CNS), 88
Cross-slip, 230, 231*f*
Crustal filtering, 80–81, 156–157, 157*f*
Cylindrical anisotropy, 195

D

Deformation mechanism maps, 239, 239–240*f*
Deformation of inner core
  creep, 223–224
  diffusion creep, 233–234, 233–234*f*
  dislocation creep
    climb, 230, 231*f*
    cross-slip, 230, 231*f*
    edge dislocation, 228, 229*f*
    Frank-Read source, 229–230, 230*f*
    glide, 228, 229*f*, 230
    Harper–Dorn creep, 230
    power law creep, 229–230
    process, 228
    screw dislocation, 229, 229*f*
    stacking faults, 230, 231*f*
    strain/work hardening, 229–230, 230*f*
    types, 228
  grain boundary sliding/superplasticity, 228, 234, 234*f*
  homologous temperature, 227–228
  mechanism, 223–224

Deformation of inner core *(Continued)*
  slip systems, 232–233, 233f
  strain rate, 228
  stress/strain
    convection, 224–226, 225–226f
    magnetic field, 227
    sources, 227, 228t
    topographical relaxation, 227
  texture, 223–224, 232
  twinning, 232
Density discontinuity, 211–212, 211–212f
Depth dependence
  attenuation, 198–199, 199f
  cylindrical anisotropy, 196
  differential travel times, 196, 196f
  radial anisotropy, 195–196
  scattered coda envelope, 198–199, 200f
Differential rotation
  differential travel times, 204–205, 206f
  mechanisms, 204, 205f
  observations, 204–205, 205–206f
  theories
    differential rotation rate, 206–207, 206f
    ray paths, 207
    super-rotation, 207, 207f
Differential travel times, 195, 196f
Dimensionless parameters, 117t
  Buckingham Pi theorem, 118
  characteristic time scales, 119
  control parameters, 118–119
  convective force, 118
  Ekman number, 119
  Elsasser number, 120
  vs. numerical models and laboratory experiments, 120, 121f
  numerical values, 118
  physical units, 118
  Prandtl number, 119
  Rayleigh number, 119, 120t
  response parameters, 118–119
  Reynolds number, 120
Dislocation creep
  climb, 230, 231f
  cross-slip, 230, 231f
  edge dislocation, 228, 229f
  Frank-Read source, 229–230, 230f
  glide, 228, 229f, 230
  Harper–Dorn creep, 230
  power law creep, 229–230
  process, 228
  screw dislocation, 229, 229f
  stacking faults, 230, 231f
  strain/work hardening, 229–230, 230f
  types, 228
Doubly diffusive convection, 122–123
Dynamic viscosity, 15
Dynamo. *See also* Geodynamo
  definition, 76
  dynamo and antidynamo mode, 87–88
  ingredients, 77
Dynamo model, 265–266, 265–266f
Dynamo model reversals
  a.c. dynamos, 164
  complex reversal, 165, 166f
  flow structure, 164
  periodic reversal behavior, 164
  polarity reversals, 164
  reversal frequency/sensitivity, 168, 168f
  reversal onset
    dynamo regimes, 167, 167f
    local magnetic Reynolds number, 167–168
    multipolar transition, 167
    reverse magnetic flux, 165–166, 166f
    thermochemical dynamo, 167
  simple reversal, 165, 165f
  types, 164

E

Edge dislocation, 228, 229f
Ekman boundary layers
  Ekman pump, 133–134, 134f
  magnetic fields, effects on, 134, 134f
  velocity component, 132–133, 133f
Ekman number, 117, 119
Elastic anisotropy, 28–29, 29f, 184, 185f, 195–196, 242–243, 244f
Elastic equation of motion, 2, 4–5, 19–20
Elastic velocity, 1
E′ layer stratification, 160, 161f, 162
Electromagnetic implantation, 274
Elliptical instabilities, 275, 275f
Elsasser number, 120
Elsasser number scaling, 159–160
Energy and entropy balance equation
  CMB heat flux, and heat and entropy on adiabat, 51–52
  compositional energy, 49–50
  conservation of energy, 51
  core-to-mantle heat flow, 52
  expression, 51
  heat of chemical reactions, 50
  Large Low Shear Velocity Provinces (LLSVPs), 52
  latent heat, 49
  mantle plume flux, 52
  Ohmic and viscous dissipation and rate of entropy, 52–53
  radioactive heat, 49
  radiogenic elements and rate of energy and entropy, 53
  rate of energy and entropy change due to secular cooling, latent heat, heat of chemical reactions, and compositional energy, 53
  secular cooling, 48–49, 48f
EPOC outer core model, 17
Equation of state, 17, 30
Equatorial undercurrent, 130–131
Equilibrium model, 259–260, 260f
Eutectic phase diagram, 37–38, 38f
Evolution of core
  buoyancy flux, 270–271
  compositional changes, 259
  convection, 268–270, 270f
  core and geomagnetic field evolution, 266, 266f
  core evolution model, 265, 265f, 276–277
  dynamo model prediction, 265–266, 265–266f
  equilibrium model, 259–260, 260f
  formative condition, 259
  inner core boundary (ICB), 260–261
  inner core growth history, 261–262, 261–262t, 262f, 276–277
  inner core nucleation (ICN) age, 262–265, 263–264f
  long-term core cooling, 266–268, 267–269f
  model inputs, 260–261
  R-factors, 276–277
  temperature profile, 277
  terrestrial planet history, 259
  thermal reservoir, 259
Exsolution convection, 269, 270f

F

Failed reversals, 88
Fe-Ni, 33, 34f
Formation of core
  calcium-aluminum inclusions (CAIs), 250
  composition, 250
  core formation process
    entrainment coefficient, 255–256
    metal-silicate separation, 254–255, 255f
    metal-silicate turbulence, 255–256, 255f
    multistage model, 254
    silicate magma ocean, 256
    single-stage model, 254
  core-forming elements
    cosmic ray spallation, 249
    Goldschmidt classification, 249, 249f
    Hubble constant, 249
    massive neutron stars, 249–250
    star formation, 249–250
    supernova event, 249
  cosmochemical and geochemical data, 247
  dust particles, 251
  early evolution, 250
  elemental abundances in Sun, 250, 251f
  energetics of Earth formation
    accretion/binding energy, 247–248
    accretion of Earth, 247, 248f
    Earth energy budget, 247–248, 248f
    energy balance, 249
    sensible heat content, 247–248
    sequestered energy, 248
    tidal friction, 248
  isotopic evidence, 256–257, 257f
  metal-silicate partition coefficient, 258
  meteorites
    achondrites, 252
    carbonaceous chondrites, 252
    chondrites, 252
    Grand Tack hypothesis, 253
    gravitational instability, 253
    metallic meteorites/irons, 252
    meter-sized barrier, 253
    pallasites, 252
    pebbles, 252–253
    solar wind, 253
    specimens, 252f
    stony irons, 252
    streaming instabilities, 253
    time line of major events, 252, 252f
  molecular clouds, 250
  Moon-forming event, 258–259, 258f

oligarchic growth, 253
pebble accretion model, 254
planetary embryos, 253
protoplanetary disk, 250
refractory and volatile elements, 250–251, 251f
siderophile elements, 257–258
solar nebula, 250–251, 250f
stellar nebula, 250
terrestrial planet formation model, 253–254
Frank-Read source, 229–230, 230f
Free inner core nutation, 204
Free oscillations, 9–11, 10–11f, 22, 22f
Free stream, 105
F region
　formation theory, 190
　seismic velocity anomalies, 189, 189f
　stability, 190
Frequency dependence and parameter trade-offs
　PKIKP waveform, 197, 197f
　pulse attenuation, 198, 199f
　velocity dispersion and attenuation, 197–198, 198f
Frozen flux
　concept, 94, 95f
　inversion, 108–109
　polar vortex, 107–108, 108f
　tracing, 105
Future research
　core dynamics
　　core-mantle boundary heat flow and age of inner core, 288–289
　　laboratory experiments on thermochemical convection, 288
　　liquid metal numerical dynamos, 286–287, 286–287f
　mineral physics
　　ab initio calculation, 285
　　analogs and processes, 285–286
　　approaches, 283
　　high-pressure experiment, 284–285, 284f
　seismology
　　ambient noise, 282
　　coherence, 282
　　numerical synthetic seismograms, 283
　　seismic instrumentation, 282
　　source-time functions, 282, 283f
　　station coverage, 281–282
　　wavefield correlation, 282

G

Geocentric axial dipole (GAD) hypothesis, 79
Geodynamo
　age, 271–272
　definition, 76
　energy pathways, 102–103, 102f
　history, 271
　homogeneous dynamo, 77
　vs. industrial dynamo, 77
　kinematic dynamo, 77
　magnetic induction
　　Alfvén's theorem, 94
　　equation, 94
　　frozen flux, 94, 95f
　　magnetohydrodynamic (MHD) induction, 110–111
　　perfect conductor limit, 94
　　Schmidt Legendre polynomials, 109–110
　　total magnetic flux, 94
　　vector diffusion equation, 94
　magnetic Reynolds number, 93
　magnetohydrodynamic (MHD) dynamo theory, 77
　numerical models of ancient geodynamo, 272–273, 272f
　Ohmic dissipation, 103–104
　poloidal and toroidal magnetic field component, 95–97, 96f
　seeding
　　geodynamo initiation, 274–275, 275f
　　geodynamo seeding times, 273–274, 277–278
　　seed magnetic fields, 273
　self-sustaining fluid dynamo (see Self-sustaining dynamo mechanism)
　simple kinematic dynamo, 99–100
　toroidal-poloidal magnetic feedback
　　α-effect, 99, 99f
　　induction mechanisms, 97
　　magnetic helicity, 98
　　poloidal magnetic field generation, 98–99, 99f
　　toroidal magnetic field generation, 97–98, 98f
　　ω-effect, 97, 97f
　von Karman Sodium (VKS) dynamo experiment, 100–101, 101f
Geomagnetic field
　archeomagnetic intensity, 86–87
　core field
　　crustal filter, 80–81
　　inner core tangent cylinder, 82
　　Mauersberger-Lowes spectrum, 80, 80f
　　reverse flux patches, 82
　　secular variation, 80, 82–83, 83f
　　South Atlantic Anomaly, 82
　　structure, 81–82, 81f
　Cretaceous normal superchron (CNS), 88
　dipole moment
　　expression, 79
　　time variations, 83–85, 84f
　dipole variations
　　on 100 kyr time scales, 87–88, 88f
　　on millennium times scales, 86–87, 86–87f
　　dynamo and antidynamo mode, 87–88
　　failed reversals, 88
　Gauss coefficient, 78–79
　geocentric axial dipole (GAD) hypothesis, 79
　as geodynamo probe, 78
　geomagnetic intensity in deep past, 90–91, 91f
　geomagnetic jerks, 85, 85f
　geomagnetic polarity time scale (GPTS), 88
　geomagnetic pole, 79
　geomagnetic potential, 78
　Jurassic hyper-reversals, 88
　Kiaman reverse superchron (KRS), 88
　long-time average core field structure, 89–90, 90f
　magnetic poles, 79
　paleomagnetic intensity, 90–91, 91f
　polarity excursions, 87–88
　polarity reversal, 88–89, 89f
　secular variation time scale, 85–86, 86f
　in spherical harmonics, 78–79
　transition field, 88–89
　virtual axial dipole moment (VADM), 86–87, 86–87f
Geomagnetic images
　columnar flow, 106
　core flow images, 106–107, 106f
　frozen flux tracing, 105, 108–109
　helical flow, 106
　length-of-day (LOD) variations, 107, 107f
　polar vortices, 107–108, 108f
　velocity constraints, 106
Geomagnetic jerks, 85, 85f
Geomagnetic polarity time scale (GPTS), 88
Geomagnetic pole, 79
Geomagnetic potential, 78
Geostrophy/geostrophic flow, 124–126, 125–126f
Glide, 228, 229f, 230
Goldschmidt classification, 249, 249f
Grain boundary, 57
Grain boundary sliding/superplasticity, 228, 234, 234f
Grain size, 239, 240f
Gravity, 1
Gruneisen parameter, 39, 67–68

H

Harper–Dorn creep, 230
Heat of chemical reactions, 50
Helicity, 98, 106
Hemispherical differences
　observations, 202, 202f
　regional complexity, 202
　theories
　　free inner core nutation, 204
　　lateral variations in heat flow, 202–204, 203f
　　translating convection model, 202–204, 204f
　travel times, 202, 202f
Heterogeneous inner core growth effects, 162–163, 163–164f
High magnetic Reynolds number dynamo, 155–156, 155f
High-pressure experiment, 284–285, 284f
Homogeneous dynamo, 77
Hubble constant, 249

I

Industrial dynamo, 77
Inner core
　attenuation and scattering
　　anisotropy of heterogeneity spectrum, 201, 201f
　　depth dependence, 198–199, 199–200f
　　frequency dependence and parameter trade-offs, 197–198, 197–199f
　　lateral variations, 200, 200f
　depth dependence, 195–196, 196f

Inner core (Continued)
  differential rotation
    observations, 204–205, 205–206f
    theories, 205–207, 206–207f
  differential travel times, 195, 196f
  elastic anisotropy, 195
  hemispherical differences
    observations, 202, 202f
    theories, 202–204, 203–204f
  shear modulus, density, and viscosity
    density discontinuity, 211–212, 211–212f
    predictions from mineral physics, 209–210, 210f
    shear modulus from seismology, 208–209, 208f
    transition region, 209, 209f
    viscosity, 210–211
Inner core boundary (ICB), 190–191, 190–191f, 260–261
Inner core dynamics
  annealing
    annihilation, 235, 236f
    coarsening, 234–235, 237, 238f
    grain boundary migration, 235–237, 237f
    paths, 234–235, 235f
    recovery and recrystallization, 235–237
    sub-grain formation, 235, 236f
  attenuation and isotropic heterogeneity, 243–244
  deformation
    diffusion creep, 233–234, 233–234f
    dislocation creep, 228–230, 229–231f, 232
    slip systems, 232–233, 233f
    stress/strain, 224–227, 225–226f, 228t
  deformation mechanism maps, 239, 239–240f
  elastic anisotropy, 242–243, 244f
  grain size, 239, 240f
  inner core viscosity, 240–242, 241–242f
  solidification
    dendritic growth, 216, 217–219f, 218
    F layer, 221–222
    microstructure and texture, 218, 220–221, 220f
    morphological instability, 215–216, 216f
    nucleation, 222–223, 223f
    snowing core model, 222
    translation of inner core, 221, 222f
Inner core growth, 261–262, 261–262t, 262f, 276–277
Inner core mineralogy
  anelastic behavior, 63–64
  creep/plastic flow, 63–64
  crystallography
    Bravais lattices, 57
    close-packed structures, 58–59, 59f
    grain boundary, 57
    hexagonal close-packed structure, 58, 58f
    microstructure, 57
    Miller indices, 59, 70–71, 70–71f
    pole figures, 59, 70–71, 70–71f
    slip system, 59–60, 60f
    spheres representation, 58, 58f
    stacking sequence, 58, 59f
  elastic properties of Fe, 62, 63f
  lattice preferred orientation (LPO), 57

  Poisson's ratio, 64
  polycrystal, 57
  quality factor, 63
  shape preferred orientation, 57
  shear modulus of Fe, 63–64
  stable phase of Fe and its alloys
    hcp ε-phase, 61, 61f
    phase diagram, 60, 61f
  textured solid, 57
Inner core nucleation (ICN), 262–265, 263–264f
Internal gravity waves, 138–140, 139f
Iron snow, 269

J
Jurassic hyper-reversals, 88

K
Kiaman reverse superchron (KRS), 88
Kinematic dynamo, 77

L
Laminar flow. *See* Steady laminar flow
Large low seismic velocity provinces (LLSVPs), 150
Large Low Shear Velocity Provinces (LLSVPs), 52
Large low-velocity provinces (LLVPs), 182, 182f
Latent heat, 49
Lateral variations of inner core, 200, 200f, 202, 203f
Lattice preferred orientation (LPO), 57, 184
Length-of-day (LOD) variations, 107, 107f
Light element, 33–35, 34f, 36f
Light element and heat transport, 122
Light element exsolution, 269–271
Liquid metal numerical dynamos, 286–287, 286–287f
Liquid outer core, 1
Long-term core cooling
  core-mantle boundary temperature *vs.* age, 266–268, 267f, 269f
  heat loss, 267
  laminar thermal boundary layer theory, 267–268
  radioactive heating effects, 267, 268f
Lowermost mantle region (D″)
  constraints
    elastic anisotropy, 184, 185f
    radiative transport, 183
    scattering, 183–184, 183–184f
  large low-velocity provinces (LLVPs), 182, 182f
  post-perovskite phase transition, 181, 181f
  slabs and plumes, 180, 180f
  structural complexity, 179, 179f
  temperature gradient, 179
  travel-time tomography, 179, 180f
  ultralow velocity zones (ULVZs), 182, 182f
Low magnetic Reynolds number dynamo
  α^2 feedback mechanism, 154
  illustration, 152–154, 153f
  mechanism, 154, 154f
  radial magnetic field, 153–154

M
Magnetic, Archimedean and Coriolis (MAC) waves, 142
Magnetic induction
  Alfvén's theorem, 94
  equation, 94
  frozen flux, 94, 95f
  magnetohydrodynamic (MHD) induction, 110–111
  perfect conductor limit, 94
  Schmidt Legendre polynomials, 109–110
  total magnetic flux, 94
  vector diffusion equation, 94
Magnetic Reynolds number, 93
Magnetic Rossby waves, 142
Magnetic wind, 128–130, 130f, 173
Magnetohydrodynamic (MHD)
  dynamo theory, 77
  induction, 110–111
Magnetostrophy/magnetostrophic flow, 128
Mantle heterogeneity effects, 160
Mantle plume flux, 52
Mauersberger-Lowes spectrum, 80, 80f
Melting temperature, 39–40
Microstructure, 57
Miller indices, 59, 70–71, 70–71f
Mode splitting, 9–11, 11f
Moment of inertia, 1, 18–19, 19f

N
Nickel, 33
Normal modes of Earth, 9–11, 10–11f, 22, 22f
Numerical dynamo model
  attributes, 152
  crustal filtering effects, 156–157, 157f
  dynamo model reversals (*see* Dynamo model reversals)
  E′ layer stratification, 160, 161f, 162
  governing equations, 152, 171
  heterogeneous inner core growth effects, 162–163, 163–164f
  high magnetic Reynolds number dynamo, 155–156, 155f
  low magnetic Reynolds number dynamo, 152–154, 153–154f
  mantle heterogeneity effects, 160
  objectives, 152
  power-law scaling, 158–160
  problem, 152
  scaling laws (*see* Scaling laws)
  3D primitive governing equations, 152
Numerical synthetic seismograms, 283

O
Ohmic and viscous dissipation, 52–53, 103–104
ω-effect, 97, 97f
Ordered alloy, 38
Outer core dynamics
  acceleration in Earth's rotating coordinates, 169–170, 169f
  convection
    discussion, 143
    fully developed convection, 146–150, 147–150f
    mantle heterogeneity effects, 150–151, 151f

onset, 143–146, 144–145f
convective dynamos, 174
dimensionless parameters, 117t
    Buckingham Pi theorem, 118
    characteristic time scales, 119
    control parameters, 118–119
    convective force, 118
    Ekman number, 119
    Elsasser number, 120
    vs. numerical models and laboratory experiments, 120, 121f
    numerical values, 118
    physical units, 118
    Prandtl number, 119
    Rayleigh number, 119, 120t
    response parameters, 118–119
    Reynolds number, 120
equations of motion, 171–172
global kinetic energy balance, 174–175
Lorentz force, 173
Maxwell stress, 173
nondimensional equations
    codensity scaling, 173
    temperature scaling, 172–173
numerical dynamo model
    attributes, 152
    crustal filtering effects, 156–157, 157f
    dynamo model reversals (*see* Dynamo model reversals)
    E' layer stratification, 160, 161f, 162
    governing equations, 152, 171
    heterogeneous inner core growth effects, 162–163, 163–164f
    high magnetic Reynolds number dynamo, 155–156, 155f
    low magnetic Reynolds number dynamo, 152–154, 153–154f
    mantle heterogeneity effects, 160
    objectives, 152
    power-law scaling, 158–160
    problem, 152
    scaling laws (*see* Scaling laws)
    3D primitive governing equations, 152
outer core environment
    characteristic time and length scales, 117, 118f
    Coriolis acceleration, 116, 169–170, 169f
    deep fluid, 115
    dimensionless parameter, 116–117, 117t
    Ekman number, 117
    fluidity of molten iron, 115–116, 116f
    lateral heterogeneity, 116
    outer core buoyancy, 115
    Rossby number, 116–117
    spatial and temporal relationship, 117, 118f
    thermochemical convection, 115
steady laminar flow
    Ekman boundary layers, 132–134, 133–134f
    geostrophic flow, 124–126, 125–126f
    magnetic wind, 128–130, 130f, 173
    magnetostrophic flow, 128
    properties, 124, 125t
    Reynolds stress, 131–132, 132f
    thermal wind, 130–131, 131f
    thin-layer geostrophy, 126–127, 127f

thermochemical transport and buoyancy
    buoyancy flux, 123–124
    codensity, 122–123
    doubly diffusive convection, 122–123
    gravitational potential energy, 121
    light element and heat transport, 122
waves
    Alfvén waves, 135–136, 136f
    inertial waves, 137–138
    internal gravity waves, 138–140, 139f
    MAC waves, 142
    magnetic Rossby waves, 142
    properties, 135t
    Rossby waves, 140–141, 140–141f
    torsional oscillations, 136–137, 137f

P

Paleomagnetic intensity, 90–91, 91f
Partition coefficient, 64–66, 65–67f
Perfect conductor limit, 94
Permanent magnetization, 76
Phase loop, 38
Planetary dynamo, 76
Planetary magnetic field, 75, 76f
Plastic deformation, 40
Plumes, 180, 180f
Poincaré equation, 137–138
Polarity excursion, 87–88
Polarity reversal, 88–89, 89f
Pole figures, 59, 70–71, 70–71f
Polycrystal, 57
Post-perovskite phase transition, 180–181, 181f
Power for geodynamo, 55–56, 55t
Power-law scaling, 158–160
Prandtl number, 119
Preliminary Reference Earth Model (PREM), 3–4, 17

Q

Quasigeostrophic flow, 126, 126f

R

Radial anisotropy, 195–196
Radial structure
    Adams-Williamson equation, 23
    anisotropy, 14–15, 15f, 28–29, 29f
    Birch's law, 26–27
    Bullen parameter, 24
    composite elastic moduli and density, 27–28
    density model, 23
    elastic velocity, 1
    equation of state, 17, 30
    geophysical evidence
        gravity, 1
        magnetic field, 1–2
        moment of inertia, 1, 18–19, 19f
        seismology (*see* Seismic wavefield)
    iron nickel alloy, 1
    liquid outer core, 1
    reference models
        AK135-F model, 3–4
        Bullen's classification, 4
        core-mantle boundary (CMB), 4f
        inner and outer core interaction, 5–6f
        PREM model, 3–4

    scattering, 12–14, 13–14f, 26–27
    solid inner core, 1
    viscoelastic attenuation, 11–12, 12t, 12f, 24–26, 25f
    viscosity, 15–16, 16f
Radiative transport modeling, 183
Radioactive heat, 49
Radiogenic element, 53
Rayleigh number, 119, 120t
Relaxation spectrum, 11–12
Response parameters, 118–119
Reverse flux patches, 82
Reynolds shear stress, 131–132, 132f
Rossby number, 116–117
Rossby waves
    barotropic Rossby waves, 140
    geometry, 140–141, 140f
    magnetic Rossby waves, 142
    potential vorticity, 140
    propagation speed, 141
    vorticity, 141, 141f

S

Scaling laws
    applications, 160
    numerical and laboratory experiments, 158
    for planetary dynamos, 158
    power-law scaling
        control parameters, 158
        diffusion-free magnetic field scaling, 159
        diffusion-free velocity scaling, 159
        diffusive magnetic field scaling, 159–160
Scattering, 12–14, 13–14f, 26–27
Schmidt Legendre polynomials, 109–110
Screw dislocation, 229, 229f
Secular cooling, 46, 48–49, 48f, 269
Seeding of geodynamo
    astrophysical batteries, 273
    Biermann battery, 273
    electromagnetic implantation, 274
    elliptical instabilities, 275, 275f
    geodynamo initiation, 274–275, 275f
    geodynamo seeding times, 273–274, 277–278
    seed magnetic fields, 273
    solar nebula seeding, 273
    tidal stirring, 274–275
    tilt-over mode, 275
Seismic wavefield
    body waves (*see* Body waves)
    elastic equation of motion, 2, 4–5, 19–20
    free oscillations/normal modes, 9–11, 10–11f, 22, 22f
    metallic core, 2, 2f
    mode splitting, 9–11, 11f
    nomenclature, 21–22, 21–22f
    outer and inner cores, 2, 3f
    relaxation spectrum, 11–12
    spheroidal modes, 9, 10f
    state of core, 2–3, 3f
    toroidal modes, 9, 10f
    viscoelastic attenuation, 11–12, 12t, 12f, 24–26, 25f
Seismology
    ambient noise, 282
    coherence, 282

Seismology (Continued)
  numerical synthetic seismograms, 283
  seismic instrumentation, 282
  source-time functions, 282, 283f
  station coverage, 281–282
  wavefield correlation, 282
Self-diffusivity, 40, 41f
Self-sustaining dynamo mechanism
  critical state, 93
  definition, 76
  energy pathways, 102–103, 102f
  external magnetic field, 92f, 93
  magnetic diffusivity, 93
  magnetic Reynolds number, 93
  magnetohydrodynamic state, 92
  physical properties, 91, 92f
  rationale, 76–77
  static state, 92
  subcritical and supercritical state, 93
Shape-preferred orientation (SPO), 184
Shear modulus
  alloy composition *vs.* temperature, 210, 210f
  nonzero shear modulus, 207–208
  pre-melting, 210
  ray paths, 208–209, 208f
  spheroidal normal modes, 208
  in transition region, 209, 209f
Shock-wave experiment, 39
Slabs, 180, 180f
Solar nebula seeding, 273
Solidification of inner core
  columnar dendritic crystals, 215
  constitutional supercooling, 216
  dendritic growth, 216, 217–219f, 218
  directionally solidified alloy, 216, 217f
  F layer, 221–222
  microstructure and texture, 218, 220–221, 220f
  morphological instability, 215–216, 216f
  mushy zone, 216
  nucleation paradox, 216, 222–223, 223f
  platelets, 218, 219f
  snowing core model, 222
  texturing, 216
  thermal undercooling/supercooling, 215
  translation of inner core, 221, 222f
Solid inner core, 1
Solid-liquid phase diagrams, 37–38, 37–38f, 64–65, 65–67f
Solid solution, 37
Source-time functions, 282, 283f
South Atlantic Anomaly, 82
Spheroidal modes, 9, 10f
Stably stratified layers in outer core, 56–57
Stacking faults, 230, 231f
Station coverage, 281–282
Steady laminar flow
  Ekman boundary layers, 132–134, 133–134f
  geostrophic flow, 124–126, 125–126f
  magnetic wind, 128–130, 130f, 173
  magnetostrophic flow, 128
  properties, 124, 125t
  Reynolds stress, 131–132, 132f
  thermal wind, 130–131, 131f
  thin-layer geostrophy, 126–127, 127f

Superlattice, 38
Supernova event, 249
Superrotation, 129, 130f, 207, 207f

T
Temperature
  adiabatic gradient, 38–39
  boundary layers, 38–39
  estimated temperature profiles, 40, 40f
  Gruneisen parameter, 39, 67–68
  melting temperature, 39–40
  shock-wave experiment, 39
  thermoelastic properties, 38–39
  T_{ICB} and T_{CMB}, 39–40, 40f
Terrestrial planet formation model, 253–254
Textured solid, 57
Thermal wind, 130–131, 131f
Thermochemical convection, 288
Thermodynamics
  age of inner core, 54, 54t, 55f
  compositional energy, 46
  conservation of energy, 47
  core as heat engine, 46, 46f
  core-mantle heat flux, 47, 47f
  energy and entropy balance equation
    CMB heat flux, and heat and entropy on adiabat, 51–52
    compositional energy, 49–50
    conservation of energy, 51
    core-to-mantle heat flow, 52
    expression, 51
    heat of chemical reactions, 50
    Large Low Shear Velocity Provinces (LLSVPs), 52
    latent heat, 49
    mantle plume flux, 52
    Ohmic and viscous dissipation and rate of entropy, 52–53
    radioactive heat, 49
    radiogenic elements and rate of energy and entropy, 53
    rate of energy and entropy change due to secular cooling, latent heat, heat of chemical reactions, and compositional energy, 53
    secular cooling, 48–49, 48f
  equilibrium thermodynamic model, 47
  gravitational potential energy, 46
  power for geodynamo, 55–56, 55t
  relations, 67–68
  secular cooling, 46
  stably stratified layers in outer core, 56–57
Thin-layer geostrophy, 126–127, 127f
Tidal stirring, 274–275
Tilt-over mode, 275
Topographical relaxation, 227
Toroidal modes, 9, 10f
Torsional oscillations, 129, 136–137, 137f
Translating convection model, 202–204, 204f
Transport properties
  electrical and thermal conductivity
    band theory, 42, 43f
    Debye model, 43, 45, 69
    electrical conductivity, 43
    electrical resistivity, 43
    electron-electron scattering, 45

  electronic thermal conductivity, 43–44
  free electron Fermi gas, 68
  phonons, 69
  resistivity saturation effect, 45
  thermal conductivity of Fe alloys, 45, 45f
  thermal conductivity values, 44, 44f
  Wiedemann-Franz law, 42–44
outer core viscosity
  activation energy, 41–42, 42f
  dynamic viscosity, 41, 41f
  F-layer, seismically inferred, 42
  low viscosity, 42
  slurry, 42
  plastic deformation, 40
  self-diffusivity/mass diffusivity, 40, 41f

U
Ultralow velocity zones (ULVZs), 150, 182, 182f
Uppermost outer core (E′)
  constraints from chemistry and MAC waves, 188
  formation theory, 188
  seismic velocity anomalies, 187–188, 187–188f

V
Virtual axial dipole moment (VADM), 86–88, 86–88f, 90–91, 91f
Viscoelastic attenuation
  applications from Q measurements, 25
  attenuation of P and S waves, 26
  elastic body waves, 11–12
  estimates, 12t
  frequency-dependent compliance, 24
  index of refraction, 24
  inner core attenuation, 12
  mechanisms, 11
  outer core attenuation, 12, 12f
  phase velocity, 24
  practical constraints, 25
  Q parameter, 24
  reference models, 12
  relaxation spectrum, 11–12, 25
  solid inner core and liquid outer core, 11
  stress and strain behavior, 24
  time-dependent rheology, 24
Viscosity, 15–16, 16f, 210–211, 240–242, 241–242f
von Karman Sodium (VKS) dynamo experiment, 100–101, 101f
von Mises condition, 232

W
Waves in outer core
  Alfvén waves, 135–136, 136f
  inertial waves, 137–138
  internal gravity waves, 138–140, 139f
  MAC waves, 142
  magnetic Rossby waves, 142
  properties, 135t
  Rossby waves, 140–141, 140–141f
  torsional oscillations, 136–137, 137f
Work hardening, 229–230, 230f

Z
Z-machine, 284, 284f

Printed in the United States
by Baker & Taylor Publisher Services